I0057300

OPTICAL MAGIC IN THE LATE RENAISSANCE

RENAISSANCE

GIAMBATTISTA DELLA PORTA'S
DE REFRACTIONE OF 1593

OPTICAL MAGIC IN THE LATE RENAISSANCE

GIAMBATTISTA DELLA PORTA'S *DE REFRACTIONE* OF 1593

A. Mark Smith

American Philosophical Society Press
Philadelphia

Transactions of the
American Philosophical Society
Held at Philadelphia
For Promoting Useful Knowledge
Volume 107, Part 1

Copyright © 2019 by the American Philosophical Society for its
Transactions series.

All rights reserved.

ISBN: 978-1-60618-071-6

US ISSN: 0065-9746

Library of Congress Cataloging-in Publication Data

Names: Smith, Mark, 1960 May 10- author. | Porta, Giambattista della,
    approximately 1535-1615.
Title: Optical magic in the late Renaissance : Giambattista Della Porta's De
    Refractione of 1593 / Mark Smith.
Other titles: Giambattista Della Porta's De Refractione of 1593 | Refractione
    of 1593
Description: Philadelphia : American Philosophical Society Press, 2018. |
    Includes bibliographical references and index.
Identifiers: LCCN 2018037988 (print) | LCCN 2018048068 (ebook) | ISBN
    9781606180761 (ebook) | ISBN 9781606180716 (alk. paper)
Subjects: LCSH: Optics—Early works to 1800. | Refraction—Early works to
    1800. | Vision—Early works to 1800. | Light—Early works to 1800.
Classification: LCC QC353 (ebook) | LCC QC353 .S64 2018 (print) |
    DDC 535—dc23
LC record available at https://lccn.loc.gov/2018037988

Also available as an ebook (ISBN: 978-1-60618-076-1)

CONTENTS

ACKNOWLEDGMENTS

Of the writing of this book (to paraphrase Ecclesiastes) there seemed no end, and surely there wouldn't have been had it not been for the all-important contributions of others. For a start, there are those who dealt with the early stages of the project, when I was beginning to establish the Latin text and, at the same time, come to grips with the English translation. For their help at that time I give special thanks to Heather McRae, Katelynn Robinson, and Matt Shaw, who proofread and edited the Latin and English texts as they developed. Furthermore, Matt and Heather were instrumental in the middle stages of the project, casting a sharp, critical eye for various grammatical glitches, syntactical errors, and terminological inconsistencies in the more crystallized version of text and translation. Matt was particularly helpful in checking the English translation against the Latin text to make sure it fit appropriately and that I hadn't strayed too far from the interpretive straight and narrow. I am also grateful to Arianna Borrelli for vetting part of my translation in its relatively early stages and offering sage advice for improving it. Heather, meantime, played a significant role at the end of this project not only by proofreading but also by helping me create both indexes.

Nor must I forget to thank my referees for their part in encouraging the American Philosophical Society Press to include the Latin text along with the English, an inclusion I was willing—albeit not happily—to forego if necessary. Thanks are due as well to that press and its director of publications, Mary McDonald, for their willingness to heed the advice of the referees and allow the inclusion of the Latin text. Thanks also to Pamela Lankas and her crew for their tireless and patient efforts to make the final, print-ready version of the book as accurate, readable, and pleasing as possible. I also want to express my appreciation to Arianna Borrelli, Giora Hon, Cesare Pastorino, Friedrich Steinle, and Yaakov Zik for organizing the workshop, "The Optics of Giovan Battista Della Porta (1535–1615): A Reassessment" held at the Technische Universität Berlin in 2014. As one of the participants in that workshop, I came away strongly encouraged to keep plugging away at this book. Finally, a nod of gratitude to the MU Research Council for funding the sabbatical five years ago that allowed me both the time and leisure to set this project into effective motion. Each in her, his, or its own way has helped bring the writing of this book to what I hope is a fitting end.

INTRODUCTION

In order to put this edition and translation of *De Refractione* into proper historical and intellectual context, I have divided the introductory portion of the book into four sections of widely varying lengths. In the first section, titled "The Making of *De Refractione*," I start with a brief account of Giambattista Della Porta's life and works and then turn to an equally brief discussion of the genesis of *De Refractione*. After that, I conclude with a fairly close examination of the textual sources on which Della Porta drew in composing the book and its analytic narrative.

The second section, "The Making of This Edition and Translation," opens with a discussion of the problems I faced in locating Della Porta's textual sources, many of which are not readily identifiable in his Latin text either because he did not cite them or because they are quoted but not marked out in any meaningful way. I then explain the why and how of my edition and translation of the original Latin text of *De Refractione*. In the process, I describe some of the conventions I have followed as well as some of the ancillary aids I have provided.

The title of the third section, "Della Porta's Optical Analysis in *De Refractione*" is pretty much self-explanatory as a topical guide to that section. To be specific, I begin by examining Della Porta's physical account of refraction, after which I deal with his overall account of vision. Then I examine his account of visual illusions and their environmental or pathological causes and conclude with a look at his analysis of the rainbow as well as of some other meteorological phenomena.

The fourth and final section, "*De Refractione* in Historical Context," addresses the historical significance of Della Porta's account of light and sight in *De Refractione* within the broader context of natural philosophy as it evolved over the late sixteenth and early seventeenth centuries. In assessing that account and its significance, I will resist the all-too-common tendency among scholars to pigeonhole Della Porta's approach to natural philosophy according to the clear-cut opposition of "medieval" and "modern." He was, after all, "modern" enough in his own day to be accepted as a peer among a select group of thinkers that included such apparently forward-thinking luminaries as Kepler and Galileo. On the other hand, he was "medieval" enough to chase after such pipe dreams as the Philosopher's Stone while

rejecting the possibility of a spinning Earth.[1] Della Porta, in short, was tied firmly to his intellectual past while yet struggling to break the bonds that tied him to it. Or, to paraphrase the notorious description of Dante in an apocryphal student essay, Della Porta stood with one foot firmly planted in medieval science while saluting the rising star of modern science with the other.[2] But pretty much every significant natural philosopher of the sixteenth and seventeenth century, including Newton, shared this awkward posture, so evaluating any of them according to an either/or "medieval/modern" dichotomy is simplistic and ahistorical.[3] I will therefore do my best to avoid it.

The Making of *De Refractione*

Giambattista Della Porta: A Biographical Sketch[4]

Although it has not yet been pinpointed, and probably never will be, the date of Della Porta's birth is now pretty much settled as autumn of 1535.[5] Scion of an aristocratic Neapolitan family of middling status but significant wealth, he was to the manner (and the manor) born. His father Nardo, or Leonardo, was well connected to the governing authorities in Spanish-ruled Naples at the time. Starting in 1541, in fact, Nardo served as royal secretary under emperor Charles V's vice-regent, Pedro Álvarez de Toledo y Zúñiga, and he was formally recognized as a court familiar in 1548. On his mother's side, Della Porta was linked to the Spadaforas, a family with patrician roots in Calabria. His maternal uncle Adriano Guglielmo was an antiquarian scholar so respected for his learning that he was appointed head of the Neapolitan archive in 1536. Learning was not limited to the Spadafora side of the family though. Della Porta's father Nardo consorted with some of the foremost intellectuals and literati of Naples, regularly entertaining them in an informal "academy" at his urban palace. During the hot, summer months, he may have shifted the venue to his villa on the coast below Naples at Vico Equense, which had been an exclusive resort for the Neapolitan social élite since the late thirteenth century. Della Porta, in short, had auspicious links through both blood and brains.

Along with his older and younger brothers Gian Vincenzo and Gian Ferrante (his oldest brother Francesco did not survive childhood, and about his sister we know virtually nothing), Della Porta was educated privately at home.[6] Such private tutelage was typical of the Renaissance aristocracy, who were eager to give their sons and daughters at least a veneer of cultivation in the arts, literature, and music. Although we are pretty much in the dark about the specific curriculum of study Della Porta and his brothers followed, it certainly included intensive instruction in Latin and Greek letters as well as in science. Music also seems to have been emphasized.[7] The quality of his education may be inferred from the quality of his tutors. One of them,

Domenico Pizzamenti, was an accomplished classicist who produced several Latin translations of Greek works.[8] Another was Giovanni Antonio Pisano. A physician noted for both his medical skill and the eloquence of his medical lectures, Pisano not only helped hone Giambattista's Latin style but also subjected him to a rigorous program of study in anatomy and physiology.[9] Meanwhile, as part of his educational program, Della Porta was apparently free to participate in the informal "academy" of literati and intellectuals convened at his father's palace. All of this took place in a city of well over 200,000 inhabitants that was emerging as one of the foremost cultural centers of Europe.

Della Porta obviously benefitted from this rich intellectual environment and the upbringing he enjoyed within it. When barely in his twenties, he published *Magiae Naturalis sive De Miraculis Rerum Naturalium Libri III (Four Books of Natural Magic or [a Treatise]) On the Marvels of Natural Things* [Naples, 1558]. This work quickly gained him renown as a natural magician, a "professor of secrets," to use William Eamon's coinage.[10] As such, he was admired for the mastery of nature and its occult qualities that allowed him to create a broad array of startling, but entirely natural effects. Five years later, in 1563 at Naples, he published a work on cryptography, *De Furtivis Literarum Notis, Vulgo De Ziferis, Libri IIII (Four Books on Understanding Secret Writing, in Common Parlance, On Cyphers)*, which added luster to his already-burgeoning reputation as an adept in natural magic. Traveling throughout Italy, then to Paris, and eventually to Spain around the same time, he was welcomed warmly by King Philip II, whose recognition of him added even more luster to that reputation. It was further burnished by the publication at Naples in 1566 of his *L'Arte del Ricordare (The Art of Memory)*.[11]

By this time, and perhaps as early as 1560, Della Porta had established the *Accademia dei Segreti (The Academy of Secrets)*. Meeting irregularly at his family estate in Naples and perhaps also at times in the villa at Vico Equense (shades of his father), this society had one fundamental requirement for membership: the ability to reveal a "secret," or startling natural effect, hitherto unknown to any of the current members. Throughout these early years, Della Porta was aided and abetted by his two brothers, the older of whom, Gian Vincenzo, had begun to distinguish himself as both a classicist and an expert in astrology and alchemy. Meantime, Della Porta himself was becoming increasingly known among select friends and acquaintances for the witty comedies he composed privately in the 1560s and 70s but left unpublished until considerably later.[12] This penchant for drama was part and parcel of the general aptitude for showmanship (and self-promotion) that made him so successful as a natural magician.[13] The path to this success was no doubt smoothed by his charismatic personality and ability to turn on the charm.

Della Porta's growing reputation was a mixed blessing, though. In 1574, Cardinal Scipione Rebiba, the chief inquisitor in Rome, sought to have him arrested and brought to Rome for trial, presumably to answer for certain

problematic claims in his *Magiae Naturalis*. This seems to have been no more than a close call, but in apparent response to Rebiba's intervention the archbishop of Naples did command that the *Accademia dei Segreti* be disbanded. By October 1577, however, things got serious when Della Porta was finally haled before the Inquisition at Rome, most likely to answer charges that at least some of his magic was demonic. Unable to prove his guilt, the tribunal enjoined him in late 1578 to undergo *purgatio canonica*, which required him to swear formally that he had never held the heretical positions with which he was charged. Because refusal to do so called for excommunication, in some cases even execution, it is hardly to be wondered that he acceded.[14]

While all this was transpiring, Della Porta married and fathered a daughter, Cinzia, and within a few years he joined a lay order of Jesuits committed to charitable works. He seems to have carried out this obligation sedulously and with good will for the remainder of his life. If this was meant to mollify suspicious ecclesiastical authorities, however, it was a failure: He spent the rest of his life under continual scrutiny by both Neapolitan and Roman Inquisitional authorities, and at times he ran afoul of others. In 1583, for instance, his *Magiae Naturalis* was placed on the Spanish index of forbidden books "until purged" (*donec expurgetur*).[15] This sort of continual surveillance by inquisitional authorities put Della Porta at constant threat of death because, according to the rules governing *purgatio canonica*, if he was subsequently found guilty of any position he forswore in that process, he would automatically be executed.

But Della Porta was not without resources. During his trial in 1577, he (like Galileo fifty-odd years later) was treated leniently for reasons of health, not only escaping actual torture but also being granted the favor of house arrest under Cardinal Flavio Orsini, a member of a leading and extraordinarily influential Roman family. He was thus spared both the indignity and discomfort of a common prison during the course of his trial. He also had the sympathetic ear of Cardinal Luigi d'Este, son of Ercole II, Duke of Modena and Ferrara, who invited him in November of 1579 to join his household in Rome, offering him both a stipend and an aegis under which he could continue his research unhampered. Della Porta accepted without hesitation, no doubt motivated as much by the protection afforded by the cardinal as by the promised subvention. In December of the next year, still under the Cardinal's aegis, he joined him in Venice, where he undertook to manufacture a large parabolic mirror, most likely with the aim of deploying it as a telescope. What, if anything, came of this project is unknown.[16]

Then, after a brief interlude at Ferrara with the Cardinal, Della Porta returned home in spring of 1581, where he entered into a long-term collaboration with his older brother, Gian Vincenzo, who had inherited his father's position as royal secretary. By this time, unfortunately, his younger brother had died. Not long after his return to Naples, rumor had it that Della Porta was hot on the trail of the Philosopher's Stone. Delighted by the prospect

of having the secret of that wonderful substance uncovered so soon, Cardinal d'Este once again invited Della Porta to join him in Rome in order to share the discovery. As it turns out, both Della Porta and his Eminence had been too sanguine about the imminence of that discovery. Realizing that it was far less close at hand than he had hoped, Della Porta turned to other things, including a study of the cultivation of fruit and trees, which eventuated in the publication of *Suae Villae Pomarium* (*The Orchard of His Estate*) in 1583 and *Suae Villae Olivetum, sive Liber Sextus* (*The Olive-Grove of His Estate, or the Sixth Book*) in 1584, both at Naples. He was also hard at work on *De Humana Physiognomonia* (*On Human Physiognomy*), which he brought to completion in 1583, although it did not see print until three years later.[17]

Hardly novel, the idea behind this book is that one's character is written in the physical features of one's face. In articulating this idea, Della Porta relied heavily on such stock authorities as Aristotle (actually, Pseudo-Aristotle), Adamantius, Polemon, and Albertus Magnus, all of whom wrote extensively on physiognomy.[18] Following their lead, Della Porta proceeds feature by feature in his analysis of physiognomic signs. A thick thatch of hair on one's head, for example, indicates a coarse, boorish nature (*De Hum. Phys.*, 42), which is presumably exacerbated when that hair is bright red, because redheads tend to be not only irascible and crafty, but also lacking in intelligence (*De Hum. Phys.*, 49). An aquiline nose, on the other hand, indicates greatness of soul and courage (*De Hum. Phys.*, 74–5), whereas a sharp, narrow one is a sign of irascibility (*De Hum. Phys.*, 78). All of these traits have their analogues in particular animals, the aquiline nose with the eagle's beak, red hair with the pelt of a fox, and so forth. At least as old as Aristotle, the

Plate 1

underlying assumption that particular kinds of humans correspond in character and nature to particular kinds of animals is exemplified in the lavish
illustrations provided throughout the book.[19] Plate 1, for example, which is
copied directly from the engraving in *De Hum. Phys.*, 55, depicts the "bulging,
high, round brow [characteristic] of stupid, heedless men," who share these
features with asses. In short, such features are a sign of asininity and the
dull-witted fecklessness that defines it. Repeated in *De Hum. Phys.*, 67 and
106, this same illustration is used to depict, respectively, the large ears and
thick, undershot lips that are also indicators of asininity.[20]

   As described here, Della Porta's approach to physiognomy was as traditional as the discipline itself, one of whose leading exponents was Albertus
Magnus, the great Dominican authority on Aristotle. There would thus seem
to be no reason for *De Humana Physiognomonia* to have aroused suspicion
among religious authorities, especially because Della Porta denied that we
lack control over the character indicated by our facial features. These features,
he says at the end of the dedicatory letter to Cardinal Luigi d'Este, "can
indicate only natural propensities but not the actions of our free wills or
those that depend on a defective or assiduous disposition, for in the case of
good and evil actions, which are in our power, virtue and vice hold sway,
but not in the case of propensities, which are not subject to our will." In
other words, character is not actually destiny because, although we cannot
control our natural inclinations, our free will enables us to control the actions
toward which they may impel us. A person may indeed be naturally irascible,
but he or she can still resist a rage-induced urge to commit violence.

   With this disclaimer added, the censors were finally persuaded to
approve publication.[21] The problem was not so much that either the book's
subject matter or the author's approach to it was tendentious as that religious
authorities, under papal pressure, had become increasingly suspicious of
anything that smacked of divination. Shortly before the imprimatur was
granted for *De Humana Physiognomonia*, in fact, pope Sixtus V issued an
official denunciation of all divinatory arts, even those that predict mere
inclinations.[22] To make matters worse, Della Porta lost a powerful religious
ally in Cardinal Luigi d'Este, who died at the very end of 1586. Nothing
daunted, however, he saw his *Phytognomonica* through to publication at
Naples in 1588. Predicated on the notion that various medicinal plants have
a particular effect on the parts of the human body they resemble, this work
was firmly rooted in the doctrine of signatures to which Della Porta was
committed throughout his life.

   In this same year, his Latin translation of the first book of Ptolemy's
Almagest, which included a translation of Theon of Alexandria's commentary,
appeared in Naples under the title *Claudii Ptolemaei Magnae Constructionis
Liber Primus (The First Book of Claudius Ptolemy's Great Synthesis)*. A year
later he published a second, greatly expanded version of the *Magiae Naturalis*
while also seeing his play *L'Olimpia* into print, a play he was said to have
written when quite young, perhaps as early as 1550.[23] Encouraged by the

warm reception of this first published dramatic piece, Della Porta issued a steady stream of plays during the remaining years of his life. The very last thing published during his lifetime, in fact, was the tragedy *L'Ulisse*, which appeared in 1614 and was the last of seventeen plays that saw print before his death.[24] Although he himself never took his plays seriously in comparison to his other works, Della Porta is regarded by modern scholars as among the most accomplished playwrights of the late Renaissance.[25]

As Della Porta's reputation grew, so did his circle of friends, acquaintances, and correspondents. Unfortunately, this circle included such controversial figures as Tomaso Campanella and Paolo Sarpi, both of whom had fallen afoul of religious authorities. Although Sarpi managed to avoid being ensnared by the Inquisition, Campanella did not. He was caught up in its trammels in 1594, charged with heresy, and eventually sentenced to prison, where he remained from 1599 to 1626. Hobnobbing with these two was bad enough, but Della Porta was suspected, perhaps rightly, of something even worse: consorting with Giordano Bruno, whose arrest by the Inquisition under the charge of heresy in May of 1592 made even the whiff of association with him potentially toxic. Tainted by these social and intellectual connections, even if they were only presumed, Della Porta was prohibited in April 1592, by order of the Roman inquisitor, Cardinal Giuglio Santorio, from publishing anything without permission from the appropriate authorities in Rome. An imprimatur from Naples would therefore no longer suffice.

The results of this constraint are evident in *De Refractione*, which was published in 1593 after two phases of censorship, the first and principal one at Rome, the second at Naples. Nonetheless, despite this restriction and despite being denied permission to publish it by the Venetian inquisition in 1592, Della Porta was able to get his Italian translation of *De Humana Physiognomonia* published under the title *Della Fisionomia dell'Uomo* at Naples in 1598, perhaps by attributing the translation to a certain Giovanni di Rosa.[26] Perhaps, too, he was given some leeway in view of his status as an international celebrity who had to be handled carefully by Church authorities in order that they not appear petty or malicious to his ever-growing public.

Whatever the reasons, religious authorities were apparently willing to relax the constraints on publication somewhat by the turn of the seventeenth century—enough, at least, to enable Della Porta to publish his *Pneumaticorum Libri Tres. Quibus Accesserunt Curvilineorum Elementorum Libri Duo* (*Three Books on Pneumatics, to Which Are Added Two Books of the Elements of Curvilinear Figures*) in 1601 at Naples. This was followed a year later by *Ars Reminiscendi* (Naples, 1602), the Latin version of *L'Arte del Ricordare*, and a year after that by *Coelestis Physiognomoniae Libri Sex* (*Six Books on Celestial Physiognomy* Naples, 1603). This last book completed what amounts to a trilogy, starting with human physiognomy, progressing through plant physiognomy, and culminating with cosmic physiognomy and its astrological entailments. Five years later, in 1608, his *De Munitione Libri III* (*Three Books on Fortification*) appeared at Naples and his *De Distillatione Libri IX* (*Nine*

Books on Distillation) at Rome. Finally, at Rome in 1610, he published an expanded version of his works on curvilinear figures under the title *Elementorum Curvilineorum Libri Tres* and a new work, *De Aeris Transmutationibus Libri IV* (*Four Books on the Transformations of Air*). These were the last original "scientific" works to be published before his death in 1615.

All the while, Della Porta was busy seeing to the republication of earlier works as well as ushering several plays into print. He also left a number of works either hanging fire or in progress when he died. Among the former was his treatise on chiromancy (i.e., palmistry), which was eventually published toward the end of the seventeenth century under the title *Della Chirofisonomia, overo Di Quella Parte della Humana Fisonomia Che Si Appartiene alla Mano, Libri Due* (*Two Books on Chirophysiognomy, or on That Part of Human Physiognomy That Pertains to the Hand* [Naples, 1677]). Among the unfinished works was *De telescopio*, which he wrote in the hope of establishing his priority as inventor of the telescope after the publication of Galileo's *Sidereus Nuncius* (*Starry Messenger*) in 1610. As we will see in short order, Della Porta's claim to priority was not entirely groundless.

If Della Porta was an indefatigable writer and publisher, he was no less indefatigable as a builder of, and participant in, scholarly networks. The abortive *Accademia di Segreti* that he founded in the 1560s is an early and obvious example. Another such network, a much later one, was that of the *Accademia dei Lincei* (the *Academy of the Lynxes*), which was founded in 1603 by the young Roman noble, Prince Federico Cesi. He and three like-minded friends, Francesco Stelluti, Anastasio de Filiis, and Jan van Heek, chose the lynx as the symbol of their association, using the illustration at the top of the frontispiece to the 1589 edition of Della Porta's *Magiae Naturalis*. Known for its keen-sightedness, the lynx had been held up by Della Porta as a proper model for the natural philosopher, who should look lynx-eyed "at whatever reveals itself to him so that, having inspected something closely, he might act on it diligently."[27] The empirical and utilitarian implications of this model are pretty clear.

In 1604 Cesi traveled from Rome to Naples to meet Della Porta and, no doubt having flattered him during the ensuing visit, got his enthusiastic approbation for the association. Hardly had he formed it, however, than Cesi was forced by his father to disband the *Accademia*, which lay dormant until 1610. During the ensuing six years of dormancy, Cesi forged a close friendship with Della Porta, who, as Clubb phrases it, "dearly loved a lord"[28] but who was also impressed by the young prince's intellect and enthusiasm.[29] Hence, when Cesi revived the *Accademia* in 1610, he invited Della Porta not only to join it as the fifth member, but also to open a Neapolitan branch, an offer that Della Porta could not refuse. This was something of a coup on Cesi's part because by this time Della Porta had risen to the very top of the top rank of intellectual luminaries of the time. It was also something of a coup for Della Porta, and even though he never got the Neapolitan branch of the *Accademia* off the ground, he took pride in his membership for the remainder

of his life, watching with approbation as Galileo joined its ranks in 1611, to be followed in subsequent years by a slow but steady accretion of new members. After having reached a total of some 32 members by 1625, the *Accademia* was effectively disbanded in 1630 with the sudden, premature death of Cesi.[30]

It was during the final six years of his life, between 1609 and 1615, that Della Porta, proud Linceian from 1610 on, become embroiled in the dispute over his priority as inventor of the telescope. As early as 1609, he complained that he was not being properly credited with its discovery, claiming that he had already described the telescope in the *Magiae Naturalis* of 1589, as well as in the ninth book of *De Refractione* (he later corrected himself, citing book 8 as the appropriate place). In fact, there is no allusion whatever to the telescope, or anything like it, in *De Refractione*. There does, however, seem to be such an allusion in book XVII, chapter 10 of *Magiae Naturalis*, where Della Porta informs us that

> concave lenses make things that are far away look quite clear, and convex lenses [do the same for] things that are near. Accordingly, you can use them advantageously for vision. With a concave lens you see small things from far away and clearly; with a convex lens you see near things magnified but blurry. *But if you know how to combine them properly, you will see both far and near things magnified and clear.* (Della Porta [1589], 269—my emphasis)

Precisely what Della Porta meant by "combine them properly" is open to interpretation, but it certainly *could* be taken to mean placing one in front of the other in an appropriate arrangement: that is, with the concave lens right in front of the eye and the convex one beyond it at an adjustable distance so as to form a primitive Galileian telescope. At any rate, this is how Kepler took it in on page 6 of his *Dissertatio cum Nuncio Sidereo* (Prague, 1610) and how Della Porta's fellow Linceian, Johann (Giovanni) Faber presented it in his dedication at the beginning of Galileo's *Il saggiatore* (Rome, 1623), which was published under the auspices of the *Accademia*.[31]

As far as the Linceians, Galileo included, were concerned, the matter was settled: Della Porta was the true father of the telescope; Kepler and Galileo, the godparents who transformed it into the wonderfully useful instrument it eventually became. But if Della Porta's priority as discoverer was more or less accepted *ab initio* among his fellow Linceians, it was still at issue outside this select group. Della Porta thus undertook to defend his claim in a new work, *De telescopio*, which he began in 1612. His intent in that work was clear enough. On the basis of his approach to convex and concave lenses in book 8 of *De Refractione*, he would prove that he had understood the theoretical foundations of the telescope and its workings long before the telescope had been brought to public light. Racked by old age and ill health, he died in 1615 still hoping in vain to bring the project to fruition.[32]

The Genesis of De Refractione

Long before 1593 and the publication of *De Refractione*, the trajectory of
Della Porta's career as a natural philosopher, experimental impresario, and,
more specific, optical thinker was plotted in his *Magiae Naturalis sive De
Miraculis Rerum Naturalium Libri IIII*. First published at Naples in 1558, when
Della Porta was only twenty-three, this compendium of "secrets" (*secreta*)
manifests the enormous breadth of his interests in natural phenomena, even
at that early date. And these interests, which occupied him for the rest of
his life, must also have occupied the thousands of readers who snapped up
the sixteen Latin, seven French, eight Italian, and two Dutch editions of
Magiae Naturalis that followed between 1558 and 1588.[33] So successful, in
fact, was the original, four-book edition that Della Porta saw fit to publish a
much-expanded version in 1589, again at Naples, under the commensurately
shortened title *Magiae Naturalis Libri XX* (*Twenty Books of Natural Magic*).
This, too, saw numerous Latin and vernacular editions, including the one
(and so far only) English version that was published in 1658, a full century
after the cycle began with the appearance of the initial Latin edition.[34]

In both the 1558 and 1589 versions of the *Magiae Naturalis* Della Porta
devotes considerable space to the discussion of optical phenomena. Well
over half of the fourth and final book of the first version and an entire book
(i.e., book 17) of the twenty comprising the second are given over to that
topic. In every case Della Porta's approach is the same: He describes a
particular optical effect, usually one that is startling or at least noteworthy
and thus "magical," and then explains how to set things up in order to
produce it. In both texts, for instance, we learn how to cast real images by
shining light onto a sheet of paper from a brightly illuminated picture or
outdoor scene through a narrow aperture in a darkened room. What Della
Porta is describing, of course, is a camera obscura, and he adds that the eye,
with its pupil and lens, is just such a device, the front surface of the lens
serving as the "paper" on which the image is cast.[35]

There is no denying that the projection of real images in a camera
obscura is eye-catching, even magical, especially with the addition of a convex
lens in front of the aperture, as suggested by Della Porta in book 17, chapter
6 of the 1589 version of *Magiae Naturalis*.[36] But there are myriad other ways
to create optical effects that are perhaps less awe-inspiring but nonetheless
eye-catching. For those of us keen to see images appear to hang in the air,
for example, Della Porta explains how to achieve this effect with a concave
spherical mirror.[37] In such a mirror, as well, we can see our face greatly
magnified if we follow his directions for placing the viewpoint correctly. Or
we might want to see ourselves approaching in one mirror while, at the
same time, walking away in another. All we need do is set up two plane
mirrors in just the right way.[38] Or perhaps we might want to see ourselves
upside down in a facing mirror. This, Della Porta assures us, is easily accomplished with a particular arrangement of plane mirrors.[39] Like virtually all

the optical phenomena discussed in both versions of *Magiae Naturalis*, these four amount to little more than funhouse effects—or *ludi* in Della Porta's parlance—a rare exception being his account of how to properly deploy a parabolic burning mirror or a spherical burning lens, which at least has practical implication.[40]

The experience and knowledge Della Porta gained during the thirty-odd years separating the two versions of *Magiae Naturalis* is evident from the wider scope of optical phenomena detailed in book 17 of the later version. Of particular note in that version is his fairly extensive discussion of glass lenses and their effects in chapters 10–13 and 19 of that book. This discussion includes several tricks with spherical and biconvex lenses that are similar to those already described using concave spherical mirrors. It is possible, for instance, to see objects through a solid-glass globe or a biconvex circular lens as if they were hanging in the air.[41] It is also possible to hold the globe or lens between object and eye, and adjust its placement with respect to the two so that the object seen through it appears magnified, diminished in size, upright, or inverted, just as can be done with a concave spherical mirror. On a more practical note, however, Della Porta describes not only how spectacles can improve poor eyesight but also how the biconvex or biconcave lenses used in them are manufactured.[42] Yet, despite the difference in breadth and depth of treatment in the two versions, they have one cardinal thing in common: In both, Della Porta's focus is on *how* the phenomena described occur, not *why* they do. In short, he was concerned not with their theoretical foundations, but with their actual, physical staging.

That Della Porta evinced no obvious interest in theoretical matters in either version of the *Magiae Naturalis* does not mean that he lacked such interest entirely. It is true, of course, that he favored an empirical, hands-on approach to natural philosophy according not only to personal observation and experiment but also to trustworthy witness. In a search for such witnesses, Della Porta tells us in the preface to the 1589 edition of *Magiae Naturalis*, he scoured the libraries of Italy, France, and Spain for descriptions of natural oddities. He also consulted with "learned men and met with skilled craftsmen" to find "new and noteworthy things they might have encountered so as to learn anything they had shown to be true and useful according to long experience."[43] What he learned, of course, were the secrets included in both versions of *Magiae Naturalis* along with those added to the second version. By its very genre as a compendium of secrets, therefore, the *Magiae Naturalis* was not an appropriate place to entertain theoretical speculation. This he reserved for *De Refractione*.

Although this work appeared in print four years after the second version of *Magiae Naturalis*, there is little doubt that Della Porta was working on both at the same time. This assumption is based on at least four allusions in that version of *Magiae Naturalis* to his "Optics," which clearly refers to the yet-to-be-published *De Refractione*. Accordingly, after claiming that the eye acts like a camera obscura, Della Porta informs us that this assertion

"has been more amply demonstrated in our *Optics*."[44] Later, having argued that a convex lens can light a fire more quickly and intensely than a concave mirror, he tells us that he "has given the reasons in the *Optics*."[45] Later yet, after briefly discussing the capacity of concave mirrors to focus sunlight to a burning point, he excuses himself from discussing the topic in more detail on the grounds that he has treated it "at greater length in the *Optics*."[46] Finally, in broaching the subject of refraction as a means of bringing sunlight to focus at a burning point, he assures us that he "will provide the explanation in the *Optics*."[47] Nor is the direction of citation only from *Magiae Naturalis* to *De Refractione*. At least four times in *De Refractione* Della Porta refers back to *Magiae Naturalis*, either explicitly or implicitly: that is, in book 2, proposition 1f, pp. 88–89; book 2, proposition 16, pp. 114–16; book 4, propositions 1 and 2, pp. 193–95 and 198–99; and book 7, proposition 8, pp. 346–47.

Such cross-referencing indicates not only that Della Porta had *De Refractione* in mind as he composed book seventeen of the 1589 edition of *Magiae Naturalis*, but also that he had this edition in mind as he composed *De Refractione*. There is virtually no question, therefore, that Della Porta wrote *De Refractione* with an eye toward explaining the optical effects described in *Magiae Naturalis* on theoretical grounds. That eye was not hyperfocused, though. Della Porta took the opportunity to look beyond the optical effects described in *Magiae Naturalis* to several topics not covered there. These include, most notably, the functioning of the eye and its parts, binocular vision and diplopia, pathological conditions leading to visual dysfunction or misperception, and the formation of the rainbow as well as certain other meteorological phenomena.

This expanded topical scope is reflected in the subdivision of *De Refractione* into nine books topically arranged according to three basic segments. Roughly speaking, the first segment, consisting of books 1, 2, and 8, deals with refraction and its physical and visual manifestations. Books 3–7 form a second segment within which vision itself and various ways it can be distorted are treated. Book 9, by far the longest of the treatise, forms a third segment by itself and focuses on various meteorological phenomena, the rainbow foremost among them.

As we will see in due course, when we examine his approach to all these topics, Della Porta availed himself of a vast array of sources in an effort to make both empirical and theoretical sense of them. We will also see that he was something of an intellectual free spirit, tied to no particular optical authority, whether ancient, medieval, or contemporary. For the most part, in fact, he took issue with such authorities, probing the weaknesses in their accounts, often on empirical or "experimental" grounds, and then offering his own alternative accounts. Some of these alternative accounts are not only innovative and ingenious but also "correct," or at least fairly close to it. Others, however, although certainly innovative, are far from correct and sometimes even self-contradictory. Consequently, it should become clear by

the end of our analysis that, as a natural philosopher, and especially as an optical thinker, Della Porta was an intriguing mixture of erudition, intellectual acuity, naïveté, and somewhat bewildering wrongheadedness. In order to bring this point home, let us begin our analysis with a look at the textual sources Della Porta used in constructing his theoretical account of light and sight in *De Refractione*.

Della Porta's Textual Sources

As is evident from the bibliography of Primary Sources for the Text on pp. 487–91, the number and variety of sources I have been able to identify in Della Porta's *De Refractione* attest to his enormous erudition. Totaling more than seventy and ranging from classical to late-Renaissance works, these sources can be parsed in several ways. At the grossest level, for example, we can distinguish printed from manuscript sources. It stands to reason of course that Della Porta consulted printed editions whenever possible, if only for comparative ease of reading and deciphering. Therefore, because the vast majority of his textual sources had been published by the 1580s at latest, we can assume fairly safely that he read them in that form. But not all of them. Four in particular stand out. Two were written by Roger Bacon: *De multiplicatione specierum*, which dates from the early 1260s, and *Perspectiva*, which was composed around 1265 as part 5 of the *Opus majus*. The latter did not see print until 1614, nearly twenty years after Della Porta published *De Refractione*, and the former was first published well over a century later, in 1733. The third work at issue is Ya'qūb al-Kindī's *De aspectibus*, whose now-lost Arabic original was written in the mid-800s and whose Latin translation was produced by Gerard of Cremona in the later twelfth century. This work remained unpublished until 1912.[48] The fourth work that Della Porta had to have consulted in manuscript form is Ptolemy's *Optics*, which did not see print until 1885.[49] Della Porta was therefore willing and able to consult manuscripts whenever necessary. Equally important, he had access to them. How often he actually needed, or chose, to consult them is impossible to determine, but in all likelihood the proportion of manuscripts to printed editions he consulted was small.

At a slightly more refined level, we can distinguish among Della Porta's sources by type, that is, according to whether they were primary or secondary. There is no doubt that most of his sources were primary, yet Della Porta's discussion of certain thinkers was clearly based on secondary authorities, even when he quoted actual passages. His references to such Presocratic philosophers as Anaxagoras (500? –428? BCE), Anaximenes (f. c. 550 BCE), Democritus (fl. c. 460/430? BCE), Empedocles (fl. c. 460 BCE), and Parmenides (c. 515?–after 450 BCE) serve as obvious examples. Only fragments of the works of these authors survived from antiquity, all of them contained in later accounts, many of which were doxographic. Della Porta therefore

had no direct access to the works of the Presocratics he cited.[50] Likewise, his allusions to such obscure classical thinkers as Artemidorus of Parium, Demosthenes Philalethes, Favorinus of Arelete, Isigonus, Metrodorus of Chios, Nicholas the Peripatetic (or Nicholas of Damascus), Nymphodorus, and Philip of Medma were based on second-hand references in such sources as Pliny's *Natural Histories*, Seneca's *Natural Questions*, and even Albertus Magnus's commentary on Aristotle's *Meteorology*.[51]

Another of Della Porta's secondary sources, this one more or less contemporary, was Lodovico Ricchieri's *Antiquarum Lectionum Libri XXX* (*Thirty Books of Ancient Readings*), a massive, topically arranged collection of textual excerpts from a wide range of classical authors.[52] How heavily dependent Della Porta was on this compendium is uncertain. What is certain is that it contains a significant number of the primary source snippets he incorporated into his own text. As a consequence, we are left somewhat unsure about how deeply he actually delved into the primary sources that he cited and, on that account, how reliant he was on secondary authorities such as Ricchieri.

At the most refined level, finally, Della Porta's sources may be parsed according to genre. This we can subdivide roughly into four categories, the first of which I characterize as "scientific." Grouped within it are works devoted to technical subjects such as geometrical optics, ophthalmology, and physiology. It also subsumes such works as Vitruvius's *De architectura* and Sacrobosco's *Sphere*. The second category, which consists of studies dealing in a more or less general way with natural philosophical topics, takes the designation "philosophical." Aristotle's *Meteorology* and *Parva Naturalia* fall naturally under this head, as do various commentaries on his works. The third category can be appropriately styled "encyclopedic." Such works as Pliny's *Natural Histories*, Aelian's *Various Histories*, and Aulus Gellius's *Attic Nights*, all of which address a wide variety of topics in somewhat scattershot fashion, are obvious candidates for this category. Last comes a sort of catchall category that I dub "literary" and that encompasses such disparate works as Virgil's *Aeneid*, the *Glossa ordinaria* of the Bible, and Leon Battista Alberti's *De pictura*.

That there are substantial overlaps among these categories is undeniable, as is the fact that the distinctions implied by them would have been at least somewhat foreign to Della Porta and his humanist contemporaries. For instance, Pseudo-Plutarch's *Placita philosophorum* (*On the Opinions of the Philosophers*), to which Della Porta appealed several times, fits fairly snugly within both the philosophical and literary categories—and possibly even the encyclopedic one. By the same token, Alberti's *De picture* can be legitimately viewed as both literary and scientific, because it addresses the ray-geometrical aspects of vision in the process of establishing the foundations of proper artistic representation through "istoria." Nonetheless, rough and inexact as it is, this fourfold breakdown of genre provides a relatively useful way to organize Della Porta's sources. With that in mind, then, let us turn to specifics, starting with the scientific sources.

At the very core of those sources, and of Della Porta's overall analysis of vision in *De Refractione*, lies a handful of treatises by four thinkers credited with establishing the so-called Perspectivist optical tradition. The first and most seminal of these thinkers was Ibn al-Haytham (d. 1048/49), whose *Kitāb al-Manāẓir* (*Book of Optics*) was translated into Latin around 1200 under the title *De aspectibus* (*On Visual Appearences*). Ascribed to Alhacen (the Latin transliteration of Ibn al-Haytham's given name, al-Ḥasan), this work is quite extensive and technically daunting, especially at the level of ray-geometry. Of the remaining three Perspectivist thinkers, all tracing their theoretical and analytic roots to Alhacen, the earliest was Roger Bacon, already mentioned as author of two optical treatises: *De multiplicatione specierum* (c. 1262) and *Perspectiva* (c. 1265). Of these two treatises, the *Perspectiva* is more heavily dependent on Alhacen for its theoretical and analytic foundations. Next came Witelo, whose *Perspectiva* (c. 1275) was so closely modeled after Alhacen's original—albeit half again as long—that Della Porta snidely referred to him as Alhacen's "ape" (*simia*), remarking that "throughout his [*Perspectiva*], whatever in it goes beyond Alhacen is utterly false."[53] And finally, bringing up the rear, was John Pecham, who intended his *Perspectiva communis* (c. 1280) as a concise introductory textbook based on Alhacen's and Bacon's more specialized and elaborate treatments of light and sight. The least technically demanding and therefore most popular of the Perspectivist works, Pecham's *Perspectiva communis* became a staple of undergraduate university instruction in optics throughout the later Middle Ages and Renaissance.[54]

That Della Porta was familiar with all of these works, including Bacon's *De multiplicatione specierum*, is evident from his explicit and implicit references to them throughout the *De Refractione*.[55] Equally evident is that he relied on them for certain theoretical principles. One clear case lies in his agreement with the Perspectivists that the front surface of the eye's crystalline lens is visually sensitive to impinging light and, as such, is where vision is inaugurated.[56] Another lies in his unequivocal acceptance that sight is due to visible radiation entering into the eye from external objects rather than to visual radiation emitted from the eye out to external objects. In short, like the Perspectivists, Della Porta strongly favored intromissionism over extramissionism.[57] Yet another lies in his adoption of the Perspectivist cone of radiation, all of whose rays originate on the surface of visible objects at the cone's base and strike the front surface of the lens orthogonally as they converge toward the vertex at the eye's center. Like the Perspectivists, Della Porta assumed that it is only within this cone that vision is veridical and undistorted.[58]

But, as mentioned earlier, Della Porta was by no means uncritically wedded to Perspectivist theory. Quite the contrary, he took issue with it in a number of fundamental and far-reaching ways. One important point of departure for him was over binocular vision and our natural tendency to see one image with both eyes. The Perspectivists argued that the image

apprehended by each eye is transmitted into and through its respective optic nerve until the nerve joins its mate from the other eye at the forefront of the brain. It is at this juncture—known as the *optic chiasma*—that the two somewhat discrepant images are fused to yield a single one that forms a sort of average of the two. Della Porta, in contradistinction, rejected binocular vision and image-fusion altogether. Instead, he insisted that each eye works independently and that what we take to be a fused image is actually what is seen by the dominant eye (the right-hand one by his account), which essentially overrides what is seen by the weaker (left) eye.[59] Another significant departure was over the viewpoint from which we perceive and judge such spatial qualities as size, location, and distance. For the Perspectivists, this point, which constituted the so-called center of sight, was the vertex of the cone of radiation at the very center of the spherical eyeball. Della Porta, on the other hand, rejected the Perspectivist center of sight in favor of a multitude of viewpoints on the anterior surface of the crystalline lens, each one constituting its own center of sight according to unique radial ties with particular points in space. Della Porta thus replaced a single point of view with an infinitude of such points clustered relatively near one another on the lens's anterior surface.[60]

In many ways, therefore, Perspectivist theory served less as a positive springboard than as a negative foil for Della Porta's analysis of vision. In this role, the Perspectivist model of sight formed a template of sorts against which Della Porta could frame his own model. The same holds for a second group of optical sources that includes Euclid's *Optics* and *Catoptrics* (c. 300 BCE), Hero of Alexandria's *Catoptrics* (c. 50 CE), and Ptolemy's *Optics* (c. 160 CE), all of which are based on extramissionism. Although he firmly rejected extramissionism in favor of intromissionism, Della Porta's main objection to these sources was not their grounding in the theory of visual radiation. Rather, he attacked them for their explanation of spatial perception, particularly their explanation of size perception. The problem, as he laid it out in book five of the *De Refractione*, was that none of those sources takes into account the refraction of light rays into the eyes, which distorts spatial perception in a variety of ways.

Still within the category of scientific sources lies a final batch of works devoted to the anatomical and physiological basis of the visual process. Among the authorities to whom Della Porta refers in this context, several date from late antiquity. Not surprising, Galen of Pergamum (fl. late second century CE) heads the list because of his enduring status as "prince of physicians," but Della Porta also refers to such secondary figures as Aëtius of Amida (c. 500 CE) and Paul of Aegina (seventh century CE). In addition to these classical and late-antique sources, two medieval Arabic authorities come to the fore. First and foremost is Avicenna, or Ibn Sīnā (c. 980–1037), whose *Al-Qānūn fī al-Ṭibb* (*Canon of Medicine*) was translated into Latin in the late twelfth century and became an authoritative text for the teaching of medicine at European universities from the late thirteenth to the seventeenth century, and even into the eighteenth century in some cases.[61]

Somewhat less significant as a medical authority was Averroes, or Ibn Rushd (1126–98). Best known for his Aristotle commentaries, Averroes also composed a medical compendium, the *Kitāb al-Kullīyāt fī al-Ṭibb* (*Book of Generalities in Medicine*), which was translated into Latin under the title *Colliget* toward the very end of the thirteenth century. This, too, was an important and influential medical source in the Latin West throughout the later Middle Ages and Renaissance. Closer to Della Porta's own time was Gentile da Foligno (d. 1348), the foremost commentator on Avicenna's *Canon* in his day and long after. Contemporary with Della Porta, finally, was Andreas Vesalius (1514–64), author of the *De Humani Corporis Fabrica*, from which Della Porta drew almost exclusively in his account of ocular anatomy in book 3 of the *De Refractione*. Della Porta was thus exceptionally well informed about the medical theory and practice of his day.[62]

Shifting from Della Porta's scientific or technical sources to those within the philosophical category, we find that the lion's share traces directly or indirectly back to Aristotle. Plato, as author of the *Timaeus*, is one exception. So is Plato's third-century CE systematizer, Plotinus, whose *Enneads* were available to Della Porta in the pathbreaking translation and commentary of Marsilio Ficino (1433–99). The first century CE philosopher/poet Lucretius is also mentioned by Della Porta, who quotes a fairly extensive passage from his long poem *De rerum natura* (*On the Nature of Things*).[63] Nevertheless, it is Aristotle and various of his commentators who figure most prominently among Della Porta's philosophical sources.

Of Aristotle's own works, it was generally to the *Meteorology*, the *Problems*, and the *Parva naturalia* that Della Porta had recourse, often quoting directly from them. Among the cited Aristotelian commentators, Alexander of Aphrodisias (fl. c. 200 CE) crops up most frequently, primarily through his *Problems* but also through his commentary on the *Meteorology*. Rounding out the list are Theophrastus (371–287 BCE), Aristotle's successor as head of the Peripatetic school; Olympiodorus of Alexandria (fl. c. 540 CE); Albertus Magnus (c. 1200–1280), the great Dominican compiler and interpreter of Aristotle's works; and Della Porta's older contemporary Agostino Nifo (c. 1470–c. 1538). Avicenna, as author of the *Liber de anima* (*Book of the Soul*), and Averroes, through his commentary on Aristotle's *On Sense*, also figure among these philosophical sources.[64]

Judged by how often they are cited and how extensively they are quoted, Della Porta's preferred sources (Aristotle perhaps excepted) were the encyclopedic ones. Cited with greatest frequency in book nine of the *De Refractione*, where Della Porta deals with meteorological phenomena, these sources include Albertus Magnus's contemporary, Vincent of Beauvais, whose immense compilation, *Speculum natural*, covers everything from cosmology to psychology. They also include a handful of Della Porta's contemporaries, most notably Girolamo Cardano (1501–76), Petrus Ramus (1515–72), and Julius Caesar Scaliger (1484–1558).

Pride of place, however, is held by a triumvirate of classical encyclopedists, the foremost of whom is Lucius Annaeus Seneca (d. 65 CE). A prolific

essayist and playwright, Seneca was author of the compendious *Natural Questions*, which Della Porta cited or drew from at least twenty times. More to the point, he quoted directly from it for a total of nearly 2,300 words in English translation. The *Natural History* of Seneca's contemporary Pliny comes in a fairly distant second, cited only nine times by Della Porta and quoted for a total of just over 370 words in English translation. Lagging even further behind is Macrobius (fl. early fifth century CE), whose *Saturnalia* Della Porta cited only four times for a total of just over 240 words in direct quotations.

In terms of sheer numbers, the fourth and final category, which consists of literary sources, is perhaps the most expansive, in great part because, as I said earlier, it forms something of a grab bag. Accordingly, it includes Leon Battista Alberti's *De pictura* and Albrecht Dürer's *Elementa Geometrica*, the 1532 Latin version of his *Unterweysung der Messung* (*Instruction for Measuring* [Nuremberg, 1525]), neither of which is "literary" in the strict sense. Likewise, such writers as Aratus, Suetonius, and Xenophon fit within this category, even though Aratus's *Phaenomena* focuses on astronomy, Xenophon's *Anabasis* on history, and Suetonius's *Lives of the Caesars* on biography. Such hybrid works aside, however, there remain several sources within this category that are indisputably literary. These include Apollonius of Rhodes and Valerius Flaccus, each responsible for a version of the *Argonautica*. They also include such distinguished classical poets as Hesiod, Virgil, and Ovid, as well as the tragic Greek playwright Euripides and his comedic Latin counterpart Plautus.

That Della Porta cast his net widely enough to capture such apparently inapposite sources as these may seem puzzling. Bear in mind, however, that in his day Della Porta was renowned as a comedic playwright and, as such, was heavily influenced by certain classical literary figures, Plautus in particular. Bear in mind, too, that Della Porta's appeal to such a wide variety of sources was not unusual for the time, especially among humanist-trained writers. A brief scan of Kepler's *Ad Vitellionem Paralipomena* of 1604 reveals much the same panoply of sources and source-types, and this work is regarded today as a scientific masterpiece. Nonetheless, it is fair to say that even by Renaissance standards Della Porta's polymathy was exceptional, even if a bit self-consciously overblown.[65]

The Making of This Edition and Translation

The Problem of Sources

Although Della Porta cited many of the authors just discussed by name—but not necessarily the relevant works—quite a few others are identifiable only by direct quotations without attribution. In most cases, such quotations consist of short passages somewhat crudely woven into Della Porta's narrative

with no indication whatever that they were borrowed, much less any intimation of the work from which they were borrowed. Worse yet, Della Porta rarely quoted his sources verbatim. More often than not he either paraphrased them, sometimes fairly loosely, or modified them to some extent by changing verbs and tenses or by altering syntax. Finding and identifying these sources has therefore been a significant challenge, and I am far from confident that I have unearthed them all.

No less significant has been the challenge of signaling these passages in the Latin text and incorporating them into the English translation. On the one hand, I had to play somewhat fast and loose with quotation marks, in some instances using them to bracket close paraphrases consisting of snippets of directly quoted text strung together by Della Porta's own verbiage. For the most part, however, the passages set off by quotation marks are fairly close to verbatim, and in a precious few cases they are truly verbatim. On the other hand, I had to identify, or at least attempt to identify, the actual, physical sources from which Della Porta drew his quoted passages. This meant ferreting out the appropriate printed works to which Della Porta would have had access and, if possible, pinpointing the particular edition or translation he might have used.

For instance, although it is certainly possible that Della Porta availed himself of the 1543 edition of Vesalius's *De Humani Corporis Fabrica* for his discussion of the eye and its anatomical structure at the beginning of book 3, I judged it likelier that he used the later edition of 1568 that was closer to the publication date of the *De Refractione*. Unfortunately, given my limited access to all possible sixteenth-century editions of the relevant works, I have not always been able to refer to the most proximate edition and have therefore had to settle for earlier ones. This is less problematic than it may seem at first blush because most sixteenth-century editions of a given work were based on earlier ones. Hence, although the specific page location of a given passage may have changed over time, the passage itself remained the same, or at least virtually the same. It is worth remembering, by the way, that not all of Della Porta's references or lifted passages came directly from the cited work. As mentioned earlier, some of his quotations came from Lodovico Ricchieri's *Lectionum Antiquarum*, although how often Della Porta actually relied on this work is uncertain. It is worth remembering as well that not all of his sources were printed; as we have seen, there is no doubt whatever that on occasion he consulted manuscripts. Nevertheless, determining by provenance what manuscripts he actually consulted is virtually impossible, and it is no less difficult to determine whether he read his Greek sources (e.g., Xenophon or Aratus) in the original language or in Latin translation. He certainly *could* have read them in the original Greek.

All things considered, therefore, we can be sure with varying degrees of certainty about several things related to Della Porta's sources and his use of them. Absolutely certain is that those sources range widely in both number and genre, adding up to more than seventy distinct works that run the gamut

from highly technical to purely literary. Somewhat less certain is how those sources vary by type, that is, according to whether Della Porta consulted them in manuscript or in printed form. Perhaps even less certain is how reliant he was on secondary as opposed to primary sources. In the first case, we can be fairly confident that the overwhelming majority of his sources were printed. In the second we can be somewhat less confident that all or at least most of his quotations from primary sources were based on actual use of those sources and not at second or third hand. Least certain, finally, is whether all of Della Porta's sources have been identified. The problem in this case is that his text is rife with unacknowledged passages, which, by their very anonymity, are difficult to identify. Nonetheless, I suspect that such unacknowledged and unidentified passages are fairly few and far between. I should add in closing that, although Della Porta was an egregious plagiarist by today's standards, he was anything but according to the far looser standards of his day, when imitation was not only a sincere form of flattery but also the mark of erudition.[66]

The Latin Edition and English Translation

Before examining the actual, optical content of Della Porta's *De Refractione*, I need to say a few words about my Latin edition and English translation, starting with an obvious question: Why include a Latin edition at all? After all, the original printed text on which it is based was rendered in eminently clear, readable Roman and Italic fonts and should therefore pose no problems of decipherment or interpretation. But the apparent readability of that text is misleading. For a start, although it is superficially clear, the original Latin text suffers from a number of deficiencies that make it hard to decipher properly and, on that account, to interpret easily. The root of the problem lies in its maladroitness. Whether the blame for this lies with the printers, Giovanni Giacomo Carlino and Antonio Pace, for failing to control the physical layout appropriately, or with Della Porta himself, for not exercising appropriate editorial supervision is an open question. Beyond question, however, is that the actual state of the text is chaotic enough to make reading it quite problematic at times.

A prime example of editorial sloppiness can be found in the sixth proposition of book 1, where the diagram that accompanies the text is inserted upside down. This is a glaring lapse, yet both the printer and Della Porta apparently failed to notice, much less correct it. Worse yet, in at least two instances (book 6, proposition 8, and book 8, proposition 6) the requisite diagrams are missing entirely. Even when appropriately provided, the diagrams are often mislabeled or misdrawn, causing them to be improperly keyed to the text, or vice-versa. Add to these lapses the occasional misnumbering of pages, and you have a clear sense that the book was compiled by the printer, or by Della Porta, or by both in greater haste than necessary or advisable.

A second set of problems pertains to the text itself. For one thing, the placement of commas, semicolons, colons, and full stops in the original Latin version tends to be idiosyncratic, even haphazard at times. Consequently, the resulting segmentation (or mis-segmentation) of text often impedes rather than facilitates interpretation by adding a layer of obscurity that the reader is forced to penetrate while trying to make sense of what is on the page. Punctuation aside, moreover, the text itself is sometimes muddled by typographical errors that reflect misunderstanding on the printer's part, sloppiness on Della Porta's part, or a bit of both on the part of both. Whatever the case, these sorts of errors can easily lead to misreading and misinterpretation.

In response to this set of problems, I have revised the Latin text in several ways. At the most basic level, I have modified the punctuation in an effort to make the text more easily accessible to a modern reader. To that end, I have not merely modernized the punctuation; I have also broken long Latin sentences into smaller, more easily digestible chunks. Such modification extends beyond mere punctuation marks or sentence segmentation to the imposition of paragraphs and other modifications of physical layout that are intended to ease the reader's way through the text. For example, I have rendered long quotations in block format in both the Latin text and English translation, even though they are not formatted that way in the original. In several instances, moreover, I have redrawn the diagrams in order to make them more accurately reflect the text to which they are tied. Conversely, I have at times revised the text (only slightly) so that it more accurately reflects the diagrams to which it is tied. Where appropriate, I have corrected the text itself, altering words or letters and signaling in footnotes whatever corrections were needed in the original. In making those alterations I have tried to be as unobtrusive as possible. Fortunately, the lion's share involves the mislabeling of diagrams, not problematic wording, so my editorial hand has been fairly light.

Strictly speaking, the Latin text that results from these modifications is not an edition at all—at least not in the diplomatic or critical sense—because it does not purport to recreate an Urtext. In short, it is not meant to accurately reflect the printed original or some ideal form from which that original derived. More loosely speaking, however, the revised Latin text approximates a critical edition insofar as it is meant to reflect Della Porta's authorial intent, or at least my interpretation of that intent. One final point: In order to facilitate comparison of my Latin text with the original, I have inserted the page numbers from the original text within brackets in boldface font in both the modernized Latin text and the English translation. I have also inserted the original diagrams, copied directly from the 1593 text, at the appropriate places in the modernized Latin text.

Like my Latin edition, the English translation based on it reflects my interpretation of authorial intent. This should come as neither a surprise nor a red flag to anyone familiar with the translating process, especially as it applies to languages as far apart in syntax and grammar as Latin and

English. My approach to translation therefore tends toward looseness rather than literalness because I am more concerned with textual content than I am with precise form. Not that I am unmindful of form. Quite the contrary, Della Porta's rather florid Latin style—perfectly exemplified in the dedicatory letter at the beginning of the treatise—is a constant reminder, as is his recherché vocabulary and orthography. Della Porta, after all, was humanistically educated and, on that account, imbued with a zest for extravagant Latinity that was fostered by his tutors, Giovanni Pisano in particular, whose mastery of the "elegant Roman style" Della Porta found especially praiseworthy. In addition to textual emendations, I have also corrected faulty diagrams whenever necessary, and in the two cases mentioned earlier I have actually constructed the missing diagrams to reflect the narrative tied to them.

Not all the stylistic peculiarities of Della Porta's text are his. Some trace back to the sources on which he drew for quoted passages. Many are classical and therefore couched in a Latin that defies easy, fluid translation into English. Whenever feasible, therefore, I have used published translations of these sources, either quoting passages verbatim from those translations or, in several cases, adapting them to Della Porta's usages. In every case of direct quotations, I have cited the Latin source according to an appropriate Renaissance edition. All such citations are keyed to the bibliography of primary sources mentioned earlier. Thus, a reference to "Plutarch (1530)" indicates *Plutarchi Chaeronei, Philosophi Historicique Clarissimi Opuscula (Quae Quidem Extant) Omnia, Undequaque Collecta, et Diligentissime Iampridem Recognita* (Basel, 1530), abbreviated in the bibliography of Primary Sources for the Text as *Plutarchi Chaeronei . . . Opuscula* (Basel, 1530). In the same vein, "Seneca (1529)" refers to *L. Annei Senecae Opera, et ad Dicendi Facultatem, et ad Bene Vivendum Utilissima, per Des. Erasmus Roterod. Ex Fide Veterum Codicum, Tum ex Probatis Autoribus, Postremo Sagaci Non Nunquam Divinatione, Sic Emendata, Ut Merito Priorem Aeditionem, Ipso Absente Peractam, Nolit pro Sua* (Basel, 1529), abbreviated in the bibliography of primary sources as *L. Annei Senecae Opera* (Basel, 1529). In typical fashion, I have supplied a running commentary for each book of *De Refractione*. Appended at the end of the given book, this takes the form of endnotes numerically keyed to the relevant passages in the text. Also, for those who are curious about some of the lesser known figures referred to in the text and commentary, I have provided a set of capsule biographies on pp. 475–86.

Finally, I should mention the format I have adopted in citing Della Porta's Latin sources in both printed and manuscript form. For the latter, I have chosen the conventional method of capitalizing the first word and leaving the rest in lower case. The manuscript version of Roger Bacon's work on the multiplication of species is therefore rendered *De multiplicatione specierum*. On the other hand, for the sake of consistent distinction, I have chosen to render the titles of printed sources in capitals throughout to match modern English usage. Consequently, even though the actual printed title

in a given Latin source may follow the format I have chosen for manuscripts (and many of them do), I will reformat the title to reflect modern English usage. Take, for instance, the 1541 collection of Galen's works, whose title is *GALENI PRIMA CLASSIS HUMANI corporis originem, formationem, dissectionem, temperaturam, facultates, facultatumque cum actiones omnes, tum instrumenta et loca singula complectitur*. This I render as *Galeni Prima Classis Humani Corporis Originem, Formationem, Dissectionem*, and so on.

Della Porta's Optical Analysis in *De Refractione*

At first glance, Della Porta's *De Refractione* seems to be mistitled because its ostensible focus is not on refraction per se but on visual perception taken broadly. However, the apparent discrepancy between titular and actual focus dwindles in the light of Della Porta's approach to vision, which rests on the assumption that how we perceive things visually is ultimately conditioned by how light rays are refracted into the eye. Such refraction, Della Porta insists, can make things look larger or smaller than they actually are. It can also make them appear displaced. Or it can make them look more distant or nearer than they actually are.

The refraction of light into the eye is therefore a crucial and consistent factor in how we visually perceive physical space and its denizens. Consequently, Della Porta's real purpose in *De Refractione* is to explore the various ways in which refraction affects vision by both determining and distorting it. Such ways include the interposition of refracting media between eye and object as well as the refractive perturbation of light that causes coloring in the rainbow and in twilight clouds. All of these points are brought out in the preface to book 1, where Della Porta assures us that the study of refraction "shines with its own light among the rest of the sciences by virtue of the nobility, exigency, usefulness, and praiseworthiness of its subject matter" (p. 9).

Della Porta also assures us that "this science springs from a mixture of mathematics and natural philosophy, and it receives and gathers together in itself whatever each of them on its own possesses in the way of delight and wonder . . . for whatever geometry produces and explores it clarifies and brings to light" (p. 11). Not surprising, therefore, Della Porta's analysis of optics throughout *De Refractione* is heavily dependent on geometrical demonstrations. In that respect his approach looks similar to that of his Perspectivist sources, Alhacen, Roger Bacon, Witelo, and Pecham. But a closer look reveals one significant difference: Whereas the Perspectivists (Alhacen and Witelo in particular) relied primarily on rigorous geometrical proofs, Della Porta was content to fall back on fairly loose geometrical illustrations. At the level of technical sophistication and exactitude, in short, Della Porta's geometrical analysis is comparatively primitive and imprecise.

The Physical Account of Refraction

Della Porta opens his study in the first book of De Refractione with a set
of seven definitions that establish the basic conceptual and terminological
grounds for his physical account of refraction. We are thus introduced to
such analytic entities as incident and refracted rays, the point of refraction,
the normal to that point, the cathetus of incidence, and the angles of incidence
and refraction, the latter consisting of what today we call the *angle of deviation*.[67] These things defined, Della Porta goes on to demonstrate a few basic
points about refraction. The first and most basic point is that when a light
ray passes obliquely from a denser medium, such as water, into a rarer
medium, such as air, it will be deflected away from the normal at the point
of refraction, which is where the light strikes the interface between the two
media.[68] By the same token, if the passage is in the opposite direction, from
the same rarer into the same denser medium, the ray will be deflected toward
the normal at the point of refraction by a commensurate amount. If, however,
the light passes through the interface orthogonally, right along the normal,
it will pass straight through without any deflection. All of these points Della
Porta confirms empirically in book 1, propositions 1–3, using an astrolabe
partially immersed in water with the sighting holes of the alidade poised to
allow a thin beam of light to shine through and strike the water's surface at
various angles or to allow light from an object below the water's surface to
pass through in the opposite direction.[69]

Having established these fundamental points, Della Porta turns to an
analysis of image location in propositions 4–7 in order to demonstrate, finally,
that the image of any spot on a luminous or illuminated object will appear
at the juncture of the refracted ray and the cathetus of incidence. For the
sake of an illustration, let BRB' in Figures 1a and 1b represent a plane or
spherical interface between air above and water below. Let R be the point
of refraction on that interface and NRN' the normal at that point. Normal
NRN' will therefore pass through the center N' of the spherical interface.
Let us start by supposing that A in the air is a viewpoint, A' a luminous point
in the water, and A'B' the cathetus of incidence, which, being perpendicular to
interface BRB', passes through center N' of the spherical interface. Instead
of continuing straight through along RI' after striking interface BRB' at point
R in both figures, the light traveling along incident ray A'R is deflected away
from normal NRN' to reach viewpoint A along refracted ray AR. Consequently,
from viewpoint A, the image I of luminous point A' will lie at the intersection
of cathetus of incidence A'B' and the rectilinear extension of refracted ray AR.

Conversely, if A' below the water is taken as the viewpoint, with A the
luminous point and AB the cathetus of incidence, then instead of continuing
straight through along RI, the light will be deflected toward normal NRN'
to reach viewpoint A' along RA'. Consequently, image I' of luminous point
A will lie where the extension of refracted ray A'R meets the extension of
cathetus of incidence AB. In both cases angles A'RI and ARI', which Della

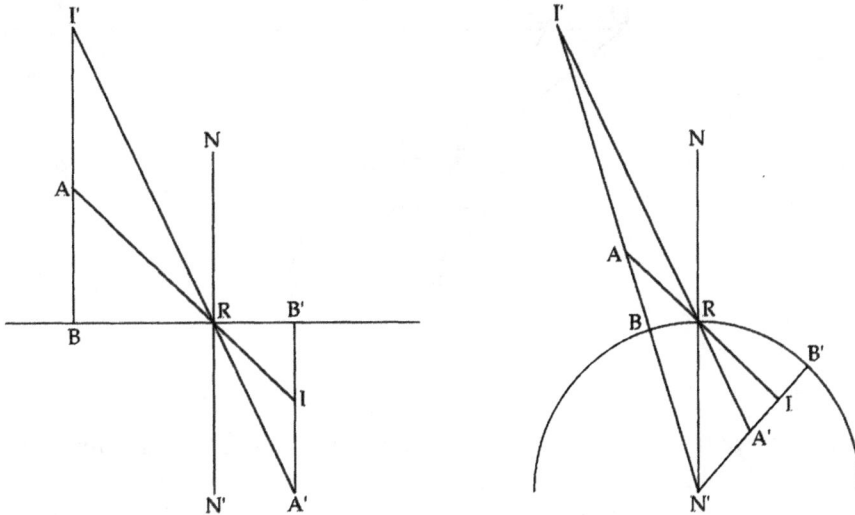

Figures 1a, 1b

Porta refers to as the *angles of refraction* (i.e., the angles of deviation according to the modern designation), are equal, because they are vertical angles. This is an important point because it underlies the principle of reciprocity, which holds that the path followed in refraction will be precisely the same no matter the direction of radiation. In other words, whether A or A' in Figure 1 is taken as the radiating point, the path will invariably be ARA'.

Della Porta's effort to establish these fundamental principles of refraction and the determination of image location that follows from them occupies the first seven propositions of book 1. In the remaining fifteen, he uses these principles to explain various visual illusions or distortions that arise from viewing things through refractive interfaces. In particular, he shows how image displacement can affect the apparent shape and size of objects seen at a slant through refractive interfaces. Such displacement, for instance, explains why large objects seen directly underwater appear concave and magnified (propositions 10 and 11), or why celestial objects not only appear to be at the horizon when they actually lie below it but also look magnified at that point (propositions 9 and 12).[70] It also explains why the bottom of a vessel partially filled with water can appear reduced in size when viewed at a slant (proposition 16).

These are just a few of the things that Della Porta accounts for on the basis of refractive image displacement, but one of the most potentially significant points he makes is in proposition 18, where he demonstrates that "to a viewer under water everything appears refracted below the water as it does to a viewer looking from air down through water" (p. 57). The proof, such as it is, is illustrated in Figure 2, where GHI is the outer, corneal surface of the eye centered on point B. Meantime, F and E are two points on the front surface of the lens inside the eye, and DA is the water's surface, beneath

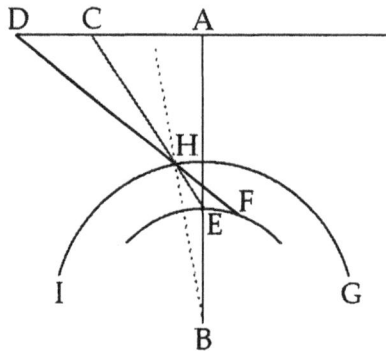

Figure 2

which the eye lies. Let C be a point on that surface, and let incident ray CH
from it strike the cornea at a slant. If there were no refractive interference
at point H, ray CH would continue straight to E, in which case point C
would be seen where it actually is from point E. However, as Della Porta
will establish later in book 3, the space between the cornea and the front
surface of the lens is filled with aqueous humor, a fluid somewhat rarer than
ordinary water according to his account. Consequently, when ray CH strikes
point H on the interface between the denser water and the rarer aqueous
humor, it is deflected away from normal BH along refracted ray HF. From
point F, therefore, the image of C will be seen along FD, and as D is displaced
to the left of C, interval AC on the water's surface will appear enlarged
along AD according to that displacement. Two key points emerge from this
somewhat peculiar and problematic analysis. First, the space between cornea
and lens is filled with a refractive medium somewhat rarer than water but
denser than air, and second, the image of C is seen from point F on the
lens's surface, not from center point B of the eye, where the Perspectivists
locate the center of sight.

Having to this point shared two basic assumptions about the anterior
portion of the eye—namely, that it is spherical and that it is filled with a
refractive medium—Della Porta is ready to undertake a study of refraction
through spherical glass lenses, to which he turns in book 2. The reason for
this study, he insists in the preface to that book, is that, because such a lens
"serves as a model of the eye, . . . the true locations of vision in the eyes
cannot be determined by us unless we touch upon a few things about that
sphere" (p. 79).

Della Porta broaches his analysis by positing a fundamental equivalence
between reflection from a concave spherical mirror and refraction through
a glass sphere. In the case of the concave spherical mirror, several things
can be determined. First, if we pass a diameter through the mirror's center
of curvature and mark the point where it intersects the reflecting surface,
and if we allow a ray parallel to that diameter to strike the surface at a point
more than 60° to the side of the intersection, the light will not reflect back

to the diameter inside the mirror. On the other hand, the closer to this intersection within an arc of 60° the ray strikes the reflecting surface, the higher up along the diameter it reflects inside the mirror until it reaches a limit at one half the mirror's radius.

These assertions are all illustrated in Figure 3a, where BGKMP is on a portion of the mirror's reflecting surface, D is the center of the sphere

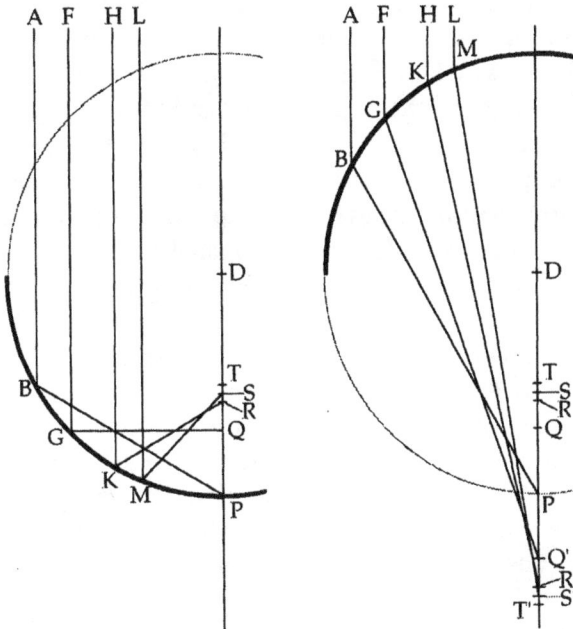

Figures 3a, 3b

containing it, and DP is a diameter of the sphere that intersects the reflecting surface at P. Let ray AB parallel to DP strike the mirror at point B, which lies 60° to the side of point P. Ray AB thus strikes the mirror's surface at an angle of incidence of 60°, and according to Della Porta's analysis, it will reflect to point P, where the diameter meets the reflecting surface. Any ray that strikes beyond B will do so at an angle greater than 60° and will therefore reflect along a line that, when continued straight, would intersect the diameter below P. Della Porta goes on to show that ray FG striking the mirror at 45° (i.e., such that arc GP = 45°) will reflect to point Q inside the mirror along ray GQ perpendicular to the diameter, whereas rays HK and LM that strike the reflecting surface at 30° and 22.5°, respectively, will reflect to points R and S.[71] Thus, the closer to P the rays strike the mirror's reflecting surface, the higher along diameter DP their reflected rays reach.

Nonetheless, no matter how close to P the incident rays strike, their reflected rays will always fall short of limit-point T, which bisects radius DP and constitutes the mirror's focal point or, as Della Porta calls it, the *point of inversion*. The reflected rays are therefore subject to spherical aberration, and the closer to P the incident rays strike, the closer their reflected rays

get to focal point T. In addition, as the intersections between reflected rays and diameter approach T, the more densely packed those intersections become. That is why sunlight gets so concentrated at a spot an infinitesimal distance below the focal point that it can ignite flammable material.

The equivalence Della Porta draws between reflection from a concave mirror and refraction through a glass sphere can be easily understood by recourse to Figure 3b, where BGKM is an arc on the glass sphere through whose center of curvature diameter DP passes. From point A in the air outside the sphere, ray AB parallel to DP strikes the refracting surface at an angle of 60° (i.e., such that the arc between B and the point at which diameter DP cuts the sphere's top surface is 60°) and is deflected toward the normal along BP. Hence, just as was the case with the concave spherical mirror, a ray incident at 60° will be diverted to the point where diameter DP meets the containing sphere. Now let ray FG parallel to the diameter strike the refracting surface at 45°. It will be refracted along GQ' such that PQ' = PQ, the distance above the mirror's surface that FG's reflected ray reached the diameter in the previous case illustrated in Figure 3a. PQ' is quite literally the mirror image of PQ. The same holds for rays HK and LM that strike the refracting surface at 30° and 22.5°, respectively. They will refract to points R' and S', respectively, such that PR' = PR, and PS' = PS. By the same reasoning, therefore, PT' = PT, from which it follows that T' is the glass sphere's focal point at a distance of half a radius from its bottom surface. It is important to note that this analysis applies to refraction *into* the glass sphere only, not *through* it, because Della Porta did not take into account the second refraction out of the sphere, even though he was well aware that such refraction occurs. In other words, his failure to acknowledge this second refraction was purposeful, not inadvertent or thoughtless.

Della Porta suggests four ways to test this theoretical model on the basis of an experimental apparatus that consists of a plaque with appropriate lines scribed on it, a glass plate cut to appropriate shape, and a filter to channel sunlight through the edge of that plate at the appropriate angles. Two examples are illustrated in Figures 4a and 4b, which reflect his analysis in book 2, propositions 4 and 5, respectively. First, Della Porta invites us to take a glass plate around 26 cm long, 13 cm wide, and 2 cm thick. Out of this plate we are to cut a semicircular section, a bit more than half of which is represented by the light-gray segment in Figure 4a. A little bit more than half of what remains after the cut is represented by the light-gray segment in Figure 4b. We then attach these chunks of glass to a plaque on which the appropriate lines have been drawn. For instance, AB, FG, HK, and LM will be scribed parallel to diameter EP and will cut off arcs BP, GP, KP, and MP of 60°, 45°, 30°, and 22.5°, respectively on the semicircular arc of both glass plates. Then, taking a thin wooden or copper strip that is somewhat wider than the glass plate is thick, we punch tiny holes in it at points corresponding with A, F, H, and L. These will form intervals such that, when we place the strip upright and parallel to the straight edge of the glass plate

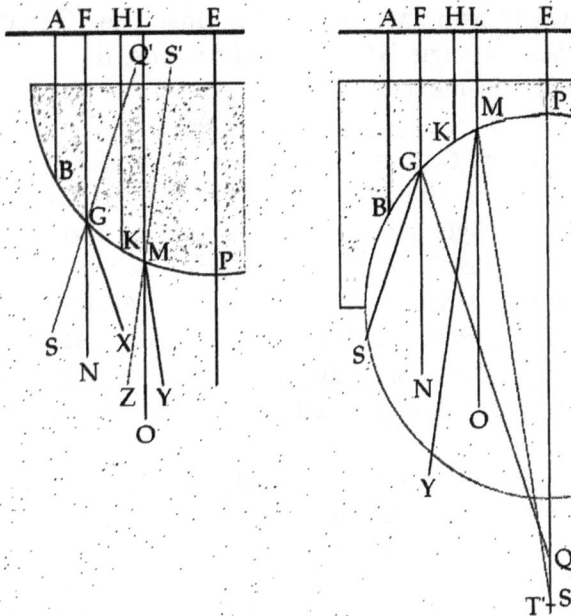

Figures 4a, 4b

on the plaque and allow thin beams of sunlight to shine through the holes to the straight edge of each glass chunk, those beams will be precisely channeled along AB, FG, HK, and LM. They will therefore reach unrefracted through the straight edge of both glass chunks to their bottom arcs, which form the refractive interface between glass and air.

Take ray FG, which strikes the arc at 45° in both cases. Starting with the case illustrated in Figure 4b, let us suppose that the space below arc PMKGB consists of glass rather than air and, conversely, that all the space above it consists of air rather than glass. According to Della Porta's previous theoretical analysis, ray FG would be deflected toward the normal on entering the glass below arc PB and would refract along GQ'. In reality, though, ray FG passes unrefracted through the straight, top edge of the glass plate, reaches G, and is deflected away from the normal on passing into the air below arc PB. By how much it is deflected is easily determined according to Figure 4a. Assume that NG, which is colinear with FG, is a ray that strikes arc BP of the glass disk at 45°. It will be refracted toward the normal along gray line GQ' at an angle determined earlier. In reality, though, ray FG, which is colinear with NG, passes through the straight edge of the glass disk unrefracted, reaches G, and is then refracted away from the normal along some line GX.

Now according to the principle of reciprocity, no matter what direction light passes along in going from one medium to the other, it will follow an identical path. Consequently, since light refracting into the glass along NG and incident at 45° refracts toward the normal along GQ', which is colinear

with GS, then light incident along FG at 45° in the opposite direction should
follow an equivalent path away from the normal along GX such that angle
FGQ' = angle SGN = angle NGX. In other words, the angle of deviation
(Della Porta's angle of refraction) from glass to air is equal to the angle of
deviation from air to glass. By the same token, in Figure 4b angle NGS
formed by refracted ray GS and continuation GN of the incident ray is equal
to angle NGQ'. Likewise, for ray LM, which strikes arc MB in both cases at
an angle of 22.5°, the light is refracted away from the normal along MY such
that angle OMY = angles OMZ and OMS', respectively in Figures 4a and 4b.
Applying the same rationale to rays AB and HK, which are incident at 60°
and 30°, respectively, we can easily determine the lines along which the light
should refract out of the glass. Thus, if we scribe those lines on the plaque
and then pass the light through each of the holes, the light should follow
those lines.

There are several problems with this experimental confirmation of Della
Porta's model, uppermost among them that the model itself does not conform
to the sine law, which is the correct governing principle of refraction. Therefore, the expected angle of refraction of 30° for an angle of incidence of 60°
according to Della Porta's model is more than 5° short of the value predicted
by the sine law. This is a major discrepancy, and so is the difference of nearly
3° between what Della Porta's model predicts and what the sine law yields
for an angle of 45°: that is, 25.5° versus 28.2°. On the other hand, for angles
of incidence of 30° and 22.5° the discrepancy dwindles to 1.8° and 1.4 °,
respectively, which can perhaps be interpreted as small enough not to constitute outright disconfirmation of the model.[72]

But inaccuracy of results is not the only problem. As it turns out, both
experiments fail to take into account the critical angle, which comes into
play when light passes from a medium that is optically denser into one that
is optically rarer. In this case, with the light passing from what amounts to
crown glass into air, the critical angle is nearly 41°, which means that, if
the light strikes interface BP between the two media at an angle of more
than 41°, all of it will be reflected internally instead of passing through.[73]
Hence, instead of refracting out of the glass along expected paths GX and
GS, respectively, in Figures 4a and 4b, ray FG incident at 45° will be reflected
internally, a fact that Della Porta should have noted had he carried out the
experiments with due care and an appropriately open mind. Nonetheless,
the experimental apparatus is crude enough that various factors, such as
diffraction, relatively poor quality of glass, inexactitude of measurement, and
so forth, could have skewed the results enough to persuade Della Porta that
they favored his model, particularly if he was so firmly wedded to it as to
be blind to all but the grossest discrepancies.

That Della Porta's model was wrong does not mean it was ill conceived
or foolish. There really are fundamental convergences between reflection
from concave spherical mirrors and refraction through glass spheres that
suggest a close kinship between the two. For one thing, both devices bring

sunlight to focus at or very near a particular point, and in both cases that point lies around half a radius from the reflective or refractive surface.[74] In both cases, moreover, the reflected or refracted light is subject to spherical aberration, and, in both cases, because of such aberration, the area on the reflecting or refracting surface implicated in such focusing is quite small. One final and important convergence between the two devices has to do with the way they produce images. Both can cast real images, and in both cases those images are inverted. In addition, what is seen in both devices can appear magnified or reduced in size as well as upright or inverted, depending on the placement of viewpoint and object. Also, depending on that placement, what is seen in those devices can appear to lie behind, on, or in front of the reflecting or refracting surface.

Although Alhacen and his Perspectivist colleagues had done a creditable job of accounting for these phenomena in concave spherical mirrors, they barely touched on spherical glass lenses, and when they did, the results were problematic.[75] Recognizing the deficiencies of the Perspectivist account, Della Porta undertook to analyze such lenses and their imaging properties more intensively and extensively in a series of fifteen propositions in book 2. The first five (i.e., propositions 7–11) analyze refraction of light through a spherical lens from a single luminous point to a single viewpoint in order to determine where the image will be located from that viewpoint. This determination is contingent on where and whether the cathetus of incidence through the lens's center meets the ray refracted to the viewpoint. Accordingly, in proposition 7·Della Porta shows that the image can lie between the viewpoint and the lens's facing surface, whereas in propositions 8, 9, and 11, he shows that the image can lie on the lens's facing surface, at the viewpoint itself, or behind the viewpoint. In proposition 10, on the other hand, he shows that if the refracted ray and the cathetus of incidence are parallel and therefore do not meet, there will be no image at all.

Having so far dealt with point objects and images, Della Porta moves to two- and three-dimensional objects and images in propositions 12–20 (proposition 15 excepted), where he shows not only how those images can appear magnified, reduced in size, inverted, or distorted in shape, but also where they might lie in respect to the refracting surface or the surface that faces the viewpoint. Several of these attempts are marked by Della Porta's bewildering failure to follow the basic rules of image location. A prime example is proposition 15, where he purports to demonstrate that "two or three images of a single magnitude can be seen in a glass sphere" (p. 113). His proof is as follows. Let A in Figure 5 be the magnitude facing spherical lens DEGF centered on L, and let I and H be the right and left eyes, respectively. The light at A is radiated along AD and AE to the top surface of the lens, then refracted along DF and EG in the lens, and finally refracted out along FH and GI to the respective eyes. Cathetus HL from the left eye through the lens's center point L will meet incident ray AE at C, and cathetus IL from the right eye through center point L will meet incident ray AD at B.

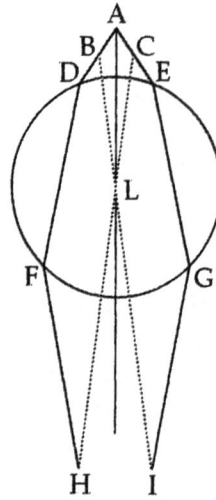

Figure 5

Consequently, Della Porta concludes, B will be A's image from the perspective of I, whereas C will be its image from the perspective of H. It follows, then, that A will appear doubled at B and C.

The flaws in this analysis are as blatant as they are puzzling. First, neither HLC nor ILB is a cathetus of incidence because neither originates at object A. The real and only cathetus of incidence is AL. Second, AD and AE are not refracted rays; they are incident rays. Earlier, however, in proposition 7 of book 1, Della Porta had established unequivocally that "the refracted image of an object will appear at the intersection of the cathetus [of incidence] and the extended line of refraction" (p. 31). Clearly, this rule has been misapplied in the proof. Nor is this misapplication a fluke; Della Porta flouts the cathetus rule three more times, in propositions 13, 17, and 19. This is all the more puzzling in light of his correct application of the rule in propositions 12, 14, 16, and 20, as well as in proposition 21, where he concludes his analysis of imaging by showing that "the image of a magnitude reaching the sphere at a slant appears to lie far away from its proper place and looks otherwise [than it actually is]" (p. 121). Della Porta then rounds out book 2 by showing in proposition 22 that light can be focused to a burning point through a glass sphere, from which it follows in proposition 23 that, by placing a concave spherical mirror at that point, we can redirect the light through the lens to a great distance.

If we continue topically rather than book by book, then the natural sequel to book 2 is book 8, where Della Porta deals with biconvex and biconcave lenses on the basis of his analysis of spherical lenses in book 2. Take proposition 6 of book 8 as an example. The point of this proposition is to prove that, when an object is placed near one surface of a biconvex lens, it will always appear upright to an eye on the other side of the lens, no matter how near or far from the lens the eye is. Della Porta starts by

assuming that the biconvex lens in Figure 6 is composed of two intersecting spheres, one of which is centered on C and contains the bottom surface of the lens, the other of which contains the top surface.

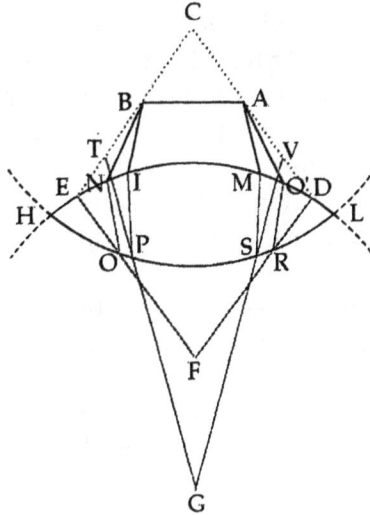

Figure 6

Let AB be a magnitude facing the lens's top surface, and let F be the eye. The light from endpoint B of the magnitude will strike the lens's top surface along ray BN, will then refract into the lens along ray NO at the angle determined by the model in book 2, and will finally exit the bottom surface along OF at an angle equal to the original angle of incidence at N, as dictated by the principle of reciprocity. According to the cathetus rule, B's image will lie at E on the lens's top surface, where the extension of cathetus of incidence CB intersects the extension of refracted ray OF. The same holds symmetrically for endpoint A of the magnitude; light from it will strike the mirror along AO', will be refracted into the lens along O'R, and will exit along RF, so its image will appear at D on the lens's top surface. Endpoints E and D of the image will therefore have the same orientation as the endpoints of magnitude AB.

Now move the eye down from F to G. In that case, the path the light from B follows in order to reach G will be along BI, IP, and PG, so B's image will lie at T, where the extension of cathetus CB intersects the extension of refracted ray PG. The same holds symmetrically for endpoint A, whose light passes to the eye at G along AM, MS, and SG so that its image V lies at the intersection of refracted ray GS extended to meet the extension of cathetus CA. Just as before, when the eye was at F, the image's endpoints T and V of the image will have the same orientation as endpoints B and A of the magnitude.

Using this same analytic technique, Della Porta shows both when and why things will appear inverted through biconvex lenses (propositions 7 and 8) as well as when and why they will appear magnified or reduced in

size in such lenses (propositions 8–10). In proposition 8, however, he violates the cathetus rule of image location by tying the wrong cathetus to the wrong refracted ray, even though his final conclusion (i.e., that the image in this case will be inverted and smaller than the object) actually follows from a correct use of the cathetus rule. Della Porta also applies the analysis of intersecting spheres to show in proposition 13 that sunlight refracted through biconvex lenses can be brought to a burning point, the same conclusion he drew for spherical lenses in book 2, proposition 22. He concludes his analysis of biconvex lenses in proposition 14 with a brief discussion of how they remedy presbyopia by concentrating the image conveyed by all the radiation passing into the lens and, in the process, brightening that image for eyes weakened by old age.

Turning to biconcave lenses in propositions 15–18 of book 8, Della Porta uses a similar analytic technique with appropriate adjustments. Take as an example proposition 16, where he sets out to prove that an object seen through a biconcave lens will appear increasingly small as the eye draws farther away from the lens. The construction for the proof is illustrated in

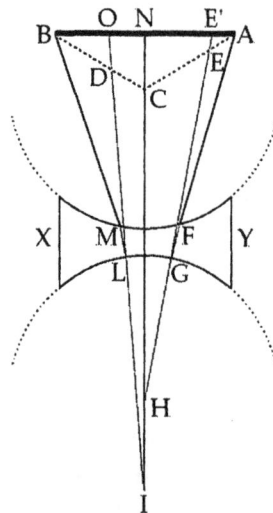

Figure 7

Figure 7. Imagine two air-filled spheres standing apart from one another, and let the top one be centered on C. Then imagine the space between the two spheres filled with glass so as to form biconcave lens XMFYLG, and let AB be an object in the air above facing its top surface.

Suppose first that the eye is at H in the air below the lens. In that case, after radiating along AF to the lens's front surface, the light from endpoint A of the object will refract toward the normal along FG, as it enters the lens, and will then be refracted away from the normal along GH on exiting. The image of the object's right-hand endpoint A will thus lie at E, where the extension of refracted ray GH intersects cathetus AC. Extended beyond E,

refracted ray HG will intersect the object at E', so the image of A will be seen along line of sight AE'. The same holds symmetrically for endpoint B; its light will follow an equivalent path to reach H. Consequently, the image of NA, which constitutes half the object, will be seen within NE', which is smaller than NA. Because an equally reduced portion of the other half of the object between N and B will be seen, it follows that the whole image will be smaller than the whole object.

Now move the eye farther down to I. The resulting image of the object's left-hand endpoint B will lie at D, and the line of sight along which it is seen will intersect the object at O, so BN, the other half of the object, will be seen within ON, which is smaller than NE'. The entire image viewed by the eye farther from the lens at I will therefore be smaller than the entire image seen by the eye nearer the lens at H.

In one key respect the demonstration as just outlined is misleading: Figure 7, which I drew according to the given parameters, does not accurately reflect its counterpart in the original Latin text, which is reproduced directly from that text in Figure 8a. Like all of the diagrams representing refraction into biconcave glass lenses in the original Latin text (i.e., in propositions

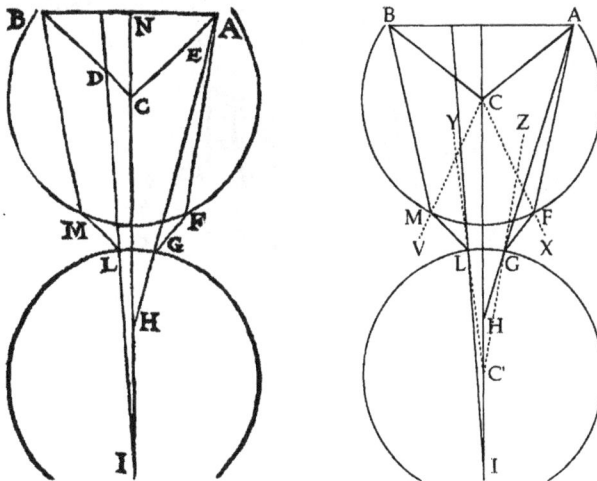

Figures 8a, 8b

15, 17, and 18), this figure misrepresents such refraction because, instead of having the rays diverted toward the normal on entering the lens, it shows them shunted away from it, as if the light were passing from glass into air. Thus, in Figure 8b, which is copied directly from the original, if we drop normals CV and CX through points of refraction M and F from center of curvature C of the lens's top surface, it is obvious that the angles of incidence BMC and AFC will be considerably smaller than angles of refraction VML and XFG formed by refracted rays ML and FG with those same normals. It should be equally clear that, if we drop normals C'Y and C'Z from center

point C' of the lens's bottom surface through points of refraction L and G, the angles of incidence MLY and FGZ formed by incident rays ML and FG with those normals will be considerably larger than the angles of refraction C'LI and C'GH formed by refracted rays LI and GH with those same normals. In short, Della Porta's original diagram represents light passing from glass through air into glass rather than from air through glass into air, which is to say that the biconvex lens in his diagram is filled with air, not glass, whereas the spheres are filled with glass, not air.

   This apparent confusion on Della Porta's part may help account for his peculiarly inept explanation of the lenticular correction of myopia, or nearsightedness, which caps his physical analysis of refraction in the nineteenth and final proposition. Correcting myopia requires two things according to him: "first, that the images dilate and somehow unite . . . and second that the light be rendered brighter." Both of these things are accomplished by a biconcave lens, he argues, "for it somehow unites the image [conveyed by incoming radiation and also] opens it out somewhat, as we see with the lines fanning out in the opposite direction, and thus the light that passes through is multiplied" (p. 383).

   What Della Porta is getting at here seems to follow from his analysis of parallel radiation into and through a biconcave lens in proposition 18,

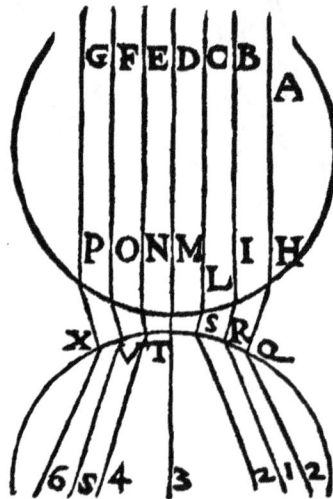

Figure 9

which is accompanied by Figure 9 in the original Latin text. The first thing to notice is that, instead of diverging, as they should when refracted into the lens, parallel rays AH, BI, CL, and so on converge along HQ, IR, LS, and so forth.[76] It is presumably this convergence that unites the image conveyed by those rays and, in the process, congregates the incoming light so as to brighten it. On the other hand, in causing the radiation to diverge along Q2, R1, S2, and so on after exiting, the lens "opens [the image] out somewhat"

in order to "dilate" it for a wider, clearer view to a myopic eye facing the bottom surface of the lens.

From this botched account, based as it is on an incorrect analysis of refraction into the biconcave lens, it is evident that Della Porta was confused about how such lenses actually work. This confusion, for its part, seems to reflect a somewhat cavalier disregard for analytic details, a disregard that may have led to Della Porta's occasional misapplication of the cathetus rule when dealing earlier with biconvex lenses. As we will see in the next section, where we discuss his account of vision, he also misapplied the cathetus rule when dealing with the refraction of light into the eye itself.

The Workings of Sight

Della Porta begins his study of vision in book 3, proposition 1, by describing the structure of the eye on the basis, primarily, of Andreas Vesalius's account of ocular anatomy in book 7, chapter 14, of *De Humani Corporis Fabrica*.

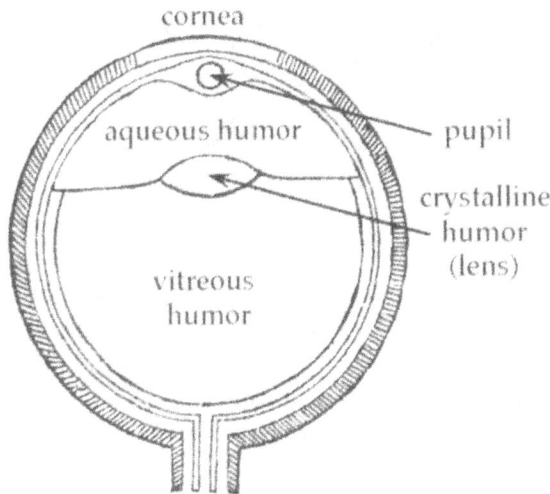

Figure 10

The resulting model is illustrated in Figure 10, adapted from the diagram that accompanies proposition 1 in the original Latin text of book 3. At the very front of the eye is the transparent cornea, which is continuous with the scleral sphere enclosing the eye as a whole. Opposite the cornea at the back of the eye is the optic nerve, which, as mentioned earlier, joins its mate from the other eye at the optic chiasma toward the forefront of the brain. Just beneath the cornea lies the pupil, which forms an aperture in the uveal membrane nesting inside the scleral sphere and concentric with it. Below the pupil, but still toward the front of the eye, lies the crystalline humor that forms the lens. The space between it and the cornea is filled with aqueous humor. Somewhat rarer than ordinary water by Della Porta's account, aqueous

humor is also rarer than the crystalline humor behind it. Finally, behind the crystalline humor lies the vitreous humor, which fills the remainder of the eye and is somewhat denser than the crystalline humor.

After describing the structure of the eye, Della Porta undertakes to show why that structure is best suited to the eye's functioning. For example, in proposition 2 he offers several reasons why the eye had to be spherical and, moreover, why it had to be convex rather than concave. One reason it could not be concave, he argues, is that if it were, the image conveyed by all the radiation passing through the cornea to the lens would be inverted. In that case we do not see things upright.[77] Della Porta's rationale for this claim is

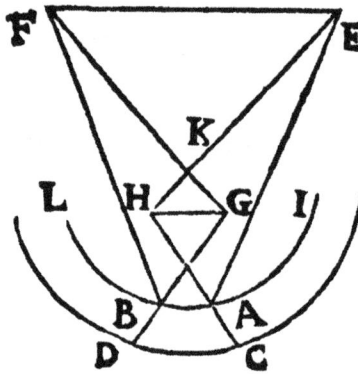

Figure 11

based on Figure 11, reproduced directly from the original Latin text. As depicted in that figure, arc IL represents the hypothetically concave cornea and DC the hypothetically concave lens behind it, both centered on K outside the eye. FE, meantime, represents an object facing the concave eye. Light from its endpoint F will radiate to point B on the cornea, will then be refracted toward the normal on entering the denser aqueous humor, and will follow refracted ray BD to point D on the lens. If that ray is extended out along BG, it will intersect the extension of cathetus of incidence FK at G. Likewise, light from endpoint E of the object will radiate to A and refract to C, so the extension of refracted ray AC will intersect the extension of cathetus EK at H. Because image G of endpoint F lies on the opposite side of the eye's center point K, as does image H of endpoint E, the object and its image will have the opposite orientation. FE's image GH will thus be inverted.

The problem with this demonstration—and it points up Della Porta's apparent disregard for analytic details—is that it requires incident rays FB and EA to refract not just *toward* the normal, but beyond it, when entering the aqueous humor. This, of course, is impossible according to the standard rules of refraction established by Ptolemy, enforced by the Perspectivists, and followed by Della Porta himself when analyzing refraction in books 1

and 2. Those rules dictate that, upon entering the eye, rays FB and EA be diverted toward the normal in such a way that extensions BG and AH of refracted rays DB and CA form fairly small angles with incident rays FB and EA. As a result, BG and AH will meet their respective catheti above center point K and on the same side of it as their generating point objects. Image H of E will therefore lie between E and K, and image G of F between F and K. Contrary to Della Porta's claim, then, the image will maintain the same orientation as its object.[78]

Della Porta identifies other factors besides its spherical shape that conduce to the eye's proper functioning. One such factor, and a crucial one, is the placement of the lens. According to Della Porta, it lies relatively close to the cornea so that it can be exposed to a wider view than would be possible were it situated further down in the eye at its center.[79] Also, as he points out in propositions 3 and 4, the aqueous humor helps widen that view by forcing rays that reach the cornea obliquely to refract toward the lens. In this way, a considerable amount of incoming radiation that would bypass the lens if it were allowed to pass straight through the cornea is diverted to its surface and thus seen. It is for this reason, Della Porta observes later, in book 4, proposition 10, that the eye's field of view is greater than 90°. Because of its exquisite clarity, moreover, the aqueous humor brightens the images conveyed by all the incoming radiation, thereby making them more clearly visible.

Although the placement of the lens and the refractive property of the aqueous humor conspire to expand the eye's visual scope, that scope is limited by the pupil, whose size varies according to the ambient light. In dazzling light, for instance, the pupil can contract to the size of a needle point, whereas in dim light it can dilate to the size of a lentil according to Della Porta's description in proposition 6. The reason the pupil reacts in this way to light or its absence is twofold. On the one hand, as Della Porta argues in proposition 6, the pupil "is set over the eye to serve as a vestibule and to be an obstacle to immoderate light during the day" (p. 155). In other words, it serves as a protective filter. Otherwise, too much incoming sunlight "would ignite a fire at the back of the eye or would generate a great deal of heat" at or near the retina (p. 159). Without the pupil to guard it, then, the lens would actually act like an optical lens and focus incoming light to a burning point in the eye. On the other hand, the pupil dilates in dim light "so that images can enter more freely through a wider opening," the result being that those images are rendered clearer and brighter (p. 155).

Della Porta offers two experimental verifications of the pupil's capacity to contract and dilate according to ambient light. The first one is described in proposition 6, where he suggests that we form ruler FDE in Figure 12 around 26 cm. long by around 2 cm. wide and scoop out a notch at E so that the ruler can be nested up to the eye. We are then to take two cylindrical pegs, G the larger and C the smaller, and place them so that both lie on the ruler's midline with smaller peg C on the side of the eye at E. Next, we are

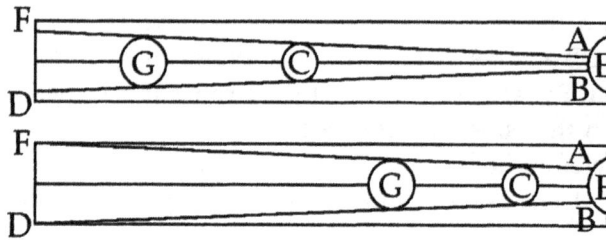

Figure 12

to hold the ruler up to a brightly lit object or landscape and move the two pegs back and forth until peg C just occludes peg G. Marking the circles at the bases of the two pegs and drawing tangents FA and DB to both of them, we measure the interval AB between the lines at the notch, as represented in the top diagram of Figure 12. This interval is commensurate with the width of the pupil that lies slightly beyond E. Repeating the experiment in dim light, we find that the interval AB, as represented in the bottom diagram of Figure 12, is greater than it was in bright light, which means that the pupil has dilated in response to the less-intense light.

A second way of verifying the pupil's reaction to brightness, this time keyed to the vividness of color, is suggested in proposition 8, where Della Porta has us start with a square board EDFG with sides of around 38 cm.,

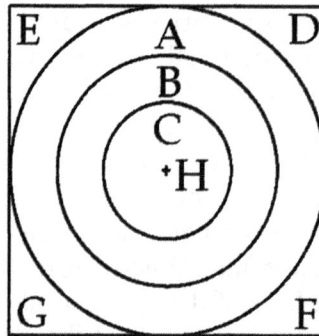

Figure 13

as represented in Figure 13. After hammering a long nail in its center H so that the head lies around 15 cm. above the surface, we are to paint the board pitch black. Then, placing our eye just above the nail and looking straight down while paying attention to our peripheral vision, we are to place small pebbles on the board as far to the edges as we can detect those pebbles. They will form the large, outer circle A. Repainting the board green and repeating the experiment, we find that the pebbles form smaller circle B. Finally, after painting the board bright white and repeating the experiment yet again, we find that the circle has shrunk to the size of the innermost

circle C. Three things thus follow from this experiment. First, the pupil is reactive to brightness, be it of light or of color. Second, the more dilated the pupil, the wider the view. And third, the field of view defined by the pupil is circular.

With these points in mind, Della Porta enters into a discussion of how the pupil's natural size and shape are adapted to the needs of particular animals. In proposition 9, for example, he argues that "those with a wide pupil see better at night . . . whereas those with a narrow pupil see more acutely during the day" (p. 165). The reason has already been given. On the one hand, a narrow pupil filters out excessive light, allowing only enough into the eye to render the images it conveys clearly visible without overstimulating the sensitive lens or overheating the retina at the back of the eye. That is why animals with normally narrow pupils, such as humans, hens, and doves, tend to be active during the day because they see well in bright light without being unduly offended by it. On the other hand, animals such as wolves and cats have naturally wide pupils, which allow more light to reach the lens and stimulate vision. These animals therefore tend to be nocturnal, avoiding daylight because of its relatively blinding light. In explaining night vision in this way, Della Porta flies in the face of the traditional explanation, which links the keenness of night vision among nocturnal animals to the relatively large amount of visual illumination they emit, as is evident from the way their eyes glow in the dark. Their eyes, in short, act as flashlights. Against this Della Porta argues that such glow is due to reflecting glitter from the highly polished crystalline lens, which is exposed to as much light as possible through the naturally wide pupil.[80]

Not only the size but also the shape of the pupil is adapted to the needs of particular animals. For instance, in the tenth proposition of the third book, Della Porta asserts that the naturally large pupils of nocturnal animals are hinged "from top to bottom, like gates, so that [they can] supply themselves with nourishment more quickly and more freely" (p. 167) according to the broad vertical field of view enabled by such pupils. On the other hand, he continues, in certain fish the pupil is hinged in the opposite way so as to provide a wide horizontal field of view, which is useful for looking to the sides. For human needs, finally, a round pupil is best suited because, as Della Porta puts it in proposition 11, such a pupil enables us to "see magnitudes facing all around without impediment" (p. 167). In other words, being endowed with round pupils gives humans the advantage of visual scope in all directions even though this scope is somewhat limited along the vertical and horizontal in comparison to that of other animals with more specialized needs. Perhaps with an eye toward skeptical Church authorities, Della Porta is emphatic throughout his discussion of the eye that it was designed in every way for the best by God, who was acting according to his supreme wisdom and benevolence.

Della Porta's denial that nocturnal animals see in the dark because of visual illumination follows from his rejection of extramissionism in favor of intromissionism. He makes a strong case for this rejection in the first proposi-

tion of book 4 and in the course of making it raises an interesting point: "Just as the light shining through a narrow window onto paper held near it represents objects illuminated by the Sun, so it depicts the images of visible things by shining through the opening of the pupil and falling onto the crystalline humor" (p. 193). The eye, in short, acts just like a camera obscura, projecting a real image onto the front surface of the lens, a point Della Porta had already made in both versions of the *Magiae Naturalis*. The problem is that such an image is necessarily inverted whereas we see things upright. Della Porta skirts this problem by supposing that the front of the lens contains an infinitude of viewpoints, each one possessing unique radial ties with particular points in space.

How this works in practice can be seen in the following demonstration, which occurs toward the end of book 4, proposition 1. A viewer is asked to look straight ahead intently along FE in Figure 14. The outer arc through

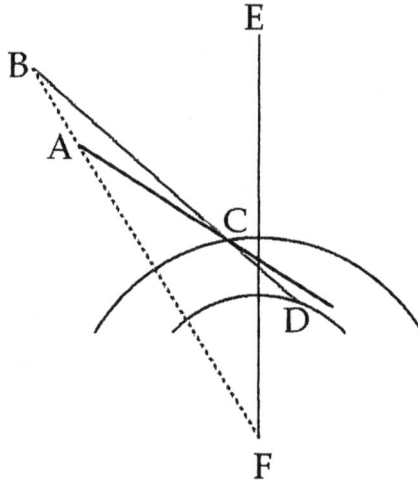

Figure 14

point C represents the corneal surface and the inner arc through D the anterior surface of the lens. Someone else is to place his finger at position A, and when the viewer attempts to touch it, he will miss, touching position B above it instead. The explanation is simplicity itself. Let ray AC from the fingertip reach point C at a slant on the cornea. Instead of passing straight through the cornea and missing the lens in this case, it will be refracted to point D on the lens along refracted ray CD. Extend that ray beyond C toward B and draw cathetus FA, extending it toward B as well, until the two meet there. B is therefore the image of A, which is where the viewer supposes the fingertip to exist in space. That of course is why he mistakenly touches this point instead of the fingertip itself.

The key thing about this analysis is that, even though the light from point A to the left of normal EF refracts to point D on the lens to the right

of that normal, the image of A will be seen to the left of EF and will therefore have the same orientation as A. In order to understand how this same analysis applies to the entire field of view and the points within it, let us turn to Figure 15, where arc DF represents the cornea, the pupil lying just below

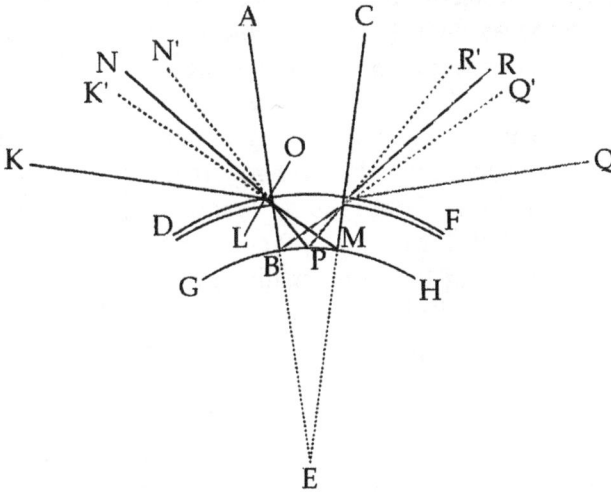

Figure 15

it. Arc GH below the pupil represents the anterior surface of the lens, and E is the center of the ocular sphere. For a start, in book 4, proposition 6, Della Porta agrees with the Perspectivists that vision takes place by means of a cone, which is represented by AEC in the figure. All of the rays within that cone, including AE and CE at the edges, reach the cornea along the perpendicular and pass unrefracted to the lens, where they are sensed at the point of incidence. For that reason, all of the points within the base of the cone outside the eye are seen where they actually are, point A, for example, being seen where it actually is from point B on the lens.

But this only accounts for part of what is seen because the cone is severely limited in scope by the pupil. In view of this fact, Della Porta goes on to argue in book 4, proposition 7 that vision is also accomplished by means of refracted rays. Take point K outside cone of direct visibility AEC. Light from it will radiate to the cornea along KL and will then refract along LM to the lens. K will therefore be seen along refracted ray MO extended toward K'. The same holds for point N outside the cone of direct visibility. Its light reaches the cornea along NO and is then refracted to the lens along OP, so it will be seen along PON', and everything described for K and N on the left-hand side of cone AEC will be mirrored on its right-hand side. Accordingly, Q and R, which are K's and N's counterparts on the left, will be seen along BQ' and PR', respectively.

Several salient points emerge from this analysis. First, although the orientation of points K and N outside cone AEC is opposite that of their

respective points of incidence M and P on the lens, their images retain their original orientation. Hence, the light from K to the left of N strikes the lens at M to the right of P, where the light from N strikes the lens, and the same holds for Q and R on the right-hand side. Yet, despite this reversal on the lens's surface, the resulting images seen along MK', PN', BQ', and PR' maintain the same orientation as their generating objects. Second, because the rays incident on the cornea's surface are refracted into the eye, the apparent field of view between K' and Q', or between N' and R', is smaller than the actual spatial separation between the respective generating points K and Q or N and R. Under normal conditions, then, the apparent field of view is somewhat smaller than the actual one. Third, the very same point on the lens can have multiple views such that, for example, P can see N along PN', while at the same time, seeing R along PR'. And finally, the farther off to the side the point viewed is, the more displaced its image will be. This becomes evident if we imagine the cathetus drawn from center point E of the eye through point K, and then we imagine extending it until it meets the extension of refracted ray MLK'. If we also imagine the cathetus drawn from E through N and extend it until it meets the extension of refracted ray PON', then it is easy to see that K's image is considerably more displaced than that of N.

Della Porta's account of peripheral vision represents a significant departure from its Perspectivist counterpart, particularly as articulated by Alhacen in book 7, chapter 6, of the *De aspectibus*. Alhacen assumes that the light from any given point to the side of the cone of direct visibility forms a sort of broken cone of radiation with its vertex at the radiating point and its base spread on the surface of the lens after refraction through the cornea. The center of sight in the middle of the eye somehow interprets the resulting composite impression on the lens's surface as emanating from its actual point of origin and on that basis "imagines" a virtual ray directed straight to that point. Thus, in Figure 15, rays from N would strike every point on the facing cornea, and many of those rays would be refracted to the lens between P and M to form a sizeable spot on it. The center of sight at E would then interpret the composite impression at that spot on the lens in such a way as to "see" N along straight, radial line EN, which passes through the opaque uvea and sclera and forms a virtual line of sight. As a result, N will appear to be where it actually is, not displaced, and the same holds for all other points to the side of cone AEC because they will all be seen within a cone of visibility consisting of both actual and virtual radial lines connected to the center of sight.[81]

That it does not depend on image displacement represents an obvious advantage of the Perspectivist account, as does its avoidance of image reversal. But these advantages are counterbalanced in Della Porta's account not only by its strict adherence to the cathetus rule of image location but also by its clear and singular association of a particular point of view on the lens with a particular point in space—all the while ensuring that those points be seen in their proper orientation within the field of view. Both accounts are

problematic, of course, because they put an extraordinarily heavy interpretive burden on either the center of sight or the lens.[82]

Visual Illusion and Visual Dysfunction

The image displacement associated with peripheral vision plays a crucial role in our perception—and misperception—of space according to Della Porta, who treats that topic in book 5 of *De Refractione* with the aim of undercutting the Euclidean approach to it. Take book 5, proposition 2, as an example. Its purpose is to explain why "parallel lines that lie in the same plane as the eye appear to converge" (p. 235). Della Porta starts with Euclid's account from proposition 6 of the Optics. Imagine the two parallel lines AG and BF in Figure 16a cut orthogonally by parallel lines RV, IL, and MC. These three lines will therefore be equal to one another. Let the large circle

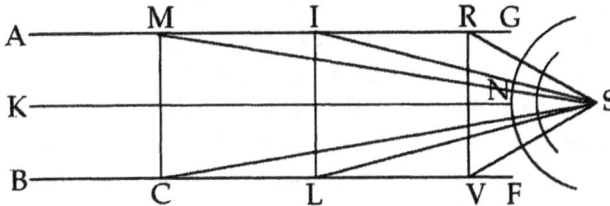

Figure 16a

centered on S be the eye, the arc inside it the lens's anterior surface and NK a line of sight that extends midway between parallels AG and BF and passes through the eye's center point S. Imagine rays SV and SR extended to the endpoints of the nearest line RV, rays SL and SI extended to the endpoints of intermediate line IL, and rays SC and SM extended to the endpoints of the farthest line MC. Because visual angle MSC under which MC is seen is smaller than visual angle ISL under which IL is seen, MC will appear smaller than IL. Likewise, visual angle ISL under which IL is seen is smaller than visual angle RSV under which RV is seen, so IL will appear smaller than RV. Consequently, since equal parallel lines RV, IL, and MC appear increasingly small with distance from the eye, parallel lines AG and FB passing through their endpoints will appear to converge toward point K as they recede from the eye.

Della Porta's alternative is illustrated in Figure 16b. Essentially the same as Figure 16a, this one also has the eye centered on S facing three equal and parallel lines RV, IL, and MC whose endpoints are connected by parallel lines AG and BF, line of sight NK lying midway between them. The difference lies in the way light from the endpoints of the facing parallel lines radiates to the lens. In the Euclidean proposition, this light—or, rather, the visual radiation from the eye—passes straight through the corneal surface along the orthogonal. In Della Porta's proposition, on the other hand, the light

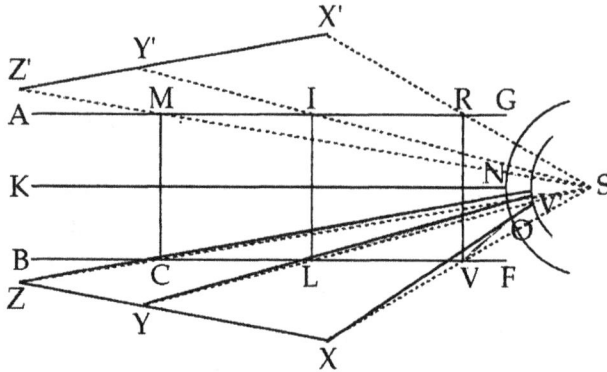

Figure 16b

reaches the cornea obliquely and is refracted to the lens. Thus, light from endpoint V of the nearest line radiates to point O on the cornea and is refracted through the aqueous humor to point V' on the lens. If, we extend the refracted ray V'O outward from the eye and also extend cathetus of incidence SV outward, the two will meet at X, which is therefore the image of V. In the same vein, the light from L and C will refract at the cornea in order to reach the lens, and their respective images will be Y and Z. The same holds symmetrically for endpoints R, I, and M of the three lines; their images will appear at X', Y', and Z'. Hence, the two lines XZ and X'Z' will appear to converge.

Della Porta uses this same analytic technique to account for various phenomena associated with linear perspective. In propositions 3 and 4, for example, he shows why the floors and ceilings in long corridors appear to converge; in proposition 5 why a plane surface viewed from on high appears to slant upward at the far end; in proposition 6 why the walls of a well appear to converge when we look down from the top, and so forth. In proposition 9, however, he turns to a different sort of illusion. "If you look at the top of a height that stands perfectly upright," he claims in the enuncia-tion to that proposition, "you will see it lean forward" (p. 247). The explana-tion is based on Figure 17, where AC is a wall facing the eye centered on K, which looks directly at point A on the top of the wall along centric ray KA. That being the case, light from C at the wall's bottom will strike the cornea at point O and will be refracted toward the lens. If we extend the refracted ray outward while also extending the cathetus of incidence KC outward, the two will meet at D, which is the image of C. By the same token, the image of B at the wall's center will be located at E, whereas point A at the top of the wall will be seen directly where it is. The entire wall AC will therefore appear to lean inward toward the eye along AD. Conversely, if the eye stares directly at the bottom of the wall, the wall will appear to lean backward (proposition 10). The same sort of analysis explains why plane magnitudes may appear curved when looked at from a particular perspective

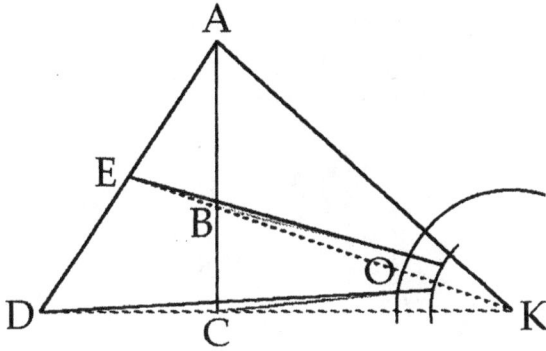

Figure 17

(propositions 11, 12, and 15) or why curved magnitudes may appear flat or else distorted in curvature (propositions 13 and 14). It also explains why, when we look directly up at the sky, it appears ovoid rather than hemispherical (proposition 17), or why an amphitheater viewed from its center appears ellipsoidal rather than circular (proposition 18).

Starting with proposition 19, Della Porta addresses the perception and misperception of size. In proposition 20, for instance, he purports to show that "there is a place from which unequal magnitudes appear equal" (p. 265). This theorem traces back to proposition 45 of the Theoninc recension of Euclid's *Optics*, which Della Porta summarizes as follows. Let BC and CD in Figure 18a be colinear, and let BC be longer than CD. Upon both lines

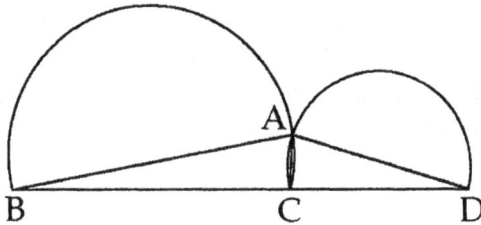

Figure 18a

erect two circular segments greater than a semicircle and proportional to one another so that the two lines form proportional chords within those segments. Given that they are greater than a semicircle, the two segments will intersect at point C on line BD and above it at some point A. Because all angles based on any given chord within a given circular segment are equal, it follows that angles CAB and CAD subtended by proportional segments are equal. Hence, lines BC and CD will be seen under equal visual angles and will appear equal on that account.

Della Porta's counter-proof, which is based in part on image displacement, takes the following form. Let AB and BF in Figure 18b be colinear

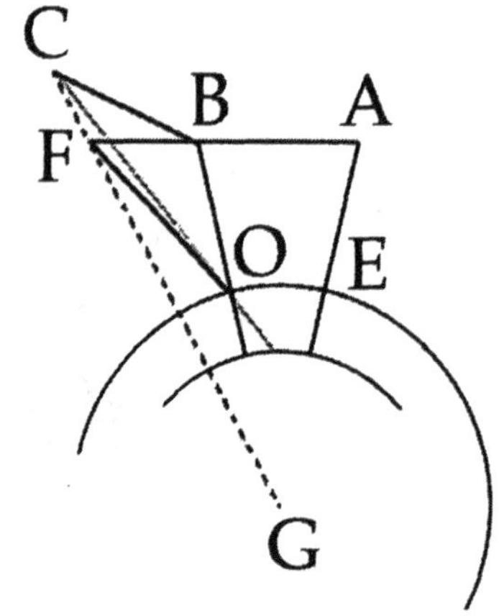

Figure 18b

magnitudes, AB being the longer, and let them face the eye centered on G. Let light from endpoints B and A of the longer line strike the cornea orthogonally at points E and O, respectively, and let the respective rays BO and AE pass directly through to the lens without refraction. Assume that rays BO and AE define the left and right edges of the cone of visibility converging on center point G of the eye. Magnitude AB will therefore be seen according to its actual size because it is viewed throughout by means of direct radiation. Meantime, let light from endpoint F of shorter magnitude FB, which lies to the side of the cone of visibility, strike O obliquely so as to be refracted through the pupil to the lens along the gray line. Extend the refracted ray beyond O, draw cathetus GF, and extend it beyond F. The refracted ray and the cathetus will meet at C, which will therefore be the image of F. Under the proper conditions, image CB of magnitude FB will be equal in length to image BA of BA, and so FB will appear to the same length as BA even though it is not.

Della Porta follows this proposition with a series of four that challenge some of Euclid's claims about shape perception in propositions 23–25, 27, 29, and 30 in the Theonine recension of his *Optics*. In book 5, proposition 21, for instance, Della Porta rejects Euclid's assertion that with two eyes we can see less than half, half, or more than half a sphere, depending on how large the sphere is. His refutation is based on assuming that we normally see monocularly, which means we always see less than half the sphere. In the next proposition, he undercuts Euclid's argument that "the closer the eye gets to a sphere, the smaller the portion of it the eye will see," and he does so by appealing to refractive displacement, which will magnify the portion of the sphere exposed to the eye when the two are close. Proposition 23 is something of an anomaly because, instead of challenging him, Della Porta supports Euclid's claim that, when viewed from directly above its center, a circle will appear true to shape. In proposition 24, however, he

attacks Euclid's assertion that, when a circle is viewed from the distance of
a radius from its center along an oblique line of sight, the circle will appear
true to shape rather than distorted. After dispensing with Euclid in that
proposition, he goes on to argue in proposition 25 that, if the pupil is
distended by disease, things will appear smaller than they actually are,
whereas in proposition 26 he shows that, if the pupil is unnaturally constricted, things will appear magnified. He then closes book 5 by explaining
in proposition 27 that older people have poor vision because their pupils
become lax and distended, which causes things to appear smaller and dimmer
than they should. As mentioned earlier, convex lenses are supposed to remedy
this condition by concentrating the incoming radiation that conveys images
and, at the same time, brightening those images.

So far we have examined how Della Porta dealt with two fundamental
kinds of visual illusion. One, covered in some detail in book 8 of *De Refractione*, involves the interposition of a refracting medium such as a biconvex
or biconcave glass lens between eye and object viewed. As we saw, this
can cause an object to appear dislocated, distorted in shape, magnified, or
diminished in size, depending on the relative placement of eye and object.
Another kind of visual illusion is due to the image displacement that arises
from the refraction of oblique radiation into the eye. This is the topic of book
5, and, as we have just seen, such refraction can lead to a misapprehension of
size, shape, or location. Yet another source of visual illusion is diplopia or
double vision. This is the topic of book 6, to which we will now turn.

As a sort of stage setting for his own treatment of diplopia, Della Porta
opens book 6 with a close examination of the Perspectivist theory of binocular
vision and its basis in image fusion in order ultimately to refute it. As Della
Porta outlines it in proposition 1, that theory can be readily understood by
means of figure 19, which is closely modeled after the diagram in book 3,
proposition 37, of Witelo's *Perspectiva*.[83] Let ABC be an object directly facing

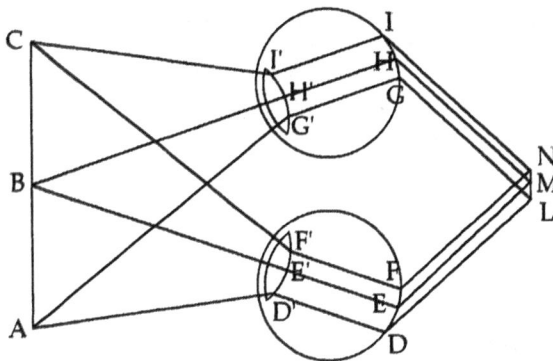

Figure 19

the two eyes. Let the top circle represent the right eye, with its associated
lens toward the front, and the bottom one the left eye, with its associated

lens toward the front, and let the axes, or centric rays, HB and EB of the two eyes converge at midpoint B of the object. Light from point C will pass orthogonally to the right eye and will enter straight through the cornea and front surface of the lens to point I' at the interface between the lens's posterior surface and the vitreous humor. There it will be refracted along I'I to the opening in the hollow optic nerve, and it will then continue through the nerve along IN to point N at the optic chiasma. Light from C will also pass orthogonally into and through the left eye to that same point along CF'FN. Meantime, light from point B will pass straight through both eyes in order to reach point M in the optic chiasma. Finally, light from point A will reach through each eye to point L in the optic chiasma. Both eyes will therefore share the same image containing points N, M, and L, so the two images will fuse naturally at the optic chiasma, and that fused image, that is, NML, will have the same orientation as its generating object ABC. Consequently, the object will appear single because it is seen by means of corresponding rays within the cones of visibility reaching through both eyes. This follows from the fact that rays AG' and AD' reach equivalent points on the front of the lens with respect to axial rays BH' and BE', and the same holds for rays CI' and CF'.

So much for image fusion and singular vision. Diplopia, or double vision, occurs when the respective rays within the two cones of visibility do not correspond properly. Figure 20 from proposition 1 serves to illustrate this

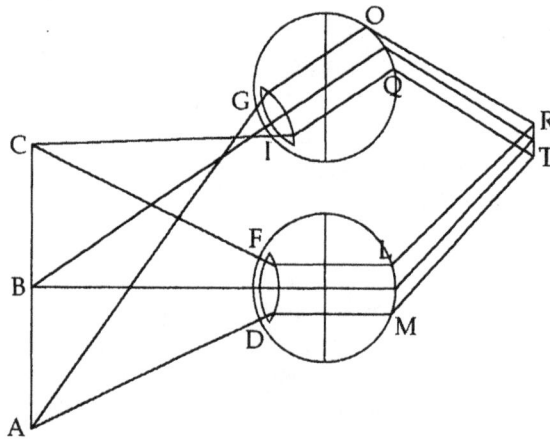

Figure 20

point. Within the cone of visibility for the left-hand eye, light from A and C will reach the optic chiasma at points T and R, respectively, along ADMT and CFLR. Through the right-hand eye, however, light from A and C will reach the optic chiasma at points T and R respectively along AGOR and CIQT. In other words, the image of C conveyed through the right-hand eye will be at T, whereas the image of C conveyed through the left-hand eye will be at R, on the opposite side of the optic chiasma. By the same token,

the image of A conveyed through the right-hand eye will be at R, whereas its counterpart conveyed through the left-hand eye will be at T. Clearly, then, CI and CF are noncorresponding rays because they reach opposite sides of both the lens and the optic chiasma, and AG and AD are noncorresponding rays for the same reason. Consequently, because one eye sees the reverse of what the other one does, the resulting images will not fuse properly at the optic chiasma, and the result will be double vision.

At least two things militate against this theory, Della Porta informs us toward the end of proposition 1. First, under normal conditions, the eyes do not actually work together. In order to confirm this, he suggests we interpose a screen or panel between the eyes, hold a book before one eye and read it while holding another book before the other eye. If the two eyes actually cooperated, we would be able to read both books at once. That we cannot indicates that one eye will always predominate. In short, we see with one eye or the other, not with both at the same time. The reason is that, under normal circumstances, virtually all of the visual power is channeled to one eye or the other. Dividing it between the eyes to achieve image fusion is unnatural. A second reason to reject the Perspectivist account of image fusion is grounded in anatomy. As noted by Vesalius and other sixteenth-century physicians, close inspection of the optic nerves reveals that, contrary to what Galen and his followers had supposed, those nerves are not hollow. Consequently, they cannot accommodate the images that supposedly pass through them to be melded at the optic chiasma. Furthermore, given that the nerves are solid, it is illogical to suppose, as do the Perspectivists and followers of Galen, that visual spirits flow through them in order not only to sensitize the lens but also to provide a supportive medium for images as they pass through the hollow optic nerves to the brain. In short, the Perspectivist theory of visual imaging and its perceptual and epistemological entailments make no sense.

Presented by Della Porta in book 6, proposition 4, an alternative account of diplopia is illustrated in Figure 21, where the two circles centered on C and B represent the right and left eyes, respectively, both facing object A. Let their axial rays (which Della Porta calls *centric rays*) be directed at point H beyond object A. According to the Perspectivist account, which closely follows that of Ptolemy, object A will be seen by the right-hand eye along CAI, whereas it will be seen along BAL by the left-hand eye. Thus, since CAI lies to the left of axis CH and BAL lies to the right of axis BH, the object is seen along noncorresponding rays, which means that A's image is seen to the left of H by the right-hand eye and to the right of H by the left-hand one.

Rejecting this explanation entirely, Della Porta bases his own account on the refraction of rays into the eyes. Accordingly, rays AO and AN strike the corneas of the left and right eyes, respectively. Were they to continue straight through, they would pass to F and D, respectively. Instead, they are refracted toward the normal so as to pass through to the lens along OG and NE respectively. If we extend those rays to meet the extension of their

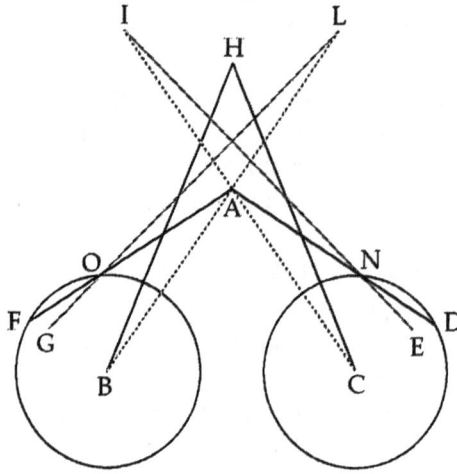

Figure 21

respective catheti BA and CA, they will converge at L and I, respectively, so the left-hand eye will see A's image at L, to the right of centric ray BH, whereas the right-hand eye will see its image at I, to left of centric ray CH. A will therefore be seen double, just as it would have been according to the Perspectivist account, which means, ironically enough, that both lines of reasoning, radically different as they are, yield the same results.

   Della Porta applies the same sort of analysis, but reversed, in an effort to demonstrate in proposition 5 how two objects can be seen as single rather than double. As before, C in Figure 22 is the center of the right-hand eye,

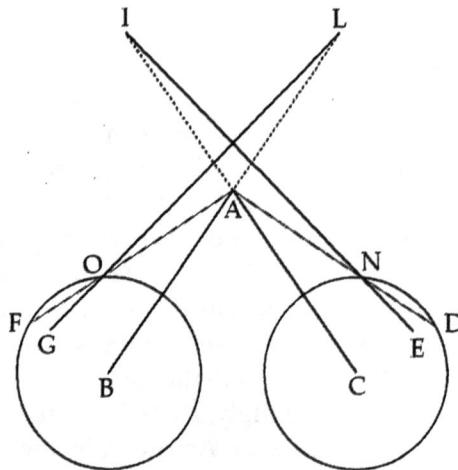

Figure 22

and B the center of the left-hand one. This time, however, points I and L are the objects, and the centric rays are focused on point A, which was the object in the previous proposition. Ray LO from point L will strike the cornea of the left-hand eye at O and, instead of continuing straight along OG, will refract to the lens along FO, which, when extended, will meet cathetus of incidence BL at point A. Likewise, ray IN from point I will strike the right-hand eye's cornea at N and, instead of continuing straight along NE, will refract to the lens along ND. When extended, refracted ray ND will intersect cathetus of incidence IC at A, so it follows that the images of I and L will be fused at A for both eyes. Moreover, if those two objects are colored differently, so that, for example, I is black and L is white, then the fused image at A will appear bicolored, white on the fused image's left side and black on its right side. Here, too, the Perspectivist account yields the same results, but it is based on the supposition that the two objects are seen through point A along axial rays BAL and BAI, so the objects share an image at A for both eyes.

The flaw in Della Porta's account, as just described, is as fundamental as it is obvious: Instead of refracting toward the normal on entering the eye, as they should, the light rays from I and L refract away from it when they enter the denser aqueous humor. Why? Earlier, in proposition 2, Della Porta attempted to justify this apparent breach of the rules by arguing that refraction into the eye can occur in two ways. One is according to the normal rules, which dictate that the incoming light refract toward the normal when entering a denser medium, as was the case in Della Porta's account of diplopia in proposition 4. The other way involves refraction away from the normal and applies "when a person divides the visual power between both eyes [and] looks with effort," for in that case "he subverts the eyes, and they do not maintain their proper disposition, as arranged by nature, but are irregularly inclined" (p. 303). This, presumably, explains why the light would refract away from the normal in proposition 5, because the divided visual power is overstrained in an effort to focus on point A.

Flimsy though it is, this rationale fully satisfied Della Porta, or so it seems if we are to judge by his attempts to explain various manifestations of diplopia in propositions 6–9, as well as in the two concluding proposition, 11 and 12, all of which appeal to this new mode of refraction. Take as an example proposition 7, where Della Porta aims to show "how each of two magnitudes might appear doubled" (p. 311). Let H and C in Figure 23 be the right and left eyes, both staring straight ahead, so that the right eye looks directly at A along centric ray AH and the left one directly at B along centric ray CB. Meantime, light from B will radiate to point G on the right eye's cornea and, instead of continuing straight along GL, will refract along GI. The same applies symmetrically to the left eye at C. Consequently, if we extend IG until it meets cathetus of incidence BH, the two will intersect at M, which will therefore be B's image from the perspective of the right-hand eye. By the same logic, N will be A's image from the perspective of the left-

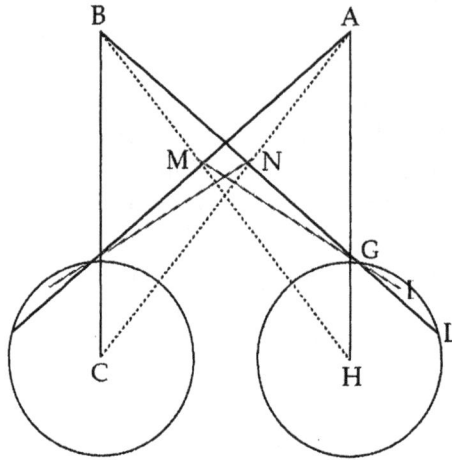

Figure 23

hand eye. Consequently, the right-hand eye will see A directly and B at M, whereas the left-hand eye will see B directly and A at N.

Bizarre as this theorem is in its reliance on anomalous refraction, even more bizarre is proposition 9, which appeals to both modes of refraction at the same time in an effort to show that three magnitudes can yield five

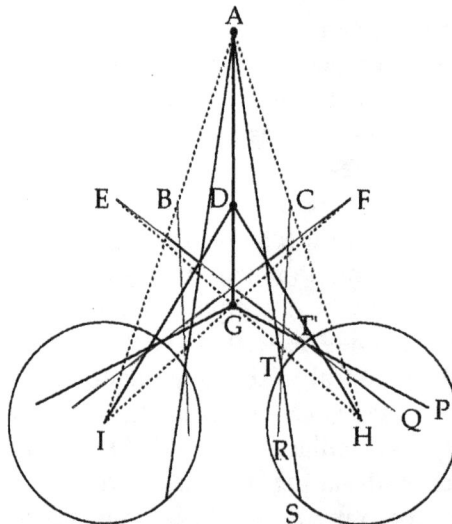

Figure 24

images. The conditions are illustrated in Figure 24, where GA is a ruler held straight out from the bridge of the nose with three small pegs placed on it at points G, D, and A. Let both the right-hand eye at H and the left-hand eye at I stare straight at peg D along centric rays HD and ID, respectively. Each eye will therefore see D where it is. Meanwhile, ray AT will reach point

T on the cornea of the right eye and, instead of continuing straight along TS, will refract away from the normal along TR. The right eye will therefore see A's image at C, where the extension of TR meets cathetus HA. From point G, on the other hand, ray GT' to the right eye will refract toward the normal along T'Q, so its image will be at E, where T'Q extended meets the extension of cathetus HG. Thus, points A, D, and G will appear at C, D, and E to the right-hand eye. Conversely, those same points A, D, and G will appear at B, D, and F to the left-hand eye. Because D is seen at its proper place by each eye, five images will be seen by both eyes at E, B, D, C, and F.

At first glance, Della Porta's willingness to posit two contrary modes of refraction in this way seems illogical, if not preposterous, especially in the light of his efforts in books 1 and 2 to put one, and only one, of those modes on a firm empirical and theoretical grounding. Bear in mind, however, that having recognized that the eyes do not act in perfect concert to produce a fused image, he was forced to reject the Perspectivist account of binocular vision and the analysis of diplopia that follows from it. What obliged him to do so was that this account is based on the assumption of corresponding radiation, which is itself based on assuming that, under normal conditions, the two eyes act cooperatively, not independently. Casting about for another explanation, Della Porta looked to the refraction of oblique radiation into the eye as the cause of double vision. Not unreasonable in itself, this move nonetheless required theoretical adjustments in response to the empirical fact that, if all the refraction occurred toward the normal, the theoretical model would not yield the observable results. It would therefore be impossible to produce the images at B and C in Figure 24 without having the rays refract away from the normal. So Della Porta had no choice but to temporize, as he in fact did.

Having explained diplopia to his apparent satisfaction, Della Porta concludes his account of visual illusions in book 7, where he discusses various pathological conditions that lead us to believe that things occurring inside the eye are actually happening outside. Floaters, for instance, appear to swim in front of the eye, as do the bright flashes of light that erupt from a blow to the head (propositions 1 and 2), and the same holds for afterimages (proposition 10). Inebriation can cause things to appear double by roiling the humors inside the eye (proposition 5), and delirium can make a person see apparitions of nonexistent things, as can a surfeit of melancholic humor (proposition 7). Drugs, such as mandrake or belladonna, can also arouse such apparitions, often fearsome ones (proposition 8). All of these visual illusions can be explained in straightforward, physical or physiological terms without recourse to occult or supernatural causes, and this is the case more generally for every type of visual illusion.

Rainbows and Other Meteorological Appearances

As Della Porta points out in the general preface to *De Refractione*, a variety of celestial phenomena, many of them considered atmospheric, or meteoro-

logical, had long inspired awe, even fear, among onlookers who did not understand them. "Who," he asks, "does not tremble at the colors in comets and eclipses of the Moon and Sun—blood-red, foul, bright, or sky-blue?" (p. 11). And who is not lost in admiration at the rainbow, a marvel that aroused such wonder in the ancients that they called it the "daughter of Thaumas," a god whose very name means wonder? Once we understand their underlying cause, however, we need not be overcome by fear or amazement at these phenomena. And what is the underlying cause? Nothing more magical or supernatural than "the refraction of light and the mixing of denser vapors with the transparent air" (p. 11). Della Porta's purpose in the ninth and concluding book of De Refractione is therefore to show how such refraction is responsible not only for halos and virgae, or atmospheric streaks, but also, and far more important, for rainbows. Indeed, the formation of rainbows is the centerpiece of Della Porta's analysis in that book.

Della Porta's effort to account for the formation of rainbows was influenced by a variety of sources running the chronological gamut from antiquity to the late Renaissance. Not surprising, a key source is book 3 of Aristotle's *Meteorology*, chapters 2–6 of which are devoted primarily to the rainbow and secondarily to halos and virgae, or atmospheric streaks. Complementing this primary source are the commentaries of Alexander of Aphrodisias, Olympiodorus, and Albertus Magnus. Apparently not among the commentaries known to, or at least used by, Della Porta were those of Jean Buridan (d. c. 1360), Nicole Oresme (d. 1382), Albert of Saxony (d. c. 1390), and Themon Judei (fl. c. 1370).[84] Nor, apparently, was he aware of the sophisticated rainbow analysis of Theodoric (or Dietrich) of Freiberg (d. c. 1310), who supposed, first, that light undergoes one or two internal reflections in individual raindrops bracketed by refraction into and out of the drops, and second, that the angles at which this happens determine the color of the exiting light.[85] Somewhat surprising is Della Porta's apparent ignorance of the rainbow theory of his older contemporary Francesco Maurolico (d. 1575), who proposed a model based entirely on internal reflections in each raindrop.[86]

From classical antiquity, meantime, Pseudo-Plutarch's *On the Opinions of the Philosophers*, Pliny's *Natural History*, and Seneca's *Natural Questions*, crop up frequently in Della Porta's account. Finally, although not cited, Witelo's extensive technical and geometrical analysis of the rainbow in book 10 of the *Perspectiva* influenced Della Porta in various ways. In typical fashion, though, he approached all these sources with skepticism, approving certain points in them, reacting against others, in order to arrive at his own, idiosyncratic theory.

Before examining that theory in detail, we need to take a brief excursus into historical background.[87] More specific, we need to look briefly at Aristotle's account of the rainbow in the third book of the *Meteorology*, for it was there that he laid the foundations for every theoretical account of the rainbow that was to follow up to Della Porta's day. Aristotle's theory is based on five key assumptions. First and foremost is that rainbows are due to reflection.[88]

Aristotle himself supposes that this process involves the reflection of visual rays from the raincloud to the Sun. Later theorists who favored intromissionism simply reversed the direction of radiation from Sun to eye via the raincloud. Aristotle's second key assumption is that the reflection occurs from single drops within the raincloud, not from the raincloud as a whole. Accordingly, as Aristotle explains, because the drops are so small, what we see in them is not an actual, formal image of the Sun, but, rather, a nebulous replication of its color.[89] Aristotle's third key assumption is that the resulting color impressions are contingent on how weak the radiation reflected from the cloud is; the weaker the radiation, the less intense the color.[90] The fourth key assumption is that the rainbow consists of three primary colors: red, green, and violet in descending order of intensity. Although the yellow between red and green is as manifest as the other three colors, it is not primary. Rather, it is due to contrast, the red being whitened by contrast to the green.[91] The color of the rainbow's yellow band, in short, is only apparent. Aristotle's fifth and final key assumption is that the rainbow and its formation are subject to certain, geometrically determined rules.[92] One such rule, for instance, is that the center of the Sun, the center of sight, and the center of the rainbow must all lie in that order on the same straight line.

Not everyone agreed with Aristotle that reflection is the mechanism by which rainbows are formed. As just pointed out, Theodoric of Freiberg supposed that rainbows are due to a combination of refractions and internal reflections within each raindrop and, therefore, that the resulting colors are somehow keyed to the angles at which this happens. Some seventy years earlier, Robert Grosseteste (d. 1253) had denied any role whatever for reflection, imputing the formation of rainbows to refraction alone—not into or through individual drops, however, but through the entire raincloud taken as a continuous refractive medium. Because "color is light mingled with a transparent medium," Grosseteste asserts in *De iride* (*On the Rainbow*), it follows that the color resulting from such mingling will vary according to the quality of the medium as well as of the light. The purer the medium, the brighter the light, and the more concentrated it is, the brighter the color. Accordingly, "if the transparent medium is pure and the light bright, its color will become equivalently white and luminous," whereas the less pure the medium and the dimmer and less concentrated the light, the darker the color will be.[93]

As we will soon see, Della Porta's explanation of the rainbow resembles Grosseteste's in certain significant respects. Whether these points of resemblance are coincidental or whether they are due to Grosseteste's influence on Della Porta is open to question. It is certainly possible that Della Porta had direct access to Grosseteste's *De iride* in manuscript form. Less likely, but not impossible, is that he inferred Grosseteste's explanation from Roger Bacon's refutation of it in part 6, chapter 11 of the *Opus majus*, since that refutation focuses on the rainbow's shape rather than on the formation of its colors.[94] Nonetheless, if Della Porta did read Bacon's account of the

rainbow, he would have realized that Bacon opposed Grosseteste in two key respects. First, unlike Grosseteste, Bacon imputes the rainbow to reflection alone, and second, his analysis is based on individual raindrops rather than on the cloud as a continuous whole. Also, unlike Grosseteste, Bacon pays close attention to the geometrical features of the rainbow. In fact, as far as we know, Bacon was the first to claim (correctly) that the maximum altitude of the primary bow, which is reached when the Sun is at the horizon, is around 42°, a figure repeated by Witelo and other sources known to Della Porta.

One other issue Bacon brings to the fore is that of the status of the colors in the rainbow according to whether they are real or apparent. Real colors, he insists in part 6, chapter 12 of the *Opus majus*, are "fixed" in things, whereas the colors of the rainbow are only apparent, arising "from imagination and a deception of sight."[95] Accordingly, although we can create the colors of the rainbow by refracting light through hexagonal crystals or water-filled glass vessels onto a dark surface, the two types of color are entirely different. Those produced from refraction are real insofar as they are not viewer dependent; every viewer sees the same spectrum because its constituent colors are all "fixed" on the surface to which they are projected. The colors of the rainbow, on the other hand, are only apparent because they cannot exist without a viewer to see them. That is why no two viewers see the same rainbow or, to put it another way, why each viewer sees his own, personal rainbow. Consequently, there are as many individual rainbows as there are individual viewers.[96]

A host of subsidiary issues also faced ancient, medieval, and Renaissance theorists. How, for instance, is the secondary rainbow formed? Why is the order of its colors reversed with respect to the primary bow? Why is there a colorless space between the two bows? Can more than two bows be formed? Is moonlight strong enough to create a rainbow? How about starlight? Are the colors of the rainbow related to differences in the substantial or material nature of the components in a raincloud? Is the red color due to heat generated at the top of the bow and the violet color to cold at the bottom? Might the colors be related to the elemental composition of the rainbow's components? How does the generation of colors in halos relate to the generation of colors in the rainbow? Are both due to essentially the same mechanism, or is there some fundamental difference in mechanism? These are some of the questions raised in Della Porta's sources, and the need to respond to them certainly informed his account of the rainbow in book nine of *De Refractione*.

Della Porta opens that account in proposition one by arguing that pure light is colorless. Pure light is the kind that radiates from the Sun and other celestial bodies "through the celestial orb." When it reaches the atmospheric shell and "flows through the elements in this sublunar world," however, the light "is tinged and tainted so that it never reaches [us] in pure form (p. 391)." In short, all the light we experience is colored, and there are three ways in which it can become so. One way, Della Porta points out in proposition 3,

is by mingling with opaque bodies and taking on their intrinsic color while, at the same time, rendering that color—and the object containing it—visible. Another way is by penetrating the outer layer of thick vapors or clouds. Thus absorbed and tinged by that outer layer, the light renders all or part of the cloud's surface red, as at dawn, or green, or even golden. The third way is by refracting through transparent media denser than air. Under the proper circumstances, such refraction can yield prismatic colors, which Della Porta characterizes as "iridescent" (*fulgidi*).

Such colors, he warns us in proposition 2, are not to be confused with those produced when light strikes the necks of doves at different angles. The difference between the two is empirically demonstrable. If you pull a clump of feathers from the neck of a dove and examine it closely, he assures us, you will see that, instead of being uniformly black, each feather is multicolored. Accordingly, "on the right side you will observe purple, on the left green, and red in between, all these areas shining upon the black color of the feathers" (p. 393). The constituent color of each part is therefore rendered visible in turn as the light strikes it from different angles, so the iridescent shimmering is only apparent because it is due to the interaction of light with opaque surfaces, not to refraction through a transparent medium.

After reviewing previous efforts to explain the formation of the rainbow's colored bands in proposition 4, Della Porta argues in proposition 5 that when intense light is refracted through a small portion of a transparent medium denser than air, it assumes a golden-yellow color (*flavus*) similar to that of "straw that is not yet mature [or] saffron that is heavily diluted in water" (p. 399). This is the color of sunlight radiated through thin, clear air at midday, and it is as close to colorless as refracted light can get. It is therefore absurd to think, with Aristotle, that this bright color somehow arises from combining the two dark colors red and green. On the other hand, Della Porta claims in proposition 6, weak light refracted through an extremely dense transparent medium produces the color blue (*caeruleus*), which approaches pure darkness, or blackness. Golden yellow and deep blue are therefore at opposite ends of the spectrum, the one verging toward the absolute colorlessness of pure light, the other verging toward the absolute colorlessness of the pure dark that arises from the absence of light.

Della Porta insists that golden yellow and blue are the only colors that are produced by the refraction of light through transparent media denser than air. These, in short, are the equivalent of Aristotle's primary colors red, green, and violet. "The rest," Della Porta argues in proposition 7, "are mixtures of neighboring colors" (p. 403). For instance, the intermediate green band of the rainbow is produced by combining the golden yellow above it with the blue below, just as we see green when we look through a combination of blue and yellow glass plates held up to the light. The successive combining of neighboring colors also accounts for the plethora of related colors other than golden yellow, green, and blue that can be seen in the rainbow, "for one color combined with another neighboring one forms a third, related

one, and from the color just produced and another one yet another one immediately forms [such that] one gradually approaches the other so as to appear indistinguishable" (pp. 401–03).

So much for the primary golden yellow and blue bands of the rainbow. So much as well for the continuum of secondary, intermediate colors formed by intermingling neighboring colors. There are two other key colors that need explaining, the red of the rainbow's topmost band and the violet of its bottommost band. These, Della Porta argues in propositions 8 and 9, are the result of contrast against an extremely dark cloud. In the case of red, the contrast is between the golden yellow of the top band and the blackness of the cloud behind. In order to explain how this arrangement yields red, Della Porta appeals to a variety of examples, many of them drawn from Aristotle's discussion of color in *Meteorology*, III, 4; *On Colours*, 1–2; and *On Sense*, 3. For instance, when the bright, golden-yellow Sun is viewed through smoke or haze, it can take on a reddish hue. So, too, candlelight seen through soot appears red. In both cases, the redness arises from the bright light's being seen through a dark veil of sorts. In the case of the rainbow, the opposite seems to be the case insofar as the reddish hue arises from the cloud's darkness being seen through the veil of the yellow band. Likewise, the violet hue at the bottom of the rainbow arises from the cloud's darkness being seen through the veil of the blue band at the rainbow's bottom. What happens in this case is much like what happened in the previous case of the golden-yellow band. Just as the cloud's darkness seen through that band is perceived as red, the cloud's darkness seen through the blue band also appears to take on a reddish cast, which, when combined with the blue, yields purple or violet.

Armed with this understanding of how the colors of the rainbow evolve from the primal couple of golden yellow and blue, Della Porta turns in proposition 11 to a physical explanation of the rainbow's primary bow. In a slight nod to Aristotle, he bases that explanation on hemisphere AEC in Figure 25, which is taken directly from the Latin text. A at the right-hand

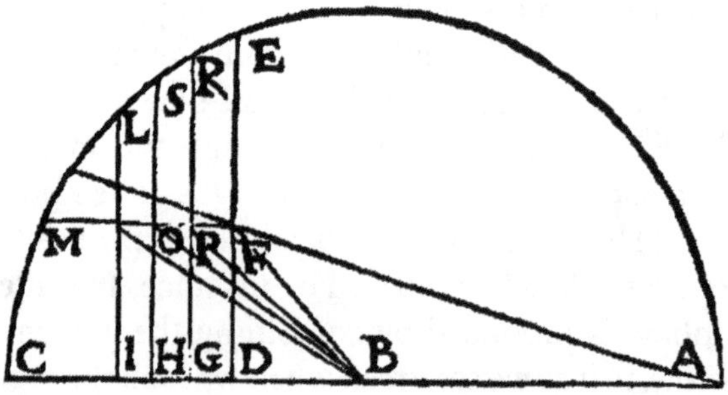

Figure 25

side of diameter ABC represents the Sun at horizon, diameter ABC the horizon line, B the viewpoint of the observer, and ELID a raincloud divided into

three equal portions by RPG and SOH, its face EFD perpendicular to the horizon. Consisting of vapor, this cloud, as a whole, forms a transparent medium denser than air. From the outset, it is clear that the rainbow cannot be formed by reflection because any ray, such as AF, that strikes face EFD of the cloud above horizon line ABC, and therefore above viewpoint B, would reflect upward, away from B, rather than toward it. This follows from the equal-angles law. Consequently, ray AF must pass into the cloud, and since the cloud is denser than the air outside, the ray will not pass straight through it but will be refracted along some line FPOM.[97]

Now, because bright light refracted into a small portion of a transparent medium denser than air, such as a raincloud, is rendered golden yellow according to Della Porta, it follows that the light at F will be that color and, moreover, that the light will remain that color for a short while as it passes farther into the cloud along FP. However, as it penetrates ever farther into the cloud, the light is progressively weakened until finally, when it reaches OM, it assumes the color blue. The combination of golden yellow along FP and blue along OM will yield green along OP in between, and the continual intermingling of ever thinner slices of color from there on will yield more and more intermediate colors.

Accordingly, from every point on refracted ray FM, color will radiate back to the eye, the darkest blue at M reaching along MB, the intermediate blue-green at O along OB, and so forth. The eye will therefore see the blue-to-bluish-green band MO under visual angle MBO, which will appear below the bluish-green-to-greenish-yellow band OP seen under visual angle OBP, which in turn will appear below the greenish-yellow-to-golden yellow band seen under the visual angle PBF. In other words, the spectrum of colors along refracted ray FPOM will appear in proper order from top to bottom of the rainbow, and if there is an adequately dark cloud behind raincloud ELID, that spectrum will include red at the top and violet at the bottom.[98] Later, in proposition 20, Della Porta agrees with Bacon, and many other classical and scholastic thinkers, that the colors of the rainbow are apparent, not real. Like Bacon, moreover, he claims in that same proposition that, although the colors produced by refracting light through a prism are the same as those of the rainbow, these colors are real, not apparent, because "they are seen on a wall [and] are discerned by every eye from every location" (p. 437).

With the physical account of the rainbow's formation out of the way, Della Porta turns to the geometry of the rainbow, with which he deals in propositions 12–19. Pretty much everything he has to say in that regard comes to ground in the supposition that the rainbow forms a circle at the common base of two cones, one of which is the cone of radiation from Sun to cloud, the other the cone of radiation from cloud to eye. This arrangement is illustrated in Figure 26, where the cone of solar radiation is ADF, with the Sun at vertex A, and the cone of radiation to the eye is DBF, with the eye at vertex B. Circle DF centered on E and common to both cones thus

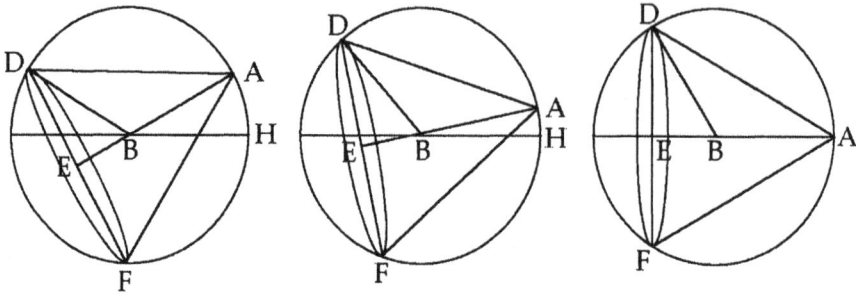

Figure 26

represents the bow (prop. 14), and axis ABE perpendicular to its center contains the Sun, the center of sight, and the rainbow's center point in that order (prop. 15).

From this figure it is clear why the rainbow reaches its maximum altitude above the horizon when the Sun is at horizon, because the higher the Sun rises above the horizon (as in the middle and left-hand diagrams of the figure), the more the combination of cones tilts about viewpoint B as a fulcrum (prop. 17). Consequently, as the tilt increases, the amount of the rainbow's arc that appears above the horizon decreases. It is also clear that, when the Sun and viewpoint are both on the horizon line, as in the right-hand diagram of Figure 26, exactly half the rainbow's circle will be seen. In all cases, therefore, the cones and their determining parameters remain constant in size and spatial relationship; only their position relative to the horizon changes. Della Porta does point out in proposition 15, however, that when a rainbow is viewed from on high, at the top of a mountain, for instance, more than half the rainbow's circle can be visible.

So far, Della Porta's geometrical model of the rainbow makes some sense qualitatively, but his attempt to quantify that model reveals a fundamental misapprehension on his part. The key to that misapprehension lies in his claim that the size of the rainbow's circle is determined by the angle formed by the edge and axis of the cone of visibility, that is, angle DBE in Figure 26. This angle, Della Porta informs us in proposition 13, can be easily determined by the measurable fact that distance EB from the center of sight to the rainbow's center and the rainbow's radius EB in Figure 26 are equal. It therefore follows that angles DBE and BDE are 45°, which, in turn, means that the maximum altitude of the rainbow at sunrise or sunset, as measured by angle DBE, will be 45°, not 42°, as given in several of Della Porta's sources. But Della Porta seems to have been aware of this latter value because he claims in proposition 21 to have measured that angle at sunset in 1590 and found it to be 42°.[99] Why, then, did he ultimately plump for a value of 45° in propositions 13 and 27? Perhaps he was appealing to some principle of simplicity and symmetry, according to which it makes sense to assume that angle at vertex B of the cone of visibility in Figure 26 would be fixed at a

nice, neat 90°. If so, then all of its constituent triangles would be isosceles, as would all the triangles like DBE within it.

Another weakness of Della Porta's account is that it fails to explain why refracted ray FM in Figure 25 is the only operative one for center of sight B. After all, the colors at F, P, O, M, and every point in between radiate in all possible directions through both the transparency of the cloud and that of the air surrounding it. Why, then, should the eye see only the color radiating from those points? Why, in fact, should it not see the color radiating from every point on every solar ray refracted through the cloud? And how can the angle at B somehow *cause* the particular selection of refracted ray FM as the only effective one in the rainbow's formation? These key, interrelated questions are left unanswered by Della Porta, who simply stipulates the answers.

However problematic Della Porta's account of the primary bow and its formation may be, it is less so than his account of the secondary bow. Granted, he had little or no useful guidance from his predecessors, as is evident from his brief overview of previous theories in proposition 21. Granted as well that, like his predecessors, he had to overcome three problems in his effort to account for the secondary bow. The first is that the order of colors in the secondary bow is reversed from that of the colors in the primary bow.[100] The second is that there is a colorless space between secondary and primary bows. And the third is that this space is of a certain, consistent size with respect to the two bows. Each of these problems poses a significant challenge in itself, and all three in concert increase that challenge exponentially. Small wonder, then, that Della Porta floundered in responding to it.

In order to make sense of the theory Della Porta proposed in response, let us turn to Figure 27, taken directly from proposition 22 in the original

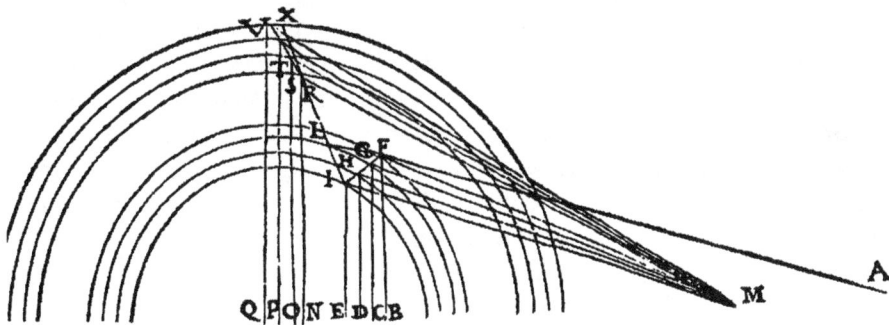

Figure 27

Latin text. In that figure, A represents the Sun, M the center of sight, FIEB a cloud facing the Sun, and RVQN another cloud facing the Sun behind cloud FIEB and separated from it by space EN. As determined in proposition 11, solar ray AF striking the front face of the first cloud will be refracted along FI to yield a golden-yellow band within FG, a green one within GH,

and a blue one within HI. The order of bands from M's perspective will therefore be from golden yellow at the top through blue at the bottom. Once the light ray refracted along FI reaches the back of cloud FIEB, however, it is refracted in the air behind it so as to reach the second cloud along IR. Instead of passing straight through that cloud along RX, it will be refracted along RV. Because the refracted light is strongest at R, it will yield a golden-yellow color as it begins to penetrate the second cloud, and by the time it reaches the back side of the cloud at V, it will have weakened to the point of yielding blue. Because of the disposition of line of refraction RV, moreover the golden-yellow band will appear at the bottom and the blue band at the top from M's perspective—hence the reversed order of colors.

Although Figure 27 is too crude to show it, Della Porta seems to have assumed that the formation of the secondary bow was dictated by some principle of geometrical symmetry. Consequently, just as the angle of refraction along ray FI through the first cloud is such that the iridescent color from every point on it radiates to center of sight M, so the angle of refraction along ray RV through the second cloud must do the equivalent thing, albeit in reverse order. The conclusion Della Porta seems to draw in proposition 23 is that the two angles of refraction into the respective clouds are perfectly symmetrical. In addition, angles VAR and FAI will be equal, so it follows that "triangles [VAR and FAI] are similar and proportional." Implicit at best in this claim is that somehow, in order to remain similar and proportional, these triangles must be spatially separated by just the right amount to yield the size of the colorless band between the two bows.

Whatever its obvious shortcomings at the qualitative level, this explanation is perhaps even more deficient at the quantitative level. Della Porta says nothing about the size of the angles at which the two rays refract into their respective clouds, nor does he make any attempt to explain how those particular angles yield the requisite spectra in the two bows. In addition, it would make more sense to base the key angles and triangles in his analysis not on the Sun at A but on the center of sight at M. After all, Della Porta's account of the rainbow is ultimately based on how the lines of refraction through the clouds are uniquely disposed with respect to the center of sight, and that in turn is determined by Della Porta's somewhat arbitrary (and incorrect) value of 45° for the maximum altitude of the primary rainbow.

Della Porta's explanation of the secondary rainbow in propositions 22 and 23 brings book 9 to something of a climax; in the remaining nine propositions he does little more than tie loose ends. In proposition 24, for example, he discusses lunar rainbows and in 25 the conditions that favor the formation of rainbows. After discussing in proposition 26 how future weather conditions may or may not be predicted from rainbows, he turns in proposition 27 to an explanation of why light refracted in a particular way through a prism yields the spectrum seen in rainbows. In the next two propositions, 28 and 29, he deals with the formation of virgae, or atmospheric streaks, and halos. Then, in proposition 30 he explains how dense clouds

are colored when light reflects from their outer surface in particular ways, after which he explains in proposition 31 how the varying colors created by light refracting into and through the atmosphere can be seen against the moon's face during the opening and closing phases of a lunar eclipse. Finally, in the thirty-second and concluding proposition he argues that the Milky Way "is formed from light and a denser portion of the heavens," which captures that light and is thus somewhat brightened by it. Strictly speaking, of course, this is not a meteorological phenomenon, since it occurs far beyond the Earth in the stellar sphere. The reason Della Porta includes it in his analysis of meteorological phenomena is to refute Aristotle and his followers, who impute the Milky Way to the combustion of vapor high up in the atmospheric shell. Thus ends book nine, and *De Refractione* with it, on a somewhat discordant note.

De Refractione in Historical Context

Unlike so many of his other works, Della Porta's *De Refractione* saw print only once. Two obvious reasons for its relative lack of popularity suggest themselves. First, being fairly technical in nature, it was not likely to appeal to the mass audience attracted by either version of *Magiae Naturalis* or by the physiognomic treatises—or, for that matter, by the seventeen published plays. Second, as we have seen in the course of analyzing *De Refractione*, the account of light and sight that unfolds in it suffers from some glaring theoretical and analytical inconsistencies. Add to these inconsistencies the lack of mathematical sophistication and rigor that marks Della Porta's analysis throughout *De Refractione*, and the likelihood that it would have appealed to anyone properly schooled in optics, especially on the basis of Alhacen's *De aspectibus* or Witelo's *Perspectiva*, dwindles sharply.

Not only was *De Refractione* unsuited to an audience among either the *dilettanti* or the *cognoscenti*, but any lingering hope it had of attracting even a fringe group among the latter was pretty much dashed in 1604, when Johannes Kepler's far superior account of light and sight appeared in *Ad Vitellionem Paralipomena*. To add insult to injury, Kepler's close and compelling analysis of lenses in the *Dioptrice* of 1611 put paid to Della Porta's corresponding account in books 2 and 8 of *De Refractione*. Worse yet, Descartes administered something of a coup de grâce in his *Discours de la Méthode* of 1637, where he offered a sophisticated and mathematically persuasive account of the rainbow that exposed the crude inadequacy of Della Porta's explanation in book 9 of *De Refractione*. Thus, within forty-odd years of its publication, *De Refractione* had been reduced from a potentially viable but highly problematic source to little more than a historical footnote.

That Della Porta's model of light and sight is both theoretically and analytically flawed should not, however, lead us to dismiss him out of hand as the intellectual lightweight described by Vasco Ronchi. "Writing a serious

book was a task beyond Della Porta's powers," he remarked well over half a century ago, adding that, being "noteworthy for the absurdities it contains, absurdities that are of course presented as mathematically demonstrated truths," *De Refractione* is proof of the point.[101] This view of Della Porta and his intellectual capacities is unduly jaundiced and anachronistic. As we have already seen, he was highly regarded by several of the leading thinkers of his day. Kepler, in particular, was unstinting in his praise, lauding Della Porta not only for demonstrating beyond doubt that vision is due to intromission alone but also for recognizing that the eye is essentially a camera obscura, an insight crucial to Kepler's own theory of retinal imaging. Kepler's praise was based on his reading of the 1589 version of *Magiae Naturalis*, though, not *De Refractione*, which he claims to have been unable to find.[102] This was probably just as well, for had he actually read *De Refractione*, Kepler's esteem for Della Porta would surely have taken a steep dive.

Some of the deficiencies of Della Porta's optical analysis in *De Refractione* may be traced to his tendency to work on several projects concurrently, often at what must have been a frenetic pace. Equally frenetic was his pace of publication. Within a span of only three years between 1583 and 1586, for instance, he published *Suae Villae Pomarium*, *Suae Villae Olivetum*, and *De Humana Physiognomonia*. Two years later he published three more works: *Phytognomonica*, *Claudii Ptolemaei Magnae Constructionis Liber Primus*, and the play *L'Olimpia*. At the same time he was engaged in these projects, he was putting the finishing touches on the much-expanded version of *Magiae Naturalis* while also hard at work on *De Refractione*. Della Porta, in short, was a full-bore multitasker. He was also gifted (or saddled) with what Clubb aptly characterizes as a "mercurial mind,"[103] its volatility evident in both the scope of his interests and the speed with which he pivoted from one interest to another. This natural tendency to shift or lose focus was undoubtedly exacerbated by the continual bouts of debilitating fever (recurring malaria?) he suffered throughout his life.[104] These are traits of a sometimes-febrile mind more inclined to jump from one big idea to another than to dwell obsessively on detail, which may help explain both the theoretical inconsistencies and the editorial sloppiness so characteristic of *De Refractione*.

For all its flaws, though, Della Porta's account of light and sight displays occasional flashes of inspired insight and originality, some foreshadowing later developments in optics and visual theory, some leading to dead ends. Among the former is his recognition that the pupil of the eye acts like the aperture of a camera obscura, shunting incoming radiation to the front surface of the lens in inverted order, an insight he had already shared in his *Magiae Naturalis*. It bears noting, however, that this insight was not original to Della Porta. Some eighty years earlier, Leonardo da Vinci had likened the eye to a camera obscura and had speculated about how the inverted radiation reaching the front surface of the lens might be refracted into and through it to reach the optic nerve at the back of the eye in upright order.[105] Della Porta, on the other hand, made no effort to trace the incoming radiation

through the lens because implicit in his account is that the visual act effectively begins and ends at the lens's anterior surface. This is more than a little
ironic because in his description of the camera obscura in *Magiae Naturalis*,
XVII, 6, Della Porta claimed that placing a biconvex lens right behind the
aperture sharpens and clarifies the image cast on a properly placed screen
within the camera.[106] He was also well aware that the eye's lens has all the
optical properties of a glass lens, including the capacity to focus incoming
light on or near the retina.

More original than the insight that the eye acts like a camera obscura
was Della Porta's effort in book 2 of *De Refractione* to quantify the radiation
of light through spherical lenses. His conclusions were incorrect, of course,
because he assumed a too-rigid correspondence between reflection in a concave spherical mirror and refraction through a glass sphere.[107] It is easy to
castigate him for this assumption and its basis in a perfect symmetry between
concavity and convexity; after all, to assume that such perfect, geometrical
symmetry extends to the physical interaction of light with entirely different
substances having those shapes, one reflective, the other refractive, is clearly
a stretch. But bear in mind that Johannes Kepler, the "father" of modern
astronomy, was persuaded throughout his career that the relative sizes and
number of the planetary orbits follow from a particular order of nesting
Pythagorean solids. Della Porta's argument from symmetry was therefore
perfectly consistent with the intellectual canons of his day and should be
judged accordingly. In addition, as we noted earlier, spherical concave mirrors
and spherical convex lenses are optically similar in a variety of significant
ways that include, most prominently, the ability to focus light and to cast
real images on a screen.

If the theoretical foundations of Della Porta's model of refraction are
problematic by today's standards, the procedures he followed in testing the
model are much less so. As we have seen, he started by calculating precisely
where parallel rays striking the glass sphere should intersect the sphere's
axis after refraction. Accordingly, he showed that the closer to the axis the
impinging rays first enter the sphere, the farther down they cross the axis
after refraction until reaching a limit at what we now refer to as the focal
point. According to the parameters of his model, which is based on parallel
rays incident at 60°, 45°, 30°, and 22.5°, he was able to specify both the
trajectory of the refracted rays and the point at which each trajectory should
intersect the axis. He was also able to show that under these conditions light
suffers spherical aberration. Suffice it to say that Della Porta's quantitative
determinations were inaccurate because his model did not conform to the
sine law of refraction, but that is beside the point because the procedure itself,
which amounts to theoretical ray tracing, was perfectly sound and sensible.

No less sound and sensible, and perhaps more inspired, was Della Porta's
effort to test the results of his theoretical ray tracing with a set of experiments
based on the apparatus described in book 2, propositions 2–5. As it turns
out, this apparatus, in its slightly varying forms, was too crude to have fully

confirmed or disconfirmed his model, but the key point is that the procedure based on it was at least valid in principle, if not in practice.[108] Another point worth mentioning is that Della Porta's two-stage procedure was a model of hypothetico-deductive reasoning, the hypothesis consisting of his particular refractive model, the observational test based on the instrumental confirmation or disconfirmation of that model. Ptolemy and the Perspectivists had also conducted experiments on refraction using fairly sophisticated equipment, especially the Perspectivists, but their primary goal was to determine the angles of refraction on the basis of that equipment, not to confirm or disconfirm an overarching hypothesis.

Interesting enough, the basic approach Della Porta took in constructing and testing his model of refraction bears some striking similarities to the one Kepler followed in his effort to determine precisely how light passes through the water-filled glass sphere at the base of a urinal flask. As recounted in chapter 5 of *Ad Vitellionem Paralipomena*, Kepler's ulterior aim was to understand how the lens of the eye works by analogy to such a water-filled sphere. Accordingly, like Della Porta, he started with a process of theoretical ray tracing based on the passage of parallel solar rays into and through the sphere, and like Della Porta, he arrived at incorrect results because his model did not conform to the sine law.[109] Like Della Porta as well, he tested the results of his theoretical ray tracing empirically by allowing a shaft of sunlight to refract through the water-filled globe and shine on a piece of paper held on the side of the globe away from the Sun. Gradually moving the paper away from the sphere along a perpendicular line allowed him not only to confirm the phenomenon of spherical aberration but also to determine the focal length of the sphere.[110]

There is no question that the procedures Della Porta and Kepler followed in developing and testing their models of refraction vary in detail, sometimes considerably so, but these differences in detail should not obscure the crucial point of similarity: Both thinkers carried out a clearly delineated, two-stage process of theoretical and empirical ray tracing. To my knowledge, Della Porta was the first to follow, or at least to explicitly describe, this particular method, and although it was with Kepler, not Della Porta, that it bore real fruit, the method itself appears to have been Della Porta's brain child. This is not to say that Kepler appropriated it from Della Porta, at least not directly, but the two-stage procedure described by Della Porta certainly foreshadowed the one followed successfully nearly a decade later by Kepler. Della Porta thus deserves credit as a methodological innovator.

The very fact that Della Porta and Kepler felt driven to study lenses so closely, and were able to do it effectively, is due in great part to developments in glass technology during the sixteenth century. One such development, whose beginnings date to the mid-fifteenth century, was the ability of Venetian glassmakers to produce *cristallo* glass of unsurpassed clarity and homogeneity. By Della Porta's day such glass was readily available at reasonable prices,

not just from Venice but from other European cities as well. Another important development, discussed in admirable detail by Vincent Ilardi, was the mass production of eyeglasses during the sixteenth century, one kind furnished with convex lenses of varying degrees of curvature for multiple stages of presbyopia, another kind furnished with concave lenses appropriately curved for two grades of myopia. The ubiquity of such lenses among spectacle makers of the later sixteenth century was crucial not only to Della Porta's study of lenses but also to the invention and evolution of the telescope in the opening years of the seventeenth century. No less important than improvements in the manufacture of glass, finally, was the mass production of glass mirrors backed by an amalgam of tin and mercury. Not only were such mirrors highly reflective, but they were also relatively cheap. In a sense, then, the sixteenth century was an age of glass, and as an optical researcher Della Porta (and Kepler after him) obviously benefitted from living in it.[111]

Della Porta's analysis of the pupil and its function represents another point of originality.[112] Of particular note in this regard is his correlation among pupil size, light intensity, and visual acuity. Accordingly, he speculated (more or less rightly) that nocturnal animals such as cats and owls have large pupils in order to allow as much light as possible into the eye, whereas diurnal animals, such as humans and hens, have smaller pupils whose function is to protect their eyes from damage by excessive light. Furthermore, Della Porta's explanation of night vision among nocturnal animals was opposed (rightly) to the standard explanation of his day: that is, that the glowing eyes of such animals act as flashlights, rendering external objects visible by illuminating them with their intrinsic light. Yet another point of originality in Della Porta's account of the pupil is his claim that the pupil shape of particular kinds of animals relates to their basic needs. Cats, for instance, have slitted pupils that open as if hinged at top and bottom because such a shape enables them "to supply themselves with nourishment more quickly and more freely."[113] To be sure, this claim is merely allusive, but in the light of recent work correlating pupil shape to animal behavior, it looks remarkably prescient.[114]

Della Porta's account of binocular vision and diplopia provides another instance of innovative thinking, this time with unfortunate results. Especially unfortunate is his explanation of diplopia, or double vision, which depends on assuming that light sometimes refracts toward the normal and sometimes away from it when entering the eye from the air outside. The direction of refraction is apparently determined not by the constraints of physical law but, rather, by the need to locate the image at a particular place in the field of view, as dictated by visual experience. Less unfortunate, but still problematic, is Della Porta's supposition that the eyes work independently and therefore that, when both eyes are open and unconstrained, only one of them is actually seeing. In other words, under normal circumstances, vision is monocular, not binocular. From this it follows that binocular vision,

as understood by everyone before Della Porta, is unnatural because it depends on full cooperation between the eyes, which by his account results in double rather than in single imaging.

In taking these problematic stands, Della Porta was reacting against the Perspectivist model of binocular vision and diplopia, which presents its own difficulties. For one thing, according to that model, proper binocular vision entails image fusion at the optic chiasma, the upshot being a melding of the separate images provided by each eye. When these images are congruent or very nearly so, they yield a single, fused image that represents an average of the two. If, on the other hand, they are not adequately congruent, they resist fusion, leading us to see double. Yet, as Della Porta correctly observed, when we stare intently at something not too far away and close each eye in succession, we notice that one eye, generally the right-hand one, provides the dominant image, which remains stationary when we close the nondominant eye. Conversely, the image shifts to the left or right when we close the dominant eye, which indicates that what we normally see when both eyes are open is not an average of what each eye sees. Della Porta was therefore clearly aware of what is currently referred to as ocular dominance or eyedness.

Another difficulty with the Perspectivist account of image fusion is that it requires the image from each eye to pass through the hollow, spirit-filled optic nerve to the optic chiasm just in front of the brain. There it supposedly fuses with its counterpart from the other eye. The problem here, as Della Porta correctly noted on the basis of close anatomical observation, is that the optic nerves are not hollow, or at least not sufficiently so to accommodate such images. This being the case, it makes no sense to assume that visual spirit passes from the brain to the eye through the optic nerves to sensitize the crystalline lens while also serving as the medium through which the images abstracted by the lens are conveyed back to the brain for perceptual processing.

In rejecting these spirits, Della Porta struck at the very foundations of the Perspectivist theory of visual perception and cognition, based as it was on the notion that visual images pass into the brain, which is suffused with animal spirits. There those images are transformed in stages by the sensitive, imaginative, and cognitive faculties within the brain's spirit-suffused ventricular system to yield fully abstract and comprehensive mental "pictures" of external reality.[115] Granted, the theory Della Porta offered as an alternative is both inchoate and incoherent, but his ability to see that an alternative was necessary in the face of mounting evidence against the Perspectivist account bespeaks a well-honed critical faculty. This faculty, as we have seen, was sporadically at work throughout his analysis of refraction and sight in *De Refractione*. And although the results of that analysis were at best mixed, they included more than enough flashes of insight and originality to belie Vasco Ronchi's characterization of *De Refractione* as a farrago of absurdities.[116]

To this point I have taken pains to characterize Della Porta's theoretical and practical insights in *De Refractione* as foreshadowings rather than antici-

pations in order to avoid the implication that later thinkers appropriated them or were even influenced by them. There is in fact no indication of such appropriation or influence, aside perhaps from the bare possibility that Kepler got wind of Della Porta's two-stage, ray-tracing procedure at second hand. It therefore follows that, in terms of actual, positive influence, Della Porta's *De Refractione* contributed little or nothing to the development of optics after its publication, except perhaps by occasionally stimulating research to correct specific errors contained in it.[117]

Despite its evident lack of positive influence on subsequent optical thinkers, *De Refractione* does reflect certain key trends in the evolving natural philosophical, or scientific, thought of its day, and that is where its true historical significance lies. Perhaps the most obvious of these trends was a strong emphasis on experiential or experimental verification, which at times devolved into naïve empiricism. At its most sophisticated, this emphasis is evident in Della Porta's effort to trace the trajectory of light rays through a glass sphere in order to determine precisely where the rays entering the sphere at various angles will intersect its axis after refraction into it. In this case, the experimental device fabricated from a carefully scribed plaque with a properly shaped glass chunk and a light filter attached to it is fairly complex. Less sophisticated is Della Porta's study of the pupil and its reaction to light, which is verified on the basis of two fairly rudimentary instruments, one consisting of a long, thin ruler and two cylinders of unequal diameter, the other of a square plank with a long nail driven perpendicular to its center. Least sophisticated are such basic observational techniques as dividing the field of view by placing an obstacle between the eyes and testing whether each eye is capable of discerning simultaneously with equal clarity within its own limited field of view. The point in this case is to offer a simple but telling demonstration that each eye works independently according to what is currently known as binocular rivalry and that the dominant eye yields the preferred point of view.

Another clear trend in the evolving scientific thought of Della Porta's day was an increasingly critical attitude toward Peripatetic thought and its scholastic avatars. This was in part due to the rise of humanism and the deep respect, even reverence, for classical sources that accompanied it. Della Porta's respect for such sources is evident in his appeal to a broad spectrum of classical authors from Aelian to Xenophon in support of various points made throughout *De Refractione*. Equally evident is his disdain for at least some scholastic sources, Witelo in particular, whom he dismissed as Alhacen's invariably wrongheaded ape. Albertus Magnus is another target of his scorn, although that scorn was not unalloyed, as it seems to have been in the case of Witelo. Likewise, despite a somewhat skeptical attitude toward Alhacen, Della Porta did not deprecate him with the same intensity that he did Witelo. Somewhat surprising, perhaps, is Della Porta's rather deferential attitude toward Aristotle. Still, this deference was only provisional; Della Porta was quick to find fault with Aristotle and his followers over key theoretical issues

on the basis, particularly, of their support of extramissionism. Nonetheless, his critical treatment of Peripatetic and scholastic sources was not a mere knee-jerk reaction against the benighted medieval past. It was born from a clear recognition of specific and potentially serious problems with certain sources, some but not all of them scholastic. Della Porta, in other words, had a sharp eye for the flaws in others' reasoning, if not always in his own.

Finally, like so many natural philosophers of his day, Della Porta was an unabashed eclectic. Wedded to no particular school of thought, albeit with Aristotelian leanings, he was willing to adopt or adapt what was useful to his purposes from thinkers representing a variety of such schools. For example, having rejected the physiology of cerebral and neurological spirits espoused by Galen and the Perspectivists, he nonetheless clung to the notion that the crystalline lens is visually sensitive. This sensitivity, however, had been explained by Galen and everyone after him in terms of a continual infusion of visual spirits arising in the brain and passing to the eyes through the hollow optic nerves—in short, on the basis of the cerebral and neurological spirits Della Porta repudiated. Likewise, while dismissing the Perspectivist center of sight as an analytic device, Della Porta nonetheless fell back on the cone of radiation with its vertex at that very same center of sight in order to account for the special clarity of vision within a limited field defined by cone's base. As a result of such partial accommodations, his analysis of sight and light often comes across as unsystematic, even incoherent, an impression bolstered by his somewhat piecemeal, topic-by-topic, problem-by-problem approach to visual theory. Not untypical of his day, in other words, Della Porta was less concerned with the overall coherence of his approach and its theoretical underpinnings than with making sense of specific phenomena within a fairly narrow empirical context.

From the foregoing discussion, it should be evident that if we evaluate Giambattista Della Porta's *De Refractione* retrospectively, looking for indicators of the "correct," straitened path of modern optical development forged by the likes of Kepler, Descartes, Newton, Fresnel, Fraunhofer, and a host of other later thinkers, then we can summarily dismiss it as historically irrelevant. This sort of retrospective evaluation has its uses, to be sure, but it ignores another important gauge of historical relevance: namely, how much light a given work sheds on the cultural and intellectual context within which it was written. In this regard, Della Porta's *De Refractione* assumes more than a modicum of historical importance. But it does more than just bring the evolving scientific thought of its day to light. It also reflects a Kuhnian crisis in that thought, at least as it applied to optics. This crisis involved a variety of by-now familiar factors. Inconsistencies in the Perspectivist visual "paradigm" and its theoretical foundations loom especially large, as does a dawning awareness of "anomalies" such as spherical aberration and the projection of real images by reflective and refractive bodies. Increasing availability of glass objects and the opportunity to study refraction and its effects ever more closely in them also played a significant role, as did the

capacity to share information (and misinformation) more widely than ever before through print. Not least significant, finally, was the proliferation of social/intellectual networks designed to promote the dissemination and critical evaluation of such information among select groups of like-minded people. A product of these unsettling conditions and the unsettled times within which they arose, *De Refractione* is emblematic of the intellectual confusion and instability that marked them and out of which modern optics began to emerge at the beginning of the seventeenth century. As such, it provides crucial insight into the state of light and sight theory just before both underwent a radical transformation during the next century.

NOTES

1. In a diatribe against William Gilbert, who accepted that the Earth rotates axially, Della Porta castigated him for holding such a foolish idea; see, for example, Edward Rosen, *The Naming of the Telescope* (New York: Henry Shuman, 1947), 27.

2. See, for example, Charles Homer Haskins, *The Renaissance of the Twelfth Century* (Cambridge, MA: Harvard University Press, 1927), 9.

3. See John Henry, "The Fragmentation of Renaissance Occultism and the Decline of Magic," *History of Science* 46 (2008): 1–48. The supposed dichotomy he addresses in this essay is between "magic" and "modern science," and his purpose is to show that the two only became oppositional during the Enlightenment, not during the sixteenth and seventeenth centuries when, in fact, magic was incorporated enthusiastically in the thought of various reformist scientific thinkers.

4. By Christian name, Giambattista Della Porta is referred to variously as Giambattista, Giovan Battista, Giovanni Battista, and Giovambattista. By family name, he is referred to as Della Porta or simply Porta. I have chosen to refer him to by Christian name as Giambattista and by family name as Della Porta because both versions are most commonly encountered in English.

5. On the dating of Della Porta's birth and its problematic aspects, see Louise George Clubb, *Giambattista Della Porta Dramatist* (Princeton, NJ: Princeton University Press, 1965), 5–6.

6. On Della Porta's siblings, see Paolo Piccari, *Giovan Battista Della Porta il filosofo, il retore, lo scienzato* (Milan: FrancoAngeli, 2007), 18.

7. For a discussion of Della Porta's possible educational background, see ibid., 19–22. Presumably, the instruction on music was part of an effort to teach the three brothers the courtly arts expected of any Renaissance Italian gentleman.

8. These include *Architae Tarentini Decem Praedicamenta Dominico Pizimentio Interprete* (Venice, 1561), *Dionysii Longini Rhetoris Praestantissimi Liber De Grandi Orationis Fenere, Dominico Pizimentio Vibonensis Interprete* (Naples, 1566), and *Democritus Abderyta, De Arte Sacra, siue De Rebus Naturalibus. Nec Non Synesii, & Pelagii in Eundem Commentaria Dominico Pizimentio Interprete* (Naples, 1570).

9. On Pisano, see notes 1 and 2 to book 1, p. 66.

10. See, for example, Eamon, *Science and the Secrets of Nature* (Princeton, NJ: Princeton University Press, 1994); see esp. 194–233 for an extensive account of Della Porta as a natural magician. Actually, Eamon did not so much coin the phrase (late Renaissance contemporaries referred to "i professori dei segreti") as popularize it in his writings.

11. In an effort (apparently successful) to curry favor with Philip, Della Porta dedicated both his *Magiae Naturalis* and *De Furtivis Literarum Notis* to him.

12. Clubb, *Magiae Naturalis* and *De Furtivis Literarum Notis*, 57–69.

13. On the possible theatrical roots of Della Porta's scientific showmanship, see Sergius Kodera, "Giambattista Della Porta's Histrionic Science," *California Italian Studies* 3 (2013): 1–27; https://escholarship.org/uc/item/5538w0qd. See also Kodera, "The Laboratory as Stage: Giovan Battista della Porta's Experiments," *Journal of Early Modern Studies* 3 (2014): 15–38.

14. On the background to Della Porta's charges and trial, see Neil Tarrant, "Giambattista Della Porta and the Roman Inquisition: Censorship and the definition of Nature's limits in sixteenth-century Italy," *British Journal for the History of Science* 46 (2013): 601–25. See also Michaela Valente, "Della Porta e l'inquisizione. Nuovi documenti dell' Archivio del Sant'Uffizio," *Bruniana & Campanelliana: Ricerche filosofice e materiali storico-testuali* 5 (1999): 415–34, and Piccari, *Della Porta*, 24–6.

15. Valente, "Della Porta e l'inquisizione," 426. Three years earlier, Della Porta had been unwittingly drawn into a dispute between Jean Bodin and Johann Wier (or Weyer) over witchcraft. Against Bodin, Wier attempted to naturalize witchcraft, thereby questioning its roots in demonic possession. In defense of his position, he cited Della Porta's account of the "witches' ointment" (*laminarum unguenta*) in book 2 of *Magiae Naturalis*, where Della Porta argued that the putative nocturnal flights and trysts with the Devil routinely engaged in by witches were simply flights of deranged fancy caused by rubbing a hallucinogenic ointment containing belladonna on their bodies; see Della Porta [1558], 101–02. Naturalization of such activities, which were emblematic of witchcraft at the time, was in direct conflict with the position of church authorities, who imputed them to demonic possession. An echo of Della Porta's "witches' ointment" can be found in book 7, proposition 8, p. 347, where he alludes to the hallucinogenic effect of belladonna.

16. See Eileen Reeves, *Galileo's Glassworks* (Cambridge, MA: Harvard University Press, 2008), 70–78. See also Valente, "Della Porta e l'inquisizione."

17. *De Humana Physiognomonia Libri IIII* (Vico Equense, 1586).

18. For background on these thinkers and their theories, see Heather McRae, "Pointing to Inclinations: Albertus Magnus's Physiognomy as a Scientific and Theological Nexus" (PhD diss., University of Missouri–Columbia, 2018).

19. In addition to the author portrait on the title page and that of Cardinal Luigi d'Este, to whom the book is dedicated, there are 79 illustrations, the vast majority of them showing human–animal pairs. Most of these illustrations are repeated at least once, often far more than once, throughout the book, so the stock of individual and individually different illustrations is far smaller than 79.

20. For a discussion of *De Humana Physiognomonia* as a literary rather than a "scientific" work, see Katherine MacDonald, "Humanistic Self-Representation in Giovan Battista Della Porta's *Della Fisonomia Dell'Uomo*: Antecedents and Innovation," *Sixteenth Century Journal* 36 (2005): 397–414.

21. For a brief account of the mechanics of church censorship in Italy at this time, see Christopher Black, *The Italian Inquisition* (New Haven, CT: Yale University Press, 2009), 158–207.

22. Clubb, *Della Porta Dramatist*, 25–6. In January of 1586, several months before the censors approved the publication of *De Humana Physiognomonia*, Pope Sixtus V issued the bull *Coeli et terrae creator Deus* in which he had roundly anathematized all divinatory arts, even those that predicted inclinations or propensities; see Ugo Baldini, "The Roman Inquisition's Condemnation of Astrology: Antecedents, Reasons and Consequences," in Gigliola Fragnito, ed., and Adrian Belton, trans., *Church, Censorship and Culture in Early Modern Italy* (Cambridge: Cambridge University Press, 2001), 79–110, esp. 91–2. That the censors apparently ignored this denunciation when vetting Della Porta's physiognomic study is somewhat surprising.

23. Clubb, *Della Porta Dramatist*, 10.

24. See Ibid., 300–01 for a listing, which includes a significant number of plays Della Porta left unpublished.

25. Ibid., xiv–xvi. This high regard for Della Porta as a dramatist is reflected in the fact that editions of his plays were among the very first of his works to be published in the Edizione Nazionale Delle Opere di Giovan Battista della Porta; see the website at http://www.edizioniesi.it/pubblicazioni/collane/letteratura_201/edizione-nazionale-delle-opere-di-giovan-battista-della-porta.html.

26. According to Clubb, *Della Porta Dramatist* 35, Giovanni di Rosa was a friend of Della Porta's. Interestingly enough, the work is still attributed to Della Porta in the full Italian title: *Della Fisionomia dell'Uomo del Sig. Gio. Battista Della Porta Napolitano Libri Sei*, yet that was evidently no obstacle to its publication.

27. *Della Porta Dramatist*, I, 2, in Della Porta (1558), 3.

28. Clubb, *Della Porta Dramatist*, 18.

29. Della Porta's esteem for Cesi was such that he dedicated three works to him: *De Distillatione* (1608), *Elementorum Curvilineorum Libri Tres* (1610), and *De Aeris Transmutationibus Libri IV* (1610).

30. For a brief account of the formation of the *Accademia dei Lincei*, see David Freedberg, *The Eye of the Lynx* (Chicago: University of Chicago Press, 2002), 65–77.

31. For a more detailed account of Della Porta's claim to priority and the circumstances surrounding it, see Rosen, *Naming of the Telescope*.

32. For an edition of Della Porta's unpublished *De telescopio*, with an introduction by Vasco Ronchi, see Maria Naldoni, ed., *Giovan Battista Della Porta: De telescopio* (Florence: Leo Olschki, 1962). The *De telescopio* itself is quite brief, but it is accompanied by a number of other short works, many of which recapitulate Della Porta's analyses of spherical, biconvex, and biconcave lenses in books 2 and 8 of *De Refractione*.

33. For a full listing, see Laura Balbiani, *La* Magia Naturalis *di Giovan Battista Della Porta: Lingua, cultura, e scienza in Europa all'inizio dell'età moderna* (Bern: Peter Lang, 2001), 212–17.

34. On the enormous popularity and broad influence of both versions of the *Magiae Naturalis*, see Laura Balbiani, "La ricezione della *Magia Naturalis* di Giovan Battista Della Porta: cultura e scienza dall'Italia all'Europa," *Bruniana & Campanelliana: Ricerche filosophice e materiali storico-testuali* 5 (1999): 277-303.

35. *Natural Magic* (1558), IV, 2, 143–44, and *Natural Magic* (1589), XVII, 6 and 7, 266–67.

36. See Della Porta (1589), 266.

37. *Natural Magic* (1558), IV, 13, 149–50; (1589), XVII, 2, 261 ("Spaerico concavo ut pensilis quoque imago videri possit").

38. *Natural Magic* (1558), IV, 8, 147; (1589), XVII, 2, 261 ("Speculum ex planis componere, in quo imago una veniens, in latero recedens conspiciatur").

39. *Natural Magic* (1558), IV, 5, 145–46; (1589), XVII, 2, 261 ("Speculis planis ut caput deorsum, pedes sursum videantur").

40. On Della Porta's account of burning by both reflection and refraction of sunlight, see *Natural Magic* (1558), IV, 15, 151–52, and *Natural Magic* (1589), XVII, 14, 271–78.

41. *Natural Magic* (1589), XVII, 10, 269 ("Lente crystallina convexa imaginem in aere pendulam videre").

42. Ibid., XVII, 21, 278–79.

43. Ibid., 3 (unnumbered): "Toto enim animo, totisque viribus maiorum nostrorum monumenta pervolui, et si quid arcani, si quid reconditi scripsissent, defloravi, dein quum Italiam, Galliam, et Hispaniam peragrassem, bibliothecas, et doctissimos quosque adii, artifices etiam conveni, ut si quid novi, curiosique nacti essent, ediscerem quae longo usu verissima, et utilissima comprobassent, agnoscerem."

44. "In nostris in opticis fusius declaratum est," *Natural Magic* (1589), XVII, 6, 267. The appropriate location in *De Refractione* is book 4, proposition 1, on pp. 192–95.

45. "Rationes in opticis reddidimus," ibid., XVII, 10, 269. See *De Refractione*, book 1, preface to the readers, pp. 10–11, and book 8, propositions 1 and 13, pp. 356-59 and 374–75.

46. "Plura non apponimus, quia latius de ea in Opticis," ibid., XVII, 14, 271. Although he does not explicitly address the focusing of sunlight to a burning point in concave mirrors, Della Porta gives an extensive account of such mirrors in book 2, proposition 1, of *De Refractione* that implies such focusing; see pp. 78–91.

47. "Rationem reddemus in opticis," ibid., XVII, 19, 277. For the apposite locations in *De Refractione*, see book 2, proposition 22, pp. 122–25, and book 8, proposition 13, pp. 374–75.

48. For details on the two Baconian treatises and their historical background, see David C. Lindberg, ed. and trans., *Roger Bacon's Philosophy of Nature: A Critical Edition, with English Translation, Introduction, and Notes, of* De multiplicatione specierum *and* De speculis comburentibus (Oxford: Clarendon Press, 1983) and Lindberg, ed. and trans., *Roger Bacon and the Origins of* Perspectiva *in the Middle Ages: A*

Critical Edition and English Translation of Bacon's Perspectiva *with Introduction and Notes* (Oxford: Clarendon Press, 1996). For al- Kindī's *De aspectibus*, see Axel Björnbo and Sebastian Vogl, eds. and trans., *Alkindi, Tideus und Pseudo-Euclid: Drei optische Werke* (Leipzig: Teubner, 1912).

49. See note 44 to book 1, proposition 11, p. 73.

50. For a full accounting of these fragments and their provenance, along with biographical sketches of the relevant Presocratic thinkers, see Hermann Diels, *Die Fragmente der Vorsokratiker* (Berlin: Wiedmann, 1903), published in a revised and expanded sixth edition by Walther Kranz (Berlin: Wiedmann, 1952). For an English translation of the fifth edition of this work, see Kathleen Freeman, *Ancilla to the Pre-Socratic Philosophers* (Oxford: Blackwell, 1948).

51. For brief, capsule biographies of some of the lesser-known figures cited here and throughout the rest of this edition, see the Biographical Sketches on pp. 475–86.

52. Also known as Caelius Rhodiginus, Ricchieri (1469–1525) was a Venetian scholar and professor of classical languages. The original version of his compendium was published at Venice in 1516 as *Antiquarum Lectionum Libri XVI* (*Sixteen Books of Readings from the Ancients*) but was reorganized and expanded from sixteen to thirty books posthumously in 1542. Both versions were enormously popular throughout the sixteenth century. For further details, see Peter G. Bietenholz and Thomas B. Deutscher, *Contemporaries of Erasmus: A Biographical Register of the Renaissance and Reformation* (Toronto: University of Toronto Press, 2003), 3: 155.

53. Book 2, proposition 22, p. 125. For the characterization of Witelo as Alhacen's ape, see book 3, proposition 6, p. 159.

54. For a detailed discussion of these four Perspectivist authors, their works, and the theory of light and sight they articulated, see A. Mark Smith, *From Sight to Light: The Passage from Ancient to Modern Optics* (Chicago: University of Chicago Press, 2015), 181–227 and 256–77.

55. The one Perspectivist work that Della Porta does not mention explicitly is Bacon's *De multiplicatione specierum*, but there are terminological reasons to think that he had read it; see note 19 to book 1, p. 69.

56. See, for example, *De Refractione*, book 3, proposition 13, pp. 168–70.

57. See ibid., book 4, proposition 1, pp. 186–99. Bacon and Pecham are exceptions insofar as they thought that vision requires radiation both into and out of the eye.

58. See ibid., book 4, propositions 6 and 7, pp. 204–13.

59. See ibid., book 6, proposition 1, pp. 288–303.

60. See ibid., book 4, proposition 7, pp. 206–13.

61. See, for example, Nancy Siraisi, *Avicenna in Renaissance Italy: The Canon and Medical Teaching in Italian Universities after 1500* (Princeton, NJ: Princeton University Press, 1987), 3–18, esp. 5–7.

62. This firm grounding in the medical theory and practice of his day was undoubtedly due to the tutorial instruction he received from Giovanni Antonia Pisano during his youth. Indeed, at one point in *De Refractione* Della Porta recalls doing "the hard work of anatomical procedures" as a boy, presumably under Pisano's tutelage; see ibid., book 7, proposition 2, pp. 329–33.

63. See book 9, proposition 2, on p. 391.

64. Avicenna's *Liber de anima*, or *Sextus de naturalibus*, is the Latin version of the sixth part of his *Shifā' (Healing)*, where Avicenna addresses psychological issues, including sensation in general and vision in particular. Della Porta was most likely also familiar with Averroes's *Long Commentary on Aristotle's* De anima, which was available to him in the sixth volume of the Giunta edition of Aristotle's works (Venice, 1550).

65. Della Porta's exceptional erudition is evident in the 1558 edition of *Magiae Naturalis*, where he cites well over 100 sources, most of them classical; see Balbiani, *La* Magia Naturalis, 41–53, and also 209-11, where she provides a complete listing of the cited sources. It should be borne in mind, however, that these include references to persons, not necessarily to actual textual sources. In short, Della Porta may have cited many of them at second or even third hand (as, for instance, reported by Ricchieri).

66. See, for example, Ari Wesseling, "Erasmus and Plagiarism," in Stephen Ryle, ed., *Erasmus and the Renaissance Republic of Letters* (Turnhout: Brepols, 2014), 203–14.

67. By current definition, the angle of refraction is formed by the refracted ray and the normal to the point of refraction, whereas the angle of deviation is formed by the refracted ray and the continuation of the incident ray. In figure 1a, p. xxxiii, if RA' is the refracted ray, then A'RN' is the angle of refraction. On the other hand, angle IRA' formed by continuation RI of incident ray AR and refracted ray RA' constitutes the angle of deviation.

68. In traditional fashion, Della Porta ties the refractivity of a given medium to its density or transparency, but neither term is to be taken literally in the physical sense; for a discussion of the problematic nature of these terms as applied to refraction, see note 9 to book 1 on p. 67.

69. For a brief discussion of how this experimental technique echoes Alhacen's more elaborately based one in book 7 of his *De aspectibus*, see note 26 to book 1, pp. 70–71.

70. The apparent enlargement of Moon and Sun at or very near the horizon, known as the Moon Illusion, is a psychological phenomenon, not a refractive one. In appealing to such refractive magnification, Della Porta is apparently following Alhacen, who points to thick atmospheric vapors as an occasional factor in the apparent enlargement of celestial bodies at or near the horizon; see note 53 to book 1, p. 74.

71. Della Porta actually defines these angles/arcs geometrically, not arithmetically. Accordingly, the arc of 60° is the arc defined by a chord equal to the side of a hexagon inscribed in the great circle on the mirror's surface, whereas the arcs of 45°, 30°, and 22° are defined, respectively, by the sides of an inscribed octagon, an inscribed 12-sided figure, and an inscribed 16-sided figure.

72. For a discussion of this point, see Smith, *From Sight to Light*, 347–49. For a full accounting of the expected vs actual angles of refraction according to Della Porta's model and the modern sine law, see note 12 to book 2, pp. 129–30.

73. The index of refraction for the glass available to Della Porta in the later sixteenth century would have been around 1.52, equivalent to that of modern crown

glass. Had he used quartz crystal instead, the refractive index would have been just a bit higher, around 1.54. In both cases, therefore, the critical angle would have been well below 45°.

74. It is important to note, however, that the limit of half the radius for the focal point of a glass lens with an index of refraction of ~1.5 is based on two refractions through the lens, not a single refraction into it, as Della Porta's analysis assumes.

75. See, for example, Smith, *From Sight to Light*, 218–22 and 333–44.

76. See Figure 8.18 on p. 381 for a proper representation of this radiation into the lens.

77. Avoiding image inversion was key to virtually every account of visual imaging in the eye until Johannes Kepler was forced to accept it as a consequence of his analysis of retinal imaging in *Ad Vitellionem Paralipomena* of 1604.

78. For clarification of this point, see note 24 to book 3 on pp. 179–80.

79. Although implicit in Della Porta's discussion of refraction into the eye, this point is not made explicit until proposition 14, pp.170–75.

80. See book 7, propositions 2 and 3, pp. 329–35.

81. For a more detailed account of Alhacen's explanation of peripheral vision, see Smith, *From Sight to Light*, 215–8.

82. For more on this point, see Alhacen (2010), lxvii–lxx, esp. lxix–lxx.

83. See Witelo (1572), 102–3.

84. These thinkers all form an intellectual "school" of sorts; for a brief discussion of their analyses of the rainbow, see Carl B. Boyer, "The Tertiary Rainbow: An Historical Account," *Isis* 49 (1958): 141–54, esp. 144–46. On Oresme and the relationship of his commentary to those of Albert of Saxony and Themon Judei, see Stephen McCluskey, ed. and trans., "Nicole Oresme on Light, Color, and the Rainbow: An Edition and Translation, with Introduction and Critical Notes, of Book Three of His *Questiones super quatuor libros meteororum*" (PhD dissertation, University of Wisconsin, 1974).

85. The classic study of Theodoric's theory of the rainbow is William A. Wallace, *The Scientific Methodology of Theodoric of Freiberg* (Fribourg: Fribourg University Press, 1959). In certain ways Theodoric's theory anticipates the "correct" theory propounded by Descartes and modified by Newton, which imputes the rainbow to particular angles of incidence into and refraction out of individual raindrops; for a brief explanation of Descartes's theory, see Smith, *From Sight to Light*, 400–403; Newton's contribution lay in having shown that "white" light is actually a composite of all colors and, accordingly, that each color refracts as a slightly different angle.

86. Maurolico's rainbow theory is contained in his posthumous *Photismi de Lumine* (1611), 49–68; An English translation is available in Henry Crew, trans., *The* Photismi de lumine *of Maurolycus: A Chapter in Late Medieval Optics* (New York: Macmillan, 1940).

87. For a general survey of the development of rainbow theory from antiquity to early modern times (and beyond) see Carl B. Boyer, *The Rainbow: From Myth to Mathematics* (New York: Thomas Yoseloff, 1959) and Raymond L. Lee and Alistair B. Fraser, *The Rainbow Bridge: Rainbows in Art, Myth, and Science* (University Park, PA: Pennsylvania State University Press, 2001). Boyer's work, which is still more or

less canonical, although a bit tendentious, is based on primary sources and is therefore the more scholarly of the two; Lee and Fraser's book on the other hand, is aimed at a more popular audience and, as such, is heavily dependent on secondary sources.

88. *Meteorology*, III, 2, 372a17–21.

89. Ibid., III, 4, 373a31–373b32.

90. Ibid., 374b20–32.

91. Ibid., 374b32–375a12.

92. On Aristotle's mathematical analysis of the rainbow and some of its problematic entailments, see Smith, *From Sight to Light*, 26–28.

93. Quotations from *De iride*, in Grosseteste (1912), 77–78.

94. See Bacon (1912), 194–95.

95. *Opus majus*, VI, 12, 196. In his *Dioptrique* of 1637 Descartes repudiated this distinction between real and apparent, or false, colors, arguing that, since it is the very nature of color to appear, it makes no sense to claim that a color appears and is yet false; see Smith, *From Sight to Light*, 403.

96. On Bacon's distinction between the real colors of the refracted spectrum and the apparent colors of the rainbow, see *Opus majus*, VI, 8, in Bacon (1912), 191–92. On the individuality of rainbows according to the individuality of viewers, see *Opus majus*, VI, 7, in Bacon (1912), 187.

97. Although it looks perpendicular to the face of the raincloud, refracted ray FPOM would actually have to refract along a line somewhere between normal FPOM and the continuation of incident ray AF straight through the cloud. The figure is therefore somewhat misleading because it is sloppily done.

98. In proposition 13 Della Porta refines this account somewhat by taking refraction of the light along the refracted ray back out of the cloud into consideration.

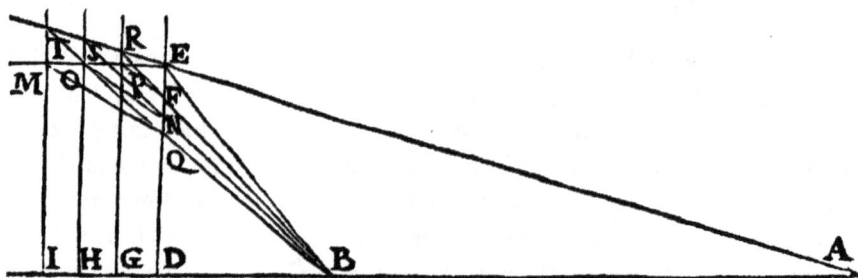

Figure 25a

Thus, in Figure 25a, lifted directly from the original Latin text, solar ray AE strikes front surface DE of the cloud and refracts into it along EM (which is misrepresented in the figure as perpendicular to ED). Instead of passing directly back to the center of sight B through ED, as represented in figure 25, the iridescent color at point P radiates to point F, where it is refracted out of the cloud to the center of sight along FB. When extended, Della Porta argues, FB intersects the continuation of G at R, which is where P's image will be seen. Applied to all the remaining points on refracted

ray EPOM, this line of reasoning leads to the conclusion that the entire spectrum along it will actually be seen along ERST. The flaw in this reasoning, of course, is that GP, HO, and IM are not catheti of incidence, so the supposed image points R, S, and T are improperly defined.

99. The actual passage reads as follows: "In the year 1590, when the Sun was setting in the west, we observed that the height of an upper rainbow was 42° and that of a lower one was 38°, so that the interval between both was 4°" (p. 437). It is not clear from this whether Della Porta was measuring the bows when the Sun was right at horizon or when it was nearing horizon. If the former, then the upper rainbow would have been primary and the lower one would have been supernumerary, in which case the separation of 4° would have been far too large. On the other hand, if the lower bow was primary and upper one secondary, 4° would have been far short of the appropriate separation of ~9°.

100. Like many other commentators before him, Della Porta viewed this reversal of colors as proof that the secondary rainbow cannot be due to reflection of the primary bow from another raincloud because in that case the colors would retain their original orientation.

101. "Mais écrire un livre sérieux était une tâche supérieure aux forces de Porta et surtout n'était pas conforme à son tempérament. Il produisit un travail [i.e., De Refractione] qui est remarquable par les absurdités qu'il contient, absurdités présentées, naturellement, comme de vérités démontrées mathématiquement," Vasco Ronchi, "Du *De Refractione* au *De Telescopio* de G. B. Della Porta," *Revue d'histoire des sciences et de leurs applications,* 7 (1954), 46–47 (reprinted as an introduction to Naldoni's 1962 edition of *De telescopio*). David Lindberg offers a more moderate, but still essentially negative, evaluation of Della Porta's abilities as an optical thinker based narrowly on his analysis, or mis-analysis, of lenses; see Lindberg, "Optics in Sixteenth-Century Italy," in Paolo Galluzzi, ed., *Novità celesti e crisi del sapere* (Florence: Giunti Barbèra, 1984), 131–48, esp. 142–48. A more recent, and more trenchantly negative evaluation of Della Porta, based on his analysis of spherical lenses in book 2 of *De Refractione* can be found in Robert Goulding, "Thomas Harriot's optics, between experiment and imagination: The case of Mr. Bulkeley's glass," *Archive for History of Exact Sciences* 68 (2013): 137–78, esp. 145, where Goulding asserts that De Refractione "as a whole is one of the most bizarre and misguided attempts on the problem of refraction ever devised."

102. On Kepler's gushingly positive evaluation of Della Porta as well as his claim to have been been unable to procure a copy of the Della Porta's "optics," see Smith *From Sight to Light*, 366.

103. Clubb, *Della Porta Dramatist,* 12.

104. Della Porta alludes to one of these bouts in book 7, proposition 7.

105. On Leonardo's treatment of the eye as a camera obscura, see Smith, *From Sight to Light*, 303–09. It is highly unlikely that Della Porta got his insight directly from Leonardo's writings because he would have to have had access to the appropriate manuscript sources which were (and still are) scattered and not easily available.

106. See Della Porta (1589), 266.

107. See book 2, props. 1 and 2, pp. ~178–95.

108. Alhacen's experimental method for measuring angles of refraction in book 7 of his *De aspectibus* provides another example of an instrumental design that was perfectly valid in principle but unfeasible in practice; for a discussion of this point, see A. Mark Smith, "Le *De aspectibus* d'Alhacen: révolutionnaire ou réformiste?" *Revue d'histoire des sciences* 60 (2007): 65–81. It is quite possible that Della Porta's experimental apparatus was at least inspired by, if not loosely modeled after, Alhacen's.

109. Although incorrect, Kepler's results were closer to the values dictated by the sine law than Della Porta's because his model was based, with minor adjustments, on Witelo's tabulations in book 10, proposition 8, of his *Perspectiva* (Witelo [1572], 412). These, in turn, were based on Ptolemy's experimentally-derived, and remarkably accurate, tabulations in the fifth book of his *Optics*; for a discussion of these tabulations and their derivation, see Smith, *From Sight to Light*, 110–14.

110. On Kepler's entire procedure, see ibid., 354–59.

111. For a discussion of these improvements in the production of glass within a broad context of technological developments during the fifteenth and sixteenth centuries, see Smith, *Optics*; for a discussion of these tabulations and their derivation, see Smith, *From Sight to Light*, 323–38. For a look at how the ready availability of glass devices influenced Della Porta's approach to optical research, see Arianna Borrelli, "Thinking with optical objects: glass spheres, lenses and refraction in Giovan Battista Della Porta's optical writings," *Journal of Early Modern Studies* 3 (2014): 39–61.

112. Although Paolo Sarpi may have recognized the pupillary response to light before Della Porta (see the testimony of Fabricius of Acquapendente in *Tractatus Anatomicus Triplex Quorum Primus Oculo . . .* [Oppenheim, 1614], 57), Della Porta nevertheless appears to have been the first to study it closely.

113. Book 3, proposition 10, p. 167.

114. See Martin S. Banks et al., "Why do animal eyes have pupils of different shapes?" *Science Advances* 1 (2015): http://advances.sciencemag.org/content/1/7/e1500391.full

115. For further elaboration on this pictures-in-the mind model of perception and cognition, see Smith, "Picturing the Mind: The Representation of Thought in the Middle Ages and Renaissance," *Philosophical Topics* 20 (1992): 149–70.

116. For a recent, more balanced re-evaluation of Della Porta's optical thought, see the set of nine essays in Arianna Borelli, Giora Hon, Yaakov Zik, eds., *The Optics of Giambattista Della Porta (ca. 1535–1615): A Reassessment* (Cham, Switzerland: Springer International, 2017); see esp. the introductory essay by Arianna Borrelli. This collection of essays is the result of a conference bearing the same title, held at the Technische Universität, Berlin, in 2014.

117. As Goulding points out in "Thomas Harriot's optics," Kepler's contemporary, Thomas Harriot, was one of very few optical thinkers of the time who actually read *De Refractione* and, on that basis, responded to the shortcomings of Della Porta's model of refraction in developing his own model, which is grounded in the sine law.

Latin and English Text

Joannis Baptistae Portae Neapolitani
De Refractione Optices Parte
Libri Novem

Giambattista Della Porta of Naples
Nine Books on the Part of Optics Dealing with Refraction

OCTAVIO PISANO
ADOLESCENTI ERUDITO

Ioannis Baptista Porta Neapolitana Salutem Dicit.

Tantus erat, Octavi carissime, Ioannis Antonii Pisani (viri nunquam sine laudum praefatione nominandi) patris tui, vel in studiorum meorum infantia, erga me amor, ut vererer an a quopiam tam vehementer adamarer. Ipse enim, quum publice doceret, scripta mea, quantulacunque fuerint, ad caelum laudibus extollebat; doctissimisque suis in scriptis mei celebrius meminerat; suaque insigni doctrina me studiis confectum; lippitudinibus, vertiginibus, et eiusmodi studiorum alumnis morbis labefactatum ex orci faucibus eripuit.

Cumque ego intelligerem a viro in excelso doctrinarum fastigio constituto adeo diligi, quique in celeberrimo Neapolitano Gymnasio duo de quadraginta annos philosophiam, et medicinam (non Actica modo, sed Romana venere) professus fuerit summo advenarum et incolarum concursu, doctorumque virorum plausu, qui etiam quintus archiatrus, tam insuper nostris regis vicem gerentibus carus, Serenissimo Ioanni Austriaco regnique proceribus, ut eum egregie coluerint et muneribus condecoraverint, ut mecum ambitiosius [2] gesserim, magnum est enim a magno viro laudari. Et piaculum mihi videretur si hominem praeteriissem, quin quoquomodo possem officia rependerem ac me tanto onere pressum reluerem.

Itaque librum hunc, qui apud posteros mutui amoris testimonia praestaret, ei nuncupare constitueram, qui tantopere meis nugis delectabatur. Ecce dum liber formis exciditur, moritur. Sed quid moritur? Non enim moriuntur qui perpetuo vivere digni sunt; sed ad superos evolavit, cuius superstes fama, et inscriptis et hominum memoriis nunquam consenescet. Aegraeferens viri iacturam, Iulio eius filio nominare decreveram, qui togatus et armatus nostri regis amore flagrans suo aere sub auspiciis Invictissimi Ducis Parmensis apud Belgas in quamplurimis bellicis functionibus ita se naviter gesserit, ut ab eo aureis torquibus honestaretur, et literas ad Catholicum Regem eius praestantiam testantes fecerit. Et dum liber sub praelo esset, fato et ipse cessit.

Mox alter filius, quare consentaneum duxi tibi superstiti ingenii et doctrine tanti viri haeredi dicare. Nec meo voto frustratum iri arbitror si

[1]
To Ottavio Pisano,
A Learned Young Man,

Giambattista Della Porta of Naples addresses respectful greetings.

My dearest Ottavio, your father, Giovanni Antonio Pisano (a man never to be named without preliminary praise), had such great love for me that even at the beginning of my studies I revered him as fervently as I was cherished by him no matter where he was.[1] For when he lectured in public, he praised my writings to the skies, however trifling they may have been, and he mentioned me in his popular and extraordinarily erudite writings; it was through his outstanding teaching that he prepared me for my studies, and he used to snatch me from the jaws of hell as I suffered from the bleary eyes, dizziness, and similar maladies of a student.

When I realized that I was thus esteemed by this man, who had attained the very pinnacle of learning and who lectured for thirty-eight years on natural philosophy and medicine at the renowned Neapolitan Gymnasium (not in the Attic but in the elegant Roman style), I carried on more ambitiously, for it is a great thing to be extolled by a great man.[2] [So distinguished was he that he lectured] to the most eminent of foreigners as well as to a gathering of local residents, and [he did so] to the applause of learned men. [Serving as] the fifth chief physician [of the Neapolitan kingdom], as well, he was so dear to the viceroys of our kingdom and to the Most Serene John of Austria and the elite of the realm that they took extraordinary care of him and lavished him with gifts.[3] [2] And so it seemed sinful to me to overlook this man rather than repay his kindnesses however I could and atone for the heavy burden he took on with me.

As a clear token to future generations of our mutual love, therefore, I had decided to dedicate this book to him, who was so delighted by my trifles. Yet while the book was in its formative stage, lo and behold, he died.[4] What is it that dies, though? In fact, they do not die who are worthy to live forever; instead, he who soared to the heights and whose reputation you stand upon will never waste away in the writings and memories of men. Painfully bearing the loss of this man, I had decided to name his [older] son Giulio [as dedicatee], a man burning with love for our king, a man who proved his worth as both a civilian and a soldier under the aegis of the Most Invincible Duke of Parma.[5] He conducted himself so diligently in numerous military exercises against the Belgians that the duke honored him with garlands of gold and composed letters to the Catholic King attesting to his superiority.[6] However, while this book was in press, he, too, yielded to fate.

Thereupon another son [took his place], for I have been drawn to you by your outstanding nature and learning and have considered it appropriate

tui patris auspiciis coeptos libros laeto animo susceperis; qui octavum vix decimum agens annum musarum sacris initiatus, non philosophica modo, sed mathematica complexus es, dum me aliquando de eiusmodi scientiis ratiocinantem audiveris. At si quid ex indole longius prospectare licebit, futurum te nulla quidem ex parte parente inferiorem iudico. Accipe igitur munus et fruere, meque ut soles ama. Vale.

to dedicate this to the heir of such a man. And I deem it not counter to my wishes for you, under the auspices of your father, to accept these books, which are on the verge of publication [and to do so] with a light heart. You, who are [intellectually] engaged at scarcely eighteen years of age, have been initiated into the sacred mysteries of the muses and have embraced not only the philosophical, but also the mathematical disciplines now that you have heard me reasoning from time to time about sciences of this sort. And if one can look ahead on the basis of natural ability, I predict a future for you that is unquestionably inferior in no way to that of your parent.[7] Accept, therefore, and delight in this gift, and love me as you are wont to do. Farewell.

AD LECTORES

PRAEFATIO

Refractionis libros, totius optices spectatissimam partem, qui sub densioribus umbrarum tenebris hucusque delituerant, lectores candidi, in lucem adducimus; quae subiecti nobilitate, necessitate, utilitate, admirabilitateque inter caeteras scientias ita sua luce coruscat ut pleraeque sine ea defectuosae et obscurae censeantur. A subiecto igitur exordientes; haec de visu tractat, sensuum omnium nobilissimo, naturalium scientiarum instrumento, quaecumque enim visu cognoscimus et experimur de luce et splendore, de syderum et coelestium corporum materie. Quomodo enim coelum ab aere diversum esse ac omnis tenuioris aeris tenuissimum cognoscemus, nisi syderum corpora alibi quam sint nobis videre viderentur? Refractionis enim lege admonemur coelum aere, vel hoc [4] aeris conspurcatae colluviei imo, lympidius esse. Vel quomodo astrorum globosum corpus ex sui orbis densioribus partibus concretum aut expolitum metallum constitutum nisi eius ope cognosceretur?

Si necessitatem contemplemur, videbimus scientias fere cunctas sine eius lumine tenebricosas esse. Ex omnibus has praecipuas eligemus astrologiam et physicam. Quid igitur de syderum motu sine refractione sentiemus, quum eorum phaenomena inter se non congruant? Nam sydera, quorum motus observantur eo loci quo videntur non vere sunt; prope enim horizonta simul contigua, eadem coeli medio a se procul divulsa conspiciuntur. Quid enim de eorum aequabili motu, quum horizontis imo tardius, in verticis culmine properantius a suis orbibus vectari concernantur, nisi refractione sancitum esset rem in nebuloso tardioris, in lympido citatioris motus videri?

Quid de syderum corporum quantitatibus, quae dum Terrae propius cernuntur se ipsis maiora, longius vero dissita minora conspiciuntur, nisi refractionis decreto constitutum esset lucem in rectum ferri, densioris medii occursu refrangi ac per varios a perpendiculari recessus distrahi, et magnitudi-

PREFACE

TO MY READERS

We are bringing to light books on refraction, fair readers, the most esteemed part of all optics, which have been hidden in the thickest shadows up to now. This is a topic that shines with its own light among the rest of the sciences by virtue of the nobility, exigency, usefulness, and praiseworthiness of its subject matter, such that many of those sciences are considered to be defective and obscure without it. Those sciences therefore take their beginning from this subject, which deals with vision, the noblest of all the senses and an instrument of the natural sciences, for it is through sight that we know and experience anything about light and brightness and about the substance of the stars and celestial bodies. Indeed, how would we know that the heavens are different from the air below and that they are rarer than the rarest air, unless the bodies of the stars seemed to appear elsewhere to us than where they actually are?[8] By the law of refraction, in fact, we are reminded that the heavens are more transparent than the air, or at least [4] than the bottommost dregs of the adulterated air.[9] Or how would it be known that the spherical body of the stars is formed from an aggregation of the denser parts of the heavenly sphere or of polished metal unless we knew it by means of this law?[10]

If we consider necessity, then we will see that, as a whole, the sciences are utterly obscure without the light of this study. Among all of these sciences, we will select astronomy and physics as especially prominent examples. What then will we perceive about the motion of the stars without a knowledge of refraction, since their appearances do not agree with one another? For when their motions are observed, the stars are not actually where they appear to be; in fact, stars that are next to one another at the horizon appear to diverge away from one another at mid-sky.[11] What about their uniform motion; [how could we assume that the stars move uniformly] when they appear to be borne more slowly by their carrying spheres down at the horizon and appear to be borne faster at zenith, unless it were established according to refraction that an object appears to move more slowly in a less transparent medium and more swiftly in a clear medium?[12]

What about the sizes of the stellar bodies; why are they perceived as larger than they actually are when near the surface of the Earth at horizon but appear considerably smaller when viewed farther above the horizon, unless it were established by the principle of refraction that light is conveyed in straight lines but is refracted when it meets a denser medium and is diverted from the perpendicular in various ways so that size, location, light,

nem, locum, lucem, et huiusmodi similia diversari? Ex paralaxis quoque observatione non solum cometas sed lacteum orbem in firmamento esse ignoraremus, si refractionis scientia fuissemus ignari. Quis non horret colores in cometis et Lunae Solisque defectibus—cruentos, tetros, lucidos, coeruleos—nisi ex refractione lucis et diaphani aere densioris mistione gigni posse doceremur? [5] Coronae, gemini Soles et tergemini, virgae per nubes extentae, irides solae et geminatae, adeo obscuris causis involutae sunt ut maiores nostri Thaumantis filiam quasi admirationis dixerint.

Refractio ancipites philosophiae quaestiones explicat; et quid vere sint dignosci largitur. Medicina quomodo irides in oculis, vertigines, et eiusmodi morbos in homine mederetur, nisi ex refractione causas cognosceret? Eius ope multa in pictura et architectura sublimia sunt inventa. Venio ad utilitatem et iucunditatem ex ipsa admirabilitate. Haec enim scientia ex mathematices et naturalis philosophiae mixtura orta est, et quicquid utraque in se iucundi et admirationis habet, in se recipit et colligit. Hinc geometricarum speculationum veritas innotescit, nam quae geometria fingit et speculatur ipsa explicat et in lucem revocat, unde ex his quasi fonte mirabilium omnium profluit scientia.

Ignis ex perspicui corporis obiectu et Solis radiorum coitu generatur; adeo validus ex refractione ut longe parabolicae sectionis vim exuperet. Nam refractio habet et radiorum unionem et multiplicitatem qua caret reflexio, unde multo remotius et validior ignis potest explicari. Specilla condit quorum causas ex refractione cognoscit quae debilibus et caecutientibus oculis accommodata longius et perspicacius dat omnia contueri, quibus si privaremur fere ipso visu orbaremur. Huius ope specilla quoque concinnantur quae ad miraculum usque elongant visum. Fiunt imagines ut in aere pendulae videantur tam clare et perspicue ut, nisi manibus [6] tangas, vix oculis credas ut oculorum fallacias et praestrigia vincant.

Multa sunt quae ne lectores taedio afficiamus omittimus. De literaria igitur republica benemeriti viri eam sequantur amplexenturque. Causa quae nos de refractione ad scribendum impulit fuit quod videremus ea de re maiores nostros breviter ac satis oscitanter perscripsisse, ut vestigia potius quaedam quam artis fundamenta iecisse viderentur. Unde indulgendum nobis erit si humiliora haec tradiderimus, necesse enim fuit ut, quantulacumque

and things of that sort diverge from the norm?[13] In addition, because of the observation of parallax, if we had not understood the science of refraction, we would have failed to know that not only the comets but also the Milky Way lie in the heavens.[14] Who does not tremble at the colors in comets and eclipses of the Moon and Sun—blood red, foul, bright, or sky blue—unless we are taught they can arise because of the refraction of light and the mixing of denser vapors with the transparent air?[15] [5] And the same holds for coronas, twin and triple Suns, streaks extending through clouds, and single and double rainbows, which up to now have been wrapped up in such obscure causes that our predecessors called the rainbow "daughter of Thaumas" as if in wonder.[16]

Refraction addresses critical questions of natural philosophy, and it grants wide latitude in determining what those questions actually are. How would medicine cure rainbows in the eyes, vertigo, and disorders of that kind in people, unless it knew the causes stemming from refraction? With its help many sublime things in painting and architecture have been discovered. On the basis of its very laudability I come to what is useful and pleasing about it. For this science springs from a mixture of mathematics and natural philosophy, and it receives and gathers together in itself whatever each of them on its own possesses in the way of delight and wonder. Hence, the truth of geometrical speculations becomes known, for whatever geometry produces and explores it clarifies and brings to light, so on this basis the understanding of all wonders flows as from a fount.

Fire is generated by the uniting of the rays of the Sun projected through a transparent body. Consequently, it is strengthened by refraction so that it can far surpass the power of light reflected from a parabolic section.[17] For refraction has a capacity that reflection lacks, both to unite and to multiply rays; accordingly, far stronger fire can be set at a much greater distance by refracted rays. One who knows the reasons based on refraction fashions lenses that are accommodated to those with weak or sightless eyes and that allow them to see everything farther and more clearly, and if we were to deprive them of these we would deprive them of sight itself.[18] With the help of this knowledge lenses are produced that extend vision out to a marvelous distance. They cause images to appear to hang in the air with such clarity and sharpness that, unless you attempt to touch those images with your hands [6], you can scarcely credit your eyes in the effort to overcome illusions and delusions of sight.

There are many things we omit in order not to afflict readers with boredom. Thus, men of merit belonging to the republic of letters should follow and embrace the subject. The reason that drove us to write about refraction was that we noticed that the things our predecessors wrote about the subject they wrote briefly and rather negligently, so that they seemed to have laid out certain traces rather than the foundations of the discipline. Therefore, we should be indulged if we deal here with quite humble things, for it was necessary that, however trifling they may have been, these things

haec fuerint, nobis placerent, quibus non nisi nova et selectiora placere possunt.

have pleased us, which cannot please us unless they are new and carefully chosen.

[LIBER PRIMUS]

PROOEMIUM

Refractionum principia, quibus cum aliis convenimus, hoc libro explicabuntur. His problemata quaedam adnectemus longe aliter quam nostris maioribus visa sunt explicata. Errata, si quae nobis visa sunt, non dissimulabimus, etiam doctissimorum virorum, ne errores in infinitum propagentur. Nova si quae habemus libenter impertimur, etsi vulgaria; doctiorum tamen animos vellicabimus ut ornatiora cultioraque proferant in lucem. Interim definitiones quasdam audies, quibus in sequentibus usuri sumus.

DEFINITIONES

1. **Linea incidens** est per quam fulgor luminosi emicat vel simulachrum per medium unius diaphani fluit.
2. **Linea refracta** est per quam radius vel imago per alterum imparis perspicuitatis diaphanum effunditur.
3. **Punctus fractionis** est utriusque lineae incidentis et refracta intercursus in utriusque tenuioris et hebetioris diaphani commissura. [8]
4. **Linea perpendicularis** est quae a fractionis puncto per corporis centrum in quo fit refractio ultra porrigitur; unde si lacus, maris, aut vasis contenta aquae observetur refractio, centrum illud universi erit, at si vitri, speculi, oculi, vel alterius consimilis, sui ipsius corporis centrum erit.
5. **Cathetus** est linea incidens per visae rei punctum et corporis centrum in rectum extensa, ut diximus.
6. **Angulus incidentiae** est incidentis lineae et perpendicularis in refractionis puncto coitio.
7. **Angulus refractionis** ubi lineae refractae et incidentis in puncto refractionis coibunt trans punctum refractionis extensae.

[BOOK ONE]

INTRODUCTION

I n this book the principles of refraction will be explained, along with other things that we include. To these we will add certain problematic things that are far different than they appear to have been as explained by our predecessors. If errors are apparent to us, even those of the most learned men, we will not neglect them in order that the mistakes not be propagated ad infinitum. If whatever we have is new, even though common, we freely offer it, and we will still pique the minds of the most learned so that they may bring more splendid and refined things to light. Meantime, you will attend to certain definitions that we will be using in what follows.

DEFINITIONS

1. An **incident line** is the line along which light shines forth from a luminous object or along which an image[19] of an object streams through a homogeneously transparent medium.
2. A **refracted line** is the line along which a ray of light or an image propagates through another medium of different transparency.
3. The **point of refraction** is where both the line of incidence and the refracted line intersect at the interface between both the rarer and the denser transparent media.[20] [8]
4. The **normal** is the line that extends out from the point of refraction through the center of the medium in which refraction occurs; hence, if refraction is observed in water contained by a lake, the sea, or a vessel, the center of that medium will be the center of the universe, but if the refraction occurs in glass, a mirror, the eye, or something similar, the center will be that of the body itself.[21]
5. The **cathetus** is the line extending straight through a point on the visible object and the center of the medium, as we just described it.
6. The **angle of incidence** is formed at the conjunction of the line of incidence and the normal at the point of refraction.
7. The **angle of refraction** is formed where the refracted and the incident lines extending through the point of refraction will meet.[22]

PROPOSITIONES

Radius perpendicularis super corpus altero minus lympidum penetrat irrefrac-
tus, obliquus vero retorquetur ad perpendicularem. PROP. 1

Id apparet experientia. Immittatur in pelvim aqua plenam vulgare astrolabium
centrotenus (ne maiori paratu refractionum avida perdiscendi ingenia remor-
emur), erigaturque perpendiculariter. Et si possibile est, transeat Solis radius
vel candelae per pinnulas perpendiculariter erectas, et radius fundo tenus
irrefractus permeabit. Si vero obliquetur regula in medio quartae astrolabii,
expecteturque usque donec Sol ascendat et per pinnularum oscula radium
transmittat, non utique recte usque pertransibit, sed ubi ad punctum venerit
ubi aqua cum aere coibit, deficit ictus, et incurvabitur et oblique defluet ad
perpendicularem radium accendo. Exemplum erit. [9]

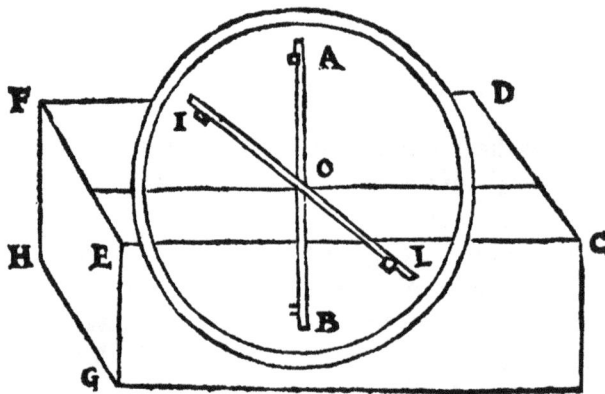

Esto pelvis CDEFGH, astrolabium ALBI; transeat aqua per medium
astrolabii, scilicet per umbilicum O. Sit regula AB perpendiculariter super
aquam erecta, et transeat lumen per A; transibit per AOB. Robustum tantum
enim est illi virium ut a vehementissimo aquae occursu deflecti non possit.
At si obliquetur regula IL, et expectes usque donec Sol occupet pinnularum
foramina, non recta meabit sed in puncto O deflectitur confinio uniusque

PROPOSITIONS

PROP. 1: *A ray [passing through one transparent medium and incident] upon a less transparent medium along the perpendicular passes through unrefracted, whereas an oblique ray is bent back toward the normal.*

This is evident through experiment. In a container full of water immerse an ordinary astrolabe up to its center (we should not hinder eager minds from thoroughly understanding refraction through greater preparation), and stand it upright. If possible, let a ray of sunlight or candlelight shine through the sighting-vanes [of the astrolabe's alidade, which is] oriented along the perpendicular, and the ray will pass through, unrefracted, all the way to the bottom.[23] If, however, the alidade[24] should be slanted within the middle of a quadrant of the astrolabe, and if it should be held up until the Sun rises and transmits a ray through the openings of the sighting-vanes, then the light will definitely not pass straight through all the way, but when it reaches the point where the water and air meet, its incidence is weakened, and it will deflect inward and will continue downward while inclining toward the normal ray. The following will be an illustration. [9]

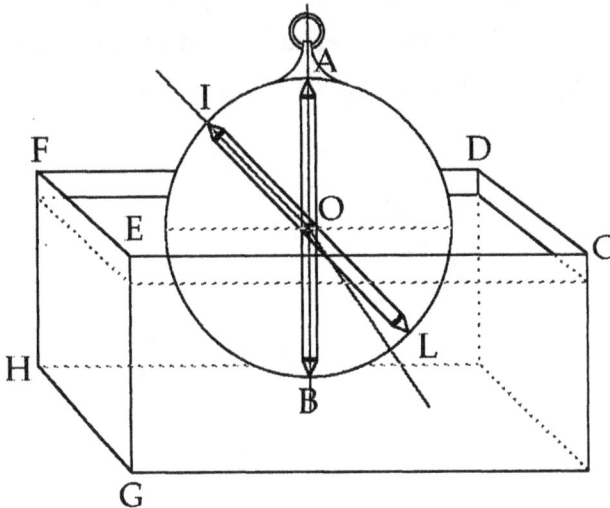

Figure 1.1

Let CDEFGH [in Figure 1.1] be the container, ALBI the astrolabe, and let the water reach the middle of the astrolabe, that is, through the center pivot at O. Let the alidade AB stand upright on the water's surface, and let the light pass through A; it will therefore pass along AOB. So forceful is it that it cannot be deflected because of the exceedingly powerful way it strikes the water. If, however, the alidade is slanted along IL, and if you hold it up until the Sun covers the openings of the sighting-vanes, the light will not pass straight through but is diverted at point O on the interface of the rarer

corporis AOI tenuioris et LOB hebetioris. Accedetque ad perpendiculum AB, quum minus virium habeat et infirmior sit. Ratio ex astrolabii fabrica pendet, nam linea IL recta meat, semperque basis fundo aequidistabit, et astrolabii manubrium, dum in libramento steterit, ad perpendiculum pendet.

Res visa sub hebetiori aeris corpore perpendicularis penetrat robusta, obliqua vero a perpendiculari recedit. PROP. 2

Haec est iam dictae contraria: nam simulachrum veniens ex hebetiori ad lucidius diaphanum—scilicet ex aquis ad aerem—non, ut prius, ad perpendicularem accedit sed reflectitur, et in contrariam partem abibit. Propositio experimento etiam probatur. Haereat magnitudo colorata vasis fundo aqua pleni, mergaturque in eo astrolabium centrotenus. Erigaturque perpendicularis regula, apponaturque oculus in summo, [10] et per utrumque foramen coloratae rei venientem formam aspicietur.[1] Recta obliquetur regula, et ex regione pinnularum foraminum res colorata aptetur sub aquis; intuentibus forma interit, et visum intercipit, videbitur enim visus aciem sublimius attolli. At si rem coloratam humilius submiseris, conspicies. Exemplo res clarior fiet.

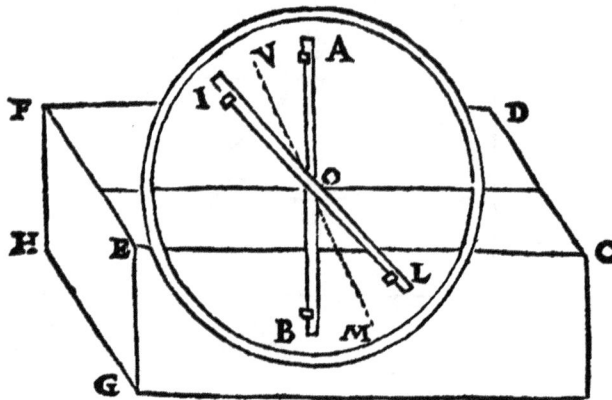

Sit pelvis aquae plena CDEFGH; mergatur in ea astrolabium AIBML[2] centro tenus O. Ponatur res colorata in B, pendeatque ex libramento astrola-

---

[1]"auspicabitur" in original text.
[2]"AILBM" in original text.

medium [containing] AOI and the denser one [containing] LOB. And so it will approach normal AB, since it has less power and is weaker than before.[25] The rationale depends on the design of the astrolabe, for line IL should continue straight, the base will always be parallel to the bottom, and while the handle of the astrolabe stands as a counterpoise, the astrolabe hangs upright along the orthogonal.[26]

PROP. 2: *When a visible object is in a medium denser than the medium of air, the light radiating from it along the normal passes straight through forcefully, but oblique light diverges away from the normal.*

This is the opposite of what was just said [in the previous proposition], for an image coming from a denser to a more pellucid transparent medium—that is, from water to air—does not, as before, incline toward the normal but is deflected away, and so it will go toward the opposite side.[27] This claim can be demonstrated experimentally as well. Affix a colored body to the bottom of a vessel filled with water, and immerse the astrolabe in the water up to its center. Orient the alidade upright, place the eye at the top, [**10**] and the image coming from the colored object will be seen through both openings. Let the alidade be tilted, and let the colored object be placed under water in a direct line with the openings in the sighting vanes; the image of the object is lost to anyone looking, and it disappears from sight, for the line of sight will seem to be raised up. If, however, you submerge the colored object farther below, you will perceive it. The point will be made clearer by an illustration.

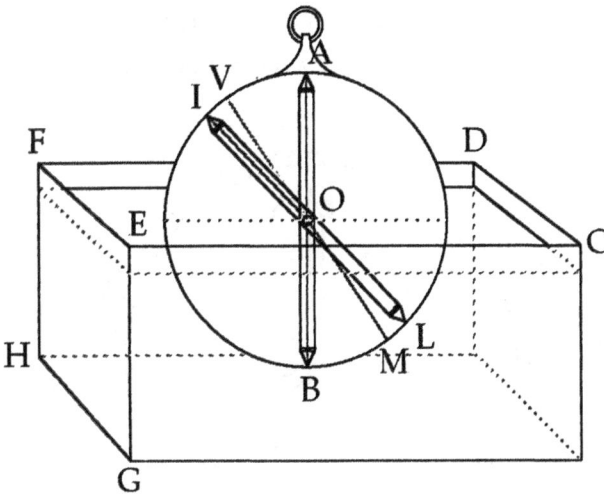

Figure 1.2

Let CDEFGH [in Figure 1.2] be a vessel filled with water, and let astrolabe AIBML be immersed in the water up to its center O. Let the colored object be placed at B, and let the astrolabe hang from its counterpoise at the

bium. Per A, B pinnulas videbitur coloratum. At si obliquetur regula, ut in
IL, coloratum existens sub L, eius visus perit, nam regulae pars altera OL
in sublime tolli videbitur, et coloratum ex repercusso oblique accepto sub-
mittitur: scilicet res colorata M recte ad O veniens non amplius in rectum
extensa in V veniet, sed oblique fertur ad I a perpendiculari se contorquendo.
Sic res existens in M ab oculo I existente conspicietur.

Anguli ingredientes egredientesque semper aequales in refractione reperi-
untur. PROP. 3

Clare etiam experientia patet angulos ex Solis radiis ingredientes egredientesque nunquam non aequales esse, ut et visus quoque anguli. Nam Solis radius
si oblique ad aquam [11] fertur, eius occursu collisus ad perpendiculum
contorquetur, et si unde resilierit, exponitur eo unde ingressus est; reflectitur
in eandem sedem a perpendiculari recedendo. Idem et de visu dicendum.
Res conspectabilis sub aquis commorans foras egrediens ubi in lympidius
diaphanum venerit, in eam partem inclinatur quae longe a perpendiculari
semota est, videlicet ad horizontem. Contra vero, si oculus sub aquis esset
res vero in aere, rei spectrum ad utriusque diaphani contactum veniens, ad
oculum contorquebitur. Unde non vaga et sine certa lege angulorum inclinatio
est, sed semper aequales vices ingrediendo egrediendoque sibi mutuo servant.
Te vero, ut rem facilius teneas, decet exemplo admonere.

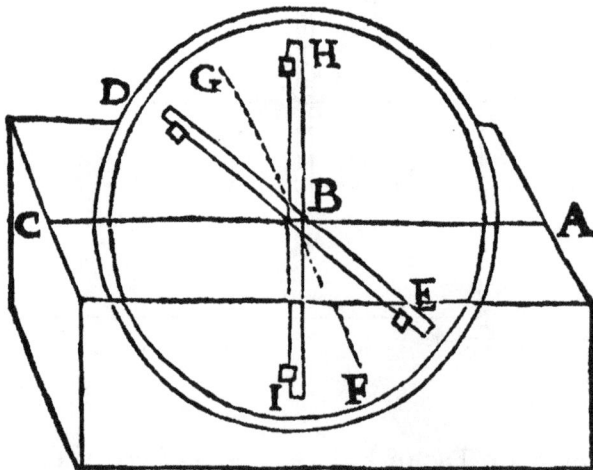

handle. The colored object will be seen through sighting holes A and B. If, however, the alidade is slanted, for instance, along IL, and if the colored object lies below [and in a direct line with] L, it disappears from view, for the other segment OL of the alidade will appear to be lifted up, and the light from the colored object is transmitted obliquely because of the deflection undergone [at the water's surface]: that is, the ray from the colored object at M that arrives straight to O will no longer extend straight to V but is conveyed obliquely to I while bending away from the normal. Thus, the object lying at M will be perceived by an eye lying at I.

PROP. 3: *In refraction the angles of entry and exit are invariably found to be equal.*

Experiment also shows clearly that the angles of entry and exit for rays from the Sun are never unequal, and likewise the angles of sight.[28] For if a solar ray is transmitted obliquely to water, [11] it will be bent toward the normal upon striking the water's surface, and if it returns back from there, it follows the same path along which it entered the water, and it is diverted along the same pathway by inclining away from the normal. The same can be said about sight. When the image of a visible object standing under water exits from the point at which it reaches the more transparent medium, it is inclined in a direction that is distant from the normal, namely, toward the horizon. On the other hand, if the eye were in the water while the object were in the air, then when the object's image reached the interface of the two transparent media, it would be deflected toward the eye. Consequently, the inclination of the angles is not ill defined or without a definite governing law; instead, they always remain equal to one another in leaving and entering the interface between media. In order for you to grasp the point more easily, however, it is appropriate to persuade you with an illustration.

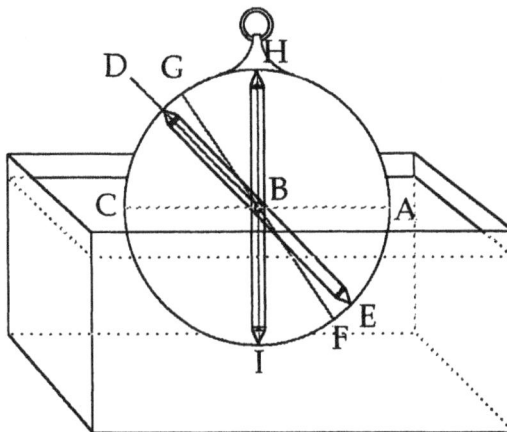

Figure 1.3

Sit aquae superficies quieta ABC; Sol D non perpendicularis sed transversalis cadat ipsi B. Non in directum labitur ad E sed alia erit quam antea fuerat, quia ex puriori ad hebetius diaphanum venit. Ideo haerebit perpendiculari HBI, et erit BF. At Sol sub aquis existens in F, sursum veniens per B, non in rectus G meabit sed in contrariam partem abibit ab ipsa perpendiculari HBI. Veniet igitur ad D, unde discesserat. Idem et de oculo dicendum. Re sub aquis existente in F, ac sursum perveniente ad B, quia ex hebetiori ad lympidius medium venit, a perpendiculari HBI separabitur, ideo non ad G sed ad oculum in D commorantem accedet. Contra oculo sub aquis in F et re extra in D existentibus, non veniet rei imago D [12] recta per DBE sed ad F accedet, et sic conspicua erit.

Quod anguli ingredientes et egredientes sint aequales. Sit Sol in astrolabio D, in medio quartae, ad gradus 45. Non veniet ad E, ubi est etiam 45 gradus, sed accedendo ad HBI in F, erit 35 gradus. At existens Sol in F, 35, non veniet ad G, ubi etiam 35, sed ad D, ubi gradus 45. Remanent ergo anguli in medio aequales GBD et EBF ex lineis se decussantibus orti ad invicem aequales.

Imago rei refracta visui occurrens suo loco non videtur. PROP. 4

Tempus iam est refractionis accidentia vestigare. Inter prima, referemus quod rei simulachrum refractum visui occurrens, nunquam suo loco videtur, et id quemadmodum fiat ratione et experientia demonstrabimus. Rationes iam supra exposuimus quod visus, ex obliquatione refracti radii ex aqua progredientis in commissuris alterius diaphani aqua tenuioris, rei simulachrum non videt in sui oppositione. Nec illud recte comprehendit, nec refracte illud se comprehendere cognoscit, sed illud nec suo loco, nec in sua forma, sed alio modo quam putat.

Experimento vulgari apparet, et a multis adducto, ex pelvi aqua plena. Capitur ergo pelvis alicuius profunditatis, in cuius fundi medio agglutinetur

Let surface ABC of the water [in Figure 1.3] be still, and let a ray from the Sun D fall not orthogonally, but at a slant on B. It does not tend straight toward E but will be inclined in a direction other than it had been following before because it comes to a denser from a more limpid transparent medium. Consequently, it will incline toward normal HBI, and its path will be BF. But if the Sun were to lie at F under the water and a ray from it were to pass upward through B, it would not continue straight to G but would incline toward the opposite side away from normal HBI. It will therefore arrive at D, from which it had originally descended. The same can also be said about the eye. If an object lies below the water at F, and if a ray from it travels upward to B, then since it passes from a denser to a more transparent medium, it will diverge away from normal HBI, so that it will not reach G but the eye stationed at D. On the other hand, if the eye lies under water at F and the object lies beyond at D, the image of object D [12] will not arrive straight along DBE but will incline toward F, and so it will be visible there.

That the angles of entry and exit must be equal [is shown as follows]. Let a ray from the Sun shine on D at 45 degrees within a quadrant of the astrolabe. It will not reach E, also at 45 degrees [in the opposite quadrant], but by approaching F toward normal HBI, it will be refracted at 35 degrees.[29] If, however, the Sun were to lie at F, at an angle of 35 degrees from the normal, a ray from it would not reach G, which is also 35 degrees from the normal, but will reach D, which is 45 degrees from the normal. Therefore, angles of deviation GBD and EBF in the middle, which are formed by the original lines of incidence that intersect each other from the source, remain equal to each other.

PROP. 4: *The image of an object that reaches the eye by refraction is not seen in its proper place.*

It is now time to investigate the characteristics of refraction. Among the foremost, we will assert, is that when the image of an object reaches the eye refracted, the object is never perceived in its proper place, and we will demonstrate how this may happen by means of reason and experiment. In our discussion above, we already laid out the reasons why, because of the slant of a refracted ray proceeding from water through the interface with another transparent medium rarer than water, the visual faculty does not see the image of an object in a directly facing position. Nor does it perceive the image correctly, nor does it recognize that it perceives the image by means of refraction or that it perceives the image neither in its proper place nor in its proper form but in another way than it judges it does.

This becomes evident through a common experiment carried out by several men on the basis of a vessel filled with water. Accordingly, take a vessel of some depth, and affix a silver coin in the middle of its bottom.

argenteus nummus. Inde a vase discedat spectator quousque latere eius obice
nummi imago intereat (nam discedendo superior, accedendo vero humilior
videtur), ibique sistatur. Mox in pelvim aquam suffundi iubeat, et videbit
iacentis[3] imaginem ad summa aquae conscendere et ibi suam imaginem
conspicuam demonstrare. Exemplum.

In vas CDEGH nummus B; visus existit in A. Radius [13] transiens
per altitudinem lateris vel marginem FD pertransibit in L, nec tangit nummum

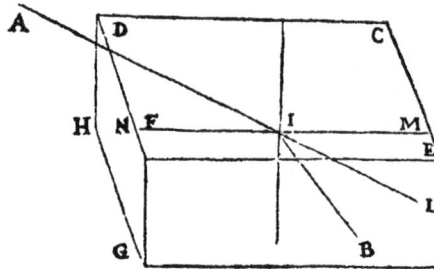

in B existentem, nam sublimior volat. Repleto aqua vase usque ad CED
summa, radius frangitur in I aquae superficie ex B, et inclinabitur ad oculum
A. Sic nummus, qui prius profunditate erat inconspicuus, refractionis ope
conspicuus evadet. Nec radius ad I frangi putabitur, sed suo loco videri,
quod nulli dubium esse potest. Urinatores et qui tridentibus venantur pisces
non ubi eos vident ictus dirigunt sed infra, nam si aliter faciunt, falluntur
et incassum laborant, ex refractione enim alibi quam sint apparent.

Imago refracta ad visum pervenit per rectas lineas. PROP. 5

Quod imago ad oculos per rectas feratur ratio est quod fractione res ad visum
veniunt temporis momento, at si in linearum fractione temporis mora aliqua
fieri deberet, et rei lationem per rectas lineas velocissime fieri non contingerit.
Pervenit igitur ad visum per linearum rectitudinem. Praeterea, in reflexionibus etiam imago ad obtutum per rectas velocissime fertur. Est enim rectitudo
lucis cognata motui quod et in aliis naturae operibus convisitur. Et id quoque
instrumentis ad id opus concinnatis rimatum est.

Mergatur astrolabium centro tenus in pelvi aqua plena, et deorsum
foramini coloratum opponatur. Si centro obstaculum aliquod opponatur,

---

[3]"iacentes" in original text.

Then have the viewer draw away from the vessel until the rim of the vessel's wall blocks the image of the coin (for when the viewer draws away, the rim appears to rise, whereas when he approaches it appears to sink), and let him stand at this spot. Then let him have water poured into the vessel, and he will see the image of the coin dropped in the vessel rise up toward the top surface of the water and reveal its image in plain sight at this point. Here is an illustration.

B is the coin in vessel CDEGH [in Figure 1.4], and the eye lies at A. The radial line of sight [from A] **[13]** that passes over the top or lip FD of

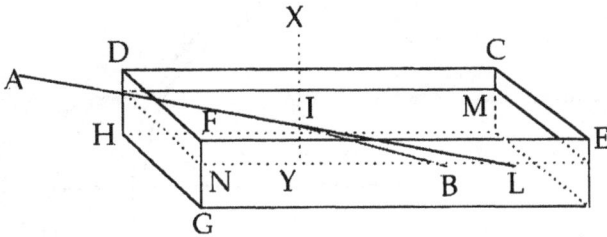

Figure 1.4

the [vessel's] rim will continue to L, and it does not touch the coin lying at B [when the vessel is empty], for it skims above. When the vessel is filled with water up to [level MN below surface] CED at the vessel's top, the ray from B is refracted at I on the water's surface, and it will be inclined toward the eye at A. Hence, the coin, which was previously invisible in the vessel's depth, will become perceptible by means of refraction. And the ray will not be judged to be refracted at I, but there can be no doubt that the object will be judged to be seen in its proper place [straight along AIL]. Divers and those who hunt fish with tridents do not aim their strike right where they see them but lower, for if they do otherwise, they err and work in vain because things appear other than they actually are on account of refraction.[30]

PROP. 5: *An image refracted toward the eye arrives along straight lines.*

The reason that the image is conveyed to the eyes along straight lines is that the images of things reach the eye by refraction in an instant of time, but if, in the refraction of lines, it had to occur during some amount of time, it would follow that the transmission of the image of the object does not occur as swiftly as possible along straight lines. It therefore reaches the eye along straight lines.[31] Furthermore, in reflection the image is also conveyed to the viewpoint as swiftly as possible along straight lines. For the rectilinearity of light transmission is akin to the motion that is also carefully observed in other operations of nature. And this motion of light transmission can be closely examined as well by means of the instruments already prepared for this task.

Immerse an astrolabe up to its center in a vessel full of water, and place a colored object directly in front of and below the sighting hole [at the lower

peribit visio, a re enim ad centrum per aquas recta fertur imago, etsi coloratum ita sub aquis locetur et per foraminulum superius visatur. Si vero in centro sit obex, iterum et imago interit, quia a centro ad oculum recte fertur. [14]

In vas aquae plenum CDEFGH mergatur astrolabium AIM centrotenus I, et signetur in ore instrumenti in oppositione visus M. Et si oculus A,

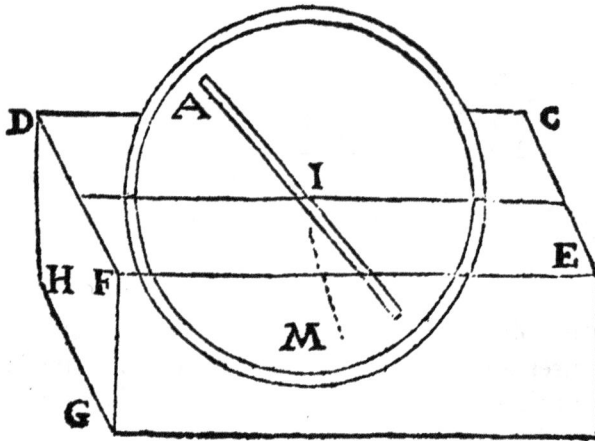

videbitur per lineas MI et IA punctus M. Si opponatur obex in puncto I, intercipietur visus rei M, nam ex M in I et ex I in A recte fertur punctus M. Si imaginetur vel signetur punctus in linea MI, et posito in aqua instrumento quovis loco obex erit in linea MI, et quaquaversus te verteris ex illius recta oppositione, nunquam M videbis, quia ex M ad I recta meat forma. Idem eveniet in linea IM. Concluditur igitur per rectas lineas venientem formam conspicari.

Rei imaginem ad aquae summum perpendiculariter ascendere. PROP. 6

Probabitur id experimento, et aliter quam a maioribus. Astrolabium, cui regula transversa sit atque aliquo puncto notata aquis paulatim mersetur, et continget res maxime mira, nam ima regulae pars ex suo loco prodire et sustolli videbitur. Nec tam male hominem feriatum reperiri arbitror quin partem perpendiculariter ascendentem confiteatur. Exemplum.

end L of the alidade]. If some obstacle is placed directly in front of the astrolabe's center, the sight of that object will disappear, for the image is transmitted from the object straight through the water to the center, even if the colored object is located in this way under water and is looked at closely through the top sighting hole of the alidade. If, moreover, the obstacle lies at the center, the image disappears again, since it is conveyed straight from the center to the eye. [14]

Let astrolabe AIM [in Figure 1.5] be immersed up to its center I in vessel CDEFGH filled with water, and mark spot M on the face of the

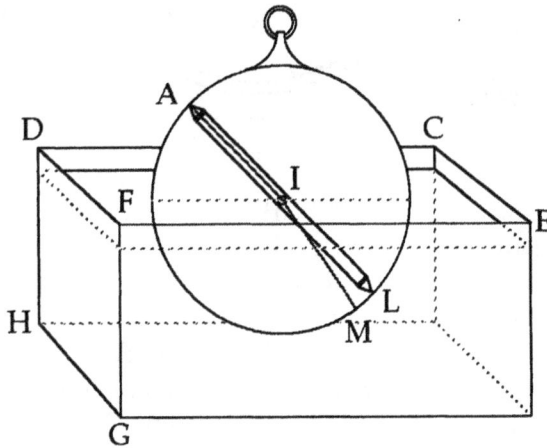

Figure 1.5

instrument facing the eye. If the eye is at A, it will see point M along lines MI and IA. If an obstacle is placed at point I, the view of object M will be blocked, for the image of point M is conveyed from M to I and from I to A along straight lines. If a point is imagined or actually marked on line MI, and if some device is placed in the water wherever the obstacle will lie on line MI, and if you situate yourself wherever you are directly facing it, you will never see M because the image travels directly from M to I. The same will happen [anywhere] on line MI. It is therefore concluded that the image is perceived to arrive along straight lines.

PROP. 6: *The image of an object rises toward the surface of the water along a perpendicular line.*

This will be demonstrated by experiment, but in a way other than that followed by our predecessors. If you immerse an astrolabe somewhat into water, and if the alidade is slanted and marked at some point, something quite wonderful will occur, for the bottom portion of the alidade will appear to move and rise up from its original location. I do not judge it such a bad thing for a person to take the time to determine that in fact the bottom part does rise up along the perpendicular. Here is an illustration.

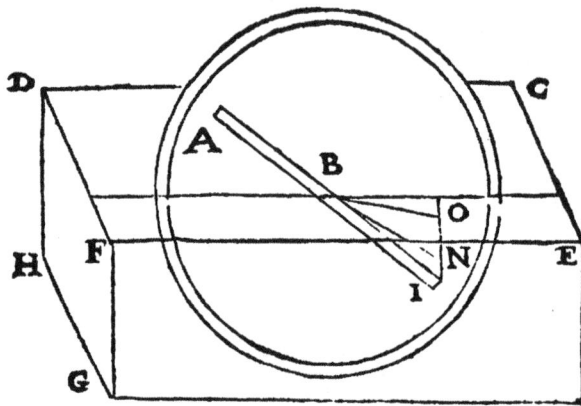

Assumatur sueta pelvis CDEFGH,[4] inclinetur regula IBA in medio quartae, et mergatur astrolabium paulatim in aquam, colore aliquo vehementer relucenti notato imo puncto regulae [15] I. Et videbitur brachium BI summa peti in N, et paulo amplius deprimendo B, semper I elevari videbitur, et tandem I ad O perveniet. Sic deprimendo sublevandoque, voti demum compos fies, quia ascendere et descendere imam partem videbis perpendiculariter.

Satis tibimet ipsi vere facies si astrolabium ex pelvi extrahes, et regulam aliquam intra vas perpendiculariter erectam constitues, et in imo repones obiectum valde relucens; inde oculos deprimendo, videbis coloratum per virgulam ascendentem perpendiculariter, sic sustollendo deprimi. Clarioris doctrinae gratia exemplum apponam. Sit vas CDEFG, regula ad rectum

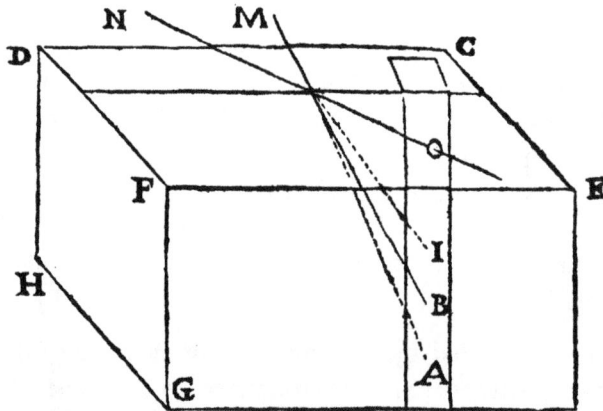

---

[4]The diagram for this demonstration is upside down in original text.

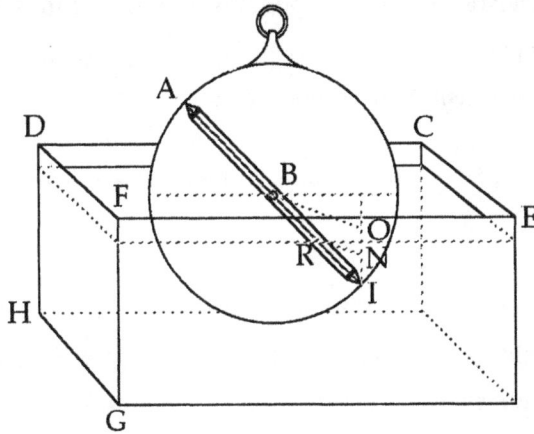

Figure 1.6a

Take the usual vessel CDEFGH [in Figure 1.6a], let alidade IBA be slanted within a quadrant of the astrolabe, and having marked point I at the bottom of the alidade with some extremely bright color, immerse the astrolabe somewhat in the water. [15] [When the astrolabe is only partially submerged, segment RI of] limb BI of the alidade will appear to move toward the surface at N, and as you lower the astrolabe farther down toward B a bit, point I will always appear to move upward, and point I will eventually reach O. By thus lowering and raising the astrolabe, you will eventually achieve the desired mastery, for you will see the bottom part of the alidade rise and descend along the perpendicular [i.e., the cathetus].

You will fully satisfy yourself about this if you remove the astrolabe from the vessel, set a ruler upright in the vessel, and reposition an extremely bright object at the bottom of the ruler. Lowering your eyes, then, you will see the colored object rise with the ruler along the orthogonal, as you continue lowering your eyes. Let me provide an illustration for the sake of clarification. Let CDEFG [in Figure 1.6b] be the vessel [within which XRY is a straight

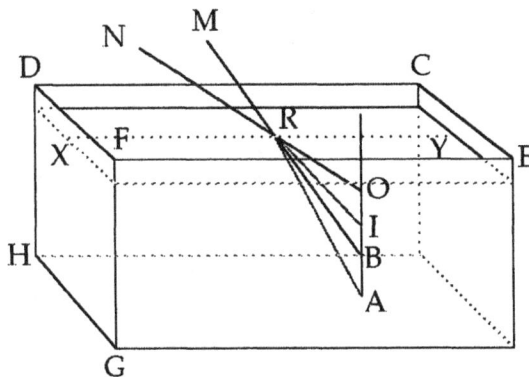

Figure 1.6b

constituta ABIO. Visus primo M, et punctus A videbitur in B. Mox elongetur a vase oculus; deprimaturque; veniat in N. Hunc punctum I in O videbitur loco sublimiori, et sensibiliter ascendere conspicietur.

Imago refracta rei videbitur in concursu catheti et lineae refractionis extensae.
PROP. 7

Si ergo res formam in summo aquae perpendiculariter enatantem remittit, eademque ad oculos perveniens refracta a densiori aere elemento per rectam lineam sui infracti speciem oculis ingerit, eamque oculus videt in linea infracta, et in rectum [16] producta, necesse est eo loco videri ubi utriusque lineae mutuus fiet concursus.

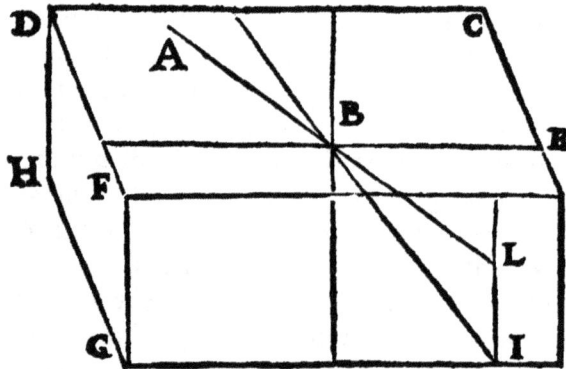

Sit vas aqua plenum CDEFGH. Res visa in aqua I pervenit ad summum per IB; refrangitur a puncto B. Veniet ad oculos per BA in rectumque producta quousque occurrerit lineae a re ipsa I perpendiculariter surgenti IL in puncto L. Ibi ergo necessario videbitur imago rei I, ut in fundo steterit.

Res sub aquis refracte visa quo magis ab oculo distat eo sublimior videbitur.
PROP. 8

Nunc visus fallacias aggrediemur quae ex refractione contingunt, et primum quod ea quae remotiora sunt in plano iacentia sublimiora videntur, et quae propinquiora humiliora, cuius rei demonstrationem a prioribus adductam afferemus.

line on the water's surface], and let the ruler ABIO be set upright in it. First, place the eye at M, and point A will be seen at B [along continuation RB of refracted ray RM]. Then draw the eye away from the vessel, let it be lowered, and let it reach N. Point I will be seen at an elevated spot, at point O [along continuation RO of refracted ray RN], and it will therefore be perceived to rise perceptibly [as the eye draws farther away from M].[32]

PROP 7: *The refracted image of an object will appear at the intersection of the cathetus and the extended line of refraction.*

Accordingly, if the object transmits an image that slips toward the water's surface along the perpendicular, and if the same object conveys its image to the eyes, reaching them refracted from the denser medium into the air along a straight line after refraction, and if the eye perceives it along the refracted line extended rectilinearly, [16] it must be seen where the two lines will intersect one another.

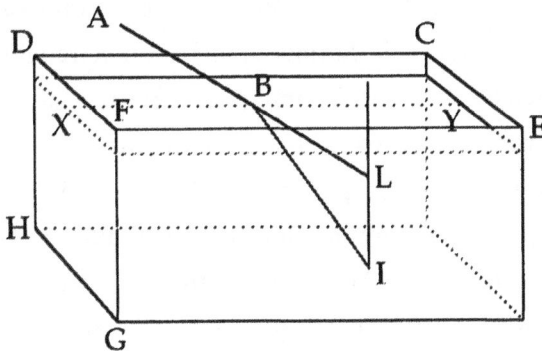

Figure 1.7

Let CDEFGH [in Figure 1.7] be a vessel full of water. The image of visible object I in the water reaches the surface along IB, and it is refracted at point B. It will reach the eyes along BA, which, when extended in a straight line, intersects line IL, along which the image of visible object I rises orthogonally toward the water's surface, at point L. The image of object I, as it actually stands at the vessel's bottom, will therefore necessarily be seen at this point.

PROP. 8: *The farther an object seen under water by refraction lies from the eye, the more elevated it will appear to that eye.*

Now we will turn to the visual illusions that depend on refraction, and the first of these is that things that lie farther away from the eye on a plane under water appear to lie higher, whereas things that lie nearer appear to lie lower; we will provide a demonstration of this point that is drawn from previous conclusions.

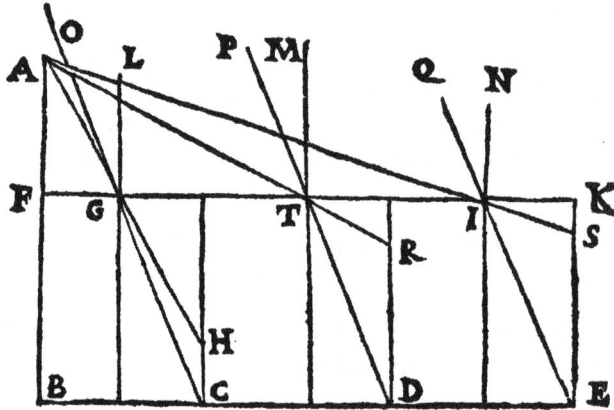

Esto visus A; puncta sub aquis visa aequidistantia B, C, D, E; quae superficiei KF perpendicularis a visu AFB. Dico quod maior refractio puncti E quam D, et maior refractio puncti D quam C, et ob id punctum E sublimius videri quam D et D quam C. Forma enim B non refrangitur, quia est in perpendiculari, sed puncta C, D, E refranguntur ex variis superficiei partibus, et quo remotior punctus a remotiori superficiei parte refrangitur.

Refragatur ergo forma E a puncto I, et D a puncto T, et [17] C a G. Et protrahantur perpendiculares a punctis refractionum, ut LG, MT, NI. Et ab A lineae[5] refractionis AG, AT, AI, et a punctis visis catheti extendantur CH, DR,[6] ES. Et protrahantur lineae refractionis ad cathetos AGH, ATR, AIS. Et quoniam trigoni AGT[7] exterior angulus AGF maior est angulo ATF, et ATF maior AIF, et quoniam anguli LGT, MTI, et NIK sunt recti et aequales, relinquitur angulus AIN maior angulo ATM, et angulus ATM maior AGL. Detrahantur ab his aequales anguli OGL, PTM, et QIN, quia ad invicem sunt aequales ab inferioribus angulis contrapositis. Relinquuntur ergo OGA minor ATP et ATP minor AIQ. Concludimus quod remotiores anguli refractionis maiores sunt, ut in septima definitione.

Quod si extendantur refractionum lineae, ut AG in H, et AT in R, et AI in S, erit S punctus sublimior R, et R ipso H. Sed Vitellio in hoc falsus est quod etsi aequaliter inter se distent in fundo iacentia colorata C, D, E, non ob id aequaliter distant in aquae summo puncta refractionum G, T, I, et ob

---

[5]"linea" in original text.
[6]"DT" in original text.
[7]"AGI" in original text.

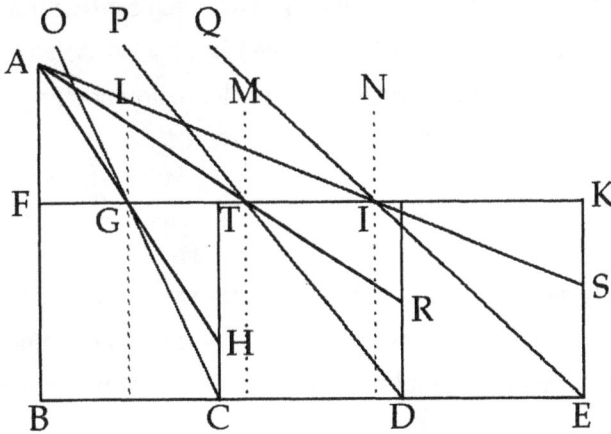

Figure 1.8

Let A [in Figure 1.8] be the eye; B, C, D, and E visible points that are equidistantly spaced under water; and AFB a normal dropped from the eye to surface KF of the water. I say that the refraction of light from point E is greater than from point D and that the refraction of light from point D is greater than from C, so the image of point E will appear higher than that of D, and the image of D will appear higher than that of C. In fact, the image of B is not refracted because it lies on the normal, but those of points C, D, and E are refracted at different spots on the water's surface, and the image of a farther point is refracted at a farther spot on the surface.

Accordingly, let the image of E be refracted at point I, the image of D at point T, and [17] the image of C at G. Let normals, such as LG, MT, and NI, be drawn from the points of refraction. From A let lines of refraction AG, AT, and AI be produced, and from the points that are seen let catheti CH, DR, and ES be dropped. Then extend lines of refraction AGH, ATR, and AIS to the catheti. Because external angle AGF of triangle AGT > [opposite interior] angle ATF, whereas [exterior angle] ATF [of triangle ATI] > [opposite interior angle] AIF, and since angles LGT, MTI, and NIK are right and thus equal, it follows that angle AIN > angle ATM and that angle ATM > angle AGL. Let equal angles OGL, PTM, and QIN be excised from them, for they are equal to the matching vertical angles below them.[33] Consequently, OGA remains less than ATP, and ATP remains less than AIQ. We conclude that, as they are described in the seventh definition, the angles of refraction that lie farther away are commensurately larger.[34]

But if the lines of refraction are extended, as AG is to H, AT to R, and AI to S, then point S will lie higher than R, and R will lie higher than H. In this regard, however, Witelo is mistaken, because even though the colored objects C, D, and E lying at the bottom are equidistant from one another [in plane BCDE], it does not follow that the points of refraction G, T, and

id anguli NIQ maior PTM,[8] et PTM[9] maior LGO. Neque ob id falsa demonstratio, nam non tam maior est NIQ ipso MTP quin QIA non sit maior ipso
PTA, ut quis facile poterit experiri vase aqua pleno, notando in aquae superficie puncta refractionum. [18]

Stellae ascendentes ab horizonte citius videntur quam vere sint. PROP. 9

Quia horizon semper densi aeris ascendentium vaporum et nebularum profundo mergitur, erigentia se astra citius nobis conspicua sunt quam vere
adsint super horizonta. Exemplo rem reddemus clariorem. Non dum oriens

astrum sit M sub horizonte BAH, transiens eius imago per densitatem sphaerae
vaporum FGL. In puncto G refrangitur et ad oculum tendit existentem in
A Terrae globi superficie DAE,[10] et venit fracta quae extendatur in rectum
quousque cum catheto erecta ex C, M occurrit in N. Unde stella quae adhuc
sub horizonte latet M, erit nobis conspicua in N, ut de nummo diximus
pelvi aquae proiecto. Ob id apud eos qui tempestates praenoscunt, quando
Sol citius solito super horizonta apparuerit, signum est futurae pluviae, quia
tunc inter oculum et stellam non pauci et crassi vapores interiiciuntur.

Plano rotundo perpendiculariter inspecto concavus videtur et maior. PROP. 10

Discus magnus sub aquis iacens et perpendiculariter inspectus concavus
videbitur, nam extremitates suae [19] circumferentiae centro sublimiores
videntur, et ob id se ipso maior. Ratio ex superiori pendet, nam pars quae
longius ab oculo distat sublimior videtur, et centrum in suo loco. Videbuntur
semidiametri obliquae et ob id maiores. Afferemus exemplum.

---

[8]"PTO" in original text.
[9]"PTO" in original text.
[10]"DAC" in the original text.

I at the surface of the water are equidistant from one another, and therefore angle NIQ > angle PTM, and angle PTM > angle LGO.[35] Nor does a false demonstration follow, for NIQ relative to MTP is certainly no greater than QIA relative to PTA, as anyone can easily test experimentally in a vessel full of water, while noting the points of refraction on the water's surface.[36] [18]

PROP 9: *Stars rising at the horizon appear earlier than they actually should.*

Since the plane of the horizon is always suffused to its depth with air made dense by rising vapors and clouds, stars are perceived by us to rise above the horizon sooner than they actually should.[37] We will clarify this point

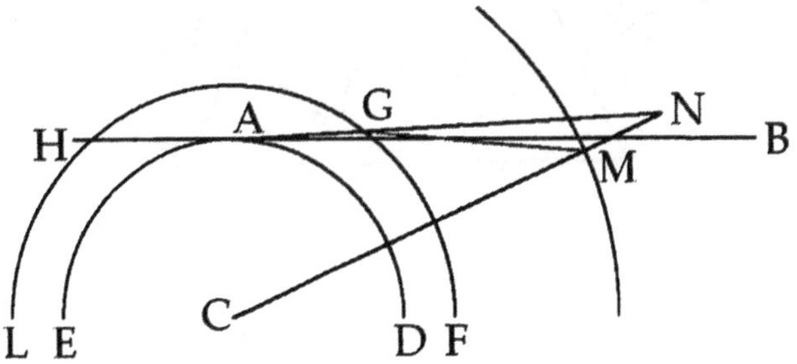

Figure 1.9

with an illustration. Let M below horizon line BAH [in Figure 1.9] be a star [on the great celestial sphere] that has not yet risen, its image passing [from the celestial aether] through the thickness of sphere FGL of [atmospheric] vapors. It is refracted at point G and inclines toward the eye lying at A on surface DAE of the terrestrial globe, and it arrives refracted along a line that can be extended straight until it meets the cathetus dropped from C through M at N. Hence, star M, which at this point lies hidden below the horizon, will be perceptible to us at N, as we said [in proposition 4] of the coin dropped in the vessel.[38] Among those who forecast storms, it follows that, when the Sun appears sooner than normal above the horizon, it is a sign of rainfall to come, because in that case lots of thick vapors intervene between the eye and the star [thus making the atmosphere denser and more refractive].

PROP 10: *When a flat, round object is looked at under water along the perpendicular, it appears concave and magnified.*

A large disk lying under water and viewed along the perpendicular will appear concave, for the edges along its [19] perimeter appear higher than the center, and for this reason it looks larger than it actually is. The explanation depends on conclusions drawn above [in proposition 8], for the portion of the disk that lies farther from the eye appears more elevated, whereas the center appears in its actual location. The radii will therefore appear slanted upward and magnified accordingly. We will provide an illustration.

Sit vas BCDE, discus in fundo iacens DHILMNE, oculus in perpendiculari respiciens A, aquae superficies BGTQZXPC, linea perpendicularis AL.

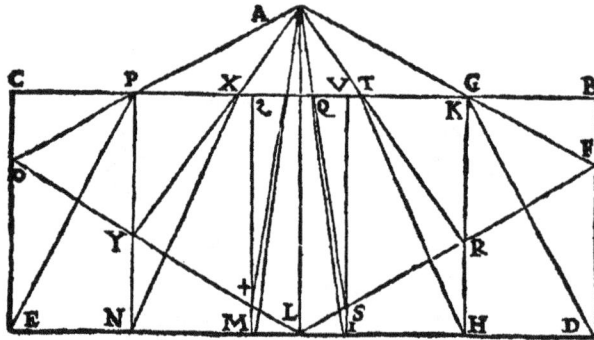

Punctus D venit ad G; refrangitur ad A oculum. Extenditur AG in F, ibi
occurrens perpendiculari DFB; punctus D videbitur in F. Sic punctus H
prope D venit ad T; refrangitur ad A. Extenditur AT in R, ibique occurrit
perpendiculari ex H erectae HRK; sic H videbitur in R. Eodem modo punctus
I per Q in A; extenditur in S semper humilior. Idem eveniet ex alia parte,
nam punctus E per P venit in A. Extendatur AP in O, et E punctus in O
videbitur. Et N per X ad A venit. Extenditur AX, pervenit in Y, et N videbitur
in Y. Sic M per Z venit ad A. Extenditur AZ in †, et punctus M erit †. Extensis
lineis per puncta F, R, S, L et L, †, Y, O, videbitur planus discus DHILMNE
obliquus FRSL†YO, et quia maior linea FL ipsa DL, et OL ipsa LE, videbitur
maior et concavus.

Ex refractione sub aquis omnia maiora sua forma videri. PROP. 11

Altera erit refractionis fallacia quod omnia quae sub aquis videntur maiora
videntur. Sed quia maiores [20] nostri multipliciter falsi sunt, priusquam
veritatem aggrediamur eorum opiniones ad stateram expendemus.

Aristoteles in *Problematis*, "Euro spirante, omnia maiora videri," dixit,
"quod eo flatu aer obscurior et caliginosissimus reddatur." Idemque in *Mete-
oris* dixit. Sed ratio reddenda erat, cur in aere obscuro et caliginoso omnia
maiora viderentur.

Let rectangle BCDE [in Figure 1.10] be the vessel, DHILMNE the disk lying at the bottom, A the eye looking down along the perpendicular,

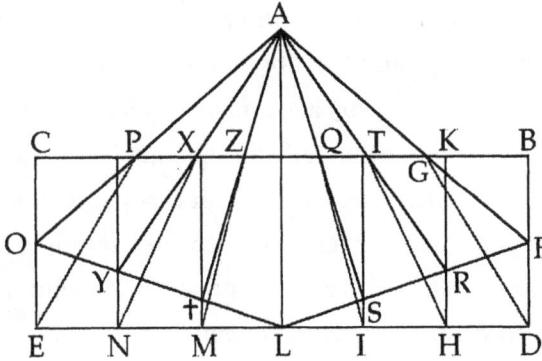

Figure 1.10

BGTQZXPC the water's surface, and AL the perpendicular line of sight. The image of point D reaches G and is refracted to the eye at A. Refracted ray AG is extended to F, and here it meets cathetus DFB, so point D will be seen at F. So, too, the image of point H next to D reaches T, and it is refracted from there to A. Refracted ray AT is extended to R, and here it meets cathetus HRK dropped from H, so H will be seen at R [which is lower than F]. Likewise, the image of point I is refracted through Q to A; and [refracted ray QA] is extended to S, which lies still lower [than R]. The same will happen on the other side, for the image of point E reaches A through P. Extend AP to O, and so point E will be seen at O. The image of N, as well, reaches A through X. AX is extended and reaches Y, and so N will be seen at Y. So, too, the image of M reaches A through Z. AZ is extended to †, and the image of point M will be †. When lines are drawn through points F, R, S, and L and L, †, Y, and O, the flat disk DHILMNE will appear slanted according to concave line FRSL†YO, and because line FL > DL, and OL > LE, the disk will appear magnified and concave.

PROP. 11: *Because of refraction everything under water looks larger than it actually should appear.*

Another visual illusion due to refraction is that everything viewed under water appears magnified. Since, however, our predecessors [20] have erred in several ways about this illusion, we will weigh their opinions before we turn to the truth.

In the *Problems*, Aristotle claimed that, "when the east wind is blowing, everything appears magnified because when it blows, the air is rendered darker and extremely misty," and he said the same thing in the *Meteorology*.[39] But the reason why everything should appear magnified in dark and misty air had yet to be given.

Ptolemaeus, in *Almagesto*, et Alfraganus Solem et Lunam caeterosque errones in oriente occidenteque maiores videri ex vaporibus et humiditatibus, ut post pluvias in diebus hyemalibus inter visum et sydera intercedentibus, immo quanto clarior et profundior ubi mersa sunt maiora conspici.

Aphrodisiensis pomum sub aqua maius videri dicit "quoniam aqua submerso adhaerens corpori eadem afficitur qualitate et colore … quo oculum fallit quasi corpus amplius sit quum undique aquae contineatur." Et hac eadem ratione, Solem, Lunam, caeterosque errones, quum vel exoriuntur vel occidunt, ampliores videri. Sed si ex accessione aquae pomum videtur maius, peto cur non videtur quantum vas illud capit?

Seneca ait omnia per aquam videri longe maiora, sic et poma per vitrum aspicientibus; "Literas quamvis minutas et obscuras per vitream pilam aqua plenam maiores clarioresque cerni, et poma formosiora vitro[11] innatantia; sydera ampliora per nubem." Rationem addit:

> Quia acies nostra in humido labitur, nec apprehendere multo fideliter potest, quod manifestum fiet si poculum adimpleveris aqua et in id conieceris annulum. Nam quum in ipso fundo iacet annulus, facies eius summo aquae redditur. Quicquid videtur per humorem longe amplius vero est. Quid mirum videri maiorem Solis imaginem quae in nube humida visitur? Quod duabus de causis id accidat: quia in nube est aliquid vitro simile, quod potest pellucere, et est aliquid aquae quam, si nondum habet, tamen iam apparet eius natura in quam extra vertatur.

Sed Seneca mathematicam artem ignorans, quid dicat non capit. Dicit quod acies nostra labitur in aquis, nec rem certam comprehendere potest. Cur igitur, si veritatem comprehendere nequit, omnia minora non videt? Sed cur sub vitro et aquis maiora videantur aliam quoque habet rationem ex rotunda vasis forma, quam reddemus quum de vitrea pila loquemur.

Macrobius hanc opinionem dilucidius aperit:

> In aqua simulachra maiora videri, quod genus apud popinatores[12] [21] pleraque scitamentorum cernimus proposita ampliora specie quam corpore. Quippe videmus in doliolis vitreis aqua plenis, et ova globis maioribus, et iecuscula fibris tumidioribus, et bulbos[13] spiris ingentibus. Et omnino ipsum videre, qua nobis ratione constet, quia solent de hoc nonnulli nec vera, nec verisimilia sentire. … Aqua est densior aeris tenuitate, et ideo eam cunctatior visus penetrat, cuius offensa percussa videndi acies scinditur, et in se recurrit. Scissa dum redit, iam non directo ictu,

---

[11]"vino" in original text.
[12]"propinatores" in original text.
[13]"belvas" in original text.

Ptolemy, in the *Almagest*, and Al-Farghānī claimed that the Sun, the Moon, and the rest of the wandering stars [i.e., planets] appear magnified at the east and west horizons on account of vapors and moisture, such as those intervening between the eye and the stars after winter rains, so that the clearer and deeper the vapors they are immersed in, the larger they look.[40]

Alexander of Aphrodisias says that an apple viewed in water appears enlarged "because the water clinging to the submerged body affects its quality and color, according to which the eye mistakenly judges that the body is larger when it is fully surrounded by the water."[41] According to this same rationale, as they either rise in the east or set in the west, the Sun, the Moon, and the rest of the wandering stars appear magnified. But if an apple appears magnified when it is surrounded by water, then I ask why the vessel that holds it does not appear magnified as well.

Seneca says that everything appears greatly magnified through water, as do apples to viewers looking at them through glass: "Writing, however tiny and difficult, is seen larger and clearer through a glass sphere full of water; fruit appears more beautiful than it is, if it is swimming in a glass bowl; the stars themselves seem larger when one looks at them through a cloud." The reason, he adds, is

> that our eyesight falters in moisture and cannot reliably grasp what it wants to; this will become plain if you fill a cup with water and drop a ring into it. For while the ring is lying on the bottom, its image is emitted at the surface of the water. Anything seen through moisture appears much larger than it is in reality. Why is it surprising that an enlarged image of the Sun is seen when it is viewed in a moist cloud? This happens for two reasons: [first], in the cloud there is something resembling glass that can transmit light, and [second], there is also some water—or even if it does not yet contain water, it is forming water, that is, it has already virtually reached the state that it is changing into from its current state.[42]

Being ignorant of the discipline of mathematics, however, Seneca does not understand what he is talking about. He says that our vision struggles in water so that it cannot perceive an object clearly. Why, then, if it cannot perceive properly, does it not see everything diminished in size? But why things should appear magnified under water and through glass has another explanation on the basis of the round form of the vessel, an explanation to which we will turn when we discuss the glass sphere [in book 2].

Macrobius sheds considerable light on this opinion [of Seneca's]:

> Images in water appear larger than the objects actually are. This is the sort of thing we find food peddlers [21] doing, when their morsels are usually presented looking larger than they really are—in water-filled glass jars we see exceptionally large, round eggs and little livers with inflated lobes and onions with huge rings. For that matter, what is the principle behind vision itself—a question on which some people hold views that are neither true nor plausible. ... Water is denser, air thinner, and so our vision penetrates the former more slowly: striking up against it and bouncing back, our line of vision becomes divided and collides with

sed undique versum incurrit lineamenta simulachri, et sic fit, ut videatur imago archetypo suo grandior. Nam et Solis orbis matutinus solito nobis maior apparet, quia interiacet inter nos et ipsum aer, adhuc de nocte roscidus, et grandescit imago eius tamquam in aquae speculo visatur.

Sed haec magis erronea sunt. Inquit enim quod, quum aquae densius aere corpus sit, ab eius densitate nostra acies refringitur et in nos denuo recurrit, et dum reddit in se, non directo ictu, sed undique ad nos refert simulachrum. Errat quod visum per extramissionem fieri putat et radios a corporibus diaphanis reflecti, quod non nisi a politis densisque corporibus evenit.

Ptolemaeus in suis *Opticis* rationem huius rei assignat quam Theon in commentariis *Almagesti*, Alhazenus, Vitellio, et caeteri sequuntur. Assumit enim pro principio quae sub maioribus angulis videntur maiora videntur. Et si hoc in simplici visione minus verum, ut videbimus, multo minus in refractione verum apparet. Est ergo sua ratio.

Sit vasculum aqua plenum DEBC, sit oculus perpendicularis [22] in A, res visa in fundo vasis BC, aer in loco ADPHE. Punctus B pervenit ad

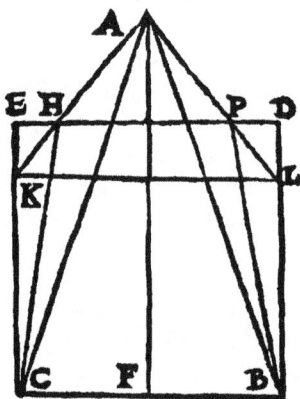

oculum A; refrangitur ab aquae superficie per punctum P. Sic punctus C per H pervenit ad A. Extenditur AP usque ad L in rectum, et AH in K. Trahaturque LK. Igitur LK est forma rei visae BC. Videtur vere, et non fracte, BC per lineas AB, AC, per refractas vero AL, AK.

Quia maior est angulus LAK ipso BAC, igitur maior videbitur ipsa LK ipsa BC. Sed hoc falsum est, quia B, ut diximus, videtur in puncto L et C in puncto K. Et KL non est maior ipsa BC,[14] quia DB et CE perpendiculares

---

[14]"BE" in original text.

itself. When it turns back after being divided, it meets the image's contours not head on but from every direction, and so it happens that the image we see seems larger than its original. The Sun's orb in the morning also appears larger than usual because the atmosphere between the Sun and us is still dewy from the night, and the Sun's image increases in size as though it were being seen in a pool of water.[43]

But these arguments are badly mistaken. For Macrobius says that, since the medium of water is denser than air, our vision is broken according to its density and then returns back to us, and when it returns on itself, it conveys the image to us not in a direct blow but in all directions. He is wrong because he assumes that vision takes place by extramission and by rays reflecting from transparent bodies, which occurs only from polished and dense transparent bodies.

In his *Optics*, Ptolemy explains this phenomenon in a way that Theon of Alexandria, in his commentary on the *Almagest*, Alhacen, Witelo, and others follow. For he takes it as a principle that things seen under larger visual angles appear larger.[44] And if this is less than true in the case of direct vision, as we will see, it seems much less true in the case of refraction. This, then, is Ptolemy's reasoning.

Let DEBC [in Figure 1.11a] be a small vessel filled with water, the eye at A on perpendicular [AF], [22] BC an object at the bottom of the vessel,

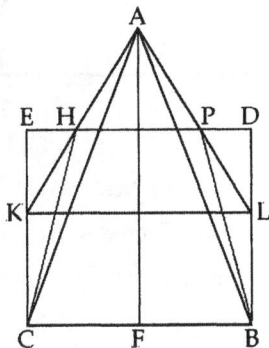

Figure 1.11a

and let air be located within the area ADPHE. The image of point B reaches toward the eye at A and is refracted through point P on the water's surface. Likewise, the image of point C reaches A through H. AP extends straight to L [on cathetus BD], and AH extends straight to K [on cathetus CE]. Connect LK. Therefore, LK is the image of visible object BC. [In air] BC is seen as it actually is, not refracted, along lines AB and AC, but [under water] it is seen by refracted lines AL and AK.

Because angle LAK > angle BAC, LK will therefore appear larger than BC. But this is false because, as we said, B appears at point L and C at point K. But KL is not longer than BC because DB and CE are perpendicular, and

sunt, et LK, BC aequidistantes. Et contra eius rationem adest experientia, quia res in fundo perpendiculariter inspecta modica maior non videtur. Etsi aliquanto maior videtur, non sua ratione sed longe subtiliori in rebus magnis, ut diximus. Sed falsam quoque eius rationem alia demonstratione confutabimus.

Sit semicirculus FAGMN, dividatur bifariam per AQ, et quartae eius FQ, QG dividantur rursusque per medium in R et I. Et erit angulus IAR

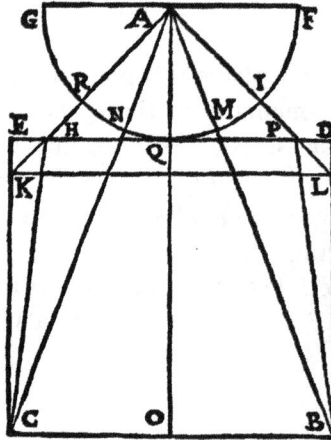

rectus. Dividantur denuo partes IQ et QR bifariam in M et N; erit angulus MAN semirectus. [23] Ex puncto Q extendatur linea DE aequidistans FG, extendaturque linea AI usque ad P et AR usque ad H. Mox extendantur lineae AM, AN in profundum infra quousque pertranseant puncta P, H perpendiculariter descendentia, et sit in B et C. Erigantur mox ex B, C perpendiculares ad DE, et extendantur AP in L et AH in K, et copulentur puncta L, K linea LK, et a puncto P et H extendantur rectae BP, HC. Dico si in hoc casu linea visa erit BC in vasis fundo iacens, vas BDEC, et per lineas BP et CH refrangerentur ad oculum HA et PA, extendanturque AP in L et AH in K, videri deberet LK duplo maior ipsa BC, quia angulus LAK duplo maior est ipsa BAC.

Nos demonstrationem ad id indicandum concinnavimus, sed quomodocunque eveniet, nunquam ad refractum angulum respondebit anguli directae visionis proportio. Sed cur res oblique in aquis inspecta per refractionem maior videatur demonstrationem meam afferam qua satis mihi ipsi facio. Quia rem in fundo iacentem rectam ex uno capite elevatam conspicimus ex

LK and BC are parallel. Moreover, experience counters Ptolemy's reasoning, for a small object at the bottom looked at along the perpendicular does not appear magnified.[45] And even if it appears slightly magnified, it does so not according to his reasoning, but according to the far more subtle reasoning about large objects, as we argued [in proposition 10]. But we will refute his mistaken reasoning with another demonstration.

Let semicircle FAGMN [in Figure 1.11b] be bisected by AQ, and then let its quadrants FQ and QG be bisected at R and I. And so IAR will be a

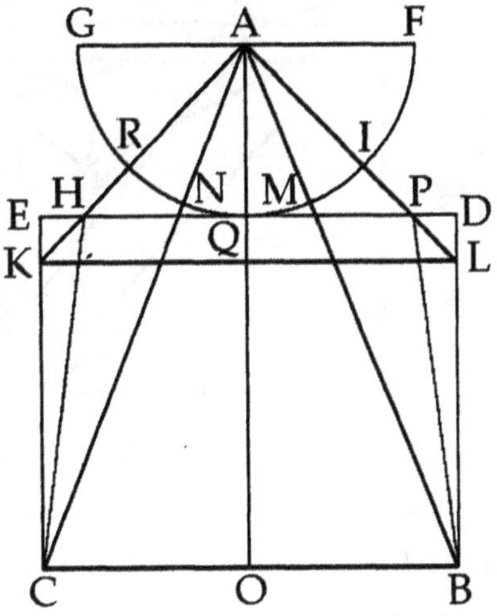

Figure 1.11b

right angle. Again, let segments IQ and QR be bisected at M and N; angle MAN will thus be half a right angle. [23] From point Q extend line DE parallel to FG, and extend line AI to P and AR to H. Then extend lines AM and AN downward along the perpendicular until they pass below points P and H, and let them descend to B and C. Next, drop perpendiculars to DE from B and C, extend AP to L and AH to K, join points L and K with line LK, and extend BP and HC straight from points P and H. In this case, I say that if BC is a visible line lying at the bottom of the vessel, if BDEC is the vessel, if [the image of BC conveyed] along lines BP and CH were refracted to the eye along HA and PA, and if AP is continued to L and AH to K, LK ought to appear twice as large as BC because angle LAK is twice as large as BAC.[46]

We have contrived a demonstration for judging this, but however it will turn out, the refracted visual angle will never correspond proportionately to the angle of direct vision. Let me, however, provide my demonstration of why an object viewed from a slant in water should appear magnified by refraction, a demonstration that satisfies me about the phenomenon. Because we perceive an object lying directly below on the bottom from an elevated perspective in refraction, the oblique diameter is longer than the one viewed

refractione, obliqua diameter maior est recta, cui si orbis circumducatur multo maior erit. Exemplum.

Infundatur in pelvim CDEFGH aqua; proiectus nummus vel pomum in fundo vasis iacens sit IL, cuius diameter IL. Perveniat L ad aquae superficiem

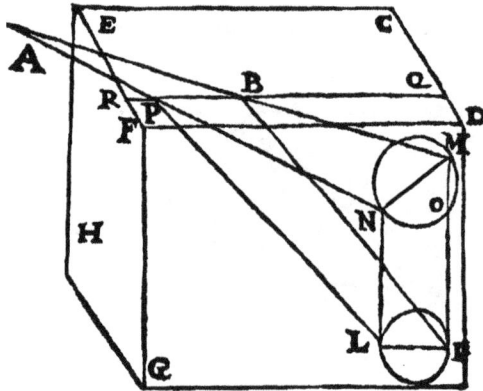

QBPR, ex P refrangatur ad oculum, extendatur AP in rectum, et ascendat cathetus ex L. Coincidet cum AP extensa in puncto N, et punctus L videbitur [24] in N.[15] Eodem modo I veniet ad aquae superficiem in B. Refrangatur in A; extendatur in rectum cathetus ex I quousque occurrerit cum linea AB, et coibunt in M. Copulentur puncta M, N per lineam MN, et eiusmodi diametro circumducatur circulus, qui maior erit circulo IL, cuius ratio est. Quia inter duas aequidistantes MI, NL diameter IL recte iacet, MN vero oblique, sit aequidistans ipsi IL ipsa ON, et ex M trahatur MO. Angulus MON rectus est, quia in semicirculo. Potentia ergo ON et OM aequalis est ipsi MN; ON aequalis est ipsi IL. Ergo potentia MN superabit potentiam ipsius ON per ipsam MO. Igitur MN ipsa IL maior erit.

Sydera maiora in oriente et occidente videri quam cœli medio. PROP. 12

Aristoteles, ut diximus, dicebat quod "astra orientia et occumbentia magis quam in medio coeli existentia" maiora videbantur. Rogerius Bacchon ex Ptolemaeo dicit ideo maiora videri orientia occidentiaque sydera quia orientia longiori intervallo videre putantur. Nam quum visus per Terrae superficiem feratur, et videat montes, turres, et civitates intercedentes, distantia maior

---

[15]"L" in the original text.

straight on, and if you draw a circle around that diameter it will be much larger. Here is an illustration.

Pour water into vessel CDEFGH [in Figure 1.11c], and let circle IL, whose diameter is IL, be a coin or a fruit lying on the bottom of the vessel.

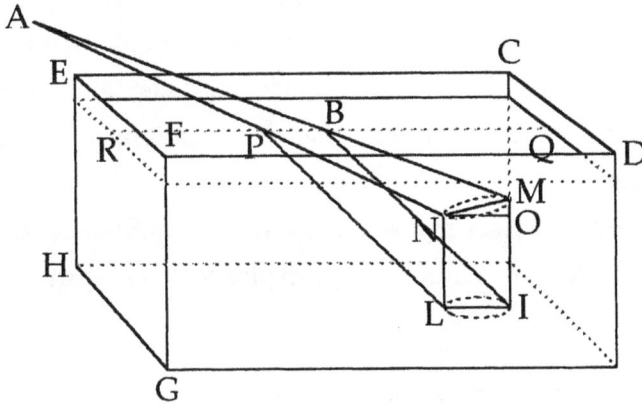

Figure 1.11c

Let the image of L reach surface QBPR of the water, let it be refracted from P to the eye, let [refracted ray] AP be extended straight, and let the cathetus [LN] extend up from L. It will intersect the extension of AP at point N, and so point L will be seen [24] at N. By the same token, the image of I will reach the water's surface at B. Let it be refracted to A, and let the cathetus be extended straight from I until it intersects [the extension of] line AB, and they will intersect at M. Connect points M and N by line MN, and draw a circle of that same diameter. It will be larger than circle IL for the following reason. Because diameter IL lies straight between the two parallels MI and NL, whereas MN is oblique [according to proposition 8 above], let ON be parallel to IL, and let MO be drawn from M. MON is a right angle because it is in a semicircle.[47] Consequently [by the Pythagorean theorem], $ON^2 + OM^2 = MN^2$, and ON = IL. Therefore, $MN^2 > ON^2$ by the added amount MO^2. Thus, MN > IL.

PROP. 12: *The stars appear larger at the east and west [horizons] than they do in the middle of the sky.*

As we said [at the beginning of proposition 11], Aristotle claimed that "the stars at rising and setting" appeared larger "than when they are in the middle of the sky."[48] Drawing on Ptolemy, Roger Bacon argues that the stars appear larger at rising and setting because on rising [or setting] they are judged to lie at a greater distance [than at zenith]. For when the visual faculty scans the surface of the Earth, and when it sees mountains, towers, and cities intervening [along the line of sight], the distance is judged to be longer,

iudicatur, secus autem in cœli medio, nam quum nullum adsit impedimentum, propinquiora putantur.

Iudicant enim philosophi distantias ex intersitis rebus, sed quae longiora putantur et maiora videntur multo maiora sunt re ipsa. Hanc eandem causam Cardanus usurpat in suis *Subtilitatibus*. Sed nescio quibus principiis haec innitantur homo putat plus distare ergo maiora ei videntur. Nos autem quae aequalia sunt et alia aliis plus distare videmus, remotiora minora et videmus et existimamus.

Alii sunt qui dicant aciem nostram in visione fatigari per densius medium illabentem, "et quemadmodum qui titubant plus occupant spatii," sic titubans acies videt maiora. Alii vero:

> sicut aqua diffunditur solidi corporis obiectione, ita species exiens atque in actum deducta per lucem, ubi in tenui aere nihil occurrit aequalis sibi prodit. Ubi vero quid sese opponit crassius, quasi retunditur atque propterea dilatatur, veluti sit in aqua, a cuius deinde superficie [25] pingitur aer eadem qua se in eam intulit quantitate.

Sed vera causa est medii densitas, ubicunque enim aer fuerit hebetior inter visum et astra, semper eorum species crassior videbitur, ratioque ex praecedenti pendet. Sed ne rudi fide oculis credas, sic ad demonstrationem procedemus.

Sit Terrae globus DAE, oculus eius in superficie situs A; sit densitas medii aeris hebetioris inter oculum et stellam LM; ipsa sphaera FIGH, horizon

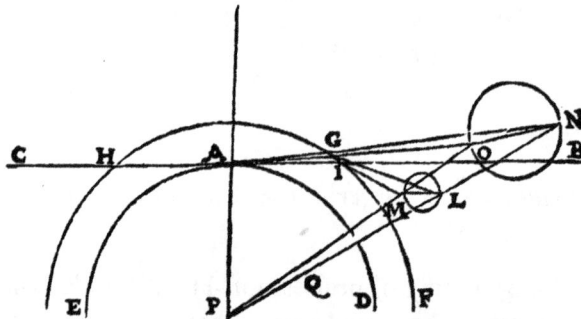

BAHC, stellae diameter LM. Punctus M venit ad I, refrangitur ad A, extenditur in O. Eodem modo punctus L venit ad G, frangitur ad oculum A; extenditur

whereas it is otherwise in the middle of the sky, for when there is nothing along the way, things are judged to be nearer.[49]

Indeed, natural philosophers [argue that we] judge distances according to intervening objects, but things that are judged to be farther away and larger on that basis appear much larger than the things themselves actually are. Girolamo Cardano appeals to the same cause in his *De Subtilitate*.[50] However, I do not know on what principles these natural philosophers base the idea that humans judge things to lie farther and thus to appear larger to them. On the contrary, we see things that are equal to one another, some of which lie farther away than others, and we both perceive and judge the ones lying farther away to be smaller.[51]

There are some who may claim that in the case of vision through a denser medium our gaze is weakened as it passes through, "and just as those who stagger take up more space," so the gaze, in staggering [through the denser medium], perceives things as larger. Others, however, argue that

> just as water permeates a solid body, so the species that emanates and is actually conveyed by light produces its equal in rare air, where it encounters nothing [to hinder it]. On the other hand, when it encounters something crasser than itself, it is hindered and therefore enlarged, just as it is in water, at whose surface the very air [25] is tinged so as to have size added to it.[52]

But the true cause is the density of the medium, for wherever the air between the eye and the stars is fairly thick, their images will always appear bulkier, and the reason depends on previous argumentation. In order for you not to believe your eyes with uninformed credulity, though, we will proceed to a demonstration.

Let DAE [in Figure 1.12] be the globe of the Earth, A an eye located on its surface, and let the density of the air between the eye and star LM be

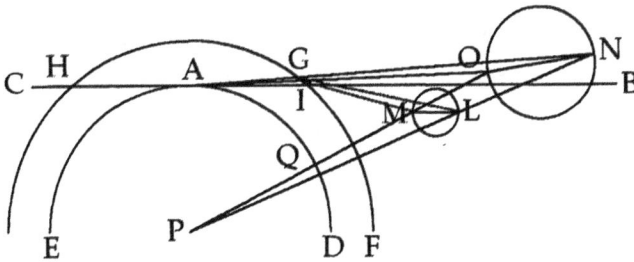

Figure 1.12

thicker [than that of the aethereal medium beyond the atmosphere]; let FIGH be the sphere of that air, BAHC the horizon, and LM the star's diameter. The image of point M reaches I and is refracted to A, [and the refracted ray AI] is extended to O. Likewise, the image of point L reaches G and is refracted to the eye at A, [and the refracted ray] AG is extended to N. Drop cathetus

AG in N. Ascendat cathetus ex MQP centro; occurret ipsi AI in O. Ergo punctus M videbitur in O, ex praecedenti.

Altitudines in aquis per refractionem breviores videntur. PROP. 13

Profunda aqua inspecta, semper fundum sublimius videbitur, quum praecipue ex obliquo conspicitur. Et id in causa fuit ut multi viri submergerentur, nam ex inspectione ipsis videre videbatur quod vix guttur aut pectus attingeret altitudo at in ipsis proiectis multo ipsis altiores aquas fuisse experti sunt. Si quis igitur id clarius conspicere velit, dimittat lignum vel columnam in aquis, cuius dimidium super aquam [26][16] emineat, et elongetur ab ea. Duplo enim fere maior superextans pars ipsa mersa videbitur, cuius causa est quod punctus imae partis ad oculos veniens sublimior videbitur, ut in subiecta figura patet.

Sit cisterna vel lacusculus CDEFGH, columnae aquae submersae pars LM, pars ima M, super aquas superextans pars altera LI aequalis. Pars ima

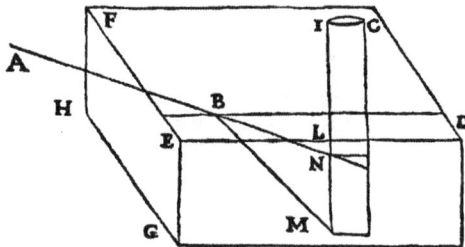

M occurrens aquae superficiei B, declinat ad oculum A, et oculo insinuatur. Occursans radius BA fractus dirigatur in rectum; perveniet in N, unde M punctus in N videbitur, et quantitas MN te latebit. Longior igitur pars IL videbitur quam LN. Sed Cardanus in *Subtilitatibus* docet profundum maris metiri; ex huius ignorantia inepte fallitur.

Ex refractione res tardius moveri videbitur. PROP. 14

Haec est et altera visus fallacia ex superiori dependens, nam stellae quae super horizonta moventur tardius ire videntur quam quae in coeli medio; cuius causa est quod duo motores aequales, qui per inaequale spatium mov-

---

[16]Misprinted as page 30 in original text.

MQP from the center of the Earth, and AI will intersect it at O. Therefore, point M will be seen at O, as is clear from an earlier proof.[53]

PROP. 13: *Heights viewed in water appear shorter than they should because of refraction.*

When one looks into the depths of water, the bottom will invariably look higher than it is, especially when it is viewed from a slant. This became an issue insofar as many men have been submerged in water, for after a close look in such water, it was noticed that the neck or chest reached virtually the same height and in cases where objects were immersed in much deeper water these phenomena were experienced. If one wants to see this more clearly, then let him stand a stick or cylinder in the water so that half of it extends above the water, [26] and let him withdraw away from it. In fact, the part extending above the water will appear fully twice as long as the submerged portion because the image of the point at the bottom that reaches the eye will appear elevated, as is clear in the illustration appended below.

Let CDEFGH [in Figure 1.13] be a water tank or a small pond, LM the segment of a cylinder submerged in water, M the bottom part of the cylinder,

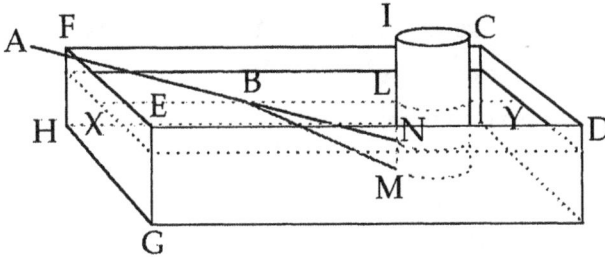

Figure 1.13

and LI the other, equal segment of the cylinder standing above the water. The image of bottom part M reaches the water's surface at B, inclines toward the eye at A, and is conveyed into the eye [along BA]. Let refracted ray BA along which [the image] reaches [the eye] be extended straight, and it will reach N, so point M will be seen at N, and the true size of MN will be concealed from you. Consequently, segment IL will look longer than segment LN. In his *De Subtilitate*, however, Cardano shows how to measure the depth of the sea, but he errs foolishly from ignorance of this refractive shrinking.[54]

PROP. 14: *A [celestial] object will appear to move more slowly on account of refraction.*

This is another visual illusion dependent on reasoning articulated above, for the stars that move just above the horizon appear to go more slowly than those in the middle of the sky. The reason for this is that when two objects

entur, ut alter brevius alter longius progressurus sit et uterque aequali tempore motum absoluturus, qui per brevius spatium se movebit tardius ire creditur quam qui per longius. Rei exemplum erit. [27][17]

Sit surgens astrum M, eadem Terrae circumferentia DAE, densiorumque vaporum FGL, idem oculus A, ac Terrae horizon BAH. Perveniat ad circulum

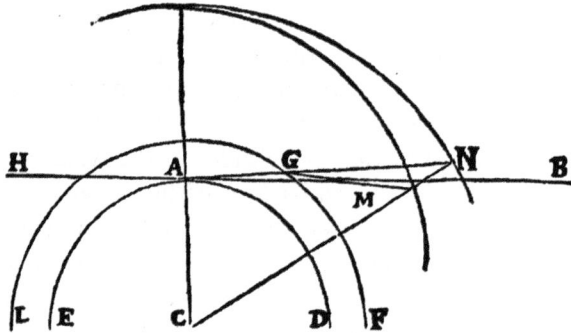

FGL ipsa MG, et refrangetur ad oculum A. Tunc oculus A videt astrum in linea AGN, ubi scilicet coibit cum ascendente catheto CMN. Spatium ergo motus erit MN brevius ipso vero; tardius ergo per spatium illud ire putabitur.

Magnitudo obliqua partim infra aquam, partim supra existens, ex refractione refracta videbitur. PROP. 15

Vitruvius de aedificiorum proportionibus loquens docet quod

> in navibus remi, quum sint sub aqua directi, tamen oculis infracti videntur; et quatenus eorum partes tangunt summam planitiem liquoris, apparent (uti sunt) directi, quum vero sub aqua sunt demissi, per naturae perlucidam raritatem, remittunt enatantes ab suis corporibus fluentes imagines ad summam aquae planitiem, atque ibi commotae efficere videntur infractum remorum oculis aspectum. Hoc autem, sive simulachrorum impulsu, seu radiorum ex oculis effusionibus (ut physicis placet) videamus, utraque ratione videtur ita esse, uti falsa iudicia oculorum habeat aspectus.

Sed falsum est id, quia species mersi remi non ad aquae superficiem summam suam [28] imaginem mittit, sed infra aquam est. Alii sunt qui remi fractam speciem dicunt. Sed hoc etiam falsum, nam species duae sunt, et quae ab

---

[17]Misprinted as page 31 in original text.

moving at equal speeds move through unequal spaces, such that the one space to be traversed is shorter, the other longer, and when both objects are to complete their motion in equal times, the one that will move through the shorter space is judged to go slower than the one moving through the longer one. An illustration of this point will be as follows. [27]

Let M [in Figure 1.14] be a rising star [on the celestial sphere], DAE the circumference of the Earth, FGL the sphere of fairly thick vapors, A the

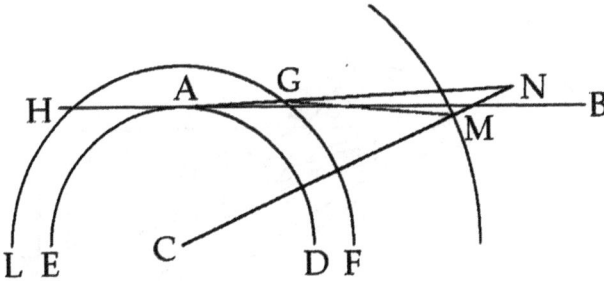

Figure 1.14

eye, and BAH the horizon line on the Earth. Let ray MG reach circle FGL, and let it be refracted to the eye at A. In that case, the eye at A sees the star along line AGN, that is, where [refracted ray AG, extended] will intersect the upward continuation of cathetus CMN. The [apparent] space MN of the star's motion will therefore be shorter than the actual space, so the star will be judged to go more slowly over that space [than it actually does in that amount of time].[55]

PROP. 15: *An object lying at a slant partly under water and partly above it will appear broken on account of refraction.*

While speaking of the proportions of buildings, Vitruvius informs us that

> in ships, when the oars are straight underwater, they look broken to the eye: the parts that extend as far as the water's surface look straight (as indeed they are), but once they are submerged underwater, they give off fluid images from their bodies; these, swimming through the shiny rarefaction of water's nature toward the upper surface of the water, and stirred up in that place, seem to create the appearance of broken oars to the eyes. Thus, either from the impact of images on our vision or by action of rays shed forth from our eyes, as the physicists would have it, for either reason it seems to be the case that the glance of our eyes may make false judgments.[56]

This, however is false because the form of the immersed oar does not emit its image [28] at the water's surface, but it is produced beneath the water. There are those who say that the image of the oar is broken.[57] But this, too, is false, for the images are double, one in the air and one in the water.

aere et quae ab aqua. Sed ratio est quod remi pars sub aquis mersa propinquior superficies aquae apparet quam sit vere, at pars ea extra aquam suo loco videtur. Igitur duae istae partes in continuum directae apparere non possunt, videntur ergo fractae. Veritatem lineis explanabimus.

Sit aquae summum FBCDE, baculus semimmersus GBH, super aquam pars extans GB, sub aquis BH. Imum baculi punctum H veniat ad aquae

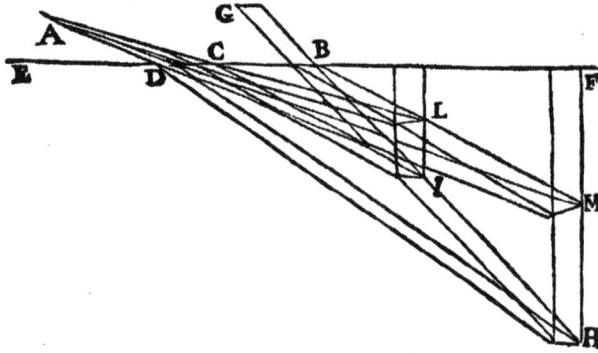

superficiem, ad A oculum per D, extendatur AD in rectum in M, et elevetur cathetus HM ab H, et occurrat AD in M punctus. Ergo H erit M. Sit altera pars baculi in medio I. Venit ad aquae superficiem per IC, ad oculum A per CA. Extendatur in rectum in L, et elevetur cathetus ab I, et occurret in L. Igitur punctus I videbitur in L. Continuentur ergo partes M, L, B. Erit pars sub aquis MLB, et sublimior videtur ipsa HIB; non continuatur in directum cum ipsa BG. Ergo fracta in B apparebit.

Vas aquis semiplena fundum minus ostendunt. PROP. 16

Sunt apud nos vasa quaedam ex semidoliis confecta ad diversos agricolarum usus parata, quorum laterales costae in [29] summo latiusculae, in imo arctiores, ut commodius ligneis circulis constringi possint. Haec eadem interdum aquis semiplena, fundum minus quam sit ostendunt. Ratio est quod latera quae oblique in aquam merguntur; pars immersa sublimior videtur. Et infracta, infractio ad centrum vergit, ut diximus. Quum igitur imae partes et sublimiores et in centrum concurrentes videantur, fundum arctius ostendunt. Exemplum.

Instead, the reason is that the portion of the oar immersed beneath the water appears to be nearer the water's surface than it actually is, and the segment of the oar outside the water is seen in its true place. These two segments can therefore not appear in a continuous straight line, so they appear broken. We will explain the truth diagrammatically.

Let FBCDE [in Figure 1.15] be the water's surface and GBH a stick submerged partway in the water, GB being the segment extending above the

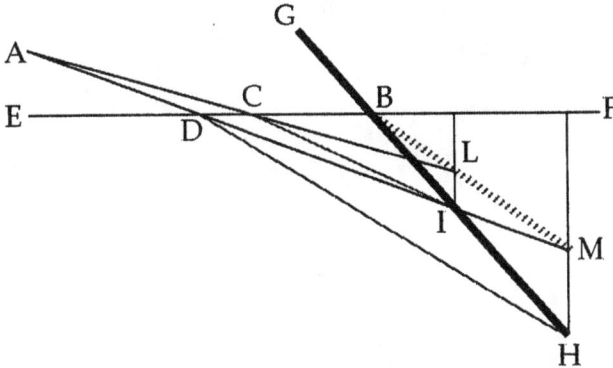

Figure 1.15

water, BH the segment extending below the water. Let the image of bottom point H of the stick reach the water's surface and then the eye at A through D, extend [the resulting refracted ray] AD straight to M, drop the cathetus HM from H, and let it intersect [the extension of] AD at point M. The image of H will thus be M. Let I be another spot in the middle of the stick. Its image reaches the water's surface along IC and refracts to the eye at A along CA. Extend CA straight to L, drop the cathetus from I, and it will meet CA extended at L. Point I will therefore be seen at L. Accordingly, join spots M, L, and B. MLB will be the image of the segment below water, and it appears to be raised above HIB, and so it does not extend in a direct line with BG. Hence, the stick will appear broken at B.

PROP. 16: *The bottom of a vessel partially full of water looks reduced in size.*

Among us there are certain vessels manufactured from half-barrels designed for various uses by farmers, the staves composing their sides being [29] wider at the top and narrower at the bottom so that they can be held together by increasingly expansive wooden hoops. From time to time, when these vessels are partially filled with water, they display a bottom that looks smaller than it actually is. The reason is that the sides are submerged at a slant in the water, so the submerged portion appears elevated. And when their image is broken by refraction, the breaking tends toward the center of the container, as we have said [in the previous proposition]. Consequently, since the lower portions appear higher and look as if they are intersecting toward the center, they show the bottom narrowed. Here is an illustration.

Sic vas ABCD aqua semiplenum usque ad EFG, et pars eius GMEF infra sit aqua plena. Sitque una eius costa EMH, et EM extra aquam MH infra.

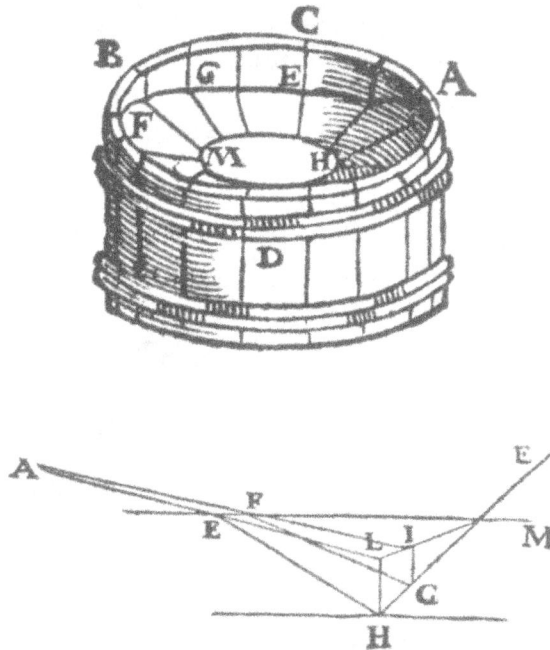

Igitur MH sublimior videbitur, et fundum ex hoc coarctabitur, et sic de caeteris intelligendum. Clarius loquemur. Sit aqua MFE, oculus A, latus obliquum semisubmersum EMGH; sit EM supra aquam, MH sub aqua. H videbitur in L, G in I sublimus. Magis ergo inclinatum latus MH sub aqua eo supra aquam ME; magis obliquum, et minus.

Quae longe in aquosis locis conspiciuntur ex refractione profundius mergi videntur. PROP. 17

In aquosis locis, palustribus, fluminibus, et mari ex longinquo conspecta omnia humiliora et quasi profundius mersa conspiciuntur. Huius rei naturalem causam reddere curantes, ut author *Sphaerae*, his probant aquae sphaericitatem indiciis. Ut [30][18] si quis abeuntis navis signum in puppi ponat, ubi non procul abierit huius signi diligentius visuro aufertur obtutus, et huius causam esse aquae tumorem. Sed miror profecto istius viri rationem, quum in tam vasta mundi magnitudine, viginti aut triginta millia passuum impedire possint navis visum; sed eius causa est medii densitas in qua navis est, ex eo namque vapores continuo elevantur ut roscidum aerem reddant, ut navis fere sub aquis videatur. Exemplum apponemus.

---

[18]Misprinted as page 26 in original text.

Let AECD [in Figure 1.16a] be the vessel partially filled with water up to GMF,[58] and let lower portion GMFH of it be full of water. Let EMH be one of its staves, with EM above water and MH below. MH will therefore appear elevated, and on that account the bottom will be narrowed, and the same should be understood for the remaining staves. We will speak more

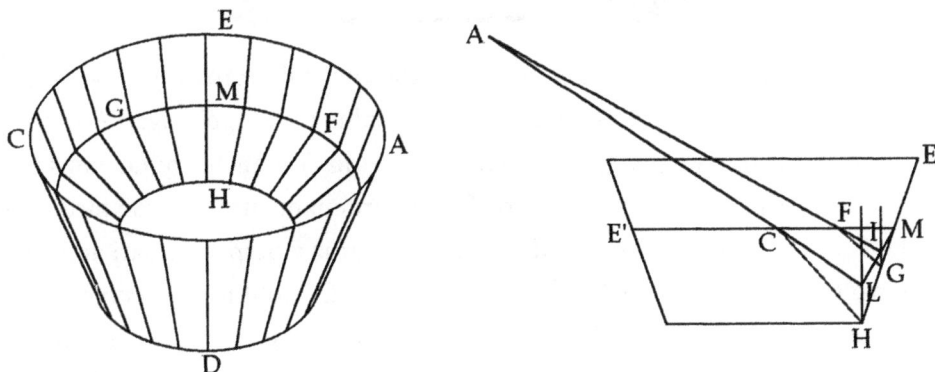

Figure 1.16 a, b

clearly. Let MFE' [in Figure 1.16b] be the [surface of the] water, A the eye, EMGH the stave that is partially submerged; and let EM lie above the water and MH below the water. H will appear elevated to L and G to I. [Image MIL of segment] MH of the stave will therefore appear more sharply slanted below the water than ME above it, and the more slanted it appears, the smaller it appears.

PROP. 17: *Things that are viewed from far away in humid locations appear sunken on account of refraction.*

Everything lying in humid places, on swampy surfaces, rivers, or the sea that is looked at from afar appears lower, as if submerged. In providing the natural cause of this phenomenon, those, such as the author [Johannes de Sacrobosco] of the *Sphere*, who pay attention to it demonstrate the sphericity of water on the basis of these appearances.[59] For instance, [30] if someone places a mark on the stern of a departing ship, then at a point where it has moved not too far away, it disappears from sight while closely watched, and the reason for this [according to Sacrobosco] is the curvature of the water. But indeed I marvel at this man's reasoning about this, for in a world of such vast magnitude, twenty or thirty thousand paces can prevent sight from seeing the ship.[60] Instead, its cause is the thickness of the medium in which the ship lies, for vapors continually rise from that place so as to render the air misty, according to which the ship may appear to lie completely below the water.[61] We will supply the following illustration.

Sit navis altitudo AB supra maris superficiem CFB. Sit roscidus et vaporosus aer EOFAB, tenuis vero EOFGHC, ubi oculus I. A puppis signo A veniat linea AO, et si in longum produceretur veniret in G. Punctus refractionis O. Ducatur perpendicularis super superficiem terminantem utrumque medium EOF, et sit perpendicularis HOL. Declinabit igitur a perpendiculari ad oculum; scilicet recedet ab OG et veniet ad OI ad oculum existentem in I. Extendatur IO; veniet ad B. Summitas ergo puppis navis A videbitur in B, maris ima superficie.

Consultius igitur Ptolemaeus in primo *Almagesti* non signum locat in navi et minima eius elongatione, sed ad quemcumque locum navigantes accedant, paulatim eorum magnitudines augeri videntur, quasi ab ipso mari emergentes, quae antea ob nimiam longitudinem submersa videbantur. Diximus quoque in 13 propositione omnia in aquis breviora videri. Posset et aliquis dicere quod densius medium de facili visum impedit, ut recte navis videri possit. Vel quod undarum motus circa navim intercipit visualem lineam, quae etsi circa navem pauca substollitur, ad oculum longe existentem. [31][19]

Sub aquis existenti omnia quae sub aquis sunt confracte videntur ut in aere sub aquis aspicienti. PROP. 18

Sub mari degentibus eadem illa accidentia eveniunt, quae si in aere steterint et sub aquis aspexerint. Causa, ut videre videtur, est nostri oculi albugo, quae adeo tenuis et pellucida est, et marina aqua hebes turbidaque, ut humor ille aeris vicem et marina aqua communis aquae sustinere videatur, ut latius postea explanabimus. Lineis clariora haec reddemus.

Sit oculus sub aquis existens IHG, mare autem infra ACD. Pars lineae AD veniat ad oculum AB. Penetrat irrefracta; suo loco videtur. C venit per pupillam H; transiret in E si recte procederet. Sed quia ex densiori ad tenuius

---

[19]Misprinted as page 27 in original text.

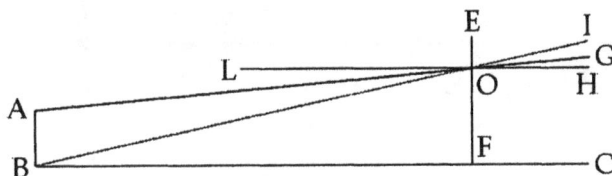

Figure 1.17

Let AB [in Figure 1.17] be the height of a ship above CFB, the surface of the sea. Let [the area] EOFAB [to the left of EF] be moist and vaporous air, [area] EOFGHC [to the right of EF] clear air in which the eye at I lies. Let line AO reach from mark A on the ship's stern; if it were extended at length, it would reach G. Let O be the point of refraction. Draw a normal to interface EOF between both media [i.e., vaporous and clear air], and let it be HOL. [Ray AO] will therefore incline away from normal [HOL] toward the eye; that is, it will verge away from OG and will reach along OI to the eye lying at I. Extend IO, and it will reach B. As a result, the top of the stern A will be seen at B, on the surface of the sea at the bottom of the ship.[62]

Accordingly, after careful consideration in the first book of the *Almagest*, Ptolemy does not locate a mark on a ship and its shortest distance [away before disappearing below the water's surface] but [notes that], as sailors approach a given location, objects there appear to grow little by little in size, as if they were rising from the very sea, whereas they had appeared submerged earlier because of their excessive distance away.[63] Also, in proposition 13 we said that everything appears shorter in water than it actually is.[64] One could also argue that the denser medium prevents the visual faculty from seeing the ship easily so that it could be perceived correctly. Or one might argue that the motion of the waves around the ship, even if it does not rise high, interrupts the line of sight from an eye located far away. [31]

PROP. 18: *To a viewer under water everything appears refracted below the water as it does to a viewer looking from air down through water.*

To those who spend time under the sea, the same things happen as would happen if they stood in the air and looked into water. The reason, as seems apparent, is the albugineous humor of our eye, which is quite thin and clear, whereas marine water is thick and turbid, so that this humor seems to hold the place of plain water between air and marine water, as we will explain more fully later. We will render this more clearly with a diagram.

Let IHG [in Figure 1.18] be the [outer surface of the] eye lying under water, and let the sea lie below surface ACD. Let the image of spot [A] on line AD reach the eye along AB. It passes through unrefracted, so it is seen in its proper place. The image of C reaches through point H on the pupil,[65] and it would pass on to E [on the crystalline lens] if it were to continue in

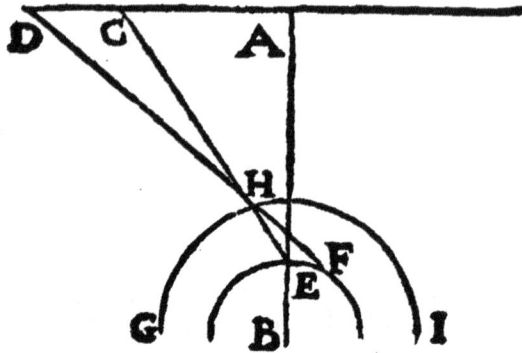

venit, fugit a perpendiculari per B, H et refrangitur in HF. Extendatur HF; pervenit in D. Igitur pars AC videbitur AD,[20] scilicet maior.

In librationibus locorum accidit fallacia ex refractione. PROP. 19

Dum aqueductus struuntur, ex longinquo libratur planities, ut inter sexagenos aut centenos pedes in sesquipedem sensim inclinetur structura, ut vim currendi habere possit aqua. Librantur autem loca dioptris, aquariis libris, aut chorobate, et quia summo mane plerunque opus aggrediuntur operarii in caliginoso aere et adhuc roscido, dum ex pinnacidiis transmittitur acies in densiori medio, quasi sub aquis existenti regulae et foramina uti non suo loco visa. Aliam atque aliam [32] demonstrabunt visionis lineam, et facillime error committitur, ut in superiori vidimus. Exemplum.

   Sit regula CB, pinnacidia A, O, oculus I, librationis linea AO progrediens in oculum quasi in tenuius. Vel si in medio densiori fuerit, AO linea visus altius progreditur in oculum I, unde quum IO radius extenditur, veniet infra in B, non in suum locum, sed inferiorem. Falluntur ergo qui mane aere

----

[20]"CD" in original text.

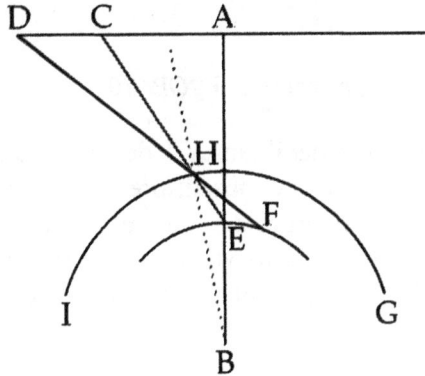

Figure 1.18

a straight line. But since it reaches from a denser into a rarer medium, it departs from the normal that passes through B and H and is refracted along HF. Extend HF, and it reaches D. Consequently, segment AC will appear as AD, that is it appears magnified.[66]

PROP. 19: *A visual illusion due to refraction occurs in the leveling of sites.*

When aqueducts are being built, their leveling is determined at a distance, so that, in order for the water to flow with adequate force, their structural framework should be inclined gradually at a rate of a foot-and-a-half every sixty or a hundred feet. Sites, however, are leveled with diopters, water levels, or a chorobates,[67] and since the laborers generally begin work in the very early morning, when the air is misty and still quite moist, then when the line of sight passes through the sighting holes into the denser medium, it is as if the leveling planes and sighting holes were under water so that what is seen is not seen in its proper location. Those instruments will instead [32] mark out one or another line of sight, and an error is most easily committed, as we saw above [in proposition 18]. Here is an illustration.

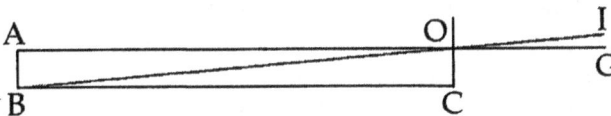

Figure 1.19

Let CB [in Figure 1.19] be the leveling plane, A and O the sighting-holes, I the eye, and AO the leveling line along which a ray reaches straight to the eye, as if it were in a rarer medium. On the other hand, if line of sight AO lies in a denser medium, [the ray passing along it] reaches higher to an eye at I, so when ray IO is extended, it will reach B below A, which will not be seen in its proper place, but lower.[68] Those who employ instruments of

caliginoso et adhuc roscido eiusmodi instrumenta exercent, ut multoties
sum expertus.

Refractione debilitari luces et colores. PROP. 20

Refractione luces et colores debilitantur,[21] de iis loquendo qui sub aquis
respiciuntur, nam excoloratiores et obscuriores videntur, quod passim clare
patet experimento. Immo si sub turbidis aquis aliqua conspiciantur, tenebros-
iora videntur. Nubes quae per lacus videntur, qui fere semper turbidi sunt
obscurique, turbidiores obscurioresque nostris oculis offeruntur. Idem Seneca docet:

> Quoties defectionem Solis volumus deprehendere, ponimus pelves, quas aut oleo,
> aut pice implemus, quia pinguis humor minus facile turbatur, et ideo quas recipit
> imagines servat. Apparere autem imagines non possunt, nisi in liquido et immoto.
> Tunc solemus notare quemadmodum se Luna Soli opponat (aut illum tanto mai-
> orem) subiecto corpore abscondat.

Stellae in aere claro videntur splendentes et clarae, unde ex his apparet
prognosticon tempestatis, nam quando obscurae sunt, medium densum, ros-
cidumque et aquosum praedicunt; clarae vero serenitatem et Boream spiran-
tem. Aratus de coloribus Lunae loquens.

> Prospice vel plenam, vel prospice dimidiatam,
> Aut hanc crescentem aut ad cornu rursus euntem,
> Mensis cuiusvis atque excipe signa colore. [33]
> Si pura omnino fuerit tu crede serenum;
> Venturum tempus, ventos, si plena rubore;
> Si nigrans, pluvias ruituras aethere ab alto, etc.

Commoto medio ex refractione res moveri videntur. PROP. 21

De stellarum scintillatione disquirentes maiores nostri multipliciter falsi sunt.
Aristoteles dicit stellas fixas scintillare, et Solem orientem occidentemque,
atque id evenire ex maxima distantia, nam visus ad sydera usque porrectus
nutat caligatque, unde scintillatio ipsa causatur; nam planetis non contingit
propter propinquitatem. Sed hoc falsum, nam Venus, Mercurius, et Mars
scintillant, etsi non ita evidenter ut stellae fixae. Praeterea, oculi facultatem
ad stellas ferri affirmat, quod est quoque falsum. Et si ex distantia fuerit,
cur Saturnus non ita scintillat, quum octavae sphaerae propinquior fuerit et
prae caeteris minus scintillet?

    Alii sunt qui scillationis causam stellarum fixarum rapidae celeritati
ascribant. Sed id coarguitur falsum, quod stellae quae iuxta polos sunt tardius
moveantur, et tamen, ut illae quae in aequinoctiali sunt, scintillant. Vel

---

[21]"deblitantut" in original text.

this sort during the early morning in misty and very moist air are therefore misled, as I have often experienced.

PROP. 20: *Lights and colors are weakened by refraction.*

Lights and colors are weakened by refraction, referring here to things that are looked at under water, for they appear sapped of color and rather dark, which is shown clearly and randomly by experience. Indeed, if various things are looked at in cloudy water, they appear quite shadowy. Clouds seen on lakes, which are continually roiled and darkened, present themselves to our eyes more muddled and obscure than they actually are. Seneca teaches this very thing:

> Whenever we want to observe an eclipse of the Sun, we set out bowls, which we fill with either olive oil or pitch, because a dense liquid is less easily disturbed and so preserves the images it receives. Images cannot be seen except in something clear and motionless. Then we regularly observe how the Moon gets in front of the Sun; by interposing its body, it eclipses the Sun (though the Sun is so much bigger).[69]

In clear air the stars appear resplendent and bright, so an omen of storms is betokened by them, for when they are obscure, they indicate a dense, misty, and humid medium, but when they are bright they indicate calm and a north wind. Speaking of the colors of the Moon, Aratus [says]:

> Inspect the Moon at full, or inspect it at half,
> Or when it is waxing or waning toward the crescent.
> Infer the forecast for each month from the color. [33]
> If it is entirely clear, assume fair weather.
> If it is full of red, assume windy weather to come.
> If it is blackening, assume that rain will fall from on high, etc.[70]

PROP. 21: *Objects appear to move because of refraction in a roiled medium.*

In discussing the twinkling of the stars, our predecessors made many mistakes. Aristotle says that the fixed stars, as well as the Sun at rising and setting, twinkle, and this arises from the vast distance, for when stretched out to the stars, the visual power wavers and becomes confused, so the twinkling is due to this; on the other hand, it does not happen with the planets because of their nearness.[71] But this is false because Venus, Mercury, and Mars twinkle, even though not as obviously as the fixed stars. Besides, Aristotle asserts that the power of the eye is conveyed out to the stars, which is also false. In addition, if the twinkling were due to distance, then why does Saturn not twinkle, since it is quite near the eighth sphere [of the fixed stars] and yet twinkles less compared to the rest?[72]

There are others who attribute the twinkling to the rapid speed of the fixed stars.[73] But that is demonstrably false because the stars that are near the poles move more slowly, and yet they twinkle, like those that lie on the

fortasse triplex octavae et multiplex planetarum motus; quippe diversis motibus moventur ad dextrum, sinistrum, sursum, deorsumque; in variis axibus titubare tripudiareque videntur, quum titubatio et tremor ex contrariis motibus gigni videantur.

Cantuarensis scribit causam esse quod stellae vices speculorum obtineant et radios Solis reflectant, et quum coelestia corpora continuo moveantur, angulus incidentiae varietur, sic et reflexionis. Haec igitur variatio facit quandam vibrationis speciem. Nos autem causam medio ascribemus. Quippe semper ventis obturbatur et inconstans est, quodque nubes densas et roridos vapores excipiat facitque ut lineae quae ad oculos perveniunt hinc inde moveantur et varia iaculatione oculos verberent, et quanto longius distant plus lineae refractionum longius et notabilius moventur.

Id conspicere licet in lapillis fluviorum fundo iacentibus, qui ob rapidum superfluentis aquae motum tremere videntur. Aqua Sole percussa et mota, quo longius lumen iaculatur, eo latius crebriusque movetur; in proximis locis [34] minus. Stellae quo maiores clarius et magis scintillare videntur, ut Venus et stellae primae magnitudinis. Accedit ad hoc quod ascendentes et descendentes stellae magis lumen vibrant, quia ibi crassiores vapores rorulentiores crassioresque. Idem evenit quo venti strenuius perflant, magis stellae crispant lumen; immo ex eo prognostica ventorum cognoscuntur, nam ea coeli parte qua plus tremere videntur et plus radios suos stellae movent et tremulae aerem quatiunt, ex ea parte venti venient et perflant. Extat etiam experimentum in nostra *Naturali Magia* stellae per fumum visae moveri videntur.

Ex refractione Solis radii in vicinia lucis et umbrae tremere videntur. PROP. 22

Aristoteles in *Problematibus* quaerit, "Cur extremum [Solis] umbrae tremere videatur?" Vel quia Sol rapidissime fertur? Sed hoc falsum, nam tremor est motus in contrarias partes, et Sol non ita movetur. Vel quia moventur corpuscula in aere, ut atomi, transmeantes enim per fenestras modo de umbra in lucem modo de luce in umbram moventur assidue, unde terminus lucis et umbrae communis continuo moveri videtur, quum non haec sed illa hunc in modum soleant agitari? Sed hoc falsum est, nam mundo in loco ubi atomus

equator. Indeed, the motions of the planets may be three times eight or more; and they certainly move by various motions to the right and left, upward and downward, and they appear to reel and dance around on different axes, since reeling about and trembling seem to arise from contrary motions.[74]

John Pecham writes that the reason for twinkling is that the stars act as mirrors and reflect the rays of the Sun, and since celestial bodies move continually, the angles of incidence vary, as do the angles of reflection. This variation thus causes a certain kind of vibration.[75] We, however, will refer the cause to the medium. Indeed, it is constantly stirred up and randomly changed by the winds, and such disturbance draws out thick clouds and dense vapors and causes the lines of radiation that reach the eyes from all sides to be moved and to strike the eyes along varying trajectories, and the farther away the light sources are, the longer the lines of refraction and the more perceptibly they move.

One can observe this in the case of pebbles lying on the bottom of rivers; they appear to tremble on account of the rapid motion of the water flowing over them.[76] When roiled water is struck by sunlight, the farther away the light is projected, the more extensively and abundantly it is moved, whereas in nearer locations it moves [34] less. Insofar as they are larger, stars such as Venus and stars of the first magnitude look brighter and appear to twinkle more. In this regard it happens that the light of rising and setting stars vibrates more than usual because at that point [on the horizon] the thick vapors [rising into the sky] are quite dewy and dense. The same thing happens where the winds blow most fiercely and cause starlight to tremble appreciably; indeed, forecasts of wind are determined from this trembling, for it is from that part of the sky, where the stars appear to tremble most and to move their rays and where the trembling appears to shake the air, that the winds will come and blow. There is also an outstanding experiment in our *Natural Magic* showing how stars viewed through smoke appear to move.[77]

PROP. 22: *Because of refraction the rays of the Sun appear to tremble in the vicinity of light and shadow.*

In the *Problems*, Aristotle asks, "Why does the edge of the Sun's shadow appear to tremble?"[78] On the one hand, is it because the Sun is carried extremely rapidly [in its orbit]? This is false, though, for trembling is motion back and forth, and the Sun does not move that way. On the other hand, is it because tiny particles in the air, such as atoms, move rapidly, at one instant transmitting shadow from light streaming through windows, at another transmitting light from shadow, so that the common boundary of light and shadow appears to move continually, since it is not the light and shadow but the particles that are customarily agitated in this manner? But this is false, for in the world, where there are no atoms or scarcely any, the edges of shadows still move.[79] The true cause, however, is the motion of the medium, which

nulla aut vix paucae, etiam umbrarum confinia moventur. Sed vera causa est medii motus, quod variis diversisque vaporibus repletum, ut diximus in superioribus, figuram omnem disturbat, aer enim qui rapitur non dat potestatem ut confinia recte possint conspici.

is replete with various kinds of vapors, as we said above, and this motion disturbs every image [passing through the medium], for air that is carried off swiftly in various directions does not permit the edges to be perceived correctly.

66

DE REFRACTIONE

NOTES

1. As Della Porta will explain in the next paragraph, Giovanni Antonio Pisano was an eminent Neapolitan physician, who served as chief physician (*archiatrus* or *protomedicus*) of the kingdom of Naples and, as such, was responsible for overseeing and regulating the full range of medical practitioners within the Neapolitan realm. This range included not only formally educated physicians, but also barber surgeons and pharmacists; for further details on this office, see David Gentilcore, "Il Regio Protomediato nella Napoli Spagnola," *Dynamis* 16 (1996): 219–36, esp, 219–20; see also Gentilcore, *Healers and Healing in Early Modern Italy* (Manchester, UK: Manchester University Press: 1998), 29–55. Pisano was among a small group of scholars commissioned by Della Porta's father to tutor him and his brothers in all the arts and disciplines suitable for a gentleman of the era.

2. The Neapolitan gymnasium was one of many such institutions founded throughout Italy during the Renaissance to prepare advanced students for entry into a university, the curricular focus being on humanistic studies, with particular emphasis on Latin and Greek. Giovanni Pisano held his teaching position until at least the mid-1580s and continued to practice medicine until his death, most likely in early 1589; on this dating, see Monica Calabritto, "Tasso's Melancholy and its Treatment: A Patient's Uneasy Relationship with Medicine and Physicians," in Yasmin Haskell, ed., *Diseases of the Imagination and Imaginary Disease in the Early Modern Period* (Turnhout: Brepols, 2011), 201–27. Della Porta's reference to the "Attic" (or Greek) versus the "elegant Roman" style of lecturing reflects a dispute over rhetoric that came to a head during the last quarter of the sixteenth century. On the one hand, there were those—sometimes characterized as anti-Ciceronian—who preferred the Attic style for its relatively straightforward simplicity and plainness, whereas, on the other, there were those, like Della Porta, who preferred the more ornate and elaborate Roman style opposed by the anti-Ciceronians. For a classic account of this dispute and its grounds, see Morris Croll, "'Attic Prose' in the Seventeenth Century," *Studies in Philology* 18 (1921): 79–128.

3. A notorious womanizer, John of Austria (1547–78) was the illegitimate half-brother of Phillip II (1527–98), King of Spain, as well as of Naples and Sicily. Starting in 1569, he served in various military capacities for Phillip and soon after distinguished himself as commander-in-chief of the Christian fleet that defeated the Ottoman Turks at the Battle of Lepanto in 1571. The viceroys to which Della Porta refers governed the Neapolitan kingdom under the aegis of the kings of Aragon and, later, of Spain, who had ruled that kingdom since 1504.

4. That the *De Refractione* was well underway by 1589 is clear from several explicit and specific references to it in his *Magiae Naturalis*, which was vetted and approved for religious orthodoxy in 1588 and granted its

imprimatur by the Vicar General of Naples in 1589. Giovanni therefore must have died right around that time.

5. This refers to Alessandro Farnese, who became Duke of Parma upon the death of his father in 1586 and served as Governor General of the Spanish Netherlands at the behest of King Phillip II from 1578 until his death in December of 1592. Giulio must have died around the same time, and certainly before March of 1593, when the book was granted its imprimatur.

6. Phillip II bore the honorific title *rex catholicissimus* ("most Christian king"), which was conferred on the Spanish monarchs by Pope Alexander VI in the very late fifteenth century.

7. Born in 1575, Ottavio moved to the Spanish Netherlands sometime in the early 1600s, at which time he was engaged in cartographic studies and, more specific, the problem of map projection. In 1612, under the rubric *Octavii Pisani Globus Terrestris Planisphericus*, he published a circular world map roughly 63 inches (160 cm) in diameter. The following year he published *Astrologia, seu Motus et Loca Sidereum* (Antwerp, 1613), a treatise of nearly ninety pages that deals with planetary motions and contains a number of elaborate volvelles. At around the same time he entered into limited correspondence with Galileo and Kepler. The last notice we have of him is a letter of 1637, which establishes the *terminus a quo* for his death. For a biography of Ottavio that includes some of his correspondence, see Antonio Favaro, "Amici e corrispondenti di Galileo Galilei: Studi e ricerche (II. Ottavio Pisani)," *Atti del Reale Istituto Veneto di Scienze, Lettere ed Arti* 54 (1895–96): 411–40.

8. Della Porta is referring here to the observational displacement of celestial bodies along the vertical caused by atmospheric refraction, especially near the horizon.

9. Della Porta is following standard usage in ascribing refraction to the relative density or transparency of the refracting medium. In neither case, however, is he referring specifically to the physical quality of density/weight or translucency. Although somewhat milky water may be significantly less transparent than clear water, it is not necessarily more refractive on that account. Likewise, clear glass may be more transparent than somewhat milky water, yet it is still more refractive. The link between refractivity and density seems less problematic, at least among the three transparent media that are standard among ancient and medieval optical thinkers—that is, air, water, and glass—air being the rarest and least refractive, glass the densest and most refractive. One might of course argue that the denser the medium is, the more it impedes light, from which it would follow that relative opacity is a function of relative density. The problem with this correlation is that many opaque substances, such as wood, are less dense than such transparent media as water or glass. In fact, what Giambattista Della Porta and his predecessors refer to as "density" in this context is called optical density in modern optical parlance.

10. According to Aristotelian cosmology, the celestial realm extending from the Moon to the outer edge of the sphere containing the universe

consists of the element aether, the fixed stars and all the other heavenly bodies being a condensed form of this element; see, for example, Aristotle, *On the Heavens*, II, 7. For the allusion to the heavens as composed of (or at least like) polished metal, see Job 37:18.

11. This is due to atmospheric refraction, which is greatest at the horizon and tapers off dramatically toward zenith. Accordingly, when a star appears right at the horizon it can actually lie a bit more than half a degree below it, the amount by which its light is refracted through the atmosphere at that point, and certain atmospheric conditions can increase that amount by a factor of nearly eight. Meantime, the light from another star lying a degree or so above it will be refracted much less. The two stars will therefore appear to lie closer to one another than they actually do, so when they rise toward zenith, where no atmospheric refraction occurs, the apparent distance between them will appear to lengthen.

12. Della Porta will explain this rather cryptic point in proposition 14 below.

13. Della Porta is adverting to the so-called Moon Illusion, according to which both the Sun and Moon appear much larger at horizon than at zenith, even though they subtend the same visual angle at both locations. As is clear from his prefatory remarks here, Della Porta will attempt to explain this phenomenon on the basis of refraction and, more specifically, refractive magnification.

14. In *Meteorology*, I, 6–8, Aristotle explains both the Milky Way and comets as phenomena occurring in the stratum of fire at the upper boundary of the sublunar atmospheric shell. Several commentators, including Olympiodorus (with whose work Della Porta was familiar), argued against Aristotle that the Milky Way is a supralunar, or celestial, phenomenon rather than a sublunar, or terrestrial, one; see Olympiodorus, *Meteors*, I, 9, in Olympiodorus (1551), 19r. As to comets, Tycho Brahe's observations of the comet of 1577 showed that it, too, was supralunar; these observations were published in *Tychonis Brahe Dani de Mundi Aetherei Recentioribus Phaenomenis* (Uraniburg, 1588). Not everyone was swayed by Tycho's argument, but evidently Della Porta was.

15. During the Renaissance and well into the Enlightenment, solar and lunar eclipses, as well as comets, were viewed as portents, generally dire ones. This is especially the case with comets, which were often associated with plague. The so-called Blood Moon is a prime example of how color could give the phenomenon an especially ominous cast. In the ninth and final book of this treatise, Della Porta will offer a general explanation for such coloring on the basis of his theory of refraction through dense vapor.

16. According to Greek mythology, the sea-God Thaumas (Θαύμα = "wonder") fathered *Iris*, the goddess of the rainbow; see, for example, Plato, *Theaetetus*, 155d. The Latin term I have translated as "streaks" in this sentence is *virgae*, which refers to the shafts of rain descending toward the ground from clouds seen at a distance, but it also includes other vapor-based phenomena that appear along more or less straight lines in the sky.

17. By Della Porta's day it was common knowledge that both concave spherical and paraboloidal mirrors can focus light to a burning point and, moreover, that paraboloidal mirrors are more effective because they bring light to an actual point focus rather than to a focal area. Later, in book 2, Della Porta will explain how sunlight reflected from a concave spherical mirror or refracted through a glass sphere can bring the light to an effective burning focus.

18. As Vincent Ilardi has shown, by Della Porta's day, and indeed well over a century before, eyeglasses with convex lenses were being produced almost to prescription for presbyopia, and eyeglasses with concave lenses were being produced for two grades of myopia; see Ilardi, *Renaissance Vision from Spectacles to Telescopes* (Philadelphia: American Philosophical Society Press, 2007), 75–95.

19. The Latin *simulachrum*, which is translated as "image" here, actually means something like "representative likeness" in the context of Della Porta's overall analysis of vision. In the very next definition, however, he uses the Latin term *imago* in precisely the same way: that is, to designate what is propagated from a physical object through the transparent medium to the eye. On occasion, Della Porta also uses the Latin terms *forma*, *species*, and even *spectrum* to denote the same thing. Because he uses these terms interchangeably, I have chosen to translate all of them with the English term "image," thus emphasizing that what passes from object to eye is a formal representation that is somehow like its originating object in the way that a portrait is somehow similar to its physical subject. Roger Bacon, to whom Della Porta refers several times in this treatise, may well be the source for his conflation of all these terms. See, for example, Bacon, *De multiplicatione specierum*, I, 1, in Bacon (1983), 2–3: *Et hec ... habet multa nomina, vocatur enim similitudo agentis et ymago et species et ydolum et simulacrum et fantasma et forma et intentio et passio et impressio et umbra philosophorum apud auctores de aspectibus* ("And this ... has many names, for it is called 'the similitude of the agent,' 'image,' 'species,' 'idol,' 'simulacrum,' phantasm,' 'form,' 'intention,' 'passion,' 'impression,' and 'shadow of the philosophers' by authors of works on vision").

20. Della Porta is following optical tradition, particularly that established by Alhacen and his Latin Perspectivist disciples, in linking transparency to the rarity or density of the medium (see note 9 above). Accordingly, the denser the medium, the less transparent it is, and the less transparent it is, the more refractive it is. Understood in these terms, then, air is more transparent than water and water more transparent than glass.

21. In short, the normal is orthogonal to the refracting surface at the point of refraction, no matter whether that surface is plane or curved, Della Porta's point about the normal in water is based on the Aristotelian cosmological scheme in which earth, water, air, and fire form nesting spheres, all of which are concentric with the center of the universe. Accordingly, because the normal must be perpendicular to the water's surface, and because the water's surface is spherical, even though it may look perfectly flat according

to our limited horizon, the normal must pass through the center of curvature of that surface—that is, the center of the universe and, coincidentally, the center of the Earth.

22. In current parlance, this is the angle of deviation, the angle of refraction being formed by the refracted ray and the normal through the point of refraction. In designating the angle of deviation as the angle of refraction, Della Porta seems to be following Alhacen and Witelo's usage; see, for example, Alhacen, *De aspectibus*, VII, iii, 10, in Alhacen (1572), 243–44, and Alhacen (2010), 41–5 (Latin) and 251–53 (English); see also Witelo, *Perspectiva*, X, 8, in Witelo (1572), 412–13. It is worth noting that in the actual medieval Latin text of Alhacen's *De aspectibus*, the Latin for *both* reflection and refraction is *reflexio*. In his 1572 edition of the *De aspectibus*, Risner "corrects" *reflexio* to *refractio* wherever "refraction" is meant. Witelo, on the other hand, does use *refractio*, but in his tables of refraction on p. 412 of Risner's 1572 edition, *angulus refractionis* refers to the angle of deviation, whereas *angulus refractus* refers to the angle of refraction as we currently designate it.

23. Shining sunlight through the sighting holes in the alidade when it is posed along the perpendicular, as represented by AB in Figure 1.1, can only be done once, or at most twice, a year, and only when the experiment is done between the Tropics of Cancer and Capricorn. In order to have actually carried out this phase of the experiment, then, Della Porta would have had to be at a latitude below that of Aswan in upper Egypt, much farther south than he ever traveled in his life. It is highly doubtful, therefore, that Della Porta carried this experiment out with sunlight; rather, he is reporting the result that would be expected according to theory.

24. Although Della Porta consistently calls this component of the astrolabe the "rule" (*regula*), it is clear he is referring to the alidade, which is unlike the rule in having sighting-holes or slits.

25. Della Porta seems to be alluding here to Alhacen's explanation in book 7 of the *De aspectibus* of why light is shunted toward the normal on passing from a rare into a denser medium. This explanation is based on the supposition that for light, as for a physical projectile, passage through a resisting interface along the perpendicular—that is, directly through—is easier and stronger than along any other path. Accordingly, when light strikes the surface of a denser medium at a slant, if it were to continue through at that slant, it would follow a weak path. It is therefore shunted toward the normal in order to recoup the potential loss of power it would suffer. See *De aspectibus*, VII, ii, 8, in Alhacen (1572), 240–42, and Alhacen (2010), 33-7 (Latin) and 244-47 (English).

26. This experiment with the astrolabe is clearly based, albeit somewhat loosely, on the experiments Alhacen conducts with a pan-shaped apparatus through whose rim a hole is pierced to allow a beam of sunlight to pass straight through. Poised upright in water so that it can be rotated freely, this apparatus will allow sunlight to pass through the hole in the rim at various

angles and to refract at various corresponding angles through the water to the other side of the pan. Unlike Della Porta's apparatus, however, Alhacen's is carefully designed to screen out extraneous sunlight so that the narrow beam passing through the holes can be clearly seen shining on the other side of the pan's rim. For the actual text that describes the design and application of Alhacen's experimental apparatus, see *De aspectibus*, VII, ii, 2–7, in Alhacen (1572), 231–40, and Alhacen (2010), 5–28 (Latin) and 220–43 (English). See also Witelo, *Perspectiva*, II, 1 and 42–6, in Witelo (1572), 61–3 and 76–81, and Witelo (1991), 235–39 and 269–84 (Latin) and 40–5 and 80–94 (English).

27. The Latin term I have translated as "denser" is *hebetior*, which connotes dimness, as opposed to lucidity, and is meant to imply a diminished transparency according to increased density as opposed to rarity. Accordingly, water is denser than air, but rarer, or less dense, than glass.

28. What Della Porta refers to as angles of entry and exit are equivalent to the angles of deviation mentioned in note 22 above: that is, the angles formed by the continuation of the incident ray and the refracted ray. Accordingly, in Figure 1.3, if DB is the incident ray and BF the refracted ray, then DBG will be the angle of entry and EBF the angle of exit, both angles obviously being equal since they are formed by the intersection of incident ray DB and its rectilinear continuation DB with refracted ray BF and its rectilinear continuation BG.

29. Where Della Porta got the figure of 35° for the angle of refraction when light passes from air to water at an angle of incidence of 45° is a mystery. According to the sine law, that angle should be just over 32°, which is just about where it should be if Della Porta had extrapolated from the tabulations provided by Witelo in *Perspectiva*, X, 7, in Witelo (1572), 412. Witelo's tabulations, which are based on 10°, not 5°, increments for the angles of incidence, were taken directly from Ptolemy's tabulations in book 5 of the *Optics*, in Ptolemy (1996), 233. It seems unlikely, therefore, that Witelo or Ptolemy was the source for Della Porta's figure of 35°. More likely is that he pulled it out of relatively thin air for the purposes of illustrating his point without regard to exactitude of measure.

30. This "floating coin" experiment is hinted at in the last postulate of Euclid's *Catoptrics* (see Euclid [1557a], 46; see also Euclid [1895], 286–87) and fully explicated geometrically by Ptolemy in book 5 of his *Optics* (see Ptolemy [1996], 230–31). Alhacen describes this same experiment without a geometrical diagram in *De aspectibus*, VII, v, 17, in Alhacen (1572), 253, and in Alhacen (2010), 70 (Latin) and 274-75 (English). See also Witelo, *Perspectiva*, X, 11, in Witelo (1572), 414–15.

31. Della Porta seems to be following the line of reasoning on which Hero of Alexandria's least-lines proof of the equal-angles law is based. In that proof, which occurs in propositions 4 and 5 of the *Catoptrics*, Hero demonstrates that, of all possible combinations of incident and reflected rays linking a center of sight, a point object, and a reflecting surface, the ray

couple that links them according to equal angles of incidence and reflection is the shortest possible. This, then, serves as proof that nature acts as efficiently as possible, which in turn requires that light travel in perfectly straight lines to get from point to point. The physical reason, according to Hero, is that the visual flux travels too swiftly to deviate from a rectilinear trajectory, which implies that it travels with maximal swiftness (i.e., in an instant) in order to be minimally curved. For the textual basis of these points, see Hero (1900), 320–30. Della Porta almost certainly got Hero's reasoning at second hand through Witelo's *Perspectiva*, V, 5 and 18–19, in Witelo (1572), 192 and 198–200, and in Witelo (1984), 195 and 207–9 (Latin) and 90–1 and 101–3 (English). See, however, Hero of Alexandria (1518), 250vb–251rb.

32. This illustration is not really to the point, because Della Porta should have shown how point A at the bottom of the ruler appears to rise as the eye is drawn ever farther away from point M. Consequently, the image O of point I is higher than image B of point A, as point I itself lies above point A. It is not clear who the "predecessors" mentioned at the beginning of the proposition are. Surely one of them is Alhacen, whose empirical demonstration that the image always lies on the cathetus is somewhat akin to Della Porta's initial demonstration in which the astrolabe is sunk little by little into the vessel full of water; see *De aspectibus*, VII, v, 18, in Alhacen (1572), 253–55, and Alhacen (2010), 70–73 (Latin) and 275–77 (English). See also Witelo, *Perspectiva* X, 15, in Witelo (1572), 416–18.

33. In other words, angles OGL, PTM, and QIN are equal to the corresponding angles of incidence formed by rays CG, DT, and EI at points G, T, and I. Thus, even though angles OGL, PTM, and QIN look equal in the diagram in the original Latin text, I take Della Porta to characterize them as "equal" in the first part of the sentence according to their equality to the angles of incidence, which increase in size according to the increasing distances of points C, D, and E from B.

34. As pointed out in note 22 above, Della Porta calls angles of refraction what we today call angles of deviation. Thus, angles CGH (= OGA), DTR (= ATP), and RIS (= AIQ) are angles of deviation, and as Della Porta concludes, they increase with the distance of the viewpoint from the point of refraction, as in fact do the actual angles of refraction as measured by vertical angles OGL, PTM, and QIN.

35. See *Perspectiva*, X, 14, in Witelo (1572), 415–16; Witelo argues mistakenly that because B, C, D, and E are equidistant from one another, angles OGL, PTM, and QIN are equal, which means that OC, PD, and QE are parallel, from which it follows that G, T, and I are equidistant from one another. Oddly enough, by representing OC, PD, and QE as parallel so that GT = TI, the figure that accompanies the proposition in Della Porta's original, printed text illustrates Witelo's point and thus contravenes Della Porta's point: i.e., that GT ≠ TI.

36. In essence, Della Porta is claiming that QIN and MTP, which are equal to the angles of incidence formed by rays EI and DT, respectively, at

points I and T, are to one another in the same ratio as the corresponding angles of deviation QIA and PTA, which is in fact false.

37. In other words, because the line of sight along the horizon passes through the atmospheric shell, which consists of air thickened by dense, rising vapors, light passing through that shell from the celestial regions beyond is refracted toward the normal, that is, the line from the center of the Earth/universe through the point of incidence/refraction.

38. This claim is somewhat inapposite because in the case of the coin experiment in proposition 4, the light passed from a denser into a rarer medium, whereas in this case the passage is from a rarer into a denser medium.

39. *Problems*, XXVI, 53, 946a33–34, in Aristotle (1552 [as XXVI, 55]), 43ra. See also *Meteorology*, II, 6, 364b25–28, in Aristotle (1550b) 196va, where Aristotle attributes the accumulation of atmospheric vapors not to Eurus but to Caecius, another east wind.

40. See *Almagest*, I, 2, in Ptolemy (1541), 3b, and al-Farghānī, *Elements of Chronology and Astronomy*, II, in al-Farghānī (1590), 18.

41. *Problems*, I, 36, in Alexander of Aphrodisias (1552), 93va.

42. Seneca, *Natural Questions*, I, vi, 5–6, in Seneca (1529), 418; English translation adapted from Seneca (2010), 152.

43. *Saturnalia*, VII, xiv, 1–2, in Macrobius (1535), 326, lines 15–27; English translation from Macrobius (2011), 274–77.

44. See *Optics* 5.13, in Ptolemy (1996), 257. See also Ptolemy, *Almagest*, I, 2, in Ptolemy (1605), 12; Theon's commentary to *Almagest*, I, 2, in Ptolemy (1605), 18; Alhacen, *De aspectibus*, VII, vii, 39–43, in Alhacen (1572), 271–74, and in Alhacen (2010), 116–23 (Latin) and 309–14 (English); and Witelo, *Perspectiva*, X, 31–35, in Witelo (1572), 431–36. Although in the *Optics*, Ptolemy does explain magnification on the basis of an enlarged visual angle when things are seen through a plane refractive interface, Alhacen and Witelo do not. Appealing instead to the weakening of light by refraction, they argue that an object seen through the interface between air and water or glass will appear dimmer than it actually is. Since the dimmer of two objects of equal size appears farther away and is thus judged to be larger, the refracted image is judged to be larger than its original object because of its apparent increased distance from the viewer. It should be noted that, since no printed version of Ptolemy's *Optics* existed until Gilberto Govi's edition of 1885, Della Porta must have had access to it through manuscript.

45. Implicitly, at least, Della Porta is appealing here to the size–distance invariance hypothesis, which assumes that a given object looks the same size at various distances despite its being viewed under various visual angles. It is because Alhacen (and Witelo following him) subscribes to this same principle that he looks to refractive dimming of the image and the resulting misperception of size according to misperceived distance in order to explain why things seen under water look magnified (see note 44 above). Moreover, Della Porta has shown in the previous proposition that, if object CB at the bottom of the vessel is sizeable enough, its image will not be straight line

KL but, rather, a concave line whose endpoints will be raised to K and L while its center lies below at F.

46. This "demonstration" hardly refutes the assumption that the apparent magnification of an object seen in water is commensurate with the increase in the size of the visual angle due to refraction. All it shows is that the disparity between the "direct" and "refracted" visual angles increases with the increasing density of the medium below surface ED because the resulting refraction brings the image commensurately closer to the eye. Implicit in this, of course, is the assumption that, even if LK may look larger than BC, it will not look twice as large.

47. As represented properly in the diagram, angle MON does not lie in the circle with diameter MN, which forms the image of the circle at the bottom with diameter LI. Consequently, the fact that it is a right angle is not due to its lying within that circle. However, since LN and IM are both constructed perpendicular to LI and therefore parallel, and since OM is constructed parallel to LI, then angle NOM = right angle LIM.

48. *Meteorology*, III, 3, in Aristotle (1550b), 206ra, lines 17–18.

49. See *Perspectiva*, III, ii, 3, in Bacon (1996), 317. Contrary to Della Porta's claim, Alhacen, not Ptolemy, is Bacon's source for this account of distance perception according to intervening objects; see Alhacen, *De aspectibus*, II, iii, 25, in Alhacen (1572), 39–42, and in Alhacen (2001), 129–44 (Latin) and 451–57 (English). Since none of Bacon's works were committed to print before Johannes Combach's *Rogerii Baconis . . . Perspectiva* (Frankfurt, 1614), Della Porta must have had access to Bacon's optical work either directly through manuscripts or indirectly through secondary sources.

50. See *De Subtilitate*, 3, in Cardano (1551), 148; see also Cardano (2013), 209.

51. I take Della Porta to be arguing here that the size–distance invariance principle does not necessarily hold so that, even when the two objects are equal, the one that appears to lie farther away may also be perceived as smaller.

52. Latin quotations from Julius Caesar Scaliger, *Exotericarum Exercitationum Liber XV*, LXIIII, 2, in Scaliger (1582), 243.

53. Della Porta leaves unsaid that N will be the image of point L, although it is certainly implicit in the narrative and, more to the point, explicit in the diagram. Consequently, it is evident that, because diameter NO of the star's image is longer than diameter ML of the star itself, the star will appear magnified. In fact, Della Porta's conclusion is incorrect; atmospheric refraction will actually cause the images of celestial objects to diminish in size along the vertical, although imperceptibly so, a point Alhacen demonstrates in book 7 of the *De aspectibus*; see *De aspectibus*, VII, vii, 52–54, in Alhacen (1572), 278–80, and in Alhacen (2010), 131–36 (Latin) and 321–25 (English). Nonetheless, later in *De aspectibus*, VII, vii, 72–73, he claims that thick atmospheric vapors can cause the celestial body to appear somewhat magnified; see Alhacen (1572), 282, and in Alhacen (2010), 143 (Latin) and 330–32 (English).

54. See *De Subtilitate*, XIII, in Cardano (1551), 461–62; see also Cardano (2013), 704–5.

55. As Della Porta presents it, this geometrical description fails to illustrate his point. What he is trying to explain can be understood by recourse to

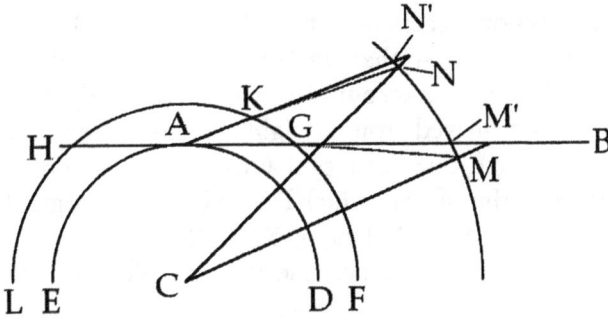

this diagram, which is based on Figure 1.14. Let A be the eye on DAE, the surface of the Earth, HAB the horizon, FGL the outer surface of the atmospheric shell, and MN' an arc on the celestial sphere. To start with, let the star lie below the horizon at point M on the celestial sphere, and let ray MG be refracted at the atmospheric shell along GA. The viewer at A will therefore see the star at M' on arc MN'. Now let the star rise to point N on the celestial sphere, and let ray NK from it be refracted at the atmospheric shell along KA. Consequently, the viewer at A will see the star at N', just above N on arc MN'. Moreover, as the angle of incidence at K is less than the angle of incidence at G, light from the star at N will be refracted less than the light from star M. As a result, the arc MN through which the star actually travels will be greater than arc M'N' through which it appears to travel. Given that the two arcs are traversed in the same time, then, it follows that the apparent motion through smaller arc M'N' will be slower than the real motion through MN. Incidentally, if we take MN as the diameter of a celestial body and M'N' as the diameter of its image, then it is clear from this demonstration that, contrary to Della Porta's claim in proposition 12, atmospheric refraction makes the body look somewhat smaller than it is along the vertical.

56. *De architectura*, VI, ii, 2–3, in Vitruvius (1586), 223; English translation from Vitruvius (1999), 78.

57. See, for example, Seneca, *Natural Questions*, I, iii, 9, in Seneca (1529), 415.

58. In order to make Della Porta's point in this proposition as clear as possible, I have taken liberties with the labeling of points in Figure 1.16a, which is why "AECD" and "GMF" do not reflect their equivalents in the Latin text.

59. See Sacrobosco, *Sphere*, I, in Sacrobosco (1578), 34. The gist of Sacrobosco's argument is that, when viewed from a ship at sea pulling away from it, the shore appears to sink below the horizon because the surface of the sea, being concentric with the Earth's surface, is spherical.

60. In essence, thirty thousand paces *millia passuum* is thirty "miles," which is often simply given by *millia*.

61. I take Della Porta's point to be that, if the curvature of the Earth alone were responsible for the apparent disappearance of a ship's stern below the horizon, then that would happen at a distance of twenty or thirty miles, whereas in reality the stern can appear to sink below the horizon at a significantly smaller distance.

62. This illustration necessitates that the viewpoint lie well above the level of point A on the ship so that ray AO will strike the refractive interface EOF at an angle. It does not account for why point A might appear to sink to the horizon when viewed from straight on—as, for instance, from the shore—unless, of course, one assumes that the ship "descends" according to the Earth's curvature, the effect of which Della Porta has essentially dismissed.

63. See *Almagest*, I, 4, in Ptolemy (1541), 4b.

64. This point is irrelevant because the ship does not lie below the refracting interface but, rather, to the left of it.

65. In general, when Della Porta refers to the pupil, he means the actual opening in the iris. At times, however, he reverts to older usage, according to which the pupil is the visually sensitive part of the eye, which Della Porta takes to be the crystalline lens.

66. Actually, the amount of refraction from seawater into the eye is fairly negligible because both the aqueous humor and seawater have nearly the same index of refraction, 1.33 for the former and around 1.34 for the Mediterranean, depending on salinity. Consequently, the magnification of things viewed by an eye submerged in seawater would be imperceptible.

67. As described by Vitruvius in *On Architecture*, VIII, v, 1, a chorobates consists of a wooden "ruler about twenty feet long and mounted on legs. Its use for leveling depends on plumb lines at both ends and a relatively short channel for water in the middle. If conditions are such that the plumb lines cannot be trusted, the channel can be filled to the top with water, and when the water fills it evenly, the chorobates is level"; for the Latin, see Vitruvius (1586 [as VIII, vi, 1]), 338–39; English translation from Vitruvius (1999), 243.

68. The diagram that accompanies this explanation does not illustrate the point because ray AO strikes the refractive interface OC along the perpendicular, so it will not be refracted. In addition, it represents a peculiar situation in which the instrument itself is in the thicker medium while the eye lies just outside it. The basic point is somewhat better illustrated by Figure 1.17, where, after being refracted at O, ray AO will give the false line of sight IO that leads to a point below A.

69. *Natural Questions* I, xii, 1, in Seneca (1529), 419; English translation from Seneca (2010), 155. The argument Della Porta makes here is inapposite to the point of this proposition, since it involves reflection rather than refraction.

70. Aratus of Soli, *Phaenomena/Apparentia*, 799–804; see Aratus (1570), 238, lines 32–37 (Greek), and 239, lines 32–37 (Latin). It should be noted, though, that the Latin text in this version of Aratus differs from that given

by Della Porta, the specific source of which I have been unable to find; cf. however, Caspar Bartholin, *Astrologia, seu de Stellarum Natura* (1612), 21–22, where the last three lines of the Latin poem as it appears in Della Porta's text are given verbatim.

71. See *On the Heavens*, II, 8, 290a17–24, in Aristotle (1550b), 61rb, lines 19–33. The term *visus*, which I have translated as "visual power," refers here to what is radiated out from the eye, whether in the form of inherent light or fire.

72. In Della Porta's day Saturn was considered to be the outermost planet, and judging by the almost thirty years it takes to orbit the Sun—as compared to twelve years for Jupiter and three years for Mars—it was assumed not only to lie commensurately far away but also to be relatively close to the sphere of the fixed stars that form the outer limit of the universe.

73. These "others" seem to be the ones against whom Aristotle argues, at least implicitly, in *On the Heavens*, II, 8, when proposing his account of stellar twinkling on the basis not of the intrinsic motion of the stars but rather on the wavering of the visual rays reaching so far out to them.

74. Della Porta is referring here to the piling of circles on circles in the detailed geocentric account of planetary motions based on the Ptolemaic system of epicycles and deferents and refined by various medieval Arabic astronomers with the addition of other more subtle circular motions. The point seems to be that, because so many of the motions on individual circles within a given planetary system are contrary, they ought as a whole to create the conditions for scintillation, whereas the stars do not partake in such complex motions. Consequently, the planets, not the stars, ought to twinkle most perceptibly, yet they do not.

75. See Pecham, *Perspectiva communis*, II, 56, in Pecham (1542), 91–92 (unnumbered); see also Pecham (1970), 208–11. The designation "Cantuarensis" in the Latin text comes from Pecham's having been Archbishop of Canterbury (Cantuaria) from 1279 to his death in 1292.

76. This example of pebbles trembling in a stream can be found in book 3 of Cardano's *De Subtilitate*; see Cardano (1551), 146; see also Cardano (2013), 207–8.

77. See *Natural Magic*, II, 18, in Della Porta (1560), 66v.

78. *Problems*, XV, 13, in Aristotle (1552 [as XV, 12]), 26vb.

79. The alternatives that Della Porta discusses to this point are all raised by Aristotle in response to the opening question of problem XV, 12. Della Porta's denial of atoms is consistent with his apparent acceptance of Aristotelian cosmology and physics, according to which the world and its constituent elements form a perfect continuum.

LIBER SECUNDUS

PROOEMIUM

De refractione iam visum est in praecedente libro subnectam speculationes quasdam de crystallina pila, scilicet quomodo solares radii in extimam eius superficiem proruentes suis statis locis refrangantur. Ab antiquis enim praecise neque loca neque rationes memoratae sunt. Haec enim pila oculi specimen refert, nec vera a nobis loca visionum in oculis decerni possunt, nisi de ea paucula quaedam praestrinxerimus, sic de visione sermonem absolvemus. Sed quia pilae refractio reciprocationem et convenientiam quandam cum concavi speculi reflexione habet, ideo de eius reflexione nonnihil attingemus; mox ad ea quae polliciti sumus revertemur, perpulchra etiam ad ignis accensionem ostensuri. Sed aliqua sunt nobis subroganda ab omnibus perspectivae authoribus accepta. (1) Solis radios aliquos ab eo emergentes sibi invicem parallelos esse. (2) Praeterea, radios veluti rectas lineas [36] se habere, et in geometricis demonstrationibus eam vim quam mathematicae lineae habent retinere. (3) Etiam solares radios in convexam, concavam, aut planam speculorum superficiem incidentes semper incidentiae angulos reflectionis aequales efficere. Iisque acceptis, iam negotium auspicemur.

Vera loca reflexionum solaris radii in speculo concavo sphaerico vestigare.
PROP. 1

Refractionum loca in crystallina sphaera exploraturis, consultius visum est vera loca in concavo speculo sphaerico determinare, unde de iis ordine referemus.

[1a] *Solis radius in speculo concavo sphaerico tangens latus*
exagoni reflectitur ad diametri calcem.

Sit concavum speculum DABC, et linea AB exprimat Solis radium B punctum tangentem circuli sextantem a diametri fine DC. Dico reflecti ad circuli

BOOK TWO

INTRODUCTION

To what has already been observed about refraction in the preceding book I will add some thoughts about the crystalline sphere, that is, how solar rays striking its outermost surface are refracted to their established locations.[1] For neither the locations nor the reasons [for the rays going there] have been accounted for with precision by the ancients. Indeed, this kind of sphere serves as a model of the eye, and the true locations of vision in the eyes cannot be determined by us unless we touch upon a few things about that sphere so as to complete our discussion of vision.[2] But since refraction in a glass sphere bears a certain reciprocal relationship and conformity with reflection from a concave spherical mirror, we will touch on this reflection somewhat; then we will turn to the things we promised and, most beautiful of all, show how it applies to the lighting of fire. But first certain things accepted by all the authors of *perspectiva*[3] must be laid down by us as follows. (1) Some of the rays emanating from the Sun are parallel to one another. (2) Rays, moreover, correspond to straight lines, [36] and in geometrical demonstrations they have the same explanatory force as mathematical lines. (3) Also, solar rays incident on the surface of convex, concave, or plane mirrors always form equal angles of incidence and reflection. With these points granted, let us broach our subject.

PROP. 1: *To trace out the precise location to which a solar ray reflects in a concave spherical mirror.*

In order for the locus of refraction in a glass sphere to be found, it seemed most prudent to determine the precise location of reflection in a concave spherical mirror, so we will account for these locations in proper order.

[1a] *A ray from the Sun that strikes a concave spherical mirror at the [endpoint of the] side of an inscribed hexagon is reflected to the bottom point of the diameter.*

Let DABC [in Figure 2.1a] be a concave spherical mirror, and let line AB represent a ray from the Sun that strikes point B one-sixth of the circle away

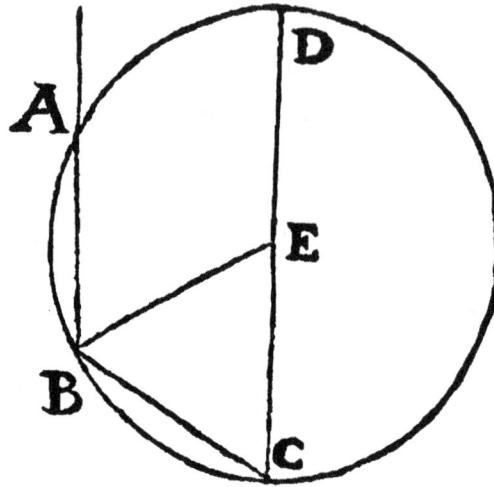

diametri calcem C. Porrigatur linea ex E ad B et ex B ad C. Quoniam triangulus EBC parium est laterum, per ea quae in 4 Euclidis libro ostensa sunt, angulus CEB par erit angulo EBC. Sed angulus CEB par est ei qui sub EBA, quia subalternus (nam incidentes solares radii cum speculi diametro in pari distantia mutuo supponuntur). Ergo CBE angulo EBA aequalis erit, et linea BC erit linea reflexionis, quia pares sunt anguli incidentiae et reflexionis. [37]

> [1b] *In reflexionibus speculorum concavorum sphaericorum,*
> *radius reflexus semper est aequalis diametri parti*
> *interceptae a circuli centro et utriusque concursu.*

Esto sphaericum concavum speculum CFBA; radius Solis incidens AB feriat punctum B; reflectatur ad diametri punctum E. Dico lineam reflexionis EB semper esse aequalem lineae DE, scilicet interceptae inter centrum D circuli et E contactum lineae reflexionis. Quoniam in reflexionibus speculorum semper incidentiae angulus ABD reflexionis[1] angulo aequalis est, et angulus ABD aequalis EDB, quia subalternus. Ergo angulus EDB aequalis est semper angulo DBE, et aequalibus angulis aequales bases oppositae sunt; igitur

---

[1]"reflexionibus"in original text.

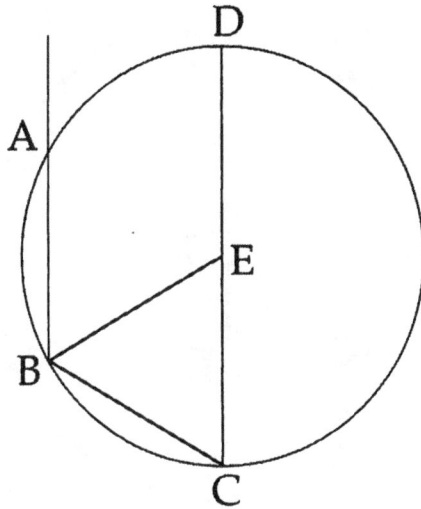

Figure 2.1a

from the endpoint of diameter DC [so that arc BC = 60°]. I say that it is reflected to bottom point C of the circle's diameter. Extend a line from [center of curvature] E to B and from B to C. Because triangle EBC is equilateral, angle CEB will be equal to angle EBC, according to what is demonstrated in proposition 4 of the [first] book [of the *Elements*] of Euclid.[4] But angle CEB is equal to the one formed by EBA because they are alternate angles (inasmuch as the incident solar rays are assumed to be equidistant from [and thus parallel to] the mirror's diameter). Therefore, angle CBE = angle EBA, and line BC will be the line of reflection, since the angles of incidence and reflection [EBA and EBC, respectively] are equal [according to supposition 3 above]. [37]

> [1b] *In reflection from concave spherical mirrors, the reflected ray*
> *is invariably equal to the segment of the diameter lying between*
> *the center of the circle and the intersection of both the reflected*
> *ray and the diameter.*

Let CFBA [in Figure 2.1b] be a concave spherical mirror; let incident ray AB from the Sun extend to B, and let it be reflected to point E on the diameter. I say that line of reflection EB will always be equal to line DE, that is, the line cut off between center point D of the circle and point E of contact between the line of reflection and the diameter. For in reflections from mirrors the angle of incidence ABD is always equal to the angle of reflection [EBD], and so angle ABD = angle EDB, because they are alternate angles. Consequently, angle EDB is invariably equal to angle DBE, and [therefore, in isosceles triangle DEB] equal bases [BE and DE] are opposite equal angles [BDE and DBE], so line of reflection BE will invariably be equal to the

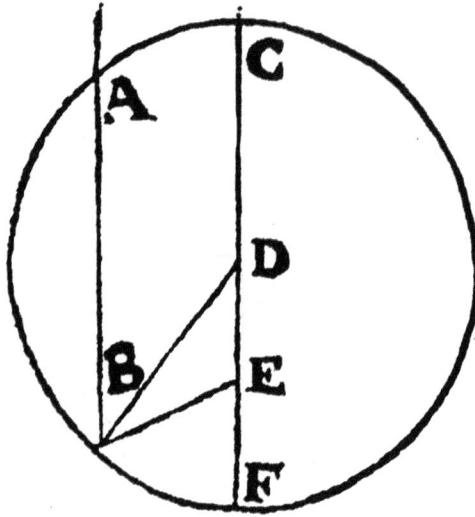

semper reflexionis linea BE par erit semidiametri parti a centro ad utriusque contactum E.

   [1c] *In reflexionibus speculorum concavorum, radius Solis incidens super latus octogoni reflexus cadit super diametrum perpendiculariter.* [38]

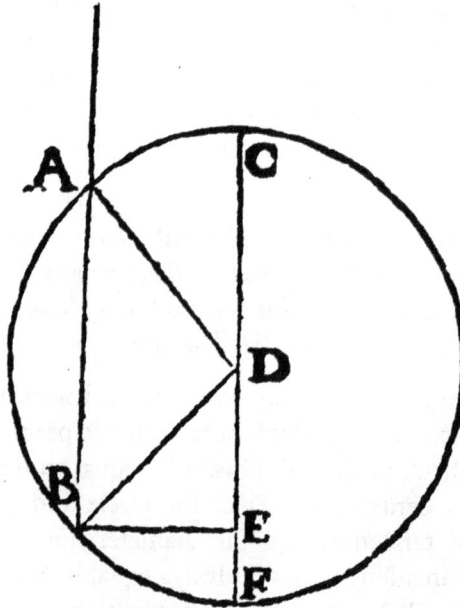

Esto Solis radius AB tangens punctum B speculi concavi, distantem a diametri fine F per octavam circuli partem. Dico radium BE reflexum super semidiame-

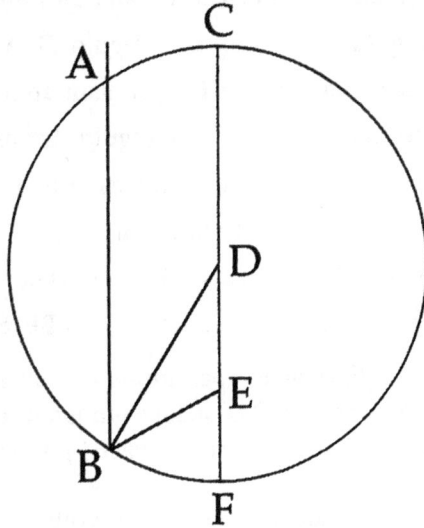

Figure 2.1b

segment of the radius from the circle's center to point E, where both [reflected ray BE and the diameter make] contact.

> [1c] *In reflection from concave spherical mirrors, a ray from the Sun that strikes [the endpoint of] the side of an inscribed octagon falls orthogonally to the diameter when reflected.* [38]

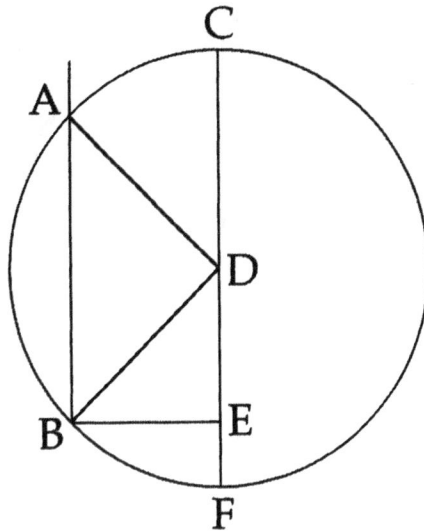

Figure 2.1c

Let AB [in Figure 2.1c] be a ray from the Sun that strikes a concave spherical mirror at point B, which lies a distance of one-eighth of the circle from end point F of the diameter [so that arc BF = 45°]. I say that reflected ray BE is

trum DF esse perpendicularem. A centro D extendito lineas ad puncta A, B, et a puncto B fiat angulus DBE aequalis angulo DBA, et extendatur BE quousque coeat cum semidiametro in E. Quoniam angulus BDA rectus est (quia quadrati diametri in centro ad rectos angulos decussantur), et DA, DB aequales sunt, ergo anguli ad A et B aequales sunt et semirecti, et DBE aequalis est DBA, quia reflexionis. Ergo semirectus, et semirectus quoque EDB, quia subalternus DBA. Remanet ergo DEB rectus per ea quae constituunt ternos trianguli angulos binis rectis aequales. Ergo BE perpendicularis.

[1d] *In reflexionibus speculorum concavorum*
sphaericorum Solis radius quanto sub latere exagoni
inferius inciderit, tanto supra a diametri fine reflectitur.

Quanto enim radius Solis incidens sub latere exagoni inferius tetigerit, tanto reflexus radius a diametri fine supra ascenderit; cuius ratio est quod incidens radius quanto magis diametro haerebit, semidiameter ex centro tangens punctum incidentiae et dividens utrinque angulum incidentiae et reflexionis tanto magis anguli angustabuntur, et reflexus radius supra ascendit.

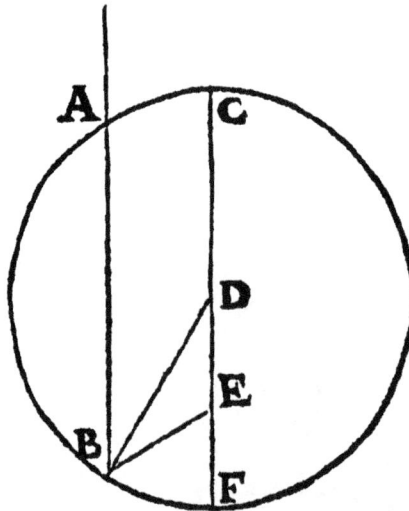

Esto incidens radius AB tangens punctum B, latus videlicet duodecagoni. Trahatur semidiameter[2] DB, et a puncto B [39] fiat angulus DBE aequali DBA. Dico lineam BE ascendere supra latus octogoni, quod erat perpendicu-

---

[2] "diameter"in original text.

perpendicular to radius DF. Extend lines from center point D to points A and B, form angle [of reflection] DBE at B equal to [angle of incidence] DBA, and extend BE until it intersects radius [DF] at E. Because BDA is a right angle (because diagonals of a square cut each other in the center at right angles), and since radii DA and DB are equal, the angles [DAB and DBA] at A and B are equal and half a right angle each, and DBE = DBA because [DBE is the angle] of reflection.[5] It is therefore half a right angle, and EDB is also half a right angle, since it is alternate to DBA. If follows therefore that DEB is a right angle insofar as the three internal angles of a triangle combined are equal to two right angles. Therefore, BE is perpendicular [to DF].

> [1d] *In reflection from concave spherical mirrors, the farther below [the endpoint of] the side of an inscribed hexagon a ray from the Sun strikes the mirror, the farther above the endpoint of the diameter it is reflected.*

Indeed, the farther below [the endpoint of] the side of an inscribed hexagon a ray from the Sun strikes the mirror, the higher the reflected ray ascends from the endpoint of the diameter. The reason for this is that the closer the incident ray lies to the diameter, the more the radius from the center that intersects the point of incidence and divides the angle of incidence and the angle of reflection [into equal halves] will narrow the angles and the higher up the reflected ray reaches [to intersect the diameter].

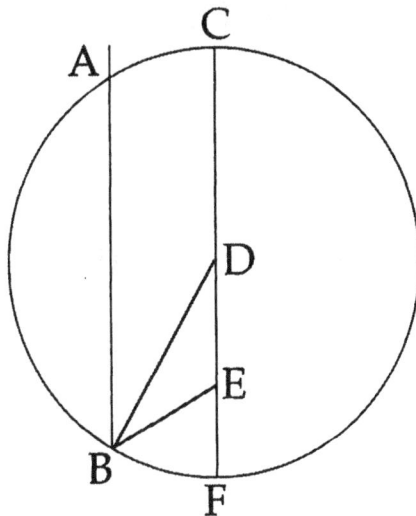

Figure 2.1d

Let incident ray AB [in Figure 2.1d] strike point B, that is, [the endpoint of] the side of an inscribed twelve-sided figure [so that arc BF = 30°]. Draw radius DB, and at point B [39] form angle [of reflection] DBE equal to [angle of incidence] DBA. I say that line BE will fall higher [on diameter CDF] than [it did when solar ray AB was incident at the endpoint of] the

lare supra DF. Angulus reflexus exagoni erat 2/3, duodecagoni vero 1/3, scilicet dimidium. Sic octogoni 1/2, sedecagoni vero 1/4, et sic angulorum cuspides semper angustabuntur.

[1e] *In reflexionibus speculorum concavorum, reflexionis linea non ascendet ultra quartam diametri.*

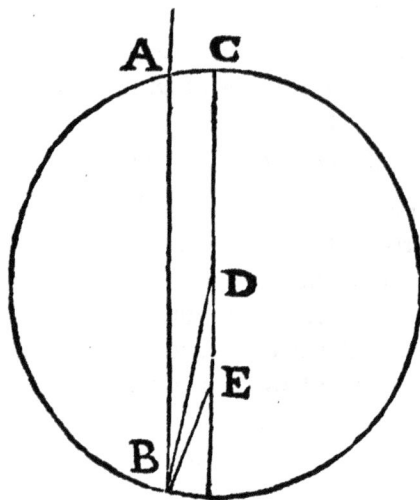

Esto speculum CAB, incidens radius Solis AB reflectatur in E. Dico E esse infra quartam partem diametri, nec unquam posse ultra quartam diametri partem ascendere; cuius ratio ex superioribus patet. Diximus enim quod reflexionis linea EB semper est aequalis diametri parti a centro ad utriusque concursum. Si reflexionis linea ultra quartam diametri partem ascenderet, nunquam esse posset aequalis illi diametri parti a centro ad utriusque concursum. Non igitur ultra ascendet. Hinc est quod in concavis rotundis speculis incensiones non fiunt ultra diametri quartam; nam puncta incidentiae infra exagoni latus a circuli axe circumvoluta circulos describent in speculi superficie, a quibus lineae reflexae in diametri quarta concurrunt, quarum concursu incensiones fiunt. Hinc patet multorum et Euclidis error, qui in sua *Catoptrica* in circuli centro incensiones fieri pronunciaverit, quod est falsum et impossibile. [40]

side of an inscribed octagon, in which [case BE] was perpendicular to DF. The reflected angle [60°] for the [inscribed] hexagon was 2/3 [of a right angle], whereas that of the twelve-sided figure is 1/3 [of a right angle], i.e., half [that of the hexagon]. Likewise, [the reflected angle] for an inscribed octagon is half [a right angle], whereas that of a sixteen-sided figure is 1/4 [of a right angle], and so the angles will continually narrow at their vertices.

[1e] *In reflection from concave spherical mirrors, the line of reflection will not rise beyond a quarter of the diameter.*

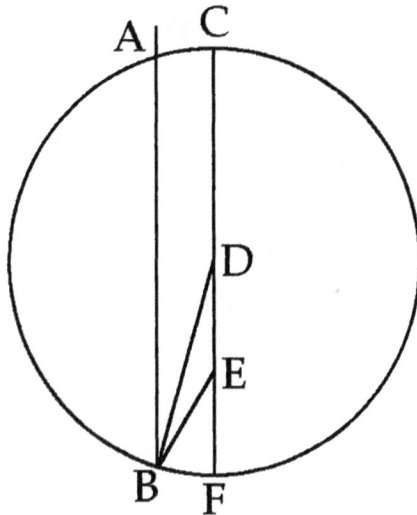

Figure 2.1e

Let CAB [in Figure 2.1e] be the mirror, and let incident ray AB from the Sun be reflected to E. I say that E lies less than a quarter of the diameter [above the bottom point F], and that it can never ascend beyond a quarter of the diameter [above that point]. The reason is clear from the things we showed above. For we have said [in prop. 1b above] that line of reflection EB is invariably equal to the segment [DE] of the diameter from the center to the intersection of both [EB and the diameter]. If the line of reflection were to ascend beyond a quarter of the diameter, it could never be equal to that segment of the diameter from the center to the intersection of both. Consequently, it will not ascend beyond [a quarter of the diameter]. This is why burning does not occur beyond a quarter of the diameter in concave spherical mirrors. For when the mirror is rotated about its axis, the points of incidence within the side of a hexagon inscribed in the mirror's circle describe circles on the mirror's surface, and from those circles lines of reflection intersect within a quarter-segment of the diameter, the intersecting rays being the cause of burning. From this, the error of many is clear, including that of Euclid, who claimed in his *Catoptrics* that burning occurs at the center of the mirror's circle, which is false and impossible.[6] [40]

[1f] *In reflexionibus speculorum concavorum, Solis radio feriente supra latus exagoni, linea reflexionis inferius ultra diametri finem cadet.*

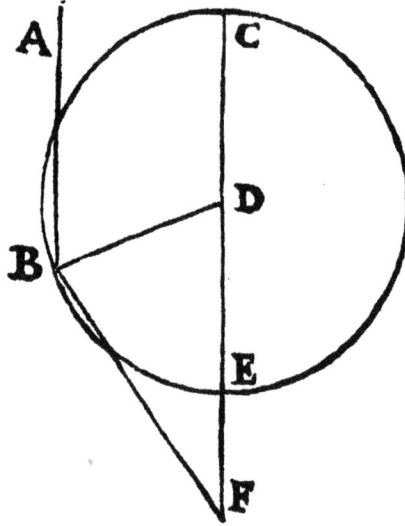

Esto speculum concavum CEB, radius Solis AB feriens B punctum, quod ab E per pentagoni latus distat, et est supra latus exagoni. Dico lineam reflexionis BF cadere extra finem diametri CDE, ultra E, inferius in puncto F. Quoniam pentagoni angulus maior est exagoni angulo per ea, quae probata sunt a Campano supra 12 primi Euclidis. Latior igitur angulus ultra diametri finem reflexionis lineam iaculabitur. Hinc specula, quae ex ea semicirculi parte parantur, quae supra exagonum est ultra speculi fundum incensionis punctum habent, ut libro *Naturalis Magiae* demonstravimus.

[1g] *Dato incidentiae puncto in speculis sphaericis concavis, punctum reflexionis in diametro reperire.* [41]

Esto speculum concavum EBAC, punctus in circumferentia speculi B. Volo reperire punctum reflexionis in diametro CE. Porrigatur ex centro circuli D ad punctum incidentiae B linea DB; dividatur per medium in puncto F. Et ex F excitato perpendicularem FE, et producito diametrum DE quousque coeat cum linea FE. Dico in E esse punctum reflexionis, quia ibi FE diametrum decussabit. Quia duorum triangulorum anguli DFE et EFB aequales sunt,

[1f] *In reflection from concave spherical mirrors, if a ray*
from the Sun reaches a point beyond the side of a hexagon
[inscribed in the mirror], the line of reflection will fall
below the bottom of the diameter.

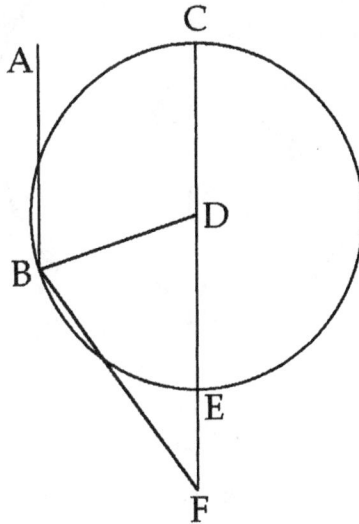

Figure 2.1f

Let CEB [in Figure 2.1f] be a concave [spherical] mirror and AB a ray from the Sun that reaches point B, which lies a distance of the side of an inscribed pentagon beyond E [so that arc BE = 72°]; it therefore lies higher than the distance of the side of an inscribed hexagon. I say that line of reflection BF falls outside the bottom of diameter CDE, beyond and below E to point F. For the angle [at any vertex] of a pentagon is greater than the angle [at any vertex] of a hexagon according to what has been demonstrated by Campanus of Novara in Euclid, book I, [theorem] 12, [proposition 19].[7] Thus, a more obtuse angle will project the line of reflection beyond the endpoint of the diameter. That is why in mirrors formed from a segment of a semicircle, the portions beyond [that defined by] a hexagon have a burning point below the bottom of the mirror, as we demonstrated in our book *Natural Magic*.[8]

[1g] *In concave spherical mirrors, if a point of incidence is given,*
to find the intersection point of reflection on the diameter.[9] [41]

Let EBAC [in Figure 2.1g] be a concave spherical mirror and B a point on the mirror's circumference [struck by solar ray AB]. I wish to find the intersection point of reflection on diameter CE. Extend line DB from center point D of the circle to point of incidence B, and bisect it at point F. Drop perpendicular FE from F, and extend diameter DE until it intersects line FE. I say that the intersection point of reflection is at E because FE will cross the diameter at this point. Since angles DFE and EFB of the two triangles

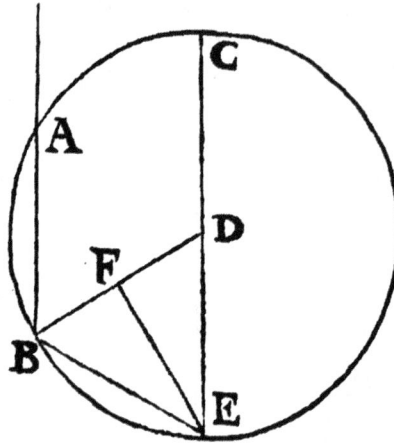

quia recti, et lineae[3] FD, FB aequales, et anguli FDE, FBE aequales, ergo DE, EB, aequales, et ubi duae hae lineae concurrunt, ibi punctus reflexionis.

Vera loca refractionis solaris radii in sphaera crystallina reperire. PROP. 2

Ad Solis refractiones in crystallina pila radiorum vestigandas, organico experimento utemur, unde oportet primo eius structuram explicare. Crystallina vel vitrea tabella paretur nitidae perspicuitatis, pedalis longitudinis latitudinisque, et digitalis crassitudinis, quae tota ex utraque parte rota quam exquisite expoliatur laevigeturque. Mox circini ope rotundetur atque exaequetur torno, marginibus rectis integris aequaliter servatis. Eius extimae superficiei ministerio ad explorandas Solis radiorum refractiones utemur. Supra tabulam ferreis anconibus firmetur; mox tabellula superponatur incisuris vel foraminulis pervia solarium radiorum suscipiendorum gratia eius circuli quarta una. Figurarum latera notentur notulisque insigniantur ex cuius regione tabellulae foramina respondeant. Et sic paratum erit instrumentum quod Solis oppositum refractiones ostendet. [42]

Esto sphaerica crystallina pila GHILMN aut cylindri pars plana explanata, supra locata tabellula AF exiguis foraminulis pervia A, B, C, D, E, F, quae radios Solis transmittat AG, BH, CI, DL, EM, FN.[4] Exploraturus igitur

---

[3]"linea"in original text.
[4]"FV"in original text.

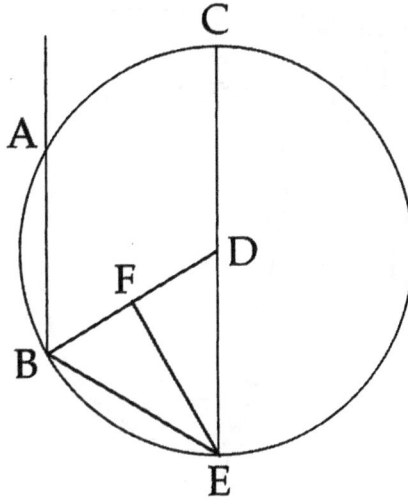

Figure 2.1g

are equal because they are right, and since lines FD and FB are equal [by construction], while angles FDE and FBE are equal, then DE and EB are equal, and where these two lines meet constitutes the intersection point of reflection.[10]

PROP. 2: *To find the precise location of refraction for a solar ray in a glass sphere.*

In order to trace out the refractions of solar rays in a glass sphere, we will use an experiment based on an apparatus, so it is necessary first to explain its construction. Get a crystal or glass plate of lustrous clarity that is a foot in length and breadth and a digit thick, and smooth and polish it exquisitely on all sides with a polishing disk.[11] Then, with the aid of a compass round it off and make it uniform on a lathe, while maintaining its straight faces parallel throughout. We will make use of the outer surface of [the resulting glass disk] in order to examine the refractions of rays from the Sun. Fasten it with iron clamps to a table; then above it place a [copper or wooden] strip punctured by incisions or tiny holes for letting solar rays through to one quarter of the disk's defining circle. Let the edges of the outlines be noted, and mark them down in line with the corresponding holes in the strip. Accordingly, the apparatus will be ready to display the refractions when facing the Sun. [42]

Let GHILMN [in Figure 2.2] be a spherical glass sphere or a flat portion of a cylinder excised from it [along a great circle], let the thin strip AF penetrated by small holes at A, B, C, D, E, and F be situated above it, and let those holes transmit rays AG, BH, CI, DL, EM, and FN from the Sun. Accordingly, whoever is to examine the refractions of rays from the Sun in

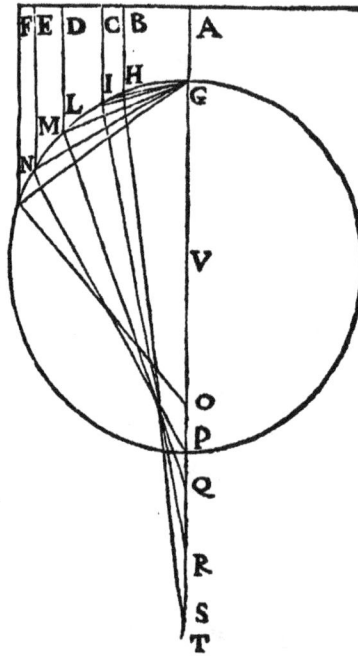

Solis refractiones in crystallina pila Soli tabulam elevet opponatque donec radius Solis per A transiens et per V centrum firmus et irrefractus permeet in infinitum et diametri vicem obtinebit.

Mox Solis radium per E foramen pertranseuntem scrupulosius observabit, nam M punctum sphaericae superficiei crystallinae feriet, et erit GM latus exagoni; refrangetur igitur radius EM ad diametri finem in puncto P, et[5] erit MP.

Deinde observabit radium CI, et erit punctus I latus duodecagoni, scilicet latus GI dimidium lateris exagoni, et [43] refrangetur eius radius infra latus exagoni, scilicet extra diametrum in puncto R, nam quantum in concavo speculo supra diametri calcem feriebat in semidiametro, tantum hic infra extra diametrum descendet.

Deinde observabimus DL tangentem L, a puncto G latus octogoni. Refrangetur eius radius ad Q punctum, scilicet LQ, et descendet infra punctum P quantum diximus.

Mox radium observabimus BH, latus scilicet sexdecagoni, dimidium octogoni. Et refrangetur ad punctum S extra diametrum inferius, et quanto diametro propius haerebit, tanto inferius descendet secundum quantitatem quam diximus.

---

[5]"ET"in original text

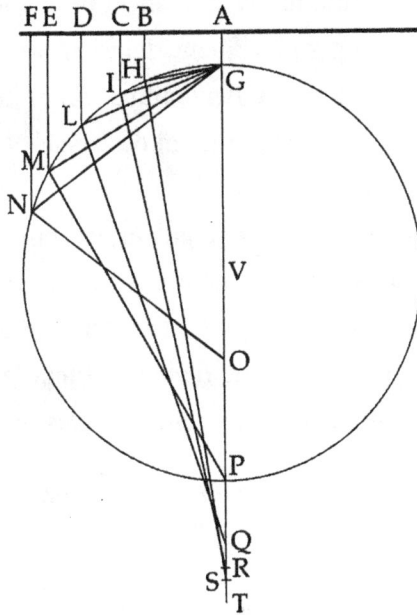

Figure 2.2

the glass sphere should raise the table and pose it toward the Sun until a
ray from the Sun that passes through both A and center point V [of the
disk] will continue straight on indefinitely without being refracted and will
thus follow the diameter.

Next he will closely observe the ray from the Sun that passes through
the hole at E, for it will be conveyed to point M on the surface of the glass
sphere, and GM will be the side of an inscribed hexagon, so ray EM will be
refracted to endpoint P of the diameter, and [the refracted ray] will be MP.

Then he will look at ray CI, and point I will be [the endpoint of] the
side of an inscribed twelve-sided figure, that is, side GI is half the side of a
hexagon, and [43] its ray will be refracted within the side of the hexagon,
that is, beyond the diameter to point R, for however far above the bottom
point of the diameter [the reflected ray] reached on the radius in the concave
spherical mirror, that is how far [the refracted ray] will descend beyond and
below [the bottom point P of] the diameter in this case.

Next we will observe DL striking L, which forms the side of an inscribed
octagon from point G. Its ray, that is, LQ, will refract to point Q and will
descend below P by the amount we just claimed [is equal to that in reflection
from the concave spherical mirror at the same angle of incidence].

Then we will look at ray BH, that is, [the ray that strikes at the endpoint
of] the side of an inscribed sixteen-sided figure, which is half the side of an
octagon. It will be refracted to point S beyond and below the diameter, and
the closer to the diameter [the incident ray] will hew, the farther down [the
refracted ray] will descend according to the amount we discussed.

Postea observabimus radium FN, qui tanget pentagoni latus GN, et quia infra latus exagoni est, ideo supra finem diametri ascendet, contra quam observavimus in concavi speculi reflexionibus, ibi enim descendebat, hic vero ascendit tantumque supra quantum ibi descendebat, nisi vitri crassities vetuerit, nam aegrius observabitur.

Sed haec irregularitas hanc habet regulam, ut contraria sint iis quae de speculo sphaerico diximus concavo, nam ubi in concavo ascendebant, hic descendunt, et ubi ibi descendabant, hic ascendunt. Solum in utroque latus exagoni metam habet, nam utrumque diametri finem ferit; cuius rationem assignat Aristoteles libro *Mechanicorum* concavum esse convexo contrarium.

Vera loca determinare refractionum solaris radii in concava sphaerali superficie. PROP. 3

Ad haec vestiganda, prius crystallinum instrumentum comminiscar parare in hunc nempe modum. Tabella crystallina paretur tantae longitudinis quanta opus erit, scilicet pedalis longitudinis et per transversum latitudinis fere semipedalis, crassitudinis digitalis, totaque exactissime explanetur ut parallelis sit superficiebus. In media longitudinis extremi marginis linea pes firmus circini statuatur, vagus circumducatur donec in eius superficie designetur semicirculus. Undique relinquatur solidum digitalis magnitudinis ut firmum integrumque permaneat instrumentum, inde rotis [44, 45] excavetur semicirculus ut latera ad rectum angulum excidantur illaesis et illibatis marginibus. Mox tabellam cupream vel ligneam supra firmabis, ut crystallino incumbat modicae crassitudinis sed tantae longitudinis quantae instrumenti. Modiceque ab eo distet, et in tot exigua foramina perforaveris quot volueris, ut solaribus radiis pervium expeditumque iter praebeat. Hactenus parato instrumento, extollatur Soli, qui per foramina illabens, tangat latera figurarum. Exemplum infra descripsi.

Esto instrumentum vitreum excavatum VXYZ, tabellula foraminibus lacessita ABCDE. Radii Solis refranguntur in contrariam partem eorum qui in circulo convexo divaricabantur, diximus enim convexo contrarium concavum, radius ergo EL per centrum transiens circuli LIHGF penetrat irrefractus EL.

Radius AF illabens per A foramen tabellae tangit circuli concavi punctum

Afterward, we will look at ray FN, which will strike [the endpoint of] side GN of an inscribed pentagon, and since it lies [beyond and] below the side of the hexagon, [the refracted ray] will ascend [to point O] above the endpoint of the diameter, which is the opposite of what we observed in reflection from a concave spherical mirror [in proposition 1f], for in that case [the reflected ray] descended [below the diameter's end point], whereas in this case it ascends as far above [the diameter's endpoint] as it descended below in the previous case, unless the thickness of the glass hinders it, for then it will be improperly observed.[12]

But this anomaly follows the rule that what happens [in this case] is the opposite of what we claimed happened in the spherical concave mirror, for in the concave mirror the rays ascended, whereas in this case the rays descend, and when rays descended in the concave mirror, in this case they ascend. In both cases there is perfect equivalence only for the side of the hexagon, for it brings both rays to the endpoint of the diameter. The reason for this, Aristotle stipulates in his book *Mechanics*, is that concavity is the opposite of convexity.[13]

PROP. 3: *To find the precise location of refraction for a solar ray passing into a concave spherical surface.*

In order to find this out, I will adapt the earlier glass apparatus as follows. Fashion a glass plate long enough to be of service, that is, a foot in length and precisely half a foot across in width, with a thickness of a digit, and flatten it throughout with precision so that it has parallel surfaces. At the midpoint of the outer edge along its length, place the point of a compass firmly and rotate it around until a semicircle is inscribed on its surface. Keep the entire apparatus a digit thick throughout so that it can remain firm and uniform, and then [44, 45] cut out the semicircle with rollers so that the sides are cut out to the right angle with the edges left undamaged and unimpaired. Then you will attach a copper or wooden strip above it so that it stands a bit higher than the thickness of the glass plate but is as long as the apparatus. Let it lie a moderate distance from the apparatus, and you will have punctured it with as many tiny holes as you wish so that it can provide easy, open passage for solar rays. From this point on, with the apparatus thus prepared, raise it toward the Sun, whose light, streaming through the holes, can strike the sides of the inscribed figures. I have provided a description below.

Let VXYZ [in Figure 2.3] be the glass instrument [with the semicircle] cut out, and let the strip be punctured by holes at A, B, C, D, and E. The rays of the Sun are refracted in the opposite direction from that in which they branched off in the case of the convex circle, for we said [in prop. 2 above] that convexity is the opposite of concavity, so ray EL that passes through the center of circle LIHGF extends unrefracted along EL.

Ray AF reaching through hole A of the strip strikes point F of the concave [semi]circle, that point forming the side of an inscribed hexagon

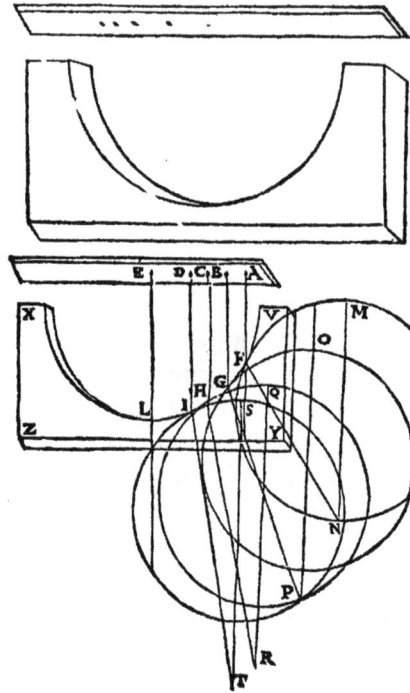

F, latus exagoni a puncto L, refrangitur ad finem diametri, ut diximus. Ideo circulum appinximus MN cuius diametri finem N; refractus radius ex F attingit.

Sic quoque radius CH tangens punctum H latus duodecagoni refrangitur ad punctum R extra diametrum, ut videtur in eius circulo QR.

Sic radius BG latus octogoni tangens G refrangitur ad punctum P, ut suo circulo apparet.

Eodemque modo DI, latus sedecagoni ad punctum T.[6]

Solus radius EL transiens per circuli centrum transit irrefractus in infinitum, ut diximus.

Vera loca determinare refractionum solaris radii egredientis ex concava sphaerali superficie. PROP. 4

Si praefatum instrumentum Soli obvertes, voti compos fies, nam refractiones Solis ex superiori parte in concavam superficiem incidentes et exeuntes

---

[6]"P"in original text.

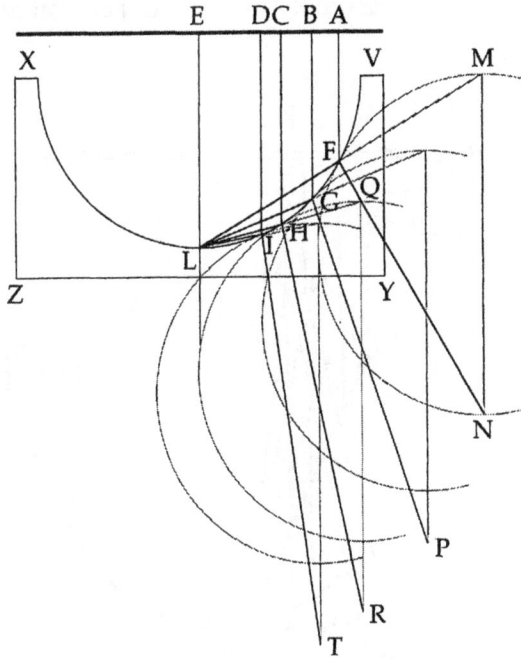

Figure 2.3

from point L, and it is refracted to the endpoint of the diameter, as we said. Accordingly, we have added circle MN, the endpoint of whose diameter is N [such that FM is equal to the side of an inscribed hexagon], and the refracted ray from F reaches that point N.

So, too, ray CH that strikes point H on the side of an inscribed twelve-sided figure is refracted to point R beyond the diameter, as is seen in its circle QR [where chord HQ is equal to the side of an inscribed twelve-sided figure].

Likewise, ray BG, which strikes G on the side of an inscribed octagon, is refracted to point P, as is apparent from its circle.

By the same token, DI [which strikes I on] the side of an inscribed sixteen-sided figure is refracted to point T.

Only ray EL, which passes through the center of the circle [containing the semicircle] continues on indefinitely without refraction, as we said.[14]

PROP. 4: *To determine precisely where a solar ray refracts on leaving a concave spherical surface.*

If you turn the aforementioned apparatus around toward the Sun, you will accomplish what you wish because you will have tested the refractions of [rays from] the Sun as they enter the top part of the apparatus and exit the

experieris. Iisdem enim angulis quibus ingrediuntur egrediuntur, ut eiusmodi schemate videre est.

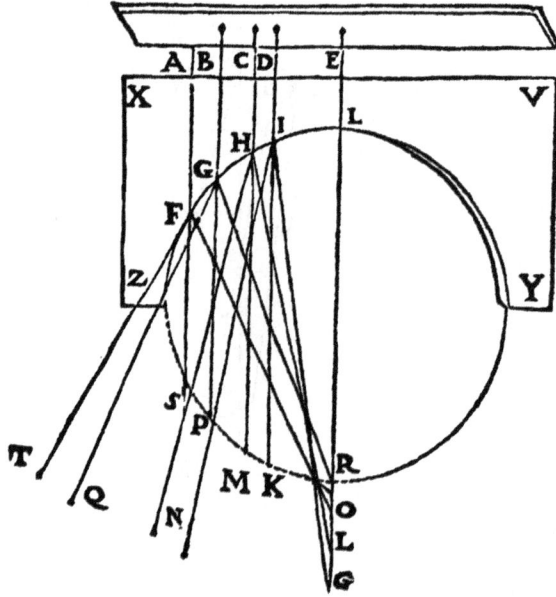

Esto instrumentum obversum VXYZ Soli oppositum, et ex tabellae foraminibus radii solares ingrediantur, refrangantur. [46] Primus ex E foramine ingrediens per LRG irrefractus labitur.

Mox radius AF, ingrediens, tangit latus exagoni. Quia ingrediens prius ad diametri calcem perrectabat; nunc in contrariam partem labitur iisdem angulis, ex F enim in T[7] labitur, et angulus[8] TFS aequalis fit angulo SFR.

Deinde observabis radium CH tangentem latus duodecagoni, et in contrariam partem pergit HN. Sic angulus NHM aequalis erit MHL.[9]

Mox observabis BG radium etiam tangentem latus octogoni; labiturque in Q, et angulus QGP aequalis erit PGO.

Postremo radium Solis DI latus sexdecagoni labiturque in H. Sic angulus HIK aequalis erit KIO—animadvertendo quod radii DI, CH, BG, AF irrefracti permeant usque ad IHGF, quia perpendiculares. [47]

---

[7]"C"in original text.
[8]"angulis"in original text.
[9]"NHL"in original text.

concave surface. In fact, they will exit at the very same angles as they enter [in the case illustrated in the previous proposition], as is evident in a diagram of the same sort.

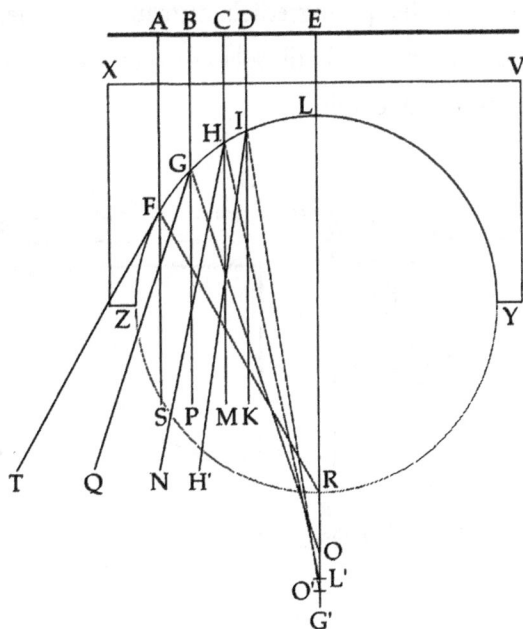

Figure 2.4

Let VXYZ [in Figure 2.4] be the reversed apparatus exposed to the Sun, and let solar rays enter through the holes in the strip and be refracted [at concave surface LIHGFZ]. [46] The first ray entering from the hole at E passes along LRG' unrefracted.

Next, ray AF strikes the [endpoint of a] side of an inscribed hexagon [i.e., chord FL] when it enters. Because it reached the bottom of the diameter when it entered [the glass sphere] earlier [in proposition 2], now it will be inclined toward the opposite side at the very same angles, for it inclines from F toward T, and angle TFS = angle SFR [according to book 1, prop. 3].

Then you will observe ray CH, which strikes the [endpoint of a] side of an inscribed twelve-sided figure and proceeds in the opposite direction along HN. Hence, angle NHM = angle MHL'.

You will next observe ray BG, which strikes the [endpoint of a] side of an inscribed octagon; it inclines toward Q, and angle QGP = angle PGO.

Finally, ray DI from the Sun strikes the [endpoint of a] side of an inscribed sixteen-sided figure and inclines toward H'. Consequently, angle H'IK = angle KIO'—bearing in mind throughout that rays DI, CH, BG, and AF reach [concave surface] IHGF unrefracted because they are perpendicular [to the top surface of the glass semicylinder].[15] [47]

Vera loca determinare solaris radii refractionis egredientis ex convexa sphaer-
ali superficie. PROP. 5

Hi anguli sunt aequales illis qui ingrediebantur in superficiem vitream convexam, attamen contrarii, nam radii illi solares in contrariam partem labuntur,
ut videbimus in subiecto exemplo.

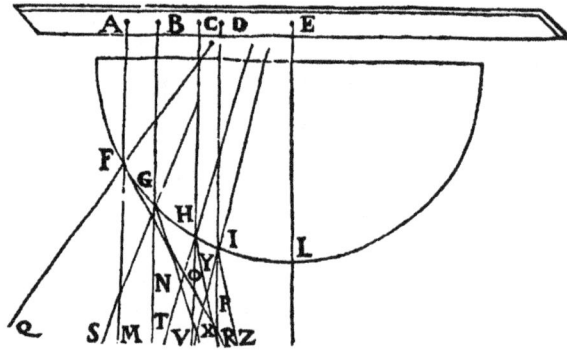

Esto tabella Soli opposita et perforata A, B, C, D, E; semicirculus vitreus
Soli oppositus FGHIL, cuius recta diameter Soli obversa, circumferentia
adversa.

Pergat radius EL transiens per semicirculi centrum; penetrat irrefractus.

Deinde AF tangens F punctum latus exagoni a puncto L[10] ab A ad F
penetrat perpendiculariter et irrefractus. Deinde foras egrediens ex F in
contrariam partem labitur in R, quando enim superficiem tangebat concavam,[11] veniebat per FQ, rectus radius FM nunc labitur in R, et angulus QFM
erit aequalis MFR.

Sic radius CH tangit latus duodecagoni. In superficie concava[12] veniebat
per HT; nunc per HY, et angulus THO[13] aequalis est angulo OHY.

Sic radius BG tangit latus octogoni; labitur in X; veniebat GS. Et DI,
latus sexdecagoni, labitur in Z, veniebat enim per IV. [48]

---

[10] "non" after L in original text.
[11] "convexam" in original text.
[12] "convexa" in original text.
[13] "DHO" in original text.

PROP. 5: *To determine precisely where a solar ray refracts on exiting a convex spherical surface.*

In this case the angles are equal to those at which the rays entered the convex glass surface [in proposition 2], although they are opposite, for these solar rays incline toward the opposite side, as we will see in the illustration below.

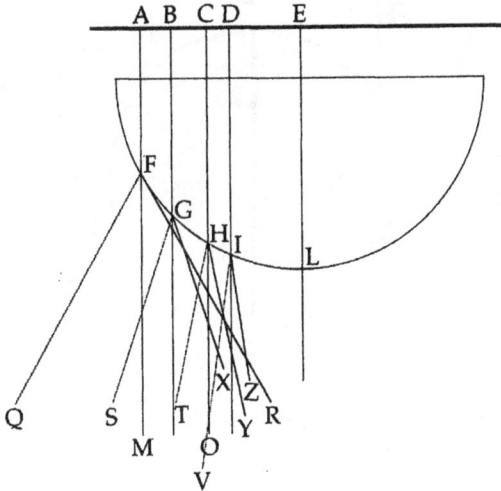

Figure 2.5

Let the strip perforated at A, B, C, D, and E [in Figure 2.5] face the Sun, let the glass semicircle facing the Sun be FGHIL, and let the flat [side of the glass semicircle along the] diameter face the Sun directly, the circumference [of the semicircle] being on the opposite side.

Ray EL should continue to pass through the center of the semicircle, and so it penetrates without refraction.

AF, in turn, which strikes point F [at the endpoint of] a side of an inscribed hexagon [FL measured] from L passes through from A to F orthogonally and unrefracted. Then it inclines in the opposite direction toward R on passing out from F, for [whereas previously, in proposition 4] when it had struck the concave surface, it exited along [line FT equivalent to] FQ, now straight ray [AF extended along] FM inclines toward R, and angle QFM = angle MFR.

Likewise, ray CH strikes the [endpoint of a] side of an inscribed twelve-sided figure. In the concave surface [in proposition 4] it had exited along [HN equivalent to] HT; now it exits along HY, and angle THO = angle OHY.

By the same token, ray BG strikes the [endpoint of a] side of an inscribed octagon, and it inclines toward X, whereas it had exited [previously, in proposition 4, along GQ equivalent to] GS. And DI, [which strikes] the [endpoint of a] side of an inscribed sixteen-sided figure, inclines toward Z, for it had [previously] exited along [IH' equivalent to] IV.[16] **[48]**

Dato Solis radio circulum vel pilam crystallinam tangente, punctum vel lineam refractionis reperire. PROP. 6

Dato Solis radio vel re quapiam pilam crystallinam invadente, quo pacto refractionis linea reperienda sit in hunc modum indicamus.

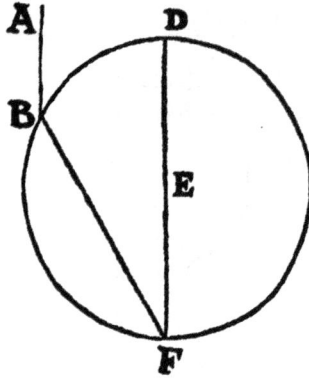

Sit circulus DBF[14]; cadat radius AB tangens circulum in B. Volo scire ubi refrangatur. Sit circuli centrum E, et dato puncto E et linea AB, ducatur linea quae aequidistans sit radio cadenti AB, per 32 primi Euclidis, et erit DEF.[15] Inventa igitur diametro aequidistanti ipsi AB, metiatur, ut diximus, quota pars erit circuli. Et si erit sextans, cadet in F, videlicet diametri fine, sic et reliquos metiemur.

Cathetus secans lineam deferentis formam inter punctum fractionis et visum, videbitur simulachrum inter visum et punctum in aere pendulum. PROP. 7

Multis modis variatur visus in pila crystallina. Id est multis variis in locis simulachrum rei videtur, idque evenit ex varietate concursus lineae deferentis imaginem cum catheto. Exemplum afferemus. [49]

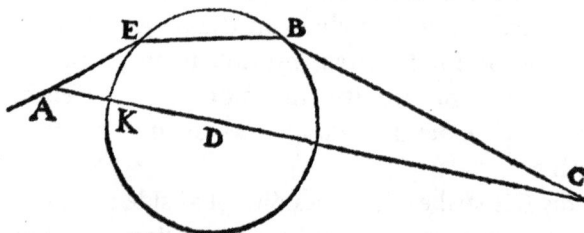

Sit res visa C; cadat super pilam in B. Refrangitur in E, per secundam

---

[14]"ABF"in original text.
[15]"EDF"in original text.

PROP. 6: *Given a ray from the Sun that strikes a circle or a glass sphere, to find the intersection point of refraction or the line of refraction.*

We show in the following way how the line of refraction is to be found for a given ray or something else from the Sun that penetrates a glass sphere.[17]

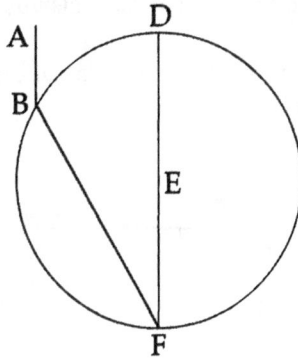

Figure 2.6

Let DBF [in Figure 2.6] be a [great] circle [on a glass sphere], and let ray AB strike the circle at B. I wish to know where it might be refracted. Let E be the circle's center, and with point E and line AB given, draw a line [through E] parallel to incident ray AB, by Euclid, [*Elements*] I, 32, and it will be DEF.[18] Having thus found the diameter parallel to AB, measure the size of segment [BD] of the circle, as we have discussed. If it is one-sixth, then [the refracted ray] will fall to F, that is, the endpoint of the diameter, and we will measure the rest of the segments in this way.[19]

PROP. 7: *When the cathetus intersects the line that conveys the image between the point of refraction and the eye, the image will appear to hang in the air between the eye and that point.*

What is seen in a glass sphere varies in several ways. That is, the image of an object appears in many different locations, and this is due to the different ways in which the line conveying the image intersects the cathetus. We will provide an illustration. [49]

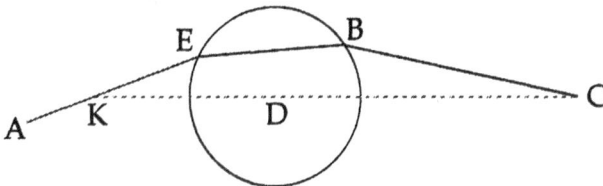

Figure 2.7

Let C [in Figure 2.7] be the visible object, and let its image strike the sphere at B. It is refracted to E, according to the second proposition of this

huius; egreditur[16] per <EA, per quintam huius, ubi firmetur oculus. Ducatur cathetus ex C per D, circuli centrum; occurret lineae refractionis in K. Ibi igitur videbitur simulachrum luminis vel alterius rei C, per 7 nostri libri primi. Videbitur igitur inter oculum A et punctum refractionis E, non sine admiratione. Sed Alhazenus et Vitellio dicunt quod clarum non videbitur, quia non suo loco. Sed hoc falsum est, nam clarius videtur quam suo loco.

Catheto secante lineam deferentem formam in superficiem pilae, ibi videbitur simulachrum. PROP. 8 [50]

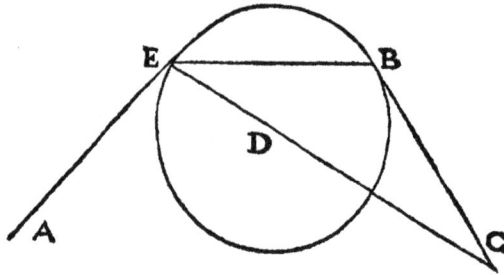

Esto res visa C tangat B pilae superficiem. Refrangitur in E, et ex E occurret oculo existenti in A. Ducatur cathetus per C et D circuli centrum; occurrit in E. In occursus puncto, qui est in pilae superficie, ibi videbitur simulachrum.

Catheto secante lineam formam deferentem in centrum visus, ibi videbitur simulachrum. PROP. 9

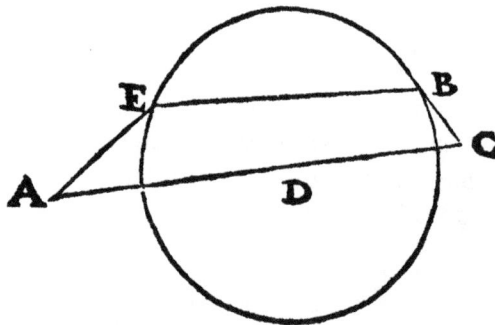

Esto res visa C. Accedit ad punctum B; transit per punctum E; venit ad oculum A. Ducatur cathetus per C punctum et D centrum. Occurret lineae EA in puncto A, ubi oculus; ibi igitur videbitur simulachrum.

---

[16]"eggreditur"in original text.

[second book], and it exits along EA to where the eye is stationed [at A], according to the fifth proposition of this [second book]. Extend the cathetus from C through D, the center of the circle, and it will intersect the line of refraction at K. Here, then, is where the image of the light or of something else at C will be seen, according to proposition 7 of our first book.[20] It will therefore be seen, not without astonishment, between the eye at A and point of refraction E. Alhacen and Witelo, however, say that it will not appear clear because it is not seen at its proper location.[21] But this is false, for it appears clearer than it would in its proper place.

PROP. 8: *If the cathetus intersects the line conveying the image at the surface of the sphere, the image will be seen at that point.* [50]

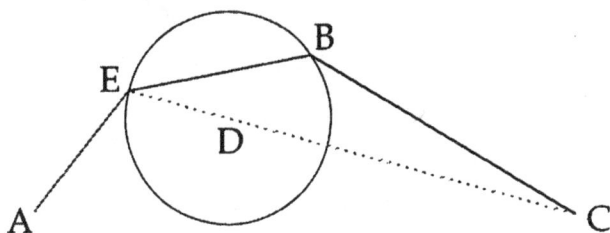

Figure 2.8

Let the image of visible object C [in Figure 2.8] strike the surface of the sphere at B. It is refracted to E, and from E it will reach the eye lying at A. Extend the cathetus through C and center D of the circle, and it intersects [refracted ray EA] at E. It is here, at the point of intersection, which is on the surface of the sphere, that the image will be seen.

PROP. 9: *If the cathetus intersects the line that conveys the image at the center of the eye, the image will be seen at that point.*

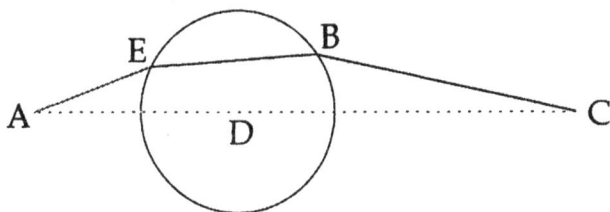

Figure 2.9

Let C [in Figure 2.9] be the visible object. Its image reaches point B, passes through point E, and reaches the eye at A. Extend the cathetus through point C and center point D of the circle. It will intersect line EA at point A, where the eye is, so the image will be seen at that point.[22]

Catheto aequidistante lineae deferenti formam, simulachrum nusquam videbi-
tur. PROP. 10 [51]

Quum cathetus non concurrit cum refractionis linea, imago nusquam videtur,
diximus enim ubi earum concursus, ibi et simulachrum. Exemplum erit.

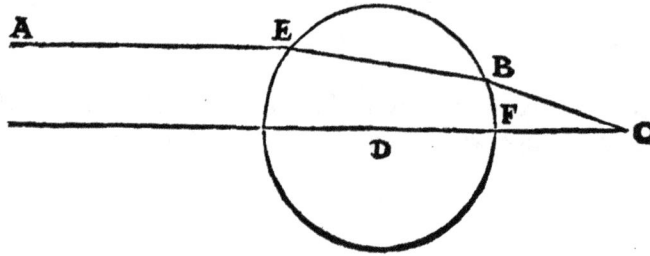

Esto res visa punctum C. Accedit ad punctum B, refrangitur ad punctum E,
inde relabitur ad F. Ducatur ex puncto rei C cathetus et per centrum D; erit
CDA. Quia CDA aequidistat EF, nunquam concursus fiet, nusquam ergo
videbitur imago.

Linea refractionis occurrens visui ante concursum cum catheto, apparebit
imago post visum, nec videbitur nisi caput retrocedat. PROP. 11

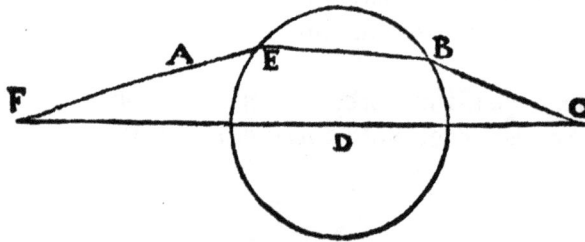

Esto res visa C. Venit ad B, refrangitur ad E, et denuo refrangitur; venit ad
A oculum. Ducta catheto per rem visam et centrum pilae D, nunquam occurret
linee EA, quia prius a visu A intercipitur; non videbitur ergo imago; at
retrocedente capite in concursu illarum in puncto F, videbitur in oculo. [52]

Superposita pila crystallina oculo, maior se ipsa videtur. PROP. 12

PROP. 10: *If the cathetus is parallel to the line conveying the image, the image will never be seen.* [51]

When the cathetus does not meet the line of refraction, no image whatever is seen, for we said that wherever there is an intersection of the two, the image will be seen at that point. This will be an illustration. Let point C [in

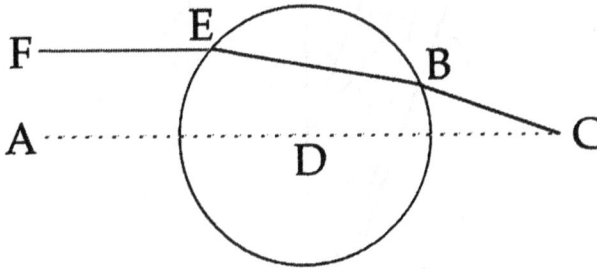

Figure 2.10

Figure 2.10] be the visible object. Its image reaches point B and is refracted to point E, from which it inclines toward F. From object-point C draw the cathetus through center point D, and it will be CDA. Since CDA is parallel to EF, it will never intersect it, so no image whatever will be seen.[23]

PROP. 11: *When the line of refraction reaches the eye before it intersects the cathetus, the image will appear behind the eye, and it will not be seen unless the viewer draws his head back.*

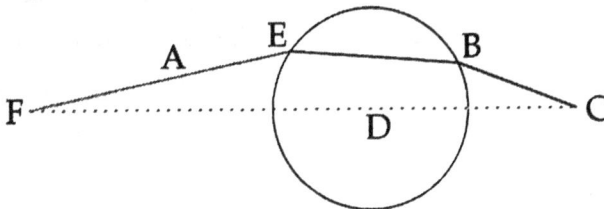

Figure 2.11

Let C [in Figure 2.11] be the visible object. Its image reaches B, is refracted to E, and is refracted again so that it reaches the eye at A. When the cathetus is extended through the visible object and center point D of the sphere, it will never intersect line segment EA, since [the extension of that line segment] is intersected ahead of the eye [at F], so the image will not be seen; but if the head is drawn back to the intersection of those lines at point F, the image will be seen at the eye [according to prop. 9]. [52]

PROP. 12: *When a glass sphere is placed right up to the eye, it will appear larger than it actually is.*

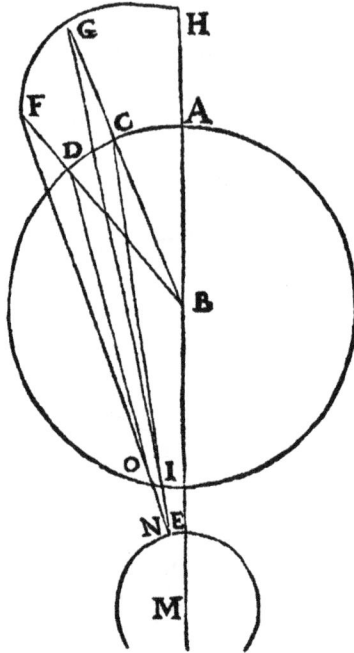

Esto crystallina pila ACDOI; opponatur oculo existenti in EN. Perpendicularis radius MEBAH irrefractus penetrabit. Mox ex parte D veniat visio ad O; exeundo veniet ad N iuxta egredientem angulum ad oculum N, et refrangendo producetur ad F. Ducta catheto ex B pilae centro per D, occurret NF in puncto F. Eodem modo punctus C veniet ad I; exeundo veniet ad E oculum; refrangetur ad G. Ductaque catheto ex BC, occurret G, et punctus C videbitur in G. Idem ex alia parte faciendo; multo maior erit circuitus HGF quam ACD. [53]

Magnitudine tangente finem diametri, parum vel longe, imago videbitur in pilae superficie quemadmodum annularis. PROP. 13

Haec est praecedentis conversa, ubi primum erat oculus EN; hic magnitudo AB. Diximus enim quando radius incidebat in sextam partem, quod in diame-

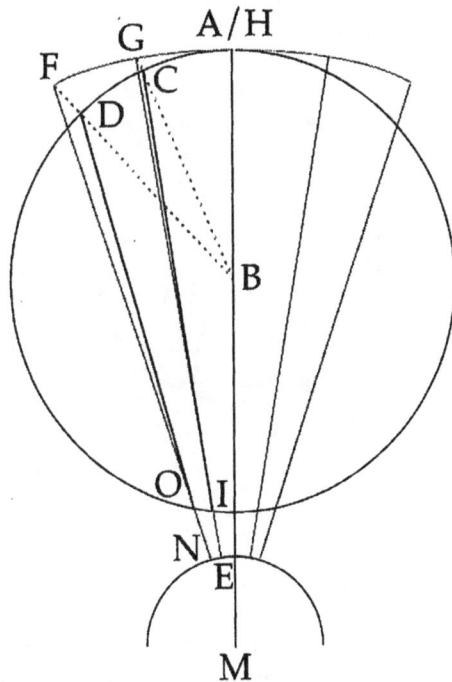

Figure 2.12

Let ACDOI [in Figure 2.12] be the glass sphere, and let it face an eye that is situated on arc EN [which represents its outer surface]. The perpendicular ray MEBAH will pass through unrefracted. Now on the side of D let a line of sight reach O; on exiting it will reach N according to the angle [of refraction] at which it exits toward N on the eye, and having refracted, it will be extended to F.[24] If the cathetus is extended from center point B of the sphere to D, it will intersect NF at point F. By the same token, the image of point C will reach I, and on exiting it will reach E on the eye, and it will be refracted [along EI, which extends] to G. If the cathetus is dropped along BC, it will reach G, and so point C will be seen at G. If the same procedure is carried out on the other side, the curve HGF [forming the image] will be considerably larger than ACD [which is the object on the back side of the sphere].[25] **[53]**

PROP. 13: *When a magnitude lies at the end of the diameter, whether near or far away [from the glass sphere], its image will be seen on the surface of the sphere in a ringlike form.*

This is the converse of the preceding theorem, where the eye was originally at EN [outside the sphere]; now it is magnitude AB [that is beyond the sphere, as in Figure 2.13].[26] For we said [in proposition 2 of this book] that when the ray fell to [the endpoint] of a segment constituting one-sixth [of

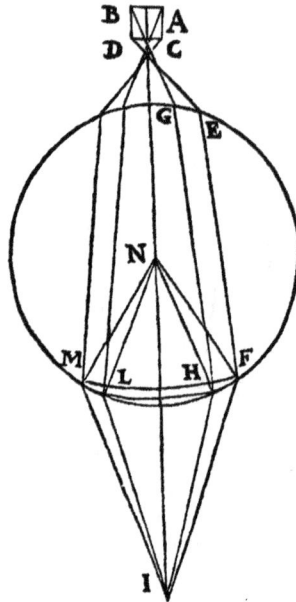

tri fine frangeretur. Nunc autem contra: quando diametri finem tangat, ad
sextam circuli partem refrangi. Magnitudo ABCD loco posita ubi omnes radii
intra FM concidant, sic ex ea procidentes radii ad oculum I, tota pilae pars
FHLM[17] obscuro colore tingetur.

Signum quod lineae moveantur ex D ad E, et ex E ad F, et ex [54] F
ad I oculum, quod tangendo partem D magnitudinis, tango F punctum, et
tangendo B ipsum H. Ex altera parte, A veniet ad L[18] et refrangetur ad I. Et
quia undique aequalis pila, uniformis erit refractio. Imaginetur ergo diametro
fixa manente eo pilam circumduci quousque ad locum suum redeat unde
discesserit; sic punctus F describit in pilae superficie circulum in MF, et
punctus H circulum HL. Unde circulus videbitur in sphaerae circumferentiae
quantitatis magnitudinis rei oppositae. Cathetus ex N ducta non potest alibi
occurrere nisi in punctis refractionis, sic veniet ad oculum magnitudo AB
in formam annuli vel armillae. Videbitur quoque directe eadem magnitudo
sine refractione suo loco. [55]

Recta linea per vitream pilam visa curva conspicietur. PROP. 14

Esto recta linea videnda ABCDEFG. Quia longior est quam possit tam prope
videri per refractionem, invertuntur puncta incidentiae, ut se possint pilae
insinuari ad visionem. Sic punctus G venit ad K, refrangitur ad P, per secundam huius, inde ad X oculum progreditur, per quintam huius. Trahatur
cathetus ex G per M centrum. Pervenit ad P, ibi igitur punctus G videbitur.

---

[17]"FALM"in original text.
[18]"F"in original text.

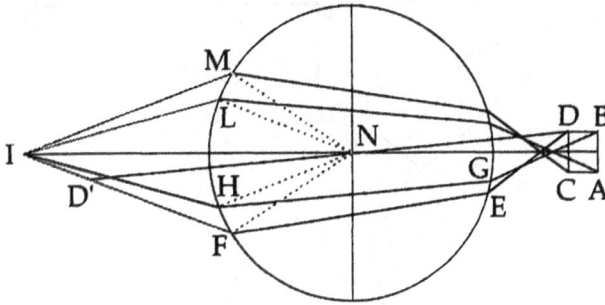

Figure 2.13

the great circle of the glass sphere], it would be refracted to the endpoint of the diameter. Now, however, the opposite obtains: when it touches the endpoint of the diameter, it is refracted to the segment constituting one-sixth [of the great] circle. When magnitude ABCD is placed where all the rays fall [from it to points] within [arc] FM, such that from there the rays fall to the eye at I, the entire portion FHLM of the sphere will be tinged with a dark color.[27]

Proof that the lines [of incidence and refraction] pass from D to E, then from E to F, and from [54] F to the eye at I is that when I touch the spot at D on the magnitude, I [appear to] touch point F, and when I touch B, [I appear to] touch H. On the other side, the image of A will reach L [after refraction into the glass sphere] and will be refracted to I. And since the sphere is even throughout, the refraction will be uniform throughout. So, imagine the diameter held firmly in place and the sphere rotated about it until it returns to the place where it started to move. Accordingly, point F describes a circle on MF on the surface of the sphere, and point H describes a circle on HL. Thus, a circle the size of the facing object will be seen on the circumference of the sphere.[28] When the cathetus is extended from N, it can intersect [the refracted rays] only at the points of refraction, so the image of magnitude AB will reach the eye in the form of a ring or hoop. The same magnitude will also be seen in its proper place, directly and without refraction.[29] [55]

PROP. 14: *A straight line seen through a glass sphere will appear curved.*

Let ABCDEFG [in Figure 2.14] be the straight line to be seen. Since it is too long to be seen by direct refraction at such a near distance, the points of incidence are inverted so that [the images associated with them] can pass through the sphere to the eye [in inverted order]. Accordingly, the image of point G reaches K, is refracted to P, according to the second proposition of this book, and then continues on to the eye at X, according to the fifth proposition of this book. Extend the cathetus from G through center point M. It reaches P, so it is here that point G will be seen [according to proposition

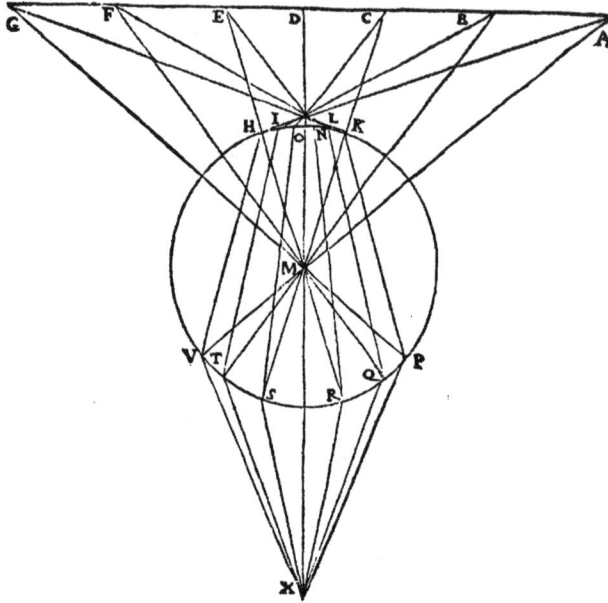

Sic F venit ad L, mox refrangitur ad Q, et ad oculum X venit. Ducta catheto ex F per M, occurret in Q lineae refractionis. Sic F[19] videbitur in Q. Sic E ad R; ad X. Idem faciendum altera parte A, B, C putavimus, et videbuntur[20] in punctis S, T, V. Quum igitur linea in superficie sphaerae videbitur, necesse est curvam videri. [56]

Possunt unius magnitudinis duae, vel tres imagines videri in pila. PROP. 15

Punctus A per D venit ad F, per secundam huius, et ad H oculum sinistrum, per quintam. Cathetus ex centro L per oculum H secat incidentiae lineam in C; ibi videbitur punctus A, nam clauso oculo H, perit punctus C, et A nusquam videtur. Idem evenit de puncto A ad oculum dextrum I, videbiturque in B. Aliquando uterque oculus etiam in A eundem punctum videt, evenit et alio modo, ut dicemus, quando res una duplicata videbitur.

---

[19]"E"in original text.
[20]"vdebuntur"in original text.

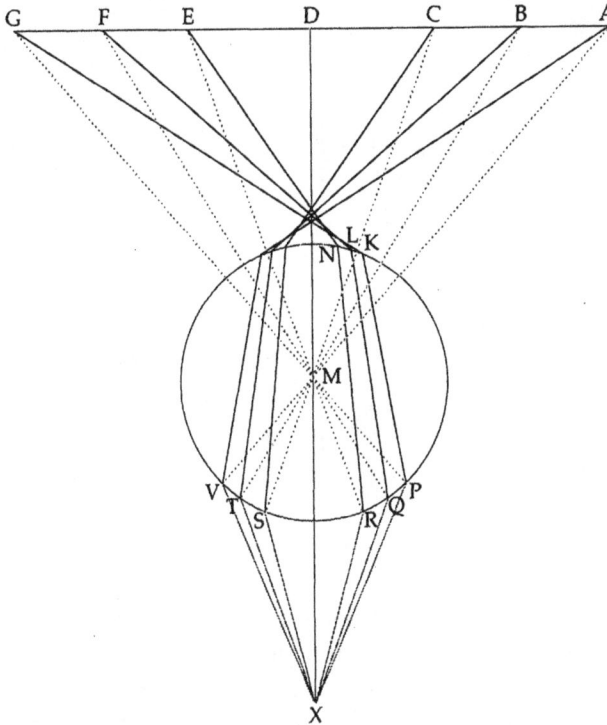

Figure 2.14

8 above]. Likewise, the image of F reaches L, then is refracted to Q, and finally reaches the eye at X. When the cathetus is drawn from F through M, it will intersect the line of refraction at Q. F will therefore be seen at Q. So, too, the image of E [refracts at N] to R, and then to X. We have supposed that the images of A, B, and C will do the same thing on the other side, and they will be seen at points S, T, and V. Therefore, since the line will be seen on the surface of the sphere, it must appear curved.[30] [56]

PROP. 15: *Two or three images of a single magnitude can be seen in a glass sphere.*

The image of point A [in the top diagram of Figure 2.15] reaches F through D, according to the second proposition of this book, and then the left eye at H, according to the fifth proposition [of this book]. The cathetus dropped from center point L through the eye at H intersects the line of incidence [AE] at C, and so the image of point A will be seen at this spot, for if the eye at H is closed, point C will disappear, and A is not seen at all. The same happens with point A with respect to the right eye at I; it will be seen at B. Also, both eyes sometimes see the same point at A, but this happens in another way, when we see a single object doubled, as we will discuss later on.[31]

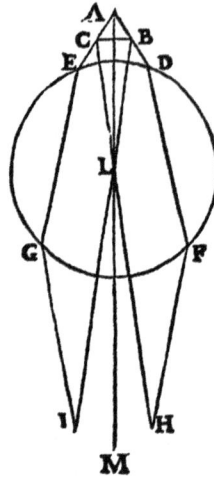

Magnitudo extra punctum inversionis posita conversa videbitur. PROP. 16

Punctus inversionis maxime in eiusmodi apparentiis vestigandus est, nam
ex eo omnes varietates nascuntur, qui punctus ex varia oculi positione et
rei variatur. Nam si oculus longe abierit, minuetur rei magnitudo, quia
anguli refractionis angustantur. Idemque eveniet magnitudinis remotione.
Contrarium eveniet utriusque rei accessione,[21] nam anguli refractionum dilatantur. De omnibus exemplo afferemus.[22] [57]

Esto magnitudo opposita GF, pila BCED,[23] oculus A. Non insinuabitur

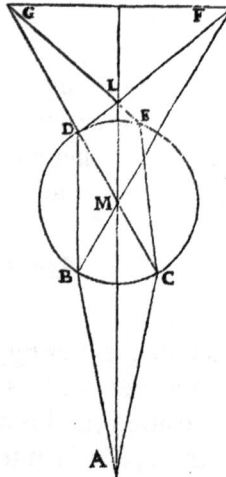

---

[21]"accersione"in original text.
[22]"A]sseremus"in original text.
[23]"BCELD"in original text.

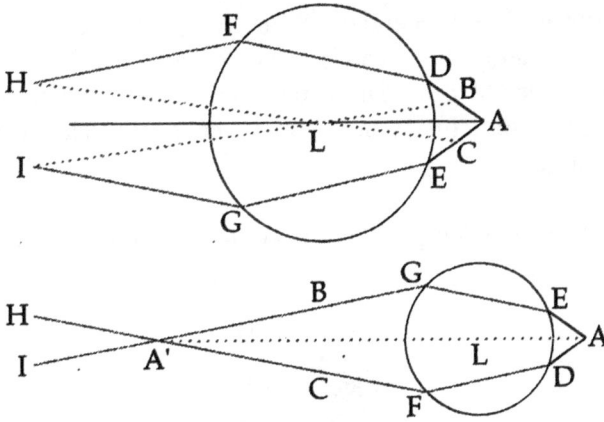

Figure 2.15

PROP. 16: *A magnitude located beyond the point of inversion will appear inverted.*

The point of inversion must be tracked down quite carefully in these sorts of appearances, for all kinds [of illusions] are due to it, and this point varies according to the placement of the eye and the visible object.[32] For if the eye lies far away, the apparent size of the object diminishes because the angles of refraction are narrowed. The same will happen according to the distance of the visible magnitude. The opposite will happen when either is brought up close, for in that case the angles of refraction widen. We will provide an illustration of all these points. [57]

Let GF [in Figure 2.16] be the facing magnitude, BCED the glass sphere, and A the eye. The image of point F will not pass through [the sphere] to

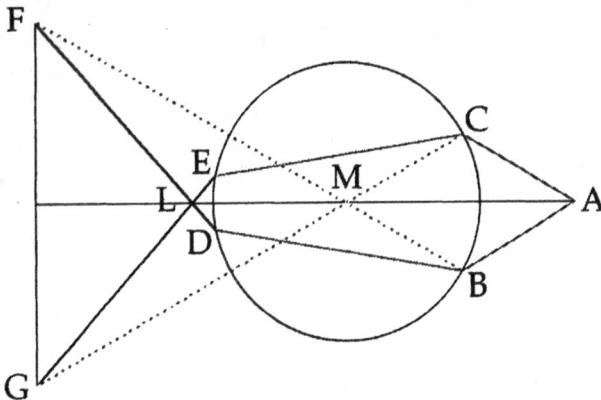

Figure 2.16

punctus F ad oculum, nisi vertatur ex sinistro ad dextrum; veniet ad D, et
ex D refrangetur ad B, per secundam huius, et egrediendo veniet ad oculum
A, per quintam. Eodem modo G ad E, mox ex E ad C, et demum ad oculum
A. Cathetus ex F per M centrum occurret lineae venienti ad oculum in B
ubi [58] videbitur punctus F. Sic G ad C, et inversa videbitur, est enim
inversionis punctus L.

Magnitudine et oculo pilae propinquis, recta videbitur, at recedente oculo
obvertetur. PROP. 17

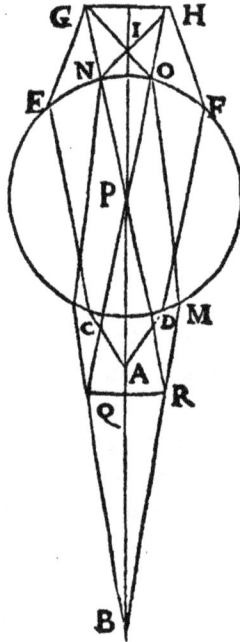

Esto primo oculus A; respiciat magnitudinem GH prope pilam, et sit quoque
A oculus ex opposito prope pilam. Punctus G insinuabitur oculo per E, ex
secunda. Refrangitur in C, et egredietur ex C, per quintam, in A. Idem ex
alia parte, H per F perveniet ad D, et ex D in A.

At elongetur oculus in B. G non insinuabitur oculo B nisi per O. Et ex
O refrangetur in M, et ex [59] M egrediendo veniet ad B. Idem ex alia parte,
H per N perveniet ad L,[24] et ex L ad B. Invertetur ergo in puncto I; H igitur
sinistra pars videbitur in dextra, et G in dextra videbitur in sinistra. Recta
igitur in CD videbitur, obversa vero in QR, at maior. Imago in QR non vere
extra videtur, sed satis obscure ubi ubi sit; et fixe intuentibus ibi apparet.

Magnitudo propinqua pilae intra punctum inversionis recta videbitur oculo
et proxime et longe se posito. PROP. 18

---

[24]"H"in original text.

the eye unless it reverses course from left to right, and so it will arrive at D and will be refracted to B from D, according to the second proposition of this book, and upon exiting it will reach the eye at A, according to the fifth proposition [of this book]. By the same token, the image of G will reach E, then continue from E to C, and finally reach the eye at A. The cathetus extended from F through center point M will intersect line [of refraction AB] reaching the eye at B, where [58] point F will be seen. The same holds for [the image of] G in reaching C, and [so image CB of object GF] will appear inverted, for L is the point of inversion.[33]

PROP. 17: *When the magnitude and the eye are close to the sphere, the magnitude will appear upright, but it will invert when the eye is drawn back.*

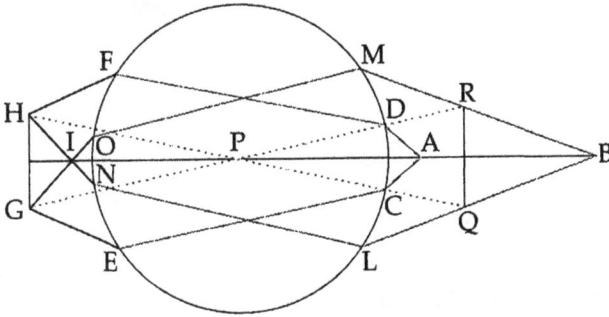

Figure 2.17

First, let A [in Figure 2.17] be the eye, let it face magnitude GH, which lies near the sphere, and also let the eye at A be near the facing sphere. The image of point G will wend its way toward the eye through E, according to the second proposition [of this book]. It is refracted to C, and it will pass out from C to A, according to the fifth proposition [of this book]. Likewise, on the other side the image of H will reach D through F, and it will continue from D to A.

   Now let the eye be moved away to B. The image of G will not wend its way to the eye at B except through O. From O it will be refracted to M, and after exiting from [59] M, it will reach B. Likewise, on the other side, the image of H will reach L through N, and B through L. The image will therefore be inverted at point I, so H on the left side will be seen on the right, and G on the right side will be seen on the left. Therefore, [magnitude GH] will appear upright at CD [to the eye at A], but it will appear inverted and larger at QR [to the eye at B]. The image at QR does not actually appear to lie outside [the sphere], but wherever it is, it appears fairly obscure; still, it does appear there to viewers who look attentively.[34]

PROP. 18: *A magnitude situated near the sphere and inside the point of inversion will appear upright to an eye placed either near or far away.*

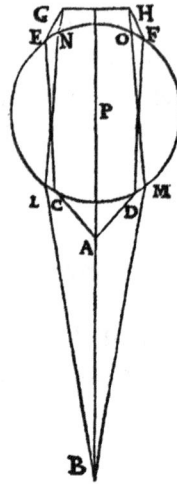

Esto conspicienda magnitudo GH, adeo prope pilam ut se oculis insinuando inverti haud possit. Veniat primo H [60] per F. Per secundam refrangetur ad D, et per quintam ad A, ubi invento oculo; videbitur recta. At se ponatur oculus in B. Idem punctus H, ut se oculo insinuet, veniet per O, refrangetur ad M, relabetur ad B per decreta angulorum egredientium. Videbitur igitur a B oculo etiam recta, nec tam prope pilam inverti valet.

Oculo pilae propinquo, magnitudine et proxime et remote posita, semper recta videbitur, sed remotior brevior videbitur. PROP. 19

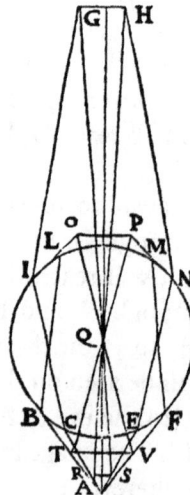

Intra punctum inversionis oculo vel magnitudine positis, semper recta videbitur magnitudo. Nunc esto oculus pilae proximus A; magnitudo proxime posita OP. Insinuet se P per [61] M. Refrangatur ad F, per secundam;

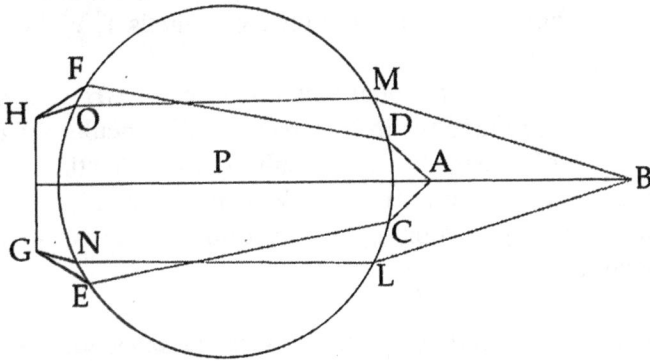

Figure 2.18

Let GH [in Figure 2.18] be the magnitude to be viewed, and let it lie close enough to the sphere that in order for its image to wend its way to the eyes, it cannot possibly be inverted. First, let the image of H [60] reach F. According to the second proposition [of this book], it will be refracted to D, and then, according to the fifth proposition [of this book], to A, where the eye is encountered, and [so the entire magnitude] will appear upright. Now place the eye at B. In order for the same point H to make its way to the eye, it will reach O, will be refracted to M, and will incline toward B by the principles governing the angles of refraction. [Magnitude GH] will therefore also be seen upright by the eye at B, and so the sphere does not have the capacity to invert it at such a close distance.[35]

PROP. 19: *With the eye close to the sphere and the magnitude situated either near or far from it, the magnitude will always appear upright, but the farther away it is, the shorter it will appear.*

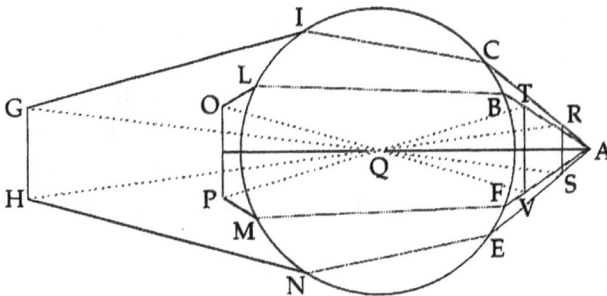

Figure 2.19

When the eye or the magnitude is placed inside the point of inversion, the magnitude will always appear upright. Now let the eye A [in Figure 2.19] be near the sphere, and let magnitude OP be placed near that sphere. The image of P will enter through [61] M. Let it be refracted to F, according to

egrediatur ad A oculum, per quintam. Ex altera parte, O per L veniat ad B, et inde ad A. Videbitur magnitudo OP recta et punctis T, V, ductis cathetis per centrum.

Longe igitur se ponatur eadem magnitudo, et sit GH. Veniet H ad N, ut se oculo insinuet, refrangitur in E, egredietur ad A oculum. Sic ex altera parte G punctus per I veniat ad C, et ad oculum labatur A; etiam recta veniet, nec datur alter concursus praeter illum. Videbitur ergo magnitudo GH, RS oculo propinquior sed brevior, ductis ex centro cathetis, quod demonstrare nitebamur.

Quanto remotior erit magnitudo a pila, eo brevior conspicietur. PROP. 20 [62]

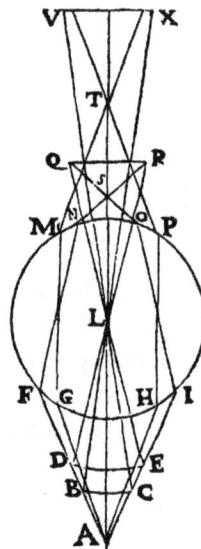

Sit primo magnitudo QR pilae proximior, et insinuans se oculo R per N. Venit ad F et mox ad A oculum. Sic Q ad O, et ad I,[25] ad A oculum. Tractis cathetis occurrent lineis deferentibus imaginem FA, IA in punctis D, E.

Sit magnitudo remotior VX, veniat X ad M, mox ad G, inde ad oculum A. Et V item ad P, mox ad H, demum ad oculum A. Ducantur catheti ex V, X per L centrum. Occurrent lineis imaginem deferentibus ad oculum GA, HA, in punctis B, C. Minor erit BC ipsa DE, quia quanto longior magnitudo abibit, tanto angustantur linearum anguli deferentium ad oculum simulachra et angustius suis occurrent cathetis.

Magnitudo oblique pilae veniens longe a suo loco et aliter videtur. PROP. 21

Sit magnitudo triangularis EFD veniens ad oculum A, et veniat oblique,

---

[25]"H"in original text.

the second proposition [of this book], and let it pass out to the eye at A, according to the fifth proposition [of this book]. On the other side, let the image of O reach B through L, and thence to A. Magnitude OP will thus appear upright and between points T and V, given the catheti [OQV and PQT] drawn through the center.

Let an identical magnitude be placed far away, then, and let it be GH. In order to make its way to the eye, the image of H will reach N, is then refracted to E, and will continue out to the eye at A. Likewise, on the other side, let the image of point G reach C through I and incline toward the eye at A, and it too will reach [the eye] upright, nor can any intersection other than this be given. Hence, magnitude GH will appear at RS, which is nearer the eye but shorter, given the catheti [GQS and HQR] extended through the center, and this is what we were at pains to demonstrate.[36]

PROP. 20: *The farther away a magnitude is from the sphere, the smaller it will appear.* [62]

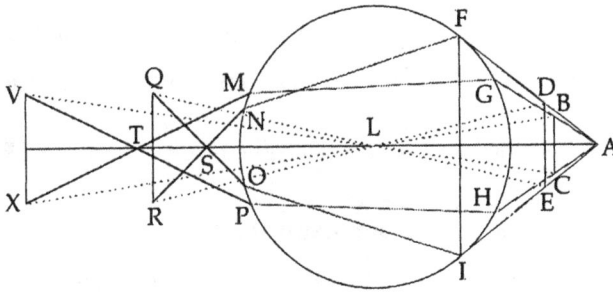

Figure 2.20

For a start, let QR [in Figure 2.20] be a magnitude nearer the sphere, and let the image of R make its way to the eye through N. It reaches F and then the eye at A. So, too, let the image of Q reach O, then I, and then the eye at A. When extended, the catheti [QLE and RLD] intersect lines FA and IA that convey the image at points D and E.

Let VX be a more distant magnitude, let the image of X reach M, then G, and then the eye at A. Also, let the image of V reach P, then H, and finally the eye at A. Draw the catheti from V and X through center point L. They will intersect lines GA and HA that convey the image to the eye at points B and C. BC will thus be smaller than DE, since the farther away the magnitude will lie, the more the angles formed by the lines conveying the images to the eye are narrowed and intersect their catheti more narrowly.[37]

PROP 21: *The image of a magnitude reaching the sphere at a slant appears to lie far away from its proper place and looks otherwise [than it actually is].*

Let the image of triangular magnitude EFD [in Figure 2.21] reach the eye at A, and let it arrive at a slant, that is, not in a directly facing position. The

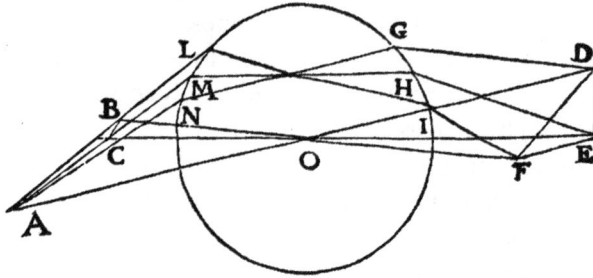

scilicet, non ex opposito. Punctus D venit ad G, refrangitur ad N per secundam, egreditur ad A per quintam. Cathetus ex D et per O centrum secabit prope oculum in A. Sic punctus E venit ad H, mox ad M, inde ad oculum A. Cathetus secabit MA in puncto C. Demum F punctus venit ad I, ac refrangitur in L, venit ad A. Cathetus occurret ex F ipsi LA in B, unde punctus F in B erit. Sic BCA et DEF obversus triangulus, ut promisimus. [63]

Pila crystallina Soli opposita ignem accendit prope pilam. PROP. 22

Radii perpendiculares supremam extimam pilae superficiem invadentes refranguntur, ut diximus, et egredientes etiam refranguntur denuo, unde decussantur prope pilam. In decussationibus igitur radiorum ibi concursus, et combustio. Exemplum damus.

Sint radii aequidistantes Solis extimam pilae superficiem invadentes A, B, C, D, E, F, G. Radii B, F, qui latera egaxoni tangunt, refranguntur ad calcem

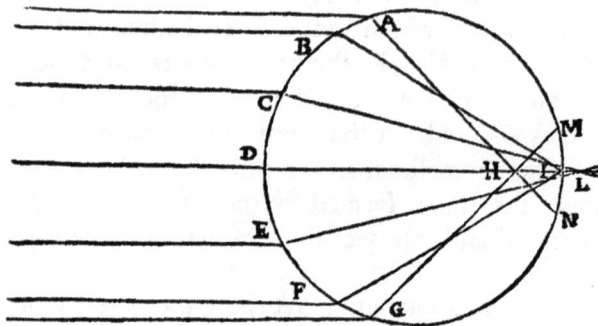

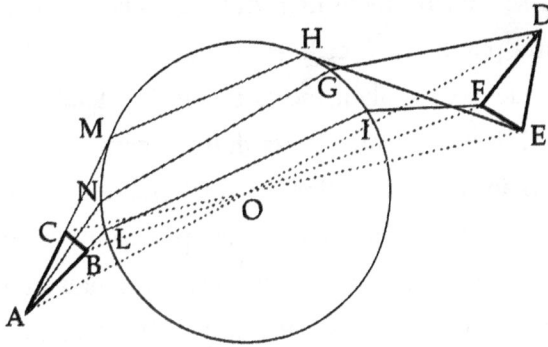

Figure 2.21

image of point D reaches G, is refracted at N, according to the second proposition [of this book], and continues out to A, according to the fifth proposition [of this book]. The cathetus extended from D through center point O will intersect [line of refraction NA] at A, the eye. Likewise, the image of point E arrives at H, then at M, and from there to the eye at A. The cathetus [EOC] will intersect [line of refraction] MA at point C. Finally, the image of point F reaches I, is refracted at L, and arrives at A. The cathetus from F will intersect this [line of refraction] LA at B, so the image of point F will lie at B. Consequently, image BCA is also an inversion of triangle DEF, as we said earlier [in the enunciation].[38] [63]

PROP. 22: *A glass sphere facing the Sun can light a fire near the sphere.*

Perpendicular rays entering the outermost surface of the sphere are refracted, as we said, and they are also refracted again on leaving the sphere, so they cross near it. Hence, there is a convergence in the crossings of the rays, and so there is combustion at that point. We provide the following illustration.

Let parallel rays from the Sun enter the outer surface of the sphere at A, B, C, D, E, F, and G [in Figure 2.22]. The rays at B and F, which strike

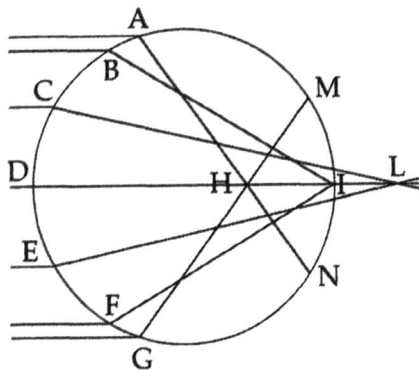

Figure 2.22

diametri DI, per secundam. Radii item A, G, quia latera pentagoni tangunt, refranguntur intra pilam M, N, et hi sunt inutiles combustioni. Radii vero, qui intra BF, omnes extra pilam incendunt foris, et ibi producunt incendium—ut radii C, E percutientes latera duodecagoni refranguntur in L.

Cantuarensis in hac propositione duplici errore notandus venit: putat enim radios a Solis centro progredientes supra pilam igne accendere, quod est falsum, nam a Solis perpendicularibus ignis excitatur, nam perpendiculares et ex centro eaedem sunt. Praeterea, nec si a Sole orirentur a centro radii, sequeretur prope pilam ignem accendi, nam tam longe dimissi radii a Sole etiam [64] longissime accenderent, quod est falsum, si vera sunt quae prius probavimus.

Maxime errat Vitellio qui putat Solis radios superficiem pilae invadentes perpendiculariter refrangi ad centrum, et quia ibi experimentum videri nequit. Optat sphaerae portionem semicirculo minorem, ut videret an in centro illius ignis excitetur, quod est falsissimum. In universo enim opere suo quidquid ex se supra illud Alhazeni est falsum fere est. Semicircularis sphaerae portio vel minor accedit loco prope finem diametri, et ex secunda refractione paulo propinquius, non autem in centro. Ratio ex his quae prius diximus deducitur.

Lumen eminus pila crystallina eiaculari. PROP. 23

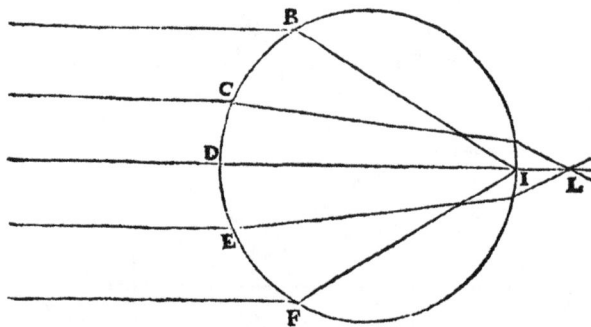

Radii, ut diximus, decussantur singuli prope pilam retro in diametro, propius et longius diffundunturque a puncto illo, ut a centro ad circumferentiam.

the [endpoints of the] sides of an inscribed hexagon, are refracted to the bottom of diameter DI, according to the second proposition [of this book]. Furthermore, since they strike the [endpoints of the] sides of an inscribed pentagon, the rays at A and G, in turn, are refracted inside the sphere to M and N, and so these rays are useless for combustion. All the rays within arc BF, however, are brought to a kindling point outside the sphere, and they produce fire at that point—for example, the rays at C and E that strike the [endpoints of the] sides of an inscribed twelve-sided figure are refracted to L.[39]

It should be noted that in handling this subject, John Pecham commits a twofold error, for he supposes that the rays reaching from the center of the Sun to the sphere light the fire, which is false because the perpendicular rays from the Sun induce fire and because the perpendicular rays and the one from the center are identical. Besides, if the rays were to originate at the center of the Sun, it would follow that they would not light a fire near the sphere, for rays sent from such a great distance by the Sun [64] would also light a fire at a great distance, which is false, if the things we have proved earlier are true.[40]

Witelo is flat wrong in supposing that the rays from the Sun that enter the surface of the sphere orthogonally are refracted to the center, and it seems he does not know how to subject this to experiment. Let him take a portion of a sphere less than a hemisphere in order to see whether it can ignite a fire at the center of curvature, which is absolutely false.[41] Indeed, throughout his work, whatever in it goes beyond Alhacen is utterly false. In a semicircular or smaller portion of the sphere [an incoming ray] reaches a point near the bottom of the diameter, and after the second refraction it reaches a point somewhat nearer, but not at the center. The explanation is based on what we have said earlier.[42]

PROP. 23: *Light can be projected to a great distance by a glass sphere.*

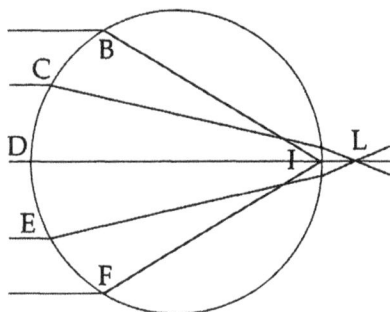

Figure 2.23

As we have said, individual rays cross one another just beyond the sphere on the diameter, and they are propagated near and far from that point, as if from a center point to a circumference. Thus, if a bright candle is placed

Si ibi igitur locetur candela ardens noctu, radii qui a candela egrediuntur a centro ad circumferentiam,[26] atque per pilam refrangentes paralleli evadunt, et longius lumen iaculabitur. Possumus et ibi concavum speculum accommodare, ut ignem non lumen, iaculetur, sed memento proportionis sphaerae et speculi. Radius C, E decussatur in L, inde spargitur. Si lumen ibi sparsum a centro, iaculabitur perpendiculariter.

---

[26]"circumferenliam"in original text.

at that point at night, the rays that are propagated from the candle are propagated as from a center to a circumference and on being refracted through the sphere pass out along parallel lines, and the light will therefore be projected to a great distance. And bearing in mind the correspondence between the sphere and [a concave spherical] mirror, we can also fit a concave mirror at this point so that fire, not light, can be projected.[43] The rays through C and E [in Figure 2.23] cross at L, and then they spread outward. If the light spreads out from this point as from a center, it will be projected orthogonally [from the sphere].

NOTES

1. Della Porta could be referring to two things with the Latin term *crystallinus*. On the one hand, he might have rock crystal, or clear quartz, in mind. On the other hand, he might be referring to the crystal glass produced at Murano since at least the mid-fifteenth century and prized for its exquisite clarity. Because he later distinguishes *vitreus* (glass) from *crystallinus*, it seems likely that in using the latter term he is referring to quartz. Technically it does not matter significantly because both quartz and the crystal glass of his day would have had an index of refraction roughly equal to that of modern crown glass, that is, ~1.5, and both would have had more or less the same clarity. Furthermore, since Della Porta earlier refers to such spheres explicitly as glass (*vitrea*) in book 1, prop. 11, I will henceforth render the Latin term *crystallinus* as glass except when the term is applied specifically to the crystalline lens of the eye.

2. In likening the eye to a glass sphere, Della Porta is appealing to the fact that, as a sphere itself, the eye is filled with transparent media and, moreover, open to incoming light through the cornea and pupil. Consequently, in order to understand how the eye works in vision, we must understand its basic optical qualities, which we can do by analogy to those of a glass sphere. In no way, however, does Della Porta mean to imply that, as the actual instrument of visual sensation, the lens should be understood by analogy to a glass sphere, although it does share optical qualities with such a sphere.

3. From the second half of the thirteenth century, the term *perspectiva* referred explicitly to the study of geometrical optics, with a focus on explaining visual perception—hence its root in the sight root *spec*. By Della Porta's day, however, the term had also taken on today's meaning, with its focus on artistic perspective as a means of depicting things in three-dimensional space naturalistically, or "realistically," on a plane surface. In this particular context, it is clear that Della Porta has the optical rather than the artistic form of the term in mind, so the "authors" to whom he is referring are the so-called Perspectivists: Alhacen, Roger Bacon, Witelo, and John Pecham, as well as later authors who follow their lead. Nonetheless, as will become clear later, Della Porta was also well aware of artistic perspective as, for instance, described by Leon Battista Alberti (1404–72) in his famous *Della pittura/ De pictura* of 1435.

4. See Euclid (1589), 82–3. That triangle EBC is equilateral follows from the fact that sides EB and EC, both being radii of the circle, are equal. Consequently, angles EBC and ECB are equal. Moreover, angle BEC is 60° by construction, so angles EBC and ECB must both be 60° in order for the internal angles of the triangle to total 180°. Because each of the angles is subtended by an equal side, all three sides of the triangle are equal.

5. Since AB is parallel to CF, then arc AC is equal to arc BF, which is 45° by construction. Hence, arc AB is 90°, which means that AB is the side of an inscribed square, from which it follows that angle ADB subtended by it = 90°. Thus, given the equality of radii AD and DB, it follows that base angles ABD and DAB of isosceles triangle ADB are equal at 45° each.

6. See *Catoptrics*, prop. 30, in Euclid (1537), 515; see also Euclid (1895), 340–43. By including the radiation within an arc of 60° on each side of the axis CE, Della Porta is including a considerable amount of radiation not involved in burning because the effective portion of the mirror is much smaller than that. Nonetheless, Della Porta's implicit recognition that the rays all undergo spherical aberration indicates that he understood both that and why the burning point in concave mirrors is actually a small area, not a point, just shy of the limit of reflection at the midpoint of the radius. On that basis, he would also have understood that the parallel rays focused to that burning area are all clustered quite close to the diameter forming the mirror's axis.

7. See Euclid (1589), 120; for Campanus's version of the proof, see Euclid (1537), 18.

8. See *Magiae Naturalis*, XVII, 14, in Della Porta (1589), 271.

9. In normal usage, the "point of reflection" (*punctus reflexionis*) designates the point at which the ray reflects from the mirror's surface. In this case, however, it designates the point at which the reflected ray intersects the diameter or axis, so I have translated *punctus reflexionis* as "intersection point of reflection" to make the intended meaning clear.

10. This rather clever construction depends on the fact that, no matter where end point B of radius DB lies on the circle within the quadrant 90° to the left of DE, dropping perpendicular FE from DB ensures that the two triangles DEF and BEF will be equal, thus making BE and DE equal. It therefore follows from proposition 1b that BE has to be the line of reflection.

11. Exactly how long Della Porta's "foot" would have been is unclear, but in all likelihood it would have been quite close to today's Anglo-American foot. The digit measures between 1.8 and 1.9 centimeters, which is nearly ³/₄ of today's Anglo-American inch (= 2.54 cm).

12. Della Porta's model in this analysis is clear: refraction through a glass sphere is perfectly symmetrical with reflection in a concave spherical mirror. Accordingly, when a ray parallel to the mirror's diameter (or axis) strikes a given point on the inner surface of such a mirror, it will reflect to a certain point *above* the diameter's end point, as long as it strikes the mirror within an arc of less than 60° from that end point. Likewise, when a ray parallel to a glass sphere's diameter (or axis) strikes the outer surface of that sphere at an equivalent point within an arc of 60° from the point at which the diameter passes through the top of the sphere, it will be refracted to a point on the diameter that lies precisely the same distance *below* the diameter's end point as its reflected counterpart struck the diameter above that end point. It is therefore crucial to bear in mind that Della Porta's analysis in this proposition is based on the ideal of a single refraction through surface GN of the sphere, not on the reality of double refraction through that surface and the surface at the bottom of the sphere. In other words, his analysis is based on the assumption that the entire area below surface GN and beyond point T is filled with crystal or glass. Accordingly, the limit of refraction below the sphere lies half a radius below the diameter's end point, just as that limit lies half a radius above in the case of reflection from a concave spherical mirror. It just so happens that the actual focal

point for parallel rays passing through Della Porta's sphere and refracting at both edges would be half the radius VP, but this fact is neither explicit nor implicit in Della Porta's analysis. Although Della Porta gives no quantitative results for this model, they are easily derived as follows. For ray EM incident at 60°, the angle of refraction is precisely 30°; for ray DL incident at 45°, the angle of refraction is ~25.5°; for ray CH incident at 30°, the angle of refraction is ~17.7°; and for ray BH incident at 22.5°, the angle of refraction is ~13.4°. The actual values according to the sine law should be: $i = 60°$, $r = ~35.3°$; $i = 45°$, $r = ~28.2°$; $i = 30°$, $r = ~19.5°$; $i = 22.5°$, $r = ~14.8°$. Della Porta's values therefore diverge fairly significantly from their modern counterparts, as they also do from Ptolemy's equivalent tabulations in book 5 of his *Optics*, which were repeated by Witelo and represented the only quantitative data available when Della Porta composed the *De Refractione*; see Ptolemy (1996), 236. Ptolemy's values are also considerably closer than Della Porta's to the values yielded by the sine law.

13. See *Mechanics*, prologue, 847b21-27, in Aristotle (1552), 54rb. Thus, the notion that spherical convexity and spherical concavity are mirror opposites provides Della Porta with a metaphysical justification for his model.

14. This analysis depends ultimately on the principle of reciprocity that was validated in book 1, proposition 3. Hence, since the light strikes the glass on a concave rather than a convex surface, it will be refracted in the opposite direction but by precisely the same angular amount. Note, by the way, that Della Porta ignores the refraction that would take place when the light leaves plane edges ZY and VY of the glass plate. In other words, as in the previous proposition, he treats rays FN, GP, etc., as if they continue to pass indefinitely through glass.

15. In fact, Della Porta's model does not work for ray BG, which is incident at 45°, because that angle of incidence exceeds the critical angle of just under 42° for crown glass. This is the angle at which (and beyond which) light passing from glass into air will undergo total reflection back into the glass. Consequently, ray BG will reflect internally at an angle of 45°, and although Della Porta's analysis does result in internal reflection along FT for ray AF incident at 60°, the actual reflection at 60° would be considerably sharper than represented by that analysis.

16. As in the previous case, so in this one, rays AF and BG incident at 60° and 45° respectively will undergo total internal reflection rather than pass from the glass into the air, and they will reflect as if from a concave spherical mirror, thus following the rules established earlier for such mirrors.

17. Aside from light, luminous color and heat, as well perhaps as other, more occult influences, can pass through glass, which is presumably what Della Porta has in mind when adding "something else" (*re quapiam*) to solar light rays refracted through the glass sphere. That such influences radiate like light and interact with physical bodies in much the same way was commonly believed during the Middle Ages and Renaissance; see, for example, Mary Quinlan-McGrath, *Influences: Art, Optics, and Astrology in the Italian Renaissance* (Chicago: University of Chicago Press, 2012). Perhaps most exemplary of this later Renaissance tendency to conflate light-

radiation with the radiation of occult influences was Della Porta's slightly older contemporary John Dee (1527–1608/09), a fellow devotee of natural magic; see for example, Urzula Szulakowska, *The Alchemy of Light: Geometry and Optics in Late Renaissance Alchemical Illustration* (Leiden: Brill, 2000), esp. 29–40; Nicholas Clulee, *John Dee's Natural Philosophy* (London: Routledge, 1988), esp. 39–72; and Wayne Shumaker, ed. and trans., *John Dee on Astronomy: Propaeduemata Aphoristica (1558 and 1568), Latin and English* (Berkeley: University of California Press, 1978).

18. The appropriate proposition is in fact I, 31, not I, 32; see Euclid (1589), 164.

19. As Della Porta has given no algorithm for directly finding where the refracted ray will intersect axis DE for any incident ray AB, we would have to turn to reflection from a concave spherical mirror according to the same parameters in order to determine indirectly where the reflected ray would intersect axis DE *above* F and then find the equivalent point at the same distance *below* F. That would be the sought-after point of refraction.

20. Presumably the "something else" to which Della Porta refers in this sentence is an actual object, as opposed to a light-source such as a candle flame.

21. Actually, both Alhacen and Witelo, following him, claim that the image appears somewhat dimmer than the object itself would appear at the same location because refraction weakens the radiation of luminous color, such weakening being proportional to the transparent medium's relative density. See Alhacen, *De aspectibus*, VII, vii, 38 in Alhacen (1572), 270, and in Alhacen (2010), 114-16 (Latin) and 308-9 (English). See also Witelo, *Perspectiva*, X, 10, in Witelo (1572), 414.

22. Della Porta's claim that the eye can see the image of something inside itself seems completely counter to experience, and it conflicts with his supposition that the front surface of the lens is where sight is initiated. According to Alhacen (and thus Witelo), when the cathetus and the refracted ray intersect at the center of sight in the middle of the eye, the resulting image formed at that point cannot actually be seen there; instead, the visual faculty makes a sort of psychological adjustment by transposing the image to the point of refraction on the refractive interface. By Alhacen's (and Witelo's) account, therefore the image of C will be seen by center of sight A not at A but at E. See Alhacen, *De aspectibus*, VII, vii, 49 in Alhacen (1572), 277, and in Alhacen (2010), 127–29 (Latin) and 318–19 (English). See also Witelo, *Perspectiva*, X, 43 in Witelo (1572), 440–41.

23. According to Della Porta's reasoning in this proposition, if F is a center of sight, then even though an image of C reaches F through the sphere, the image will not be seen unless the cathetus of incidence intersects FE at or in front of that point. In following this line of reasoning, Della Porta is completely in accord with such optical thinkers as Euclid, Ptolemy, Alhacen, and Alhacen's Perspectivist followers.

24. In other words, ray DO will be refracted away from the normal along ON when it enters the air between the glass sphere and the eye, and so refracted ray ON can be extended to F and beyond.

25. Because ray ABEM passes to the eye unrefracted, A will be seen at its actual location, not at some location H beyond A as represented in the diagram provided

in the original text. Consequently, A and H actually coincide, which is why they are labeled A/H in Figure 2.17.

26. In characterizing this analysis as the converse of the one in the preceding proposition, Della Porta presumably means only that in both cases equivalent arcs on the sphere are implicated, one being the back surface seen through the sphere, the other being the front surface of the sphere seen from directly in front of it.

27. What Della Porta is getting at in this paragraph is far from obvious. My interpretation is based on this figure, which is in turn based on Figure 2.13. Start

by assuming that the direction of radiation is reversed, so that I is taken as a point-source of radiation rather than as a center of sight. Let ray IF be parallel to diameter MNE, and let arc MF be 60°. Accordingly, chord MF would be the side of an inscribed hexagon, so it follows from proposition 2 that the ray refracted at F into the sphere will intersect diameter MNE at its end point E. From there, finally, it will be refracted along ED. Now let ray IH be parallel to diameter LNP, and let arc HL be 45° so that chord LH is the side of an inscribed octagon. According to proposition 2, then, the ray refracted at H will intersect diameter LNP at some point P outside the sphere. Before reaching that point, however, it will be refracted at point G along GB. Meantime, rays from I to M and L will refract symmetrically. Presumably, Della Porta's ulterior point is to show that his quantitative analysis of refraction in proposition 2 can be applied to analyses of this sort.

28. What Della Porta intends by this cryptic claim is anyone's guess; clearly, the ring bounded by the two circles through MF and LH will be considerably larger than the object on the other side of the sphere.

29. This proposition harks back to Alhacen, *De aspectibus*, VII, vii, 49 (in Alhacen [1572], 277, and in Alhacen [2010], 127–29 [Latin], and 318–19 [English]), but with some significant adjustments. For a start, Alhacen shows that the annular image is created by a line on the sphere's diameter, not by a magnitude straddling that diameter. Thus, rays from line AB in this figure will refract through the sphere to form an annular image on the sphere's surface from the perspective of center of sight I. Alhacen's conclusion that the image is annular and lies on the sphere's surface rests on the fact that the diameter through the line and the center of sight forms the cathetus of incidence for all the points of radiation. Consequently, I is the image location for the entire line, and in such a case, Alhacen contends, the visual faculty transfers the image of each point on AB to its corresponding point of refraction, all

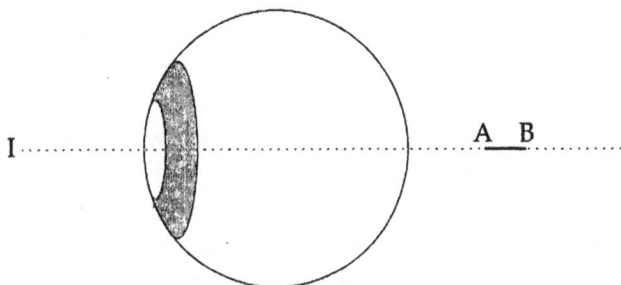

of which form a circle on the sphere's surface (see note 22 above). The total of all such circles thus forms the ring. Della Porta, on the other hand, seems either to have misunderstood or to have consciously rejected Alhacen's explanation. Instead, he attempts to explain the annular image according to the intersection of normals MN, NL, and so on, with their respective refracted rays at points M L, H, and F of refraction. The problem is that, contrary to his assumption, these normals are not the catheti of incidence for points A, B, C, and D. For instance, according to the cathetus rule, as articulated by Della Porta in book 1, proposition 7 above (see also book 1, definition 5), the image of point D in Figure 2.13 would appear not at F, on the sphere's surface, but at D', beyond the sphere, where the extension of cathetus ND intersects refracted ray FI. This, then, would be a case covered by book 2, proposition 7 above. Consequently, being based on a false assumption, the analysis in this proposition is logically invalid.

30. In this proposition Della Porta's analysis is based on the correct use of the catheti of incidence GM, FM, etc., so the conclusion is logically valid. The correct use of the cathetus rule here makes Della Porta's misuse of it in the previous proposition (and in subsequent ones) all the more puzzling.

31. In this analysis, Della Porta is doubly wrong insofar as he links false catheti to the wrong rays. Thus, according to his account, as illustrated in the top diagram of Figure 2.15, the image C of point A seen by the eye at H lies at the intersection of HLC (which is not a cathetus of incidence, since it does not intersect object A) with incident ray AE, not refracted ray FH. Furthermore, according to his analysis in proposition 9, the image should be at center of sight H. Della Porta's point is illustrated properly by the lower diagram of Figure 2.15. Accordingly, the extension of cathetus AL intersects both refracted rays FH and GI at A', so A' is the image for both eyes. Nonetheless, from H it is seen along HA'CF, whereas from I it is seen along IA'BG, so it will appear to the right from center of sight H, as seen along line of sight HF, and to the left from center of sight I, as seen along line of sight IG. Della Porta deals more fully with diplopia—that is, seeing multiple images of a single object—in book 6, where, as he claims, his mode of explanation is entirely different from the one he uses here.

32. Della Porta's reference to the point of inversion (*punctus inversionis*) here, as applied to spherical lenses, has a counterpart in book XVII, chapter 4 of his *Magiae Naturalis*, where he applies it to concave spherical mirrors; see Della Porta (1589), 264. This is the point at which the rays are brought to focus as well as

where the real image cast on a screen by those rays will be formed upside-down. It is not, however, the point at which an inverted virtual image is formed, which depends upon both the placement of the object and the placement of the eye with respect to the reflecting surface.

33. This is just a stripped-down version of proposition 14, the emphasis in this case being on the inversion of the image according to how the rays passing from the endpoints of the object to the surface of the glass sphere cross before reaching it.

34. Although Della Porta's analysis of the formation of inverted image QR for the eye at B is logically valid, his analysis of the formation of upright image CD for the eye at A is not. The problem in this case, as in propositions 13 and 14, is that he links the wrong cathetus with the wrong refracted ray. Consequently, the image of G cannot be at C, as Della Porta concludes, because that is where the extension of cathetus HP from point H on the object intersects refracted ray CA from point

G. This diagram provides one way of explaining the formation of an upright image when both the object and eye lie close to the sphere. Accordingly, let the image of the object's end point H enter the sphere along HF, refract internally along FD, and refract out symmetrically along DA. Likewise, the form of G enters along GE, refracts internally along EC, and reaches the eye along CA. The resulting image G'H' lies at the intersection of extensions PG' and PH' of catheti PG and PH with extensions AG' and AH' of refracted rays AC and AD. Note, however, that image H'G' lies beyond the object and is larger than it.

35. In this proposition, the analysis is too cursory to include an actual determination of image location, although on the basis of the previous proposition it seems evident that D will be the image of H, as seen from A. By extension, then, CD on the mirror's surface will be the image of GH, as seen from the same center of sight. Presumably, then, the image of GH as seen from B will lie either between arc LM on the surface of the sphere and the center of sight or on the surface itself. In both cases, however, Della Porta is wrong because he associates the wrong catheti with the wrong rays. This becomes clear if we compare this figure with the one in note 34 to the previous proposition. There, the eye lies somewhat farther away from the sphere than it does in this figure, so image H'G' of HG lies farther beyond the sphere. Moreover, according to the logic of both figures, as the eye draws farther away from

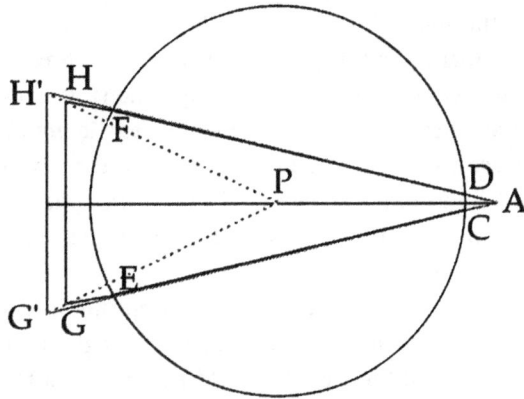

the sphere, while the object remains close to it, the image maintains the same orientation as the object but gets increasingly larger.

36. This entire demonstration, for both objects OP and GH, is vitiated by Della Porta's having again linked the wrong catheti to the wrong refracted rays. Consequently, the images are incorrectly located. The accompanying figure provides

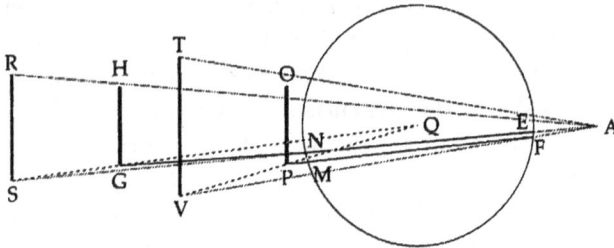

a proper analysis of the phenomenon, according to which the image of point P on nearer object OP is refracted at M, then at F, so as to arrive at A along FA, whose extenstion AV intersects cathetus QPV at V. The same holds by symmetry for the other side, so point O will be seen at T along extension AT of the refracted ray. Likewise, the image of G refracts at N, then at E, so as to arrive at A along EA, whose extension AS intersects cathetus QGS at S. Consequently, the two images TV and RS are upright, and image TV of nearer object OP is larger than image RS of farther object HG.

37. This analysis depends on assuming arbitrarily that the rays from the end-points of the two objects intersect before reaching the surface of the sphere. Accordingly, while rays VP and XM from distant object VX in the figure for proposition 20 intersect, rays HN and GI at the endpoints of distant object GH in the figure for proposition 19 do not. Yet, like VX in Figure 2.20, GH in Figure 2.19 lies well beyond the point of inversion, at least according to the placement of A in the figure as it is actually rendered in the original text. The core problem here, as well as in the previous proposition, is that Della Porta does not specify clearly enough where the eye should be placed. In proposition 19, for instance, the eye should be located

right up against the sphere. In proposition 20, on the other hand, it should lie moderately far from the sphere.

38. Because the radiation in this proposition occurs in three dimensions rather than in a single plane, as in all the previous propositions, it is extremely difficult to represent diagrammatically. The main point to bear in mind is that the two triangles essentially face each other through the sphere, so the three ray-paths DGNA, EHMC, and FILB all lie in different planes of refraction. One clear problem with this analysis is that, since point A on image ABC lies inside the eye, it will not be seen properly at that point.

39. By failing to make the conclusion of this proposition explicit, Della Porta leaves open the inference that all the rays that pass through and out of the sphere are implicated in the burning. Indeed, he could be construed to mean that every set of rays, such as those entering points C and E, and all the other rays entering the circle formed by them when the sphere is rotated about axis DL, come to a burning point outside the sphere. It is fairly clear, however, that he understood that the burning point lies immediately in front of point T in Figure 2.2 above and that this point is actually a spot where rays entering the sphere fairly close to the diameter form a maximum aggregate of intersection points after refraction through the sphere.

40. For Pecham's discussion of the refraction of solar rays through a glass sphere to a burning point, see *Perspectiva communis*, III, 16, in Pecham (1562), 105 (unnumbered); see also Pecham (1970), 228–31. Della Porta misrepresents Pecham insofar as Pecham does not claim that the perpendicular ray from the Sun lights fire; rather, he claims that all the rays emanating from the point where the diameter of the sphere intersects the Sun along the perpendicular are refracted to an equivalent point on the other side of the sphere, where they will light a fire. Della Porta is right, however, in interpreting Pecham to locate that point at a great distance from the sphere, contrary to what Della Porta has established both theoretically and empirically in book 2, proposition 2.

41. See *Perspectiva*, X, 48, in Witelo (1572), 443–44. As Della Porta points out, Witelo does claim that a spherical glass lens smaller than a hemisphere may (*forte*) bring rays that strike it along the perpendicular to focus at the center (*Forte tamen portio sphaera crystallinae minor hemisphaerio fortius inflammaret in loco centri sui posita re inflammabili: quoniam omnes radii totali illi superficie sphaericae perpendiculariter incidentes concurrerent in centro*). It is clear, however, from Witelo's appeal to *Perspectiva*, I, 72 (in Witelo [1572], 29–30), that the incident rays he has in mind strike the spherical surface along the perpendicular and thus pass toward the center unrefracted. Della Porta seems to interpret Witelo as claiming that the rays entering the sphere along parallels (as in proposition 2 of this book) are refracted to the sphere's center, a claim that Witelo would have known full well to be false.

42. What Della Porta seems to be getting at here can be best explained according to the accompanying figure. Let FIGD represent a glass sphere with its center at C. Let section FGD, which is less than a hemisphere, be cut from it. Let arc AD = 60°, and let a ray parallel to diameter DCI strike the section at A. According to proposition 2, the light will refract toward endpoint I of the diameter when it passes from air into the glass. On re-entering the air through the flat face of the section, however,

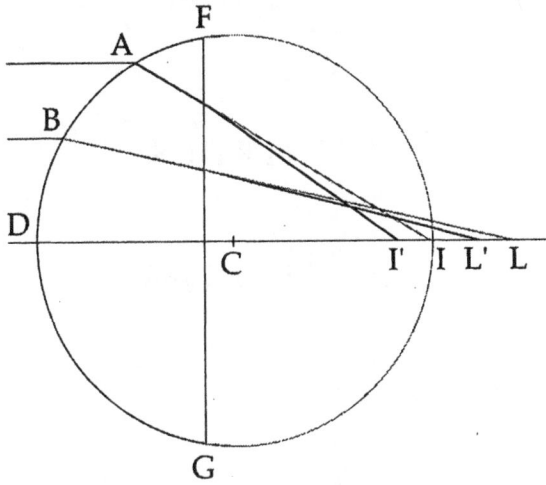

the light will be refracted away from the normal so as to arrive at I', which is closer to center point C than I. Likewise, if arc BD is 30°, a parallel ray striking B will be refracted toward point L outside the sphere. On meeting flat face FG, though, the light will be refracted away from the normal to arrive at L', which is closer than L to center point C. Accordingly, all the parallel rays passing through arc AD (and, indeed, the full arc DF) of the section will approach center point C after the second refraction at face FG, and the higher above point D they strike, the closer their approach to C. Nevertheless, none of those rays will ever reach that point.

43. Della Porta seems to be suggesting that a concave spherical mirror be placed so as to capture light at the sphere's focal point, refocus it in order to augment it, and then project it back through the sphere in the form of the heat that accompanies the light.

LIBER TERTIUS

PROOEMIUM

A refractione eiusque passionibus initio libri pro nostri ingenii captu, ut decuit auspicati sumus, nunc antequam ad visionem accingamur, operae pretium visum est oculorum constructionem eorumque munia aggredi. Frustra enim alicuius rei cognitionem assequi tentamus cuius partium constitutionem ignoramus; unde insimulandum aut praetereundum non duximus. Sic iis praecognitis ad ulteriora gradatim ascendemus, ne quid ad exactam visionis cognitionem deesse videretur; quorum maiorum quaepiam correximus, alia de medio sustulimus, et nonnulla addidimus nova. Nunc rem aggrediamur.

Oculum quatuor tunicis et tribus humoribus compaginatum. PROP. 1

Stupenda ac admirabilis oculi constructio haec est. Oculus, ut ex concava calvariae ossis sede exemptus est, exactae [66] sphaerae orbitam imitatur, et tuniculis et humoribus simul compactus est. Extima eius tunica oculum obducens et orbiculatim convestiens dura est et a dura substantia cerebri orta. Graeci σκληρότην dixerunt; neoterici (barbare quidem) consolidativam quod ossi oculum consolidet uniatque, et universa oculi constitutio praestantissime confoveatur.

Haec non sibi semper persimilis, nec simplici donata est nomine, sed posteriorem oculi sedem constituens crassa est, quae in anteriori ad pupillae regionem pervenit illic tenuis evadit, et in cornu faciem abit, non solum quod pelluciditate, verum quod cornu instar dura in laminas desquametur. Graeci κερᾱτὸειδῆς, nobis cornea dicitur. Tenuis ut cornu ne eius crassitie praepediretur visus; dura loricataque ut si quid extrinsecus occursurum

BOOK THREE

INTRODUCTION

Having been concerned at the beginning of this treatise with refraction and its effects in order to commence our study appropriately according to our intellectual capacity, it has seemed worthwhile at this point to deal with the structure of the eyes and the functions of their parts before we touch on vision. It is in fact pointless for us to try to attain knowledge of anything when we are ignorant of the structure of its parts; hence, we have carried on in a way not to be blamed or passed over. With these things known beforehand, then, we will ascend to higher [more complex] things incrementally so that what is necessary for a precise understanding of vision does not seem to be wanting. Some of these things, which are provided by our predecessors, we have corrected, others we have put forth as common, and we have added not a few new ones. Now let us broach our subject.

PROP. 1: *The eye is composed of four tunics and three humors.*

The structure of the eye, which is amazing and wonderful, is as follows. When it is removed from its concave socket in the skull, the eye resembles the orbital shape of a perfect [66] sphere, and it is composed altogether of thin tunics and humors. Its outer tunic, which covers over and cloaks the eye all around, is hard and arises from the dura mater of the brain. The Greeks called it σκληρότην[1]; the moderns (barbarously, to be sure) call it *consolidativa* because it can hold the eye together and unite it with the bone [of its socket], and it can sustain the entire structure of the eye most perfectly.

This tunic is not homogeneous throughout, nor is it endowed with a single name, but the portion comprising the posterior seat of the eye is thick, whereas the portion reaching toward the front in the area of the pupil becomes thin and transforms into something resembling horn, not only because of its clarity but also because it is hard like horn and can be peeled apart into layers. In Greek it is called κερᾰτόειδής[2] and cornea[3] by us. It is thin like horn so as not to hinder vision by its density, and it is hard and armored as if to fend off anything obtruding from outside that might lead

expelleret cum cuique oblaesioni obvia futura esset. Haec in fine visorium nervum convestit, et tandem se explicans cerebrum ex dura meninge ortum habet.

Interna huius tunicae superficies tota ea sede quae non pellucet (qua scilicet oculi album ab nigro distinguitur) alteram tunicam ambit ipsique adcubat, qua vero parte lucida cernitur ab ea subducitur. Id ei peculiare accidit quod mediam fenestravit pupilla (nam fenestram illam pupillam dicimus), comprimiturque usque ad acus tenuitatem, dilatatur vero usque ad lentis magnitudinem, instar loculorum, ut lucis accessu recessuve commodius visibilium se species insinuarent. Universam hanc tunicam fuscedo quaedam obnubilat, ne incertam aciem vagari sinat, qua parte corneae substernitur varie tingitur glauca, nigra, caesia, fulvaque.

Totam hanc tunicam a conterminis vulsam si contempleris, uvae nigro folliculo simillima est a quo pediculus exemptus sit. Foramen quo pediculus visebatur ipsi pupillae comparabitur, si ex eo evacuari intelligas, ob id Graecis $ῥαγοειδής$ nobis uvea dicitur. Membrana haec usque ad visorii nervi ortum comes producitur, mox nervum subiens, in tenuem meningem definit, quae quia $χοροειδής$ dicitur, Graeci quoque illam $χοροειδῆ$ vocant.

Sed spatium quod a concava corneae superficie continetur usque ad crystallini humoris sedem liquore quodam completur, qui quum tenuis, lympidus, fluidusque sit instar aquae, a Graecis $ὑδᾰτοειδής$ a nobis aqueus dicitur. Antiquiores medici albuginem nominarunt, non quod ovi albo collatus par sit sed qualis cernitur [67] in recentibus ovis transudare quum igni admoventur. Uvea in medio natat, ac hiatu quo in medio pertusa est humor in anteriorem corneae sedem adlabitur ut liberius spectra obiectorum commeent. Hic globi portionem explicat, anteriori parte gibba corneae parte concluditur. Posterior tantum a plana superficie se subducit, quantum crystallini pars protuberans in eum se insinuat.

Paulo supra oculi centrum humor, qui a nimia similitudine quam cum mediocriter concreta glacie obtinet, Graeci $κρυσταλλοειδῆ$ vocant. Construitur non corporis habitu soliditateve sed luce potius, qui etsi humor dicatur, non defluit sed molliculus caerae substantiam refert. Globuli speciem, sed non adamussim refert; ac si duabus sphaerae sectionibus compactus fuerit, lentis formam retinet, unde etiam Graecis $φᾰκοειδῆ$ vocari arbitror, superficie constans laevi, lubricaque. Ipsi membranula circumtecta est instar

to its being injured. This [entire outer] tunic sheaths the optic nerve at the back end [of the eye], and in spreading out finally to the brain, it has its origin in the brain's dura mater.

The whole [portion of the] inner surface of this tunic that is not transparent in its place (i.e., according to which the white is distinguished from the black of the eye[4]) surrounds and nests with another tunic, a portion of which is excised from it and thus open to light. Specifically, it happens that the pupil has opened a window in its middle (indeed, we call that window the pupil), and it [sometimes] shrinks to the size of a needle point, and [sometimes] dilates to the size of a lentil, much as purses [do in opening and closing], so that with an increase or decrease in light the images of visible objects can pass through quite conveniently. A certain darkening covers the entirety of this tunic so that it does not allow the gaze to lapse into confusion, and the portion of it that is stretched under the cornea [i.e., the iris] is variously tinged gray, black, blue, or amber.

If you look at this entire tunic stripped apart from its surroundings [in the eye], it is very much like the dark skin of a grape from which the stem has been removed. If you think of the stem as removed, the opening through which it was viewed will be comparable to the pupil, whence the tunic is called ῥᾱγοειδής[5] by the Greeks, and uvea[6] by us. This membrane extends all the way to the companion origin of the optic nerves, and then extending along the nerve, it terminates in the [brain's] pia mater, which is why it is called χοροειδής,[7] and the Greeks also call it χοροειδῆ.[8]

Now the space that is enclosed by the concave surface of the cornea up to the seat of the crystalline humor is filled with a certain fluid that is called ὑδᾰτοειδής[9] by the Greeks and aqueous by us because it is like water in being thin, clear, and serous. Ancient physicians called it albugineous, not because it is comparable to the white of an egg, but because of how it is seen [67] to exude from newly laid eggs when they are brought near a fire.[10] The [front of the] uvea is submerged in this medium, and this humor seeps through the opening with which the uvea is perforated up to the anterior seat of the cornea so that the images of objects can pass through it quite freely. This fluid fills up a portion of the ocular globe, and it is enclosed at the front by the anterior, bulging portion of the cornea. Its posterior side falls short of a plane surface by the amount that the bulging part of the crystalline humor reaches into it.[11]

Somewhat above the center of the eye is a humor that the Greeks call κρυσταλλοειδῆ[12] on account of its great similarity to what happens with moderately congealed ice. It is composed of a body qualified not by its solidity but rather by its partaking in light, and even though it is called a humor, it does not flow but is soft and gives the impression of a waxy substance. It looks like a kind of globe, but not exactly; it has the shape of a lentil, as if it were composed of two spherical sections, according to which I judge that it is called φᾰκοειδῆ[13] by the Greeks, its surface firm, smooth, and polished throughout. This is covered by a thin membrane that is like

tenuissimae caeparum tunicae (quae inter earum orbes reperitur) exacte transparens; quam a telarum araneae similitudine *ἀραχνοειδῆ* vocant Graeci, sed non robore et duritia, sed tenuitate; expanditurque undequaque ad latera oculi ad uveam, at veluti interseptum humorem aqueum a vitreo disterminat, ne laxioribus sese meatibus commisceretur.

Caeterum posterior crystallini sedes alio humore fulcitur, qui non lucidus ut crystallinus, neque eiusdem consistentiae duritiaeque, nec ut aqueus diffluens, sed quia liquati ab igne vitri consistentiam et colorem referat, ob id a Graecis *ὑάλοειδής* a nobis vitreus nominatur; ex inferiori parte dum in membranis continetur, semiglobi formam refert, sed ubi crystallino adiungitur planam superficiem habet, si in eius medio sinus non imprimeretur tantus quantus crystallini humoris globus in eo mergitur innatatque, nil aliter inter ipsum et vitreum intercedens.

Reliqua pars quae a crystallino non occupatur ex tenui membranula (si membranula et non mucor dici meretur) quemadmodum ovi pellicula albuginem continet, vitreus humor circumsepitur, quae Graecis *ἀμφιβλησ-τροειδής* dicitur, eo quod *ἀμφιβλήστρω*, idest reti similis non formam sed plexum referens. His tunicis hisque humoribus natura oculum machinata

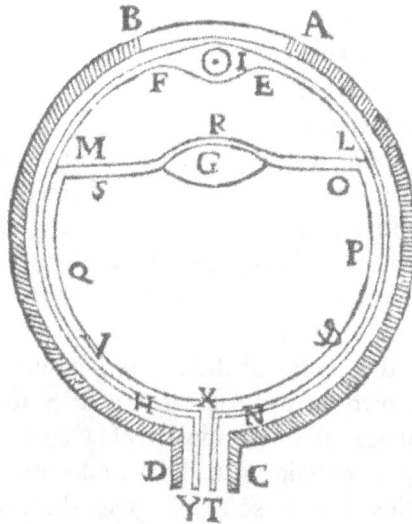

est, qui ut lucidissime nostris oculis exponatur, ipsum depingemus, ac si per medium secuissemus cum medietate nervi optici eo modo quo astrologi

the thinnest of onionskins (which are found between their rings) and that is exquisitely transparent. The Greeks call it ἀραχνοειδῆ[14] because of its similarity to spider webs, not for its strength and hardness but for its tenuousness; and it expands on all sides to the uvea at the edges of the eye, and just like a barrier it separates the aqueous from the vitreous humor so that they may not intermingle by passing through the space [between the crystalline humor and the uvea].

The rest of the space behind the seat of the crystalline humor is sustained by another humor that is not as pellucid as the crystalline humor, nor of the same consistency and hardness; nor is it fluid like the aqueous humor, but since it presents the consistency and color of glass melted by fire, it is called ὑάλοειδής[15] by the Greeks and vitreous by us. Although it is enclosed by [the uveal and retinal] membranes on its lower side, it takes the form of a hemisphere, but where it is adjoined to the crystalline humor, it has a plane surface, if that surface were not indented in the middle with a cavity just the right size for the globe of the crystalline humor to be fitted into it and float in it, with nothing else lying between it and the vitreous humor.

The remaining portion not occupied by the crystalline humor is surrounded by a very thin membrane (if it merits being called a membrane and not a mold) that envelops the vitreous humor in the way that a thin skin surrounds the albumen of an egg, and it is called ἀμφιβληστροειδῆ[16] by the Greeks because it is ἀμφιβλῆστρῳ, that is, similar to a net, not referring to its shape but to its interwoven structure.[17] Nature has devised the eye with these tunics and humors, and in order for it to be revealed in a perfectly

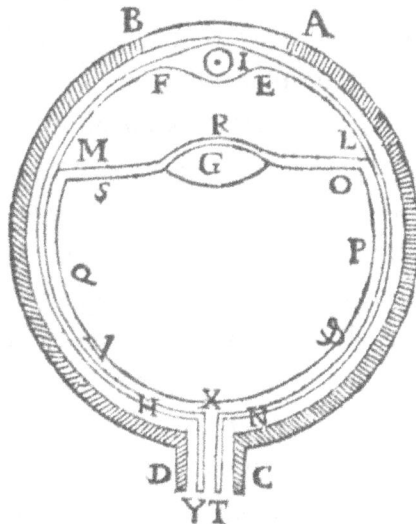

Figure 3.1

clear way to our eyes, we will depict it [in Figure 3.1] as if we had cut it in half, along with the middle of the optic nerve, in the way that astronomers

caelestes [68] sphaeras visui obiectant, ut unico intuitu totius fabrica contempletur.

ABCD dura tunica, cuius pars AB instar cornu pellucet et degenerat in crassam meningem CD.

EFNH uvea, cuius pars I fenestrata pupilla est, et degenerat in tenuem meningem TY.

ORS aranea tunica.

Et VX retiformis.

LRM humor aqueus.

G humor crystallinus.

PQ humor vitreus.

CD nervus opticus.

Oculum necessario fuisse spaericum. PROP. 2

Nunc oculi singulas partes exactius perpendamus, nam quae intus a natura conclusa sunt, ita locata sunt, ut nil eorum supervacaneum sit, nihil ad perfectam visionem non [69] necessarium. Sed demonstrabimus primo necessario fuisse rotundum. Primo ex capacitate, aliter enim capiendae quantitati suffecturus non erat; est orbicularis figura isoperimetrorum omnium capacissima. Quum etiam ex omnium partium corporis multis particulis ad multa munia obeunda, indigeret humoribus et tunicis, easque angustissimo spatio coarctarentur, ne caput gravarent; ex nunquam satis laudatae naturae munere largitum est ut orbicularis constitueretur.

Altera ratio ex necessitate, oportebat enim oculos sursum adducti, deorsum abduci, illuc huc circumvolvi, et deflecti. Nam si alterius figurae triangularis vel quadrangularis fuissent, horum motu oportebat partes mutuo contundi dilacerarique, et interceptas partes vacuas relinqui, unde quo visus aboleretur insignes noxas motus his partibus inferret. Vel si immoti essent, caput illuc, huc circumvolvendo, maxime hominem defatigari contingeret, quod erat inconcinnum. Addimus praeterea rotundationis illam rationem, ut liberiori motionis munere fungerentur, ut nostro arbitratu celerrime multa visu apprehenderemus, quum omnium figurarum sphaerica minus revolutionibus obnoxia, globosisque summa inest revolutionis agilitas.

Necessarium quoque erat oculum circumacta globataque figura esse ut rectae lineae super eum orientes perpendiculares et in centro concurrentes, quanto longius distraherentur, maiorem basim interciperent, quae solae veri-

portray the heavenly [68] spheres to the viewer so that the structure of the whole can be contemplated at a single glance.

ABCD is the hard tunic [i.e., sclera or *consolidativa*] whose portion AB [forming the cornea] is transparent like horn and reverts to the thick membrane CD [of the optic nerve at back].

EFNH is the uvea, whose open portion I forms the pupillary opening, and it reverts into the thin [inner] membrane TY [of the optic nerve].

ORS is the cobweb-like tunic.

VX is the retiform tunic

LRM is the aqueous humor.

G is the crystalline humor.

PQ is the vitreous humor.

CD is the optic nerve.[18]

PROP. 2: *The eye had to be spherical.*

Let us now consider the individual components of the eye more precisely, for those components that have been confined inside by nature are located there so that nothing more is needed beyond them and nothing more is needed for vision [69] to be accomplished. But we will start by demonstrating that the eye had to be round. First, it is round for the sake of capacity, for it could not be constructed other than by taking up space, and of all shapes that are bounded by equal surface areas, the spherical is the most capacious.[19] Also, since it consists of a body, all of whose parts comprise several smaller components that fulfill several functions, the humors and tunics needed to be compressed into the smallest space in order not to burden the head. There has never been sufficient praise for nature in ensuring that the eye is constructed in a spherical shape.

Another reason is rooted in necessity, for the eyes had to be drawn upward, pulled downward, revolved to one side or the other, and deflected. Now if they were of another shape, for example, triangular or quadrangular, the components, in moving in these ways, would have to be bruised and lacerated by one another and would have to leave the intercepted components behind in unfilled spaces, so the fact that vision would be destroyed indicates the injury such motion might bring to these components. On the other hand, if it were immobile, a person would then be excessively fatigued by moving his head back and forth, which would have been awkward. Furthermore, to this reason for roundness we add that the eyes would function with great freedom of motion so that we could quickly apprehend many things consciously by sight, since of all shapes the spherical is least affected by rotation, and ease of motion is greatest in the rotation of spheres.

The eye also had to be of a round and globular shape so that the farther away the straight lines that are perpendicular to it and that intersect at the center [to form the radiative cone] are extended out, the larger the base they intercept, according to which those lines alone reveal the truth, and this

tatem ostendunt, quod nulli, nisi sphaericae figurae contingit. Si planus enim
esset, non perciperet unico intuitu nisi aequale sibi spatium. Sit visus AB; a

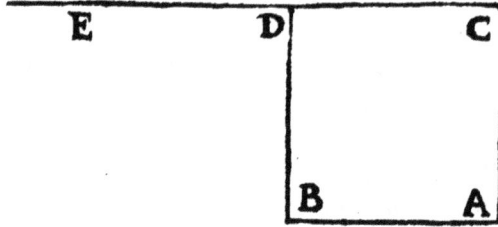

puncto A et B oriantur binae perpendiculares AC, BD. Interceptum solum
spatium CD videret aequale ipsi AB, per 23 et 25 primi, et si CD longior
esset usque ad E, oporteret ut oculus ad dextram moveretur, et si propius[1]
vel longius superficies obiectaretur. [70]

At si concavus, contrariae functionis opifex fuisset, nam multo minus
quam ipse est videret. Esto oculus ABCDLI, et occurrat ei res videnda EF.

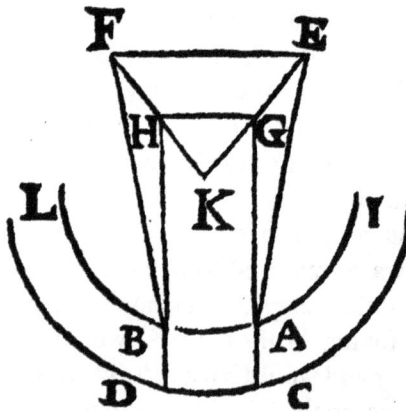

Cadit radius EA; refrangitur in C, per tertiam primi nostri. Extendatur;
perveniet in G. Eodem modo F venit in B, refrangitur in D, relabitur in H.
Sit oculi centrum foris K; ex E et F trahantur catheti. Occurrent lineis formam
referentibus in G, H, erit igitur GH forma minor re visa EF.

_____

[1]"proprius" in original text.

occurs with no shape other than the sphere.[20] For if the eye were flat, it would only perceive a space equal to itself with a single glance. Let AB [in

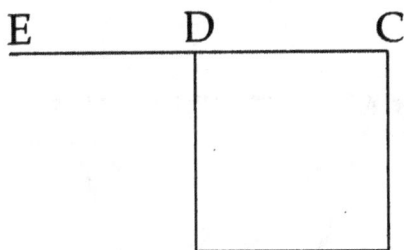

Figure 3.2a

Figure 3.2a] be the [surface at which] sight [occurs], and drop the two perpendiculars AC and BD from points A and B. All that would be seen is the intercepted space CD, which is equal to AB, according to I, 33 and 34 [of Euclid's *Elements*],[21] and if CD were extended farther to E, the eye would have to move to the left [in order to see E], whether the surface were to face the eye nearer or farther away.[22] [70]

If the eye were concave, however, the Creator would have been operating in an unfavorable way, for the eye would see things as far smaller than they

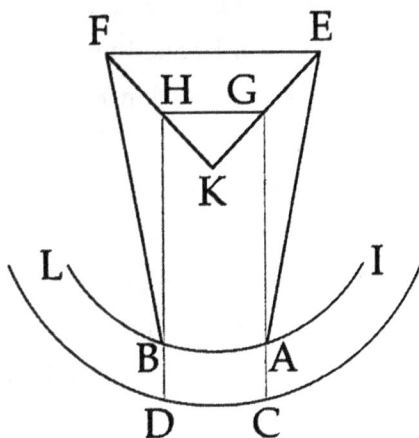

Figure 3.2b

are. Let ABCDLI [in Figure 3.2b] be the eye, and let the image of visible object EF reach it. Ray EA falls [to the cornea] and is refracted to C [on the crystalline humor], according to the third [proposition] of our first book.[23] Extend it [i.e., AC], and it will reach G. Likewise, the image of F reaches B and is refracted to D [on the crystalline humor], and [when BD is extended] it inclines toward H. Let K be the eye's center point, which is outside the eye, and drop the catheti [EK and FK] from E and F. They will intersect the lines conveying the image [of EF to the crystalline humor] at G and H, so the image GH will be smaller than visible object EF.

Plerunque etiam dextra sinistra et sinistra dextra, et quae sursum deorsum, et quae deorsum sursum viderentur. Esto oculus IABLCD, visibile

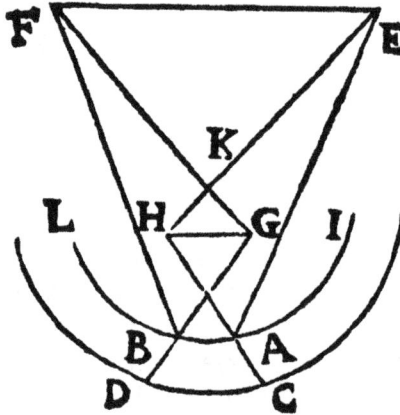

obiectum EF. Veniant ad pupillam AB lineae EA et FB. Ipsa EA refrangitur ad C, per tertiam primi; relabitur in H. Sic B refrangitur in D, relabiturque in G. Occurrent cathetis ex F, E per centrum K transeuntibus in G, H. Sic videbitur punctus F dexter in sinistrum G. [71]

Unde in sphaericam figuram animadversus sis, nullam nisi sphaericam exclamares, quae tot muniis sola praestat, ex centro enim per per pupillam, etsi per angustissimum spatium, longius exporrectae lineae totum fere hemisphaerium intercipiunt. Neque est, quod huic corroborando prolixius insistam. Gentilis Avicennae interpres medicam oculi rotundationis rationem affert, ob id globosum fuisse quod figura ipsa sola quicquid impedimenti vel noxii se ingerentis declinare potest. Sed Vitellionis opinionem non probo, oculum inquientis sphaericum propterea esse quia aquae naturae est, cuius proprietas est rotundari semper, sententiam Ioannis de Sacrobosco sequuti, qui aqueum elementum rotundum dicit quia gutta in terram cadens rotundatur.

Organum visus necessario aqueum futurum. PROP. 3

Organum visus utrum ex igne vel aqua esset haud parva fuit inter principes

For the most part, moreover, right would appear left and left right, and what lies above would appear below, and what lies below would appear

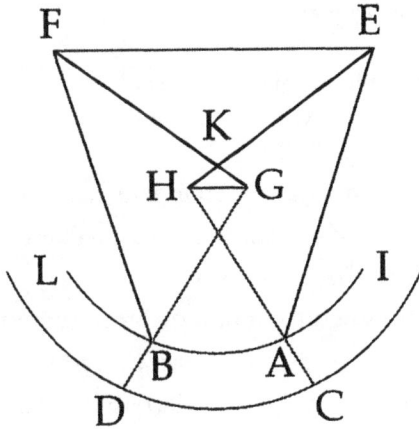

Figure 3.2c

above. Let IABLDC [in Figure 3.2c] be the eye and EF the visible object. Let lines EA and FB reach pupil AB. EA is refracted to C, according to the third proposition of the first book, and [the line of refraction AC] inclines toward H [when extended]. So, too, [ray FB reaching] B is refracted to D, and [the line of refraction BD] inclines toward G [when extended]. These lines of refraction intersect the catheti passing from F and E through center point K at G and H. Consequently, right-hand point F will be seen to the left at G.[24] [71]

Thus, when you have closely considered the spherical shape, you should exclaim that nothing but the spherical shape stands out on its own for having so many functions, for even though the lines of sight [are funneled] through the narrowest space in passing from the center through the pupil, when they are extended outward at great length, they intercept nearly an entire hemisphere.[25] But I do not need to continue reinforcing this point at greater length. Gentile da Foligno, the commentator on Avicenna['s *Canon*] offers a medical reason for the eye's roundness, according to which it was spherical because only that shape can avoid whatever impediments or injuries befall it.[26] But I do not approve the opinion of Witelo, who says that the eye is spherical because it is of the nature of water, which has the property of always being rounded, basing his claim on the opinion of John of Sacrobosco, who says that the element of water is rounded into drops on falling to the ground.[27]

PROP. 3: *The organ of sight had to be watery.*

Whether the organ of sight was composed of fire or water was an issue of no small controversy among the major natural philosophers. Empedocles

philosophos controversia. Empedocles igneum oculum existimavit, quatuor enim elementis quatuor sensus assignavit, tactui terram, gustui aquam, auditui aerem, visui vero ignem, inquiens:

> Conspicimus terram tellure, liquore liquorem,
> Aere naturam aeream; igni cernimus ignem.

Galenus etiam igneum dixit oculum Platonem sequutus et opticos sui temporis, qui per emissionem visionem fieri existimabant. Tantos viros decepit splendor in oculis conclusus, ut videbimus.

Democritus aqueum oculum dixit, et hunc sequutus Aristoteles libro *De animalium partibus*, et *De sensu et sensili*. Tradit enim oculum aqueum potius quam aereum ab rerum opifice solertissime constructum, quod aqua non solum diaphana sed euphana esset; nam etsi pellucidior sit aer, aqua tutior vallari conservarique potest, quum aer sua tenuitate facile diffletur. Oportebat oculum densis solidioribusque tunicis construi; quod si ita factum esset, transmissioni fuisset inopportunus, neque constructionem suo muneri idoneum obtineret, nam animalia densam oculi tunicam habentia obtusi sunt visus, ut locustae. Praeterea euphana est, quo visibiles formae melius insinuari, suscipi, conservarique possint, quod est oculo proprium, nam si aer suscipit, non retinet. Additque melius terminari [72] aquae diaphanitatem ex sua profunditate sine alterius opaci oppositione, quam aer, ut in puteis et aquae plenis vasis videtur, quae imaginem reddunt.

Sed haec isti nimis oscitanter aggressi sunt. Terminatio aquae in oculo necessaria non est, nec color magis aquae quam aeri inhaeret, nam si quis sub arbore decubuerit viridi colore perfundetur[2]; eadem lux colore intincta aquam et aerem tingit, qua sublata, uterque absque colore est. Sed cur oculos sagacissimus artifex ex aqua eleganter extruxerit explicabimus. Potissimae binae sunt rationes, prima in refractione. Quia sola in oculo pupilla pervia est, et visio per eius arctum et angustum foraminulum ducenda erat, nisi oculus aqueus fuisset, in eo refractio fieri non posset; obiectarum rerum minimam partem videret: nil nisi quae intra perpendiculares lineas intercluderetur. Res linearum ductibus dilucidior fiet.

Sit oculus KEL, rectae lineae huic perpendiculariter incumbentes sint AI,[3] BI; hic recta visio intercipitur. Linea vero quae his impensius obliquatur

---

[2]"perfudetur" in original text.
[3]"A" in original text.

assumed the eye consists of fire, for he assigned four senses to the four
elements, earth to touch, water to taste, air to hearing, and fire to vision,
saying:

> We perceive earth by earth, water by water,
> By air what is of an airy nature; by fire we perceive fire.[28]

Following Plato and the optical thinkers of his day, who thought vision
occurs through extramission, Galen also claimed that the eye is composed
of fire.[29] As we will see, the brightness contained within the eye deceived
many men.

Democritus claimed that the eye consists of water, and Aristotle followed
suit in his book *On the Parts of Animals* and in *On Sense*.[30] He in fact proposes
that the eye was fashioned most adroitly by the Creator of things out of water
rather than air because water was not only translucent but also exquisitely
transparent. For even though air is clearer, water can be more securely
protected and conserved, since air is easily dispersed because of its tenuousness. The eye had to be constructed from thick and very solid tunics.
But if it were made that way [exclusively], it would be unsuitable for the
transmission [of light and images], and it would not possess a structure
conducive to its function, for animals, such as lobsters, that have thick tunics
in their eye are weak-sighted. Furthermore, water is exquisitely transparent
so that visible forms can be better received by it and make their way through
it while also being conserved, which is appropriate for the eye, for if air
receives those forms it does not retain them.[31] He adds that the transparency
[72] of water is limited better than air by its depth without any opaque
object in the way, as is apparent in wells and vessels full of water, which
convey images.

But these points were put forward by him quite carelessly. There is no
necessary limit to water in the eye, and color does not inhere to a greater
degree in water than in air, for if someone reclines under a tree, he will be
saturated with a green color, and the same light tinged with color tints water
and air, each of which is colorless when it is removed [from the coloring
agent]. We will explain, however, why the Creator, who is most wise, constructed the eye from water so elegantly. There are two extremely compelling
reasons, the first of which is grounded in refraction. Since only the pupil
allows passage into the eye, and since sight had to be conducted through
its close and narrow opening, it follows that unless there were aqueous
humor, refraction could not occur in it, and so it would see a very small
portion of exposed objects: that is, nothing but what would be included
between the perpendicular lines [comprising the cone of direct radiation].
This point will be made clearer by a drawing.

Let KEL [in Figure 3.3] be the eye, and let the straight lines incident
on it along the perpendicular be AI and BI; this is where direct vision is
intercepted.[32] On the other hand, let EF be a line at a considerable slant to

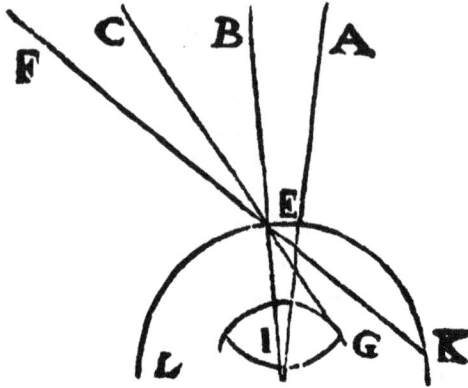

sit EF; per pupillam EF[4] inseratur. Si oculus inanis esset, progressus fieret ad K, et non tangeret crystallinum; nil ergo videret. Sed densioris corporis obiectu flexus faciet ad G per secundam primi nostri, et relabitur ad C. Videbit igitur quae intra CEB continentur.

Altera ratio est quod rerum imagines per pellucidissimum aqueum humorem pertranseuntes luculentiores crystallino defferentur, nam et claritate se induentes, fulgentiores conspiciuntur, ut apparet in Sole crystallum pertranseunte. Et nos specillis crystallinis multo clarius quam vitreis cernimus. Multiplicatur enim lux per refractionem, ut iam dicemus; aquea igitur claritate oculus donatus est, qua probe suum munus obire posset. [73]

Humorem aqueum tenuem et pellucidissimum esse. PROP. 4

Aqueus humor intaminata pelluciditate tenuitateque visendus venit; intervallum intra crystallinum et corneam implet, nec aliter fieri poterat. Oportebat enim visus instrumenta lucida futura, quia colorum dignotionem habitura erant. Natura enim comparatum est ut sola eiusmodi corpora a coloribus afficerentur. Diximus lucem omnium colorum expertem, quia omnibus coloribus tingi oportuerat, ut veros colores oculi repraesentaret. Sic aqueum vel albugineum humorem (quod guttae ex ovo igne transudanti par dictus est) nullum colorem habere necesse erat; alioquin difficulter alienos admitteret, neque crystallino colores et lineamenta visibilius ostendere potu-

---

4"EC" in original text.

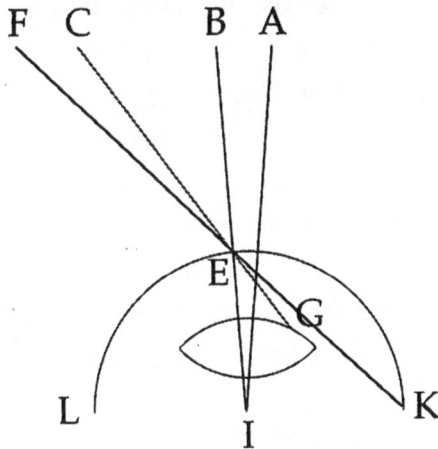

Figure 3.3

these lines, and let it be introduced through the pupil along EF. If the eye [up to the crystalline humor] were empty [of aqueous humor, EF] would continue to K and would not strike the crystalline humor, so the eye would see nothing. But the trajectory through the denser body [of aqueous humor] will cause it to be bent toward G, according to the second proposition of our first book, and [refracted ray EG] is inclined toward C. The eye will therefore see what lies within angle CEB.[33]

Another reason is that in passing through the extremely clear aqueous humor, the images of things reach the crystalline humor in a very bright state, for when they also assume such clarity they appear brighter, as is apparent when sunlight passes through crystal. And we perceive much more clearly through crystalline lenses than through glass ones.[34] For light is augmented by refraction, as we will soon discuss; consequently, the eye is endowed with the clarity of water by means of which it can fulfill its function properly. [73]

PROP. 4: *The aqueous humor is thin and extremely clear.*

The aqueous humor is noteworthy for its unblemished clarity and delicacy; it fills the space between the crystalline humor and the cornea, nor could it be otherwise. For the instrument of sight had to be perspicuous because it needed to be appropriately receptive of colors. Indeed, it was constituted by nature in such a way as only bodies of its kind are affected by colors. We have said that light is affected by all colors because it had to be tinged by all colors in order to present true colors to the eye. Thus, it was necessary that the aqueous or albugineous humor (which was said to be like the drops exuding from an egg near fire) have no color; otherwise it would admit external colors with difficulty, and it would not be able to reveal colors and outlines with utmost visibility to the crystalline humor. By way of showing

isset. Argumento, icterici esse possunt, quibus bilis vaporibus oculis sese ingerentibus, omnia videntur citrina, vel si nubeculis et fumositatibus conspurcaretur, fumosa et defaedata obiecta cerneremus, ut in senibus et defluxione laborantibus est videre. Imitatur tenuitate sua roridum laticem illum Maio mense olerum foliis illabentem irrorantemque, et in argenteas guttulas coeuntem. Argumentum suae tenuitatis est ut statim dissecto oculo in tenues auras facile diffletur. Purissimus igitur et ex omnibus defaecatior a rerum opifice excolatus est, et aeri fere par, ut per eum lux deferens simulachra et intromittens rerum veritatem inhibere non possit.

Per pupillam rerum simulachra intromitti. PROP. 5

Diximus de aqueo humore; sequitur de pupilla dicamus, per eam namque simulachra crystallino intromittuntur. Quasi fida ianitrix oculo adsidet, ibique sua munia exercet. Sed veritatem medicis experimentis probabimus. Morbus est unguis Latinis, pterigion Graecis dictus; carnosa excrescentia est corneae vel durae quam diximus membranae obveniens. Haec ab oculi cantho exoritur et pupillam ad usque pervenit, quam ubi attigit tenebris offunditur et visus orbamur, sic chyrurgi [74] manum passa redit amissus visus. Hypochima supra pupillam est humor congelatus quo visu homo destituitur; acum admittendo ex oculi angulo per obliquum usque ad pupillam, compactum humorem illum in infima parte deducit, et statim visus restituitur.

Pupillam accessu recessuve luminis arctari laxarique. PROP. 6

Visus instrumentum fulgidum, transparens, et tenuissimum diximus. Ob id immoderato lumini obnoxium, oportebat igitur tunica quadam circummoeniri,[5] foraminulo pervia, cuius famulatu ad lucis accessum obtenderetur, remitteretur vero ad eiusdem abscessum, ut in incerta luce rectius quid conspicare valeremus. Pupilla igitur in die in oculi vestibulo huic muneri praeficitur, immoderato lumini obstaculo futura, in crepusculis vero retinacula relaxaret ut spectra per latiorem fenestram intro liberius commearent. Unde uvea circa foramen rugosa est, ac loculorum instar comprimitur laxaturque, noctu vero post obitas functiones quiesceret.

Sed prius rei veritatem post artificium a natura excogitatum enarremus. Si amici oculos apertos intentosque vehementius Solis lumini obiectos con-

---

[5]"circummonire" in original text.

this, these humors can be jaundiced so that, when vapors from bile are forced into the eyes, everything appears yellow, or if those humors are adulterated by small cloudy or smoky spots, we perceive objects as cloudy or blemished, as can be seen in old people and those suffering from discharge.[35] In its delicacy it is like the water of falling dew that is sprinkled upon the leaves of herbs during the month of May and that consolidates into silvery droplets. Proof of its delicacy is that as soon as the eye is dissected the aqueous humor easily dissipates into thin air. It has therefore been rendered the purest and most refined of all the humors by the Creator of things, and it is virtually like air so that light conveying images and passing into it cannot mask the reality of things.

PROP. 5: *The images of things are transmitted inward through the pupil.*

We have talked about the aqueous humor, so it follows that we should discuss the pupil, for images are transmitted through it to the crystalline humor. It stands by like a faithful doorkeeper for the eye, and here it exercises its functions. But we will demonstrate the truth by medical experiments. There is a disease called *unguis* by the Latins and *pterigion* by the Greeks; it is a fleshy growth on the cornea or sclera, which we said is the membrane that comes up to meet the cornea. This growth arises at the corner of the eye and extends all the way to the pupil, where, on reaching it, it overspreads it with shadow and we lose vision, so if it is subjected to the hand of a surgeon, [74] the lost sight returns.[36] *Hypochima* is a fluid congealed over the pupil that deprives a man of sight, but by introducing a needle at a slant from the angle of the eye up to the pupil, a surgeon pushes the compacted fluid out of the way into the lower portion [of the basin in front of the pupil], and sight is immediately restored.[37]

PROP. 6: *When light approaches or draws away from it, the pupil is constricted or dilated.*

We have said that the organ of sight [i.e., the crystalline humor] is bright, transparent, and extremely delicate. Because it is subject to immoderate light, it was necessary that it be fenced round by a particular tunic penetrated by a small hole through whose service it would be covered up when light is brought close but would be reopened when the light is drawn away so that we would be capable of perceiving something correctly in uncertain light. Therefore, the pupil is set over the eye to serve as a vestibule and to be an obstacle to immoderate light during the day, whereas at dusk what held it closed loosens up so that the images can enter more freely through a wider opening. Consequently, the uvea around the pupillary opening is wrinkled, and it is drawn tight and loosened in the manner of a [drawstring] purse, but at night, after its duties are done, it can rest.

But first we will expose the truth behind a device we reasoned out from nature. If you look closely at the open eyes of a friend who is staring intently

templaberis, adeo pupillam coarctari videbis, ut per angustissimum[6] foramen vix tenuis acus aciem admitteret, eosdem si in obscuro cubiculo convertat, parvo tempore curriculo foramen adeo dilatari conspicies, ut fere lentem capiat. Huius rei instrumento certius fies compos quod Archimedes in dignoscenda Solis quantitate usus est.

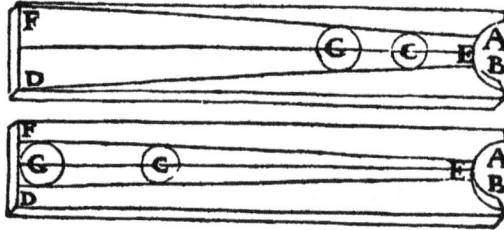

Sit regula lignea pedalis longitudinis et digitalis latitudinis ABFD; in medio linea EC, excavetur ex capite uno AEB, ut oculum commode excipiat. Eligatur cylinder ligneus torno politus minor C prope oculum, maior G longius. Appositoque uno oculo aperto, altero clauso, in AB converso contra Solem, tantisper cylindri accedant recedantque quousque radii per pupillam introeuntes latera utriusque cylindri contingant. Hoc facto, remove oculum a regula AB, et protractis lineis contingentibus cylindrorum circulos, iterum eandem regulam in obscuro loco [75] convertat. Appositoque oculo eodem loco, iterum cylindros appropinquet, removeatque donec visuales lineae latera contingant. Sic videbis lineas circulos contingentes in obscuro latiores esse.

Causas constrictionis et dilatationis nos plures excogitavimus. Prima in Luna est, nam huius syderis auctu, nostri corporis humores augescunt crassescuntque. Oportet igitur ut luculentius oculus illuminetur maioris luminis invasione, in defectu vero, quia humores tenuiores defaecatioresque fiunt, minori lumini indigi sumus. Unde si quis eodem lumine constitutus, utraque tempora observaverit, veritatem inveniet. Hoc clarius in aeluro perspicere licet, nam lunaribus quotidianis incrementis decrementisque pupillas maiores, minoresque habet. Ob id Isidis numen totam in se complecti naturam Aegyptii credentes, illi hoc animal dicarunt quod Lunae imaginem quandam in oculis repraesentet, ut ex *Oro Apolline* patet. Cardanus vero pro arbitrio felem contrahere ac relaxare pupillas dixit sine ratione, si enim ex libitu contrahere aut relaxare possemus, essent a natura musculi ad id opus constituti, at nulli apparent ex dissectione, non igitur ex voluntate aut constringere possumus aut relaxare.

---

[6]"angustismum" in original text.

at objects in bright sunlight, you will see that his pupil is thereby constricted so that his gaze can pass through the narrowest of openings that is hardly the size of a sharp needle point, whereas if he shifts the same eyes to a dark room, in a short span of time you will see the opening dilate so that it assumes virtually the size of a lentil. You will gain a more certain grasp of this phenomenon by means of an instrument that Archimedes used in determining the size of the Sun.[38]

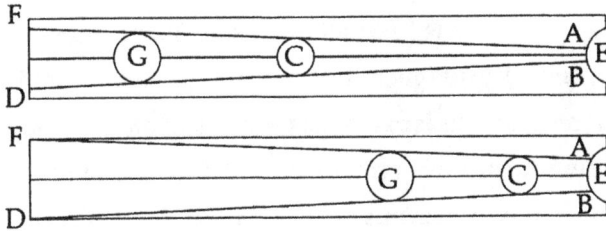

Figure 3.6

Let ABFD [in Figure 3.6] be a wooden ruler a foot long and a digit wide, and at one end of midline EC cut out nook AEB so that the ruler can be conveniently fitted to the eye. Take smaller wooden cylinder C smoothed in a lathe and place it near the eye while placing larger cylinder G farther away. With one eye closed and the other open at AB and facing toward the Sun, move the cylinders back and forth until the rays entering the pupil are tangent to the edges of both cylinders. When this is done, remove the eye from AB on the ruler, and having drawn the lines tangent to the circles [at the bases] of the cylinders, take the same ruler to a dark location. [75] With the eye placed at the same spot [on the ruler], the viewer should again move the cylinders to and fro until the lines of sight are tangent to the edges of both cylinders. Accordingly, you will see that the lines tangent to the [cylinders' base] circles are wider in the dark [as in the lower diagram of Figure 3.6].

We have thought a great deal about the reasons for the pupil's constriction and dilation. The first such reason is found in the Moon, for when this celestial body is full, the humors of our body increase in quantity and thicken. It is therefore necessary that the eye be illuminated more brightly by the introduction of more light, but in an eclipse, since the humors are rendered thinner and more refined, we need less light.[39] Therefore, if someone standing in the same light undertakes the observation at both times, he will find the truth. This may be seen quite clearly in a cat, for during the daily lunar waxings and wanings, it has larger and smaller pupils. For that reason, the Egyptians, believing that it embraced the full nature and will of Isis within itself, used to say that this animal portrays a particular image of the Moon in its eyes, as is shown in *Orus Apollo*.[40] Girolamo Cardano, in turn, claimed that cats purposely constrict and relax their pupils, but he did so without reason, for if we could constrict or relax our pupils at will, then there would be muscles designed by nature for that function, yet none are revealed from dissection, so we are unable to constrict or relax our pupils voluntarily.[41]

Altera ratio est aetatis. Senibus humores crassescunt et faeculentiores fiunt; maiori lumine indigent ad inspectionem, ob id eis relaxatur pupilla; iuvenibus vero, quia tenues purissimique, arctius coarctatur pupilla, ne validae lucis intromissu laederentur. Praeterea quod nimio Solis vel alicuius fulgidi obtutu laedatur visus patet in iis qui diu in Sole deambulant vel locis nive obsitis, nam tenebricosi evadunt et diu illatam oculis affectionem servant. Tandem si quis apertis oculis [76] aliquantisper Solem intueretur, et pupilla panderetur, ex Solis illapsu ignem in oculi fine accenderet vel ibi maximum ardorem ingereret (causam iam explanavimus superiore libro, prop. 22[7]).

Sed maiorum nostrorum errata dissimulanda non sunt, ne et nos ipsi posteritati imponamus. Alhazenus doctissimus Arabs dicit corneam et uveam sphaeras eccentricas esse, ac se mutuo secari in contact, Et si sphaera sphaeram secet, sectio circulus erit, ex Theodosio II; ob id pupillam sphaericam esse. Ab eo eadem simia Vitellio refert. Sed uterque fallitur. Horum fallaciam coarguit oculi anatomen, ut diximus, nam externa uveae superficies internae corneae adamussim adcubat usque ad corneae pellucidum; unaque fit in eo loco. Ab illa recedit et plana distenditur, et in medio parvo foramine scinditur; nec uspiam intersecantur, neque eccentricae sunt. Difficultates ponentes ubi non sunt ut scientias admirabiliores faciant.

Subitam pupillae dilatationem et constrictionem non fieri sine dolore. PROP. 7

Animadversione quoque dignum videtur hiulcas aut obstrictas pupillas repentino lucis occursu non fieri sine dolore. Repentina luce illata iis qui in tenebris commorati sunt, non sine laesione evadunt. Sic per Solem deambulantes, si ad tenebricosos locos illico divertant, caecutiunt, nam pupilla Solis praesentia coarctata, repentino tenebrarum occursu dilatari nequit.

> Dionysius Siciliae tyrannus domum super carcerem construxit clarissimam ac longe splendidissimam calce illitam ut homines qui diutissime carcere fuissent conclusi vinctos ipsos sursum educebat, qui ex multa profunda caligine in splendidiorem lucem egressi, quam cupide erant intuituri, occaecabantur.

Quod etiam oculi praepropera coarctatione oblaedantur, argumento sunt sternutationes, nam subita pupillae restrictio ad se telam trahit intenditque

---

[7]"21" in original text.

Another cause stems from age. In old people the humors thicken and become increasingly adulterated, and so it takes more light for those people to see, which is why the pupil is relaxed in them; in young people, on the other hand, since the humors are thin and exceedingly pure, the pupil is more narrowly constricted so that they will not be offended by the admission of strong light. Furthermore, the fact that looking too long in sunlight or any bright light offends sight is clear in the case of those who walk around in the Sun for a while or in places thickly covered in snow, for in darkened places they escape the bright light yet for a while maintain the effect in the eyes. Finally, if someone were to look for a while with open eyes [76] at the Sun, and if the pupil were dilated, the Sun shining down would ignite a fire at the back of the eye or would generate a great deal of heat (we already explained why in proposition 22 of the previous book).[42]

The errors of our predecessors, however, must not be ignored so that we do not inflict them on future generations. That most learned Arab, Alhacen, says that the cornea and uvea are eccentric spheres, and being in contact, they cut one another.[43] If a sphere cuts a sphere, the section will be a circle, according to Theodosius [in *Sphaerica*, book] 2, for which reason the pupil is a spherical section.[44] Witelo, Alhacen's ape, reproduces the same account as drawn from him.[45] But both are wrong. The anatomy of the eye refutes their error, for, as we said, the outer surface of the uvea corresponds perfectly to the inner surface of the cornea all the way to the transparent part of the cornea, and it forms a unified whole at that point. From there it draws away and is stretched flat, and it is perforated in the middle by a small hole. And the cornea and uvea do not intersect anywhere, nor are they eccentric.[46] Those who pose difficulties where they do not exist do so in order to render scientific subjects more awe-inspiring.

PROP. 7: *A sudden dilation or constriction of the pupil does not happen without pain.*

It also seems worthwhile to note that light suddenly striking dilated or constricted pupils does not happen without pain. When light is suddenly shone on those who have been lingering in darkness, they do not escape without distress. The same holds for people walking around in the Sun; if they suddenly wander into darkened places, they are blinded, for the pupil, having been constricted in the presence of the Sun, cannot be dilated immediately when darkness occurs.

> Dionysius, the tyrant of Sicily, built a bright white and exceptionally dazzling house of limestone above his prison so that he might take men defeated by him and shut up for a long time in prison up to it. Having come out of the profound gloom into the brilliant light, and having avidly gazed at it, they were blinded.[47]

That the eyes are also offended by a sudden constriction of the pupils is demonstrated in sneezing, for a sudden constriction of the pupil draws

quae oculi latera ad se abducit lancinatque quibus adhaeret tunica, latera intra nasum respondent, et in vicinia sunt loco qui delicatulus et pruriginosus est, unde si quis filo vel molliculo aliquo paxillo scalpet locum, ex ea tentingine sternuimus. Unde obiter adnotanda [77] venit Aristotelis ratio in *Problematis*: nos ideo talis occursu sternutare, "quia Sol calescens movet; perinde atque qui penna nares sollicitant . . . eumque calefacientes ocius spiritum ex humore eliciunt, cuius spiritus excursio[8] sternutatio est."

Sed Aristotelis propositum falsum est, nam non qui Sole ambulant sternutant sed qui ex tenebris exeuntes splendenti Solis lumini fiunt illico obviam, ubi nulla calefactio intercidit, quod non alio evenit quam subita pupillae constrictione. Mox quae calefactio ex fili impositione in naribus? Multi sunt qui Solis eclipsim inconniventibus oculis observantes ut effectum illum assequerentur, imprudenter occaecati sunt. Ob id summus rerum opifex diei et nocti crepuscula intermedia constituit, ut constricta diurnae lucis pupilla tenebrarum adventu in crepusculis dilatetur, ut noctu satis hiulca sit. Sic adventu solaris lucis in matutino crepusculo iam iam ex tenebris arctata, dilatari condiscat.

Qui colores pupillam referent, qui constringant, quive mediocriter arctent.
PROP. 8

Nec solum lux pupillam coarctari cogit, sed colores fulgidi et qui maxime visibiles, ut albi, crocei, sic et pulita metalla. Laedunt maxime candidae res visum, candidum enim luci affine, unde quum lux cum multo candido se offert visui, oblaedit illum et lancinat: "Xenophontis milites laesi oculis fuere, quum per multam nivem[9] iter fecere." Fabri et aurifabri saepe caecutiunt ex continua horum ignium inspectione. Sed fabulosos puto "philosophos Indiae, quos Gymnosophistas vocant, qui ab exortu ad usque Solis occasum contentis oculis orbem candidissimi syderis intuentur, in globo illo igneo rimantes secreta quaedam."

Contra vero nigri colores vel ad nigram tendentes, ut caerulei, halurgi, visum hebetant pupillam relaxando. At medii colores, ut viridis, qui ex flavo et caeruleo constat aequis partibus, visum refocillat et confovet, quia neque constringunt neque dilatant. Ideo permagnus ille rerum opifex arbores, herbam, et Terram ipsam amaeno virore amicit et decorat, ut continua horum inspectione affectus visus refocilletur.

---

[8]"excussio" in original text.
[9]"lucem" in original text.

in and stretches the web of membrane that leads to the edges of the eye and disrupts it where the tunic adheres to the corresponding sides within the nose, and they are near a spot that is extremely delicate and ticklish, so if someone touches that spot with a thread or a small, soft tendril of some sort, we sneeze from the stimulation. From this, incidentally, [77] comes a need to address Aristotle's reasoning in the *Problems*: according to this reasoning, we sneeze when this sort of thing happens "because heating by the Sun moves us to sneeze, and just like those who stimulate their noses with a feather, those being heated draw their breath very quickly from the humidity, and the rapid expulsion of breath is sneezing."[48]

But Aristotle's account is wrong, for it is not those who walk around in the Sun who sneeze but rather those who walk out of shadow and suddenly encounter the brilliant light of the Sun, at which point no heating occurs because nothing happens but the sudden constriction of the pupil. Also, what heating occurs from inserting a thread into the nostrils? There are many who have observed an eclipse of the Sun with eyes wide open in order to ascertain the effect, and they have foolishly been blinded. For that reason the supreme Creator of things established twilight between day and night so that the pupil constricted by daylight might be dilated in twilight by the advent of darkness in order to be sufficiently exposed at night. Just as the pupil can become thoroughly accustomed to constricting during morning twilight as soon as passes from darkness at the approach of sunlight, so it can become thoroughly accustomed to dilate [during evening twilight].

PROP. 8: *When colors are presented to the pupil, some may constrict it, and some may narrow it moderately.*

Not only does light lead the pupil to be constricted, but so do bright colors and ones that are highly visible, such as white and yellow, and the same holds for polished metal. White things offend the sight most, for white is akin to light, so when light is presented to sight along with something extremely white, it injures and disrupts sight: "Xenophon's soldiers suffered damage to their eyes when they made their way through deep snow."[49] Blacksmiths and goldsmiths are often blinded by continually looking at their fires. But I deem stories about "the philosophers of India whom they call gymnosophists" to be fables, according to which, "from the rising to the setting of the Sun, they gaze at the globe of that extremely bright star with fixed eyes, prying into certain secrets in that fiery globe."[50]

On the other hand, black or colors tending toward black, for example, sky blue or violet, deaden sight by relaxing the pupil. The intermediate colors, such as green, which is composed of yellow and sky blue in equal parts, revive and nurture sight because they neither constrict nor dilate the pupil. For that reason, the paramount Creator of things imbues and adorns trees, grass, and the Earth itself with a delightful green, so that by continually looking at these things our sight may be revived.

Smaragdo nihil iucundius, nihil utilius vident oculi, in primis virent ultra irrigua [78] gramina, ultra amnicas herbas. Defatigatos obtutus coloris laevitate reficiunt visus, quos alterius gemmae fulgor retunderit. Smaragdus recreat et exacuit; nec aliam ob causam non scalpitur ne offensum decus imaginum lacunis corrumperetur.

Recte igitur Aristoteles se habere oculum in coloribus, sicut auris in consonantiis, nam sicut ex proportione sesquialtera nascitur diapente, sic colores ex proportionalibus compositi adinvicem sunt delectabilissimi, ut ex croceo et caeruleo viridis. Sed id aliquo experimento comprobemus.

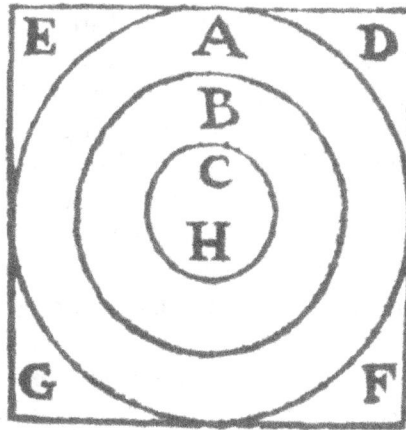

Sit quadrata tabula cubitalis FDGE; eam nigerrimo colore tinges. Centro H infige clavum, ut semipede a tabula distet. Pone oculum supra H, et immota acie, quousque visus extendatur lapillos circumpones. Mox amove visum et in circumactu lapillos dispositos invenies, quia pupilla rotunda est, ut videbimus. Eandem tabulam albissimo colore illines, eodemque loco oculum accommoda. Et iterum quousque acies extenditur lapillos dispones, et amovendo oculum, parvum invenies circulum C. Demum colore viridi tabulam superlines. Et oculum eodem loco constitutendo eodemque modo operando, circulum tertium B inter utrosque extremos medium cadere conspicies.

Sed interim falsa eliminemus ex mathematicorum familia. Stoici referente Plutarcho tenebras cogere visum aggregareque, et propterea hebetare dicunt; contra lumen disgregare. Sed satis improprie loquuti sunt, ut ocu-

The eyes see nothing more pleasant and nothing more useful than emerald, foremost among greens, greener than well-watered [78] grass and greener than plants that grow by rivers. Emeralds restore tired gazes by the soothing nature of their color, whereas the glittering of another gem would deaden them. Emerald revives and sharpens sight, and it is not engraved in order that its beauty not be corrupted by the damage of incised images.[51]

Aristotle therefore rightly holds that colors are to the eye as harmonies are to the ear[52]; for just as the interval of a fifth is based on the proportion of 2:3, colors that are compounded together according to proportions are most delightful, green, for example, being compounded of yellow and sky blue. But we will demonstrate this with a specific experiment.

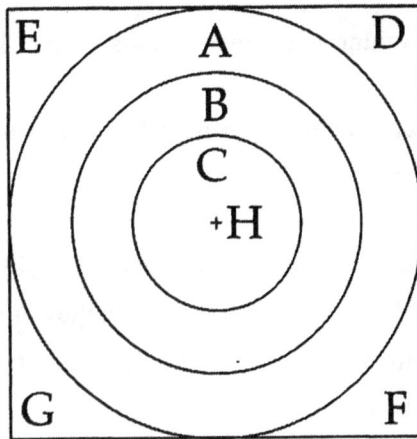

Figure 3.8

Let FDGE [in Figure 3.8] be a square board a cubit on each side, and color it pitch black.[53] At center point H drive a nail so that it[s head] lies half a foot above the board. Position an eye right above H, and, keeping your gaze steady, place pebbles all around as far as your sight extends. Then remove your eye and you will find the pebbles arranged in a circle [designated A], since the pupil is round, as we will see. You will paint the same board with a bright-white color, and position your eye at the same spot as before. Again, place the pebbles as far as your gaze extends, and having removed your eye, you will find [the pebbles forming] small circle C. Next, you will cover the board over with a green color. Placing your eye at the same spot as before and following the same procedure, you will notice that a third circle [of pebbles] B falls between the two extremes.

Meantime, however, let us dispense with falsehoods among the sect of mathematicians [who favor the visual ray theory]. The Stoics, according to Plutarch, claim that darkness collects and compresses the visual radiation, thereby weakening it, whereas light spreads it out.[54] But they spoke rather

lorum acies disgregari possit ut gallina pedibus tritici acervum, et congregari, ut gallina pullos suos. Aristoteles in *Problematibus* dicit oculos ad omnium rerum colores fatiscere et deteriores reddi, sed quum ad virides declinant, [79] optime se habere. Albi enim colores et nigri viribus suis officiunt, at viridi et mediis coloribus refocillantur corroboranturque.

> Et quemadmodum quum corpora nostra laborant deterius se habent, quum mediocriter optime efficiuntur, ita oculi ad res solidas offendentes laborant; ad humidas, quia nil obstat, oberrant, quum vero ad virides, quia solidae mediocriter sunt et aliquid humoris continent faciliter acquiescunt.

Bone Deus! Quid commune habet visus cum solido et liquido?

Qui lata pupilla sunt noctu melius, interdiu hebetius; qui arcta interdiu acutius, noctu caligant. PROP. 9

Sunt qui arcta, et qui lata donati sunt pupilla, quorum *Physiognomoniam* suo loco diximus. Sed qui lata sunt pupilla noctu melius, interdiu hebetius, qui vero arcta interdiu acutius noctu vero nil penitus vident. Noctivaga animalia, ut bubones, noctuae, feles, et lupi lata pupilla donati sunt, ut et vespertiliones; at gallinae, columbae, et passeres, et eiusmodi nil penitus noctu, interdiu acutissime vident. Et illa in die caecutiunt, et haec parvis pupillis praedita sunt. Aphrodiseus ob id

> animalia quaedam interdiu vident, noctu nequeunt, dicit, quia spiritum usui visus accomodatum habent crassiorem, quam ut aerem permeare ad rem usque videndam valeat, crassescit enim per noctem quum Solis absentia refrigeretur, interdiu extenuatur aeris beneficio qui tenuis et calidus est. Feles tamen, hyenae et vespertiliones contra se habent, quum enim interdiu nil fere videant; acute omnia per noctem conspiciunt. Spiritum quidem illum visorium et tenuissimum et dilucidissimum habent, ita ut per noctem hic modice incrassescens idoneus ad rerum conspectum reddatur; per diem luce praeclara supramodum extenuatus expandatur et evanescat. Leones per acriorem Solem ingredi non patiuntur, sunt enim oculorum acie acutiuscula, et ignem interdiu eamobrem fugiunt. Constat talem leonis aciem esse argumento pupillae quae praefulget, atque etiam partis humidae oculorum quae dilucidissima est.

Haec eadem ab eo Rogerius Bacchon. Sed haec mihi falsa videntur, qui enim lata sunt pupilla interdiu lumine multo splendoreque offenduntur, ac [80]

poorly in saying that the gaze of the eyes can be spread out in the way that a hen disperses a mound of wheat with her feet, or that it can be collected in the way a hen gathers her chicks. In the *Problems*, Aristotle says that our eyes weaken and deteriorate by looking at the colors of everything, but when they look toward green things, [79] they maintain themselves best.[55] For white and black colors hinder their powers, whereas they are revived and strengthened by green and the intermediate colors. [As Aristotle puts it],

> Just as our bodies may be poorly maintained when they work hard but are kept at their best when working moderately, so the eyes exert themselves when their sight is thrust against solid objects, whereas they operate more freely when directed at moist things because nothing blocks them, and they operate freely when directed at green things because those things are only moderately solid and contain a modicum of moistness.[56]

Good God! What does sight have in common with solid and liquid?

PROP. 9: *Those with a wide pupil see better at night and more dimly during the day, whereas those with a narrow pupil see more acutely during the day but are blind at night.*

There are those who are endowed with a narrow pupil and those with a wide one, and we discussed these in the appropriate place in *On Human Physiognomy*.[57] Those who are of a wide pupil see better at night and more weakly during the day, whereas those with a narrow pupil see more acutely during the day but scarcely anything at night. Nocturnal animals, such as horned owls, night-owls, cats, and wolves, are endowed with a wide pupil, as also are bats; however, hens, doves, sparrows, and the like see almost nothing at night but see quite clearly during the day. Whereas the former are blinded in daylight, the latter have already been said to have narrow pupils. In view of this, Alexander of Aphrodisias says that

> certain animals see in daylight but cannot see at night because they have a thicker spirit devoted to the functioning of sight, and in order for it to be forceful enough to permeate the air out to a visible object, it thickens at night when it is cooled by the absence of the Sun, whereas during the day it is attenuated with the help of air that is thin and warm. Cats, hyenas, and bats, however, operate in an opposite way, for during daylight they see almost nothing but see everything clearly at night. They of course possess an extremely tenuous and bright visual spirit so that at night, when it is somewhat thickened, it may be rendered capable of viewing things, but in bright daylight it can become so attenuated that it spreads out and dissipates. Because the gaze of their eyes is quite sharp, lions cannot bear to go out in piercing sunlight, and for this very reason they flee fire during the day. It is held that this is proof that such sight shines forth from the pupil of a lion and also from the moist portion of its eyes, which is extremely bright.[58]

Roger Bacon says the same thing on this basis.[59] But these things strike me as false because those who have a wide pupil are bothered during the day

noctu ob sui amplitudinem rerum simulachra facile subeunt et videntur. Contra vero qui arcta sunt oculorum acie, quia lumine non offenduntur, acutius vident; noctu rerum simulachra penetrare nequeunt.

Pupillas ad usum animalium accommodatas varias esse. **PROP. 10**

Ob id provida natura secundum animalium usum et necessitatem pupillas construxit; noctivagis, ut a summo ad imum referentur, quasi fores, ut citius et liberius suo fungi munere possint. Oportebat enim ut universa pupilla operiretur et oculi semiglobus foras prominens undique apertus et fulgentissimus foret. Non enim in eorum oculis albugo est, sed quicquid intra palpebras continetur id totum pupilla sit.

Contra se habent homines et diurna animalia, columbae, gallinae, et huiusmodi, nam quum die ab eis res gerendae sint, parvo pupillae hiatu contentae sunt; nam si laxius aperirentur, hebesceret potius nimia luce visus, et caecutiret. Noctivaga haec, quia sursum deorsumque conspiciendi usus erat, ab imo ad summum pupilla scinditur. Contra autem in quibusdam piscibus, quorum usus ad latera erat, a dextris ad sinistra hiatus erat. Sic pro uniuscuiusque officio pupilla et eius scissura animalibus est accommodata.

Pupillam necessario rotundam fuisse magnitudines sine impedimento oppositas rotundas videri. **PROP. 11**

Pupilla nostra rotunda est, et ex omni parte radii exeuntes et libere vagantes rotundum spatium intercipiunt, quod crystallini centrum in medio pupillae sit. Unde si quis humo supinus iacens et uno oculo caelum contueatur immoto, intuitum spatium rotundum erit, a puncto enim crystallini exeuntes radii et oppositam superficiem ferientes non nisi rotundum spatium intercipient. Sed si prope res contuenda erit, hoc experimento te ipsum docebis, ut diximus in 8 huius. In plana tabula fac oculus in centro immotus maneat, et dextra manu lapillos [81] constitues ubi radii oculares plus extendi non possint. Sic remoto oculo, circumactum lapillis spatium reperies. Idque hoc modo demonstrabimus.

Sit crystallini centrum I, et pupilla ABCD rotunda, opposita ei superficies EFGH. Linea IM erecta perpendiculariter super superficiem EFGH, et transiens per centrum crystallini I. Protrahantur lineae a centro crystallini I ad pupillae extremitates ad subiectum planum, et erunt IBF, ICG, IAE, et

by a great deal of light and brightness, but [80] at night the images of things enter easily and are seen because of the width of their pupil. On the other hand, those with eyes of a narrow gaze see clearly in daylight because they are not bothered by light, but at night the images of things cannot penetrate [their vision].

PROP. 10: *Pupils are adapted in various ways for the utility of animals.*

For this reason provident nature designed pupils according to the practice and need of animals. For nocturnal animals she designed them to open up [as if hinged] from top to bottom, like gates, so that those animals could supply themselves with nourishment more quickly and more freely.[60] Indeed, it was necessary that the entire pupil be covered over and that the hemisphere of the eye protrude outward while being open and shining brilliantly everywhere. For there is no white in their eyes; instead the entire space contained between their eyelids is filled by the pupil.

The opposite holds for men and diurnal animals, such as doves, hens, and the like, for when things are being carried out by them during daylight, their pupils are possessed of a small opening. If their pupils were opened wider, their sight would weaken appreciably from too much light, and it would be blinded. Because it was useful for nocturnal animals to look up and down, their pupil is perforated from the bottom to the top. On the other hand, in certain fish, for which it was useful to look to the sides, the opening was from the right to the left. Thus, in order to be of use to any of them, the pupil and its opening have been accommodated to animals.

PROP. 11: *The [human] pupil had to be round so that it could see magnitudes facing all around without impediment.*

Our pupil is round, and the radial lines extending out and deviating freely on all sides intercept a round space because the center of the crystalline humor lies in [line with] the middle of the pupil. Consequently, if someone lies supine on the ground and stares at the sky with a single unmoving eye, the space viewed will be round, for the radial lines extending out from the [center] point of the crystalline humor and reaching the facing surface will only intercept a round space. But if something near is to be looked at, you will learn this yourself by means of an experiment, as we discussed in proposition 8 of this book. Hold your eye steady over the center of the flat board, and with your right hand place pebbles [81] where the lines of sight cannot extend farther out to the edges. When the eye is drawn back, then, you will perceive the space enclosed in a circle by the pebbles. We will clarify this point as follows.

Let I [in Figure 3.11] be the center of the crystalline humor, ABCD the round pupil, and EFGH a plane facing it. Line IM stands perfectly upright upon plane EFGH, and it passes through center I of the crystalline humor. Extend lines from center I of the crystalline humor to the edges of the pupil and on to the plane below, and they will be IBF, ICG, IAE, and IDH. IB, IA,

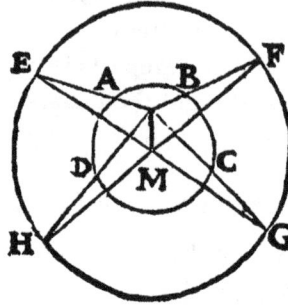

IDH. Sunt ergo aequales IB, IA, ID, et IC, et quia subiectum planum aequidistat ei, erunt quoque lineae AE, BF, CG, et DH aequales, et quoque simul iunctae IBF, IAE, IDH, et ICG. Trahantur mox a centro plani M ad linearum extremitates lineae MG, MH,[10] ME, MF, quae quoque aequales erunt, nam anguli omnes ad M planum recti, et equales quoque qui ad I, erunt quoque iam dictae MH,[11] ME, MF, et MG pares, ergo M centrum circuli EFGH.

Planum pupillae non aequidistans ovale videbitur. PROP. 12 [82]

Secus autem eveniet in plano oblique pupillae opposito, nam etsi lineae IB, IA, IC, et ID aequales erunt, quia planum ex una parte longius distrahitur,

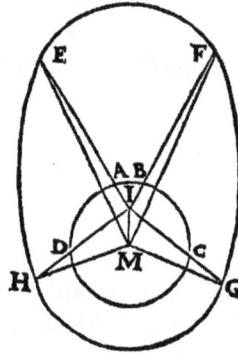

ut FE, et altera pars proprius accedet GH, lineae productae inaequales erunt, unde AE,[12] BF longiores, breviores vero CG, DH. Sic subtensa latera inaequalia erunt, nam MH, MG[13] breviora, ME, MF longiora. Igitur M non erit centrum FEGH figurae circularis.

Crystallinum humorem praecipuum esse in visu. PROP. 13

Aristoteles non inefficaciter probavit cor primariam esse corporis partem quod natura illud in animalis medio locaverit; sic nos crystallinum humorem,

---

[10]MN in original text.
[11]MN in original text.
[12]AC in original text.
[13]MH in original text.

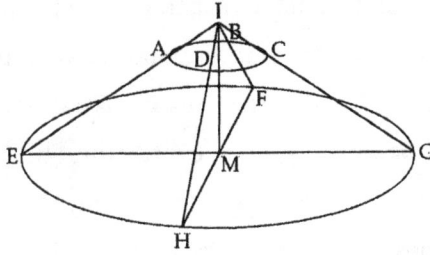

Figure 3.11

ID, and IC are therefore equal, and since the plane below is parallel to the pupil, lines AE, BF, CG, and DH will also be equal, as are IBF, IAE, IDH, and ICG combined together. Now extend lines MG, MH, ME, and MF from center point M of the plane to the endpoints [of lines ICG, IDH, IAE, and IBF]; these will also be equal because all the angles at M on the plane are right while those at I are equal, and the aforesaid lines MH, ME, MF, and MG will be equal, so M is the center of circle EFGH.

PROP. 12: *A plane not parallel to the pupil will appear oval.* [82]

Something different, however, will occur in the case of a plane facing the pupil at a slant, for even though lines IB, IA, IC, and ID [in Figure 3.12]

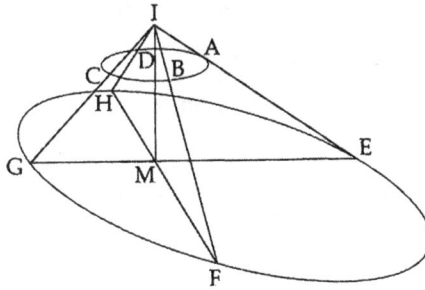

Figure 3.12

will be equal, the plane extends farther on one side, for example, on side FE, whereas the other side GH will come closer, so the lines extended [to those sides] will be unequal, such that AE and BF will be longer while CG and DH will be shorter. The subtended sides will thus be unequal, for MH and MG will be shorter, ME and MF longer. M will therefore not be the center of the curvilinear figure FEGH.

PROP. 13: *The crystalline humor is the primary instrument in sight.*

Aristotle demonstrated not ineffectively that the heart is the chief organ of the body because nature placed it in the middle of the animal[61]; in the same

quia suam regiam in oculi fere medio habet, non inepte praecipuum visionis instrumentum dicemus. Medius locus declarat quod ab omnibus custoditur iuvaturque; aliae circumadsitae partes eius gratia factae sunt.

A vitreo namque tamquam gemma continetur, [83] committiturque in eo, nam sanguis a cerebro veniens ad eius nutrimentum a vitreo excipitur, et in vitrum primo convertitur. Mox crystallino ministratur, nam proximior est vitri natura crystallo quam sanguis; si enim illico in crystallum abiret, eius pelliciditas cruore inficeretur, unde eius colluvias recipit.

Aqueus humor eum humectando prohibet ne calore, aut rigore, vel eum ambientis araneae siccitate oblaederetur. Ipse humectus est, ut suscipere possit visibiles formas eximia perspicuitate, et affici a luce possit. Est alicuius spissamenti, ut in eo figi sistique simulachra queant; aliter evanida aboleretur.

Sed verissima ratio est—quod constans et firma est medicorum opinio—visionem in crystallino fieri, quia si coit densaturque humor quem intra corneam et crystallinum statuunt, illico visus aboletur, sed ubi in alias partes adlabitur et liberum praestat a pupilla ad ipsum aditum, liberabitur. Sed quoties ipse ex aliqua intemperie laborat, omnis aboletur oculi functio. Tandem mathematicorum nos admonent rationes quod ubi lineae non directae advenerint, refractionis ratione illum contingunt; visionem causant.

Crystallinum humorem extra oculi centrum esse. PROP. 14

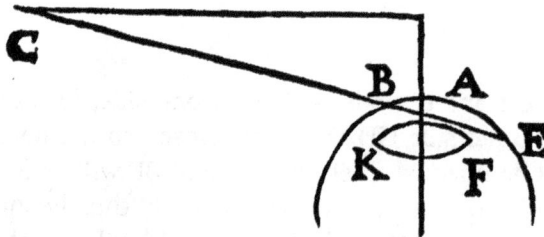

Humor crystallinus extra oculi centrum locatus est, prope pupillam, ut clarior et uberior fieret visio. Clarior, quia simulachra, si profundius mergerentur,

vein, we may say not ineptly that the crystalline humor is the chief instrument of vision because it has its residence virtually in the middle of the eye. Holding the middle position indicates that it is protected and aided by all [the rest of the eye's components]; the other parts that surround it were made for its sake.[62]

It is encased just like a jewel by the vitreous humor [83] and is conjoined with it because the blood reaching from the brain for the purpose of nourishing it is received by the vitreous humor and is initially transmuted in the vitreous humor. Then it is furnished to the crystalline humor, for the nature of the vitreous humor is closer to that of the crystalline humor than blood; indeed, if the blood were to pass immediately into the crystalline humor, its clarity would be compromised by the blood insofar as it receives its adulterants.

By moistening the crystalline humor, the aqueous humor keeps it from being injured by heat, chilling, or the dryness of the aranea surrounding it. It is moist so that it can take on visible forms and be affected by light according to its special transparency. It is slightly dense so that the images may be imprinted and remain in it; otherwise, they would vanish [without effect, and vision] would be destroyed.

The most compelling reason for assuming that sight occurs in the crystalline humor—and this is the unyielding and firm opinion of physicians—is that if the [aqueous] humor that they determine to lie between the cornea and the crystalline humor coagulates and thickens, vision is immediately destroyed, but when it is slid off to the side and opens free access to the crystalline humor from the pupil, vision will be untrammeled.[63] By however much the eye is strained by some disorder, though, the entire function of the eye is lost. Finally, the rationale of the mathematicians tells us that when lines of light cannot reach the lens directly, they strike it by means of refraction, and so they cause vision.[64]

PROP. 14: *The crystalline humor lies outside the center of the eye.*

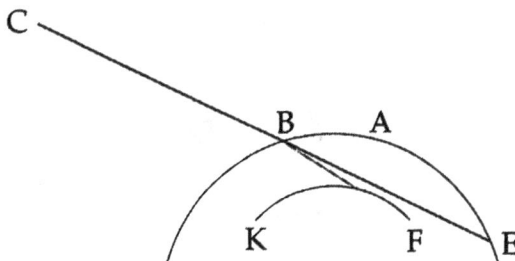

Figure 3.14a

The crystalline humor is situated outside the center of the eye, near the pupil, so that vision might occur more clearly and with a wider scope. It occurs more clearly because if the images were to sink more deeply into the

obscuriora viderentur, ut ea quae in aquis mersa videmus. Fit et uberior visio, nam si in centro esset, maior pars eorum non videretur, quae nunc videtur. Huius rei exemplum erit: sit oculi pupilla AB, cadat punctus C in B, prodibit directe ad E; lex refractionis facit ut crystallinum tangat FK, qui prope est. [84]

Praeterea, si in centro esset, et difficilius lineae illuc pervenirent, et maior pars visionum amitteretur. Punctus H perveniat ad pupillam B. Lex

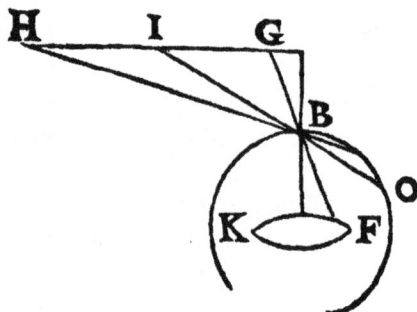

refractionis cogit eum in O, satis longe a crystallino FK. Concedamus ut tangat eum in puncto F; relabitur in G. Totum spatium GH non videbitur. Sic eorum maior pars non videretur.

Praeterea, si in centro crystallinus esset, pyramidales lineae ad eum pervenientes perpendiculariter et in centro se decussantes; dextra viderentur sinistra, et sinistra dextra, et quae sursum deorsum, et quae deorsum sursum,

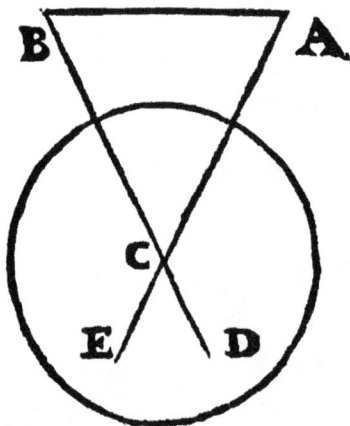

et demum omnis visio conturbaretur. Rei visae dexter punctus B, pervenient ad centrum C, laberetur ad D, et A ad E, sic dextra[14] sinistra viderentur.

---

[14]"destra" in original text.

eye, they would appear darker, as do things we see sunk in water. It also makes vision wider in scope, for if it were in the center, most of what is now seen would not be seen. This will suffice as an illustration of this point: let AB [in Figure 3.14a] be the pupil of the eye, let the image of point C fall to B, and it will continue directly to E [if not refracted]. But the law of refraction causes it to strike crystalline humor FK, which is near [the pupil]. [84]

In addition, if the lens were in the center, the lines would reach it with greater difficulty, and a significant portion of visible things would be missed.

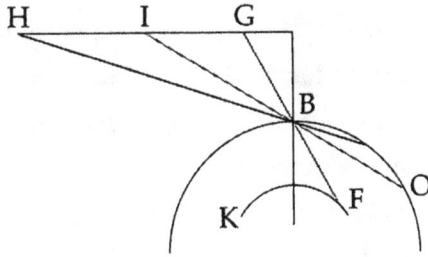

Figure 3.14b

Let the image of point H [in Figure 3.14b] reach the pupil at B. The law of refraction diverts it to O, which is fairly far away from crystalline humor FK. Let us concede that it strikes it at point F, and so [line of refraction FB] inclines toward G. The entire space GH will therefore not be seen. Hence, the major portion of those things [facing the eye] would not be seen.[65]

Furthermore, if the crystalline humor were at the [eye's] center, the lines of the radiant cone reaching it orthogonally would cross each other in the center, and right would appear left and left right, and what is up would appear down and down up, so vision would be completely confused. The

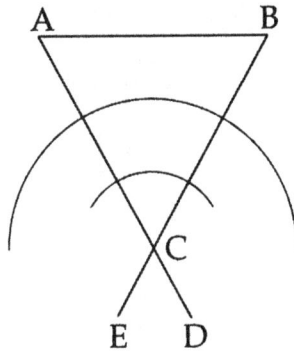

Figure 3.14c

image of leftward point B on a visible object [in Figure 3.14c] will reach center point C and continue toward D, and the image of A would continue [straight through C] to E, so right-hand things would appear to the left.[66]

Sed Vessalius suis in anatomicis administrationibus et plerique medicorum humorem crystallinum in centro oculi locatum dicunt, putantes ut nil referat medicis, ita et mathematicis vel aliis artificibus, quod est valde erroneum. [85]

Utramque corneae superficiem, humorem albugineum, aqueum, uveam, et crystallinum concentricos esse omnes. PROP. 15

Diximus utrasque corneae superficies mutuo aequidistantes esse; unde et si extima circularis quoque interior circularis erit, nam eiusdem soliditatis orbis perpetuus est. Continet interior cornea albugineum humorem; contenti pars cum suo continente eandem superficiem habet. Impletque oculum usque ad crystallinum; sic vitreus humor concavam quoque corneam replet, ut una sphaera cum albugineo sit.

Sic uvea quae interiorem etiam tunicam corneae convestit, sequitur eius formam et illa quoque concentrica, nisi tantum ea pars quae a lucida corneae parte se subducit, ubi fenestratur. Crystallinus etiam humor, etsi longe a centro abest, eius tamen suprema superficies oculo est concentrica, oportebat enim et concentricam esse et prope uveam, ob id summa Dei providentia factum est satis utrique, ut et concentrica, et longe a centro absit. Pyramis igitur visiva per corneam, et pupillam, per aqueum humorem usque ad crystallinum infracta pervenit hoc modo.

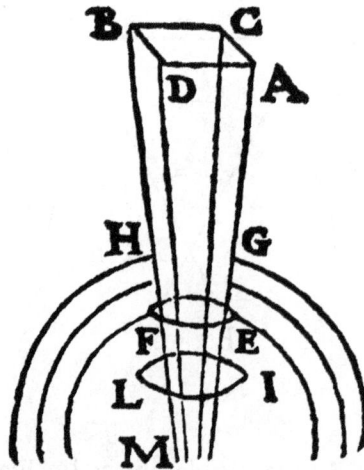

Esto visiva pyramis quadrilatera CBAD dirigens se ad centrum oculi M. Transit corneam GH, pupillam EF, et aqueum humorem EF, sic usque ad

In his book on anatomical procedures, however, Vesalius, along with most physicians, claims that the crystalline humor is located at the center of the eye, but they suppose that nothing contrary has been claimed by physicians, mathematicians, or other skilled practitioners, which is grossly in error.[67]
[85]

PROP. 15: *Both [the inner and outer] surfaces of the cornea, the albugineous humor, the aqueous humor, the uvea, and the [anterior surface of the] crystalline humor are all concentric.*

We said that both the surfaces of the cornea are parallel; and so, if the outer surface is circular, the inner surface will also be circular, for the orb [of the eye] is of the same solidity throughout. The inner surface of the cornea envelops the albugineous humor, and the portion that is enveloped shares the same surface with what envelops it. The albugineous humor also fills the eye up to the crystalline humor, just as the vitreous humor fills the space up to the concave surface of the cornea so that it forms a single sphere with the albugineous humor.

Likewise, the uvea, which also nests with the inner tunic of the cornea, takes its shape and is also concentric with it, except for the [anterior] segment that is split off from the transparent portion of the cornea, where it is perforated [by the pupil]. Even though the crystalline humor lies far from the eye's center, its anterior surface is still concentric with the eye [as a whole], for it was necessary that it be both concentric and near the uvea, for which reason it was made to be sufficiently concentric and far from the center by God's supreme providence. As a result, a cone of vision extends unbroken to the crystalline humor through the cornea, the pupil, and the aqueous humor in the following way.[68]

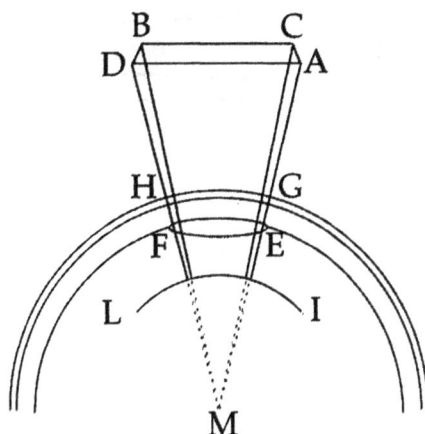

Figure 3.15

Let quadrilateral CBDA [in Figure 3.15] be [at the base of] a visual cone directed toward center point M of the eye. The cone thus passes through

crystallinum IL. Unde tota pyramis irrefracta et incolumis pervenit, et rem ut est repraesentat. Non conturbatur visio; servatur in ea partium ordo et concinnitas sine confusione rei visae.

Haec videtur germana visio, et ita ex anatome [86] apparet. Unde me nescire fateor quae insania vel furor sit Alhazeni, Vitellionis, et reliquorum perspectivorum inquientium uveam sphaeram eccentricam esse eiusque centrum ad anterius oculi plus accedere. Sic vitreum eccentricum et centrum eius ad anterius oculi plus adhaerere; ut astrologi non tot eccentricos et epicyclos in magno mundo confinxerint quot isti in tam parvo oculi corpusculo. Cur in re tam clara et perspicua tot ambages, ut potius ad puerorum terriculamenta quam ad veritatem indagandam conficta esse videantur, ut tyrones scientiam prius odisse, quam amasse incipiant?

In crystallino visionem non fieri nisi aliqua rei latitudine intercepta. PROP. 16

Oculus nunquam rem visam comprehendit in suo esse nisi quum pyramidis cuspis ad eum perveniens depingit in eius superficie rem ut est ordinata in suo esse. Igitur ut crystallinus recte diiudicare possit, oportet ut pyramidis conus alicuius sit in eius superficie quantitatis, aliter enim nil videret aut satis perturbate. Quomodo autem id animadvertere possimus, instrumento quod in sexta huius propositione descripsimus utemur.

Assumatur regula decem vel octo pedem, quae sit AB, sint duo cylindri inaequales CD minor, EF maior. Erigatur minor circa oculum regulae pars, quae oculo accommodabitur; excavetur ut commode oculum excipiat. Sic tandiu cylindrum accedes, et removebis, quousque pyramidales oculi lineae per eius latera excurrant. Tunc amove regulam ab oculo, et per extrema cylindrorum lineae protrahantur, et videbis circa oculum tantae latitudinis quantae ad crystallinum productae super eius superficiem ad rem depingendam sint suffecturae.

cornea GH, pupil EF, and the aqueous humor [below] EF to crystalline humor IL. The entire cone reaches the crystalline humor unbroken and whole, therefore, and it presents the object as it is. The resulting visual impression is not disordered, and the arrangement and order of the visible object's parts is appropriately preserved in it without confusion.

This seems to be vision in its truest sense, and so it appears from anatomy. [86] I therefore confess that I do not know what mental disorder or madness drove Alhacen, Witelo, and the rest of the Perspectivists to claim that the uveal sphere is eccentric and that its center lies toward the front of the eye.[69] Accordingly, they argue that the vitreous humor is eccentric and that its center lies toward the front of the eye, and they are like astronomers who [nonetheless] do not pack as many eccentrics and epicycles in the huge universe as these thinkers pack in the tiny body of the eye. Why, in a matter that is so clear and self-evident, import so much ambiguity, which seems designed more to frighten children than to lead to truth, so that beginners start to hate rather than to love rigorously derived knowledge?

PROP. 16: *Vision does not occur in the crystalline humor unless the object apprehended is of a certain width.*

The eye never apprehends an object as it actually is except when the angle at the vertex of the radiant cone reaching it depicts the object on its surface as it is actually arranged. In order for the crystalline humor to be able to distinguish objects correctly, therefore, it is necessary that the [angle at the] vertex of the cone form an area of some size on the lens's surface, for otherwise it would see nothing or would see in a rather confused way. We can observe how this is so by using the instrument we described in the sixth proposition of this book.

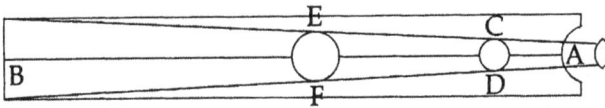

Figure 3.16

Take a ruler eight or ten feet long, let it be AB [in Figure 3.16], and let there be two cylinders of unequal size, CD being smaller and EF larger. Stand the smaller one near the eye on the end of the ruler that will be accommodated to the eye, and let that end be scooped out so that it can fit the eye comfortably. Accordingly, you will draw the cylinder toward you and move it back until the lines of the radiant cone of the eye are tangent to the edges of the cylinders. Then remove the ruler from your eye and draw lines through the edges of [the bases of] the cylinders, and you will see an area near the eye that is the size of the area produced upon the surface of the crystalline humor and that corresponds to the [image of the] object to be depicted there.

NOTES

1. The likeliest source for the Greek ocular terminology here and throughout the rest of this proposition is Andreas Vesalius, *De Fabrica*, VII, 14, in Vesalius (1568), 497–99 (see esp. 499, line 8, for this specific term). That Della Porta was familiar with this work is clear from proposition 14, below, where he cites Vesalius ("Vessalius") by name. For a recent English translation of *De Fabrica*, VII, 14, see Vesalius (2009), 221–35.

2. Vesalius (1568), 499, line 14.

3. That is, hornlike, from *cornu* ("horn") in Latin.

4. Although Della Porta may be referring to the pupil as the "black" part of the eye, it is more likely that he is referring to the iris, which is not necessarily black but is invariably blacker, or darker, than the white surrounding it. Also, the iris is part of the uvea, the majority of which lies beneath the scleral tunic and was considered to be black.

5. Vesalius (1568), 498, line 29.

6. That is, grapelike, from *uva* ("grape") in Latin.

7. Vesalius (1568), 498, line 30.

8. Because of its extreme vascularity, this posterior portion of the uvea is called "choroid" after the chorionic layer of the placenta.

9. Vesalius (1568), 499, line 24.

10. See ibid., lines 24–27 for the likely source for this explanation of "albugineous" as applied to the aqueous humor.

11. Tempting though it is to translate "crystalline humor" (*crystallinus*) as "crystalline lens" or simply "lens," I have resisted doing so because the term "lens," with all its modern connotations, did not come into general use until after Kepler's *Dioptrice* (see Kepler [1611], 9). I have therefore reserved "lens" for Della Porta's *specillum*, which refers explicitly to crystalline or glass lenses.

12. Vesalius (1568), 497, line 1.

13. Ibid., line 12.

14. Ibid., line 19.

15. Ibid., line 33.

16. Ibid., 498, line 2.

17. Della Porta is of course referring here to what is known today as the retina (from Latin *rete* = "net").

18. As is clear from the diagram below, which is drawn from Vesalius's *De Fabrica*, Della Porta and Vesalius were in general agreement about the basic structure and individual components of the eye, differing only in where they placed the lens. Following the medieval tradition, Vesalius placed it at the center of the eye, whereas Della Porta shifted it toward the front.

19. This is simply an extension to three dimensions of the fact that, of all plane figures bounded by a perimeter of the same length, the circle occupies the largest area.

20. Here Della Porta is following the dicates of Perspectivist theory, according to which clear and proper vision occurs by means only of rays that strike the front surface of the crystalline lens along the perpendicular. Taken as a whole, these rays,

which are bounded by the pupil, form a cone with its base defining the field of clear

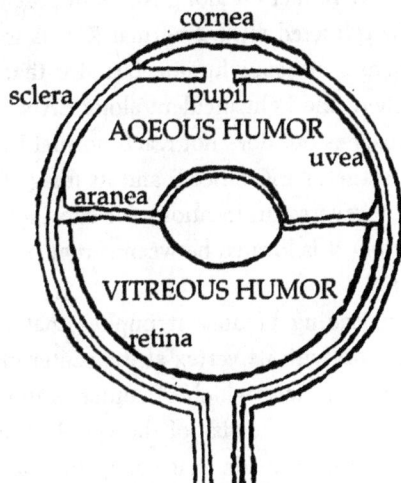

view and its vertex at the center of the eye, which constitutes the so-called center of sight.

21. See Euclid (1589), 176–78, where the propositions are listed as problem 23, proposition 33, and theorem 24, proposition 34. Presumably it was to the problem/theorem numbers rather than the proposition numbers that Della Porta was referring in the Latin text, although his citation of 25 instead of 24 was inapposite.

22. For the likely source of this analysis, see Witelo, *Perspectiva*, III, 3, in Witelo (1572), 85, and in Witelo (1991), 293–94 (Latin) and 103–5 (English).

23. This reference is inapposite because the cited proposition does not deal with the refraction of oblique rays into the eye; its primary purpose is to validate the principle of reciprocity in refraction: that is, that when light follows a given path from a rarer into a denser medium, it will follow precisely the same path if the direction of radiation is reversed. The third proposition of this book, however, does deal with refraction into the eye.

24. Based on a faulty diagrammatic representation, this conclusion is false. In the accompanying figure, EA and AC represent the incident and refracted rays from point E in Figure 3.2c, the extension of refracted ray AC intersecting the extension

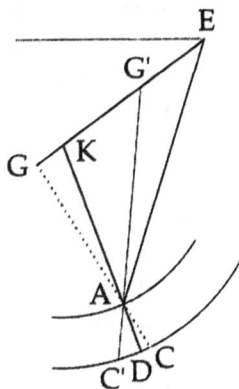

of cathetus EK at G, to the left of center point K. Accordingly, there will be a right-left reversal in location. But refraction along AC is impossible because it requires not only that the light be refracted *toward* normal KD, as indeed it must be, since it passes from air into denser aqueous humor, but also that it be refracted *beyond* that normal. In reality, then, the light incident along EA would be refracted along some path AC' that approaches but does not reach normal KD, and when extended, it will reach point G' on cathetus EK. Since E and its image G' both lie on the right side of K, there will be no reversal in location, nor can there be for any object EF, no matter its size or whether it is located between center point K and surface KABL or beyond center point K.

25. This claim is misleading because it implies that the perpendicular rays forming the cone of radiation with its vertex at the center of the eye comprehends nearly a hemispherical field of view, which of course is impossible because of the small size of the pupil relative to the size of the eye. If, however, we take all the radiation entering the eye from outside that cone, then, as Della Porta argues in proposition 3 below, and later in book 4, proposition 10, the actual field of view is considerably greater than what is seen within the cone of radiation.

26. See Gentile's commentary on Avicenna's *Canon*, III, iii, 1, 1, in Gentile da Foligno (1522), 135va, lines 12–14; my thanks to Joël Chandelier for pointing me toward this source.

27. See Witelo *Perspectiva*, III, 4 in Witelo (1572), 85, and in Witelo (1991), 105 (English) and 294 (Latin); see also Sacrobosco, *De sphaera*, 1, in Sacrobosco (1578), 35.

28. The probable source of this Latin quotation is Galen, *On the Doctrines of Plato and Hippocrates*, VII, in Galen (1535), 103, lines 39–40. It also occurs in a slightly different form in Aristotle's *De anima*, I, 2, 404b13–15, in Aristotle (1550c), 112vb, lines 22–25.

29. See, for example, Galen, *On the Doctrines of Plato and Hippocrates*, VII, in Galen (1535), 103, line 36.

30. See esp. *On the Parts of Animals*, II, 10, 656a35–656b3, and *On Sense*, 2, 438a5–24.

31. In *De anima* 3, 12, 435a9, Aristotle likens the reception of images by the eye to the stamping of a seal in wax, hence his insistence that the watery nature of the eye is dictated in part by water's capacity to retain images.

32. In other words, BIA is the cone of direct radiation, and the arc on the eye's surface cut by it is commensurate with the pupil, so only the radiation within this cone that converges on I will be perpendicular to the eye and its lens.

33. Implicit in this analysis, therefore, is that the interposition of aqueous humor between the cornea and the crystalline lens facilitates peripheral vision by refracting light that enters the pupil obliquely toward the lens, whereas if that refraction did not occur, this incoming radiation would miss the lens entirely.

34. Presumably the glass to which Della Porta refers here is not the fine, Venetian crystal glass of his day (not to be confused with later "lead crystal" glass), but the poorer common glass out of which cheap spectacles were routinely manufactured.

35. Also known as *lippitudo*, the condition known as *defluxio oculorum* ("discharge from the eyes," or blepharitis, as it is known in modern terminology) involves an inflammation of the eyelids. Often accompanied by a viscous discharge that can form a crust on the eye, this condition can cause blurry vision as well as light sensitivity.

36. In current opthalmological practice, this condition is referred to as pterygium, after the Greek *pterugion* ("wing"). The Latin term *unguis*, which denotes this same condition, literally means "nail." Suffice it to say, having a surgeon pass his hand over such a growth (i.e., scrape it off) would have been a last resort in the most extreme of cases.

37. From the Greek *hupochuma*, the latin term *hypochima* refers to cataracts in pre-modern medical practice, as also does *suffusio*—and sometimes *glaucoma* as well. Believed to develop between the cornea and the lens, not in the lens itself, cataracts were supposedly "couched" according to the procedure described by Della Porta. In actuality, however, what was couched was the defective lens itself, not the imagined congealed humor in front of it. The very restoration of sight, albeit with diminished acuity, constituted proof that the lens itself could not have been been the source of the problem and therefore could not have been couched because it was universally assumed to be the true seat of visual sensation.

38. Archimedes describes this device early in *The Sand-Reckoner*. For a discussion and English translation of his description, see Alan E. Shapiro, "Archimedes's Measurement of the Sun's Apparent Diameter," *Journal for the History of Astronomy* 6 (1975): 75–83, esp. 82.

39. On the relationship between the phases of the Moon and an increase or decrease in bodily humors, see, for example, Ptolemy, *Quadripartitum*, I, 4, and *Centiloquium*, 56, in Ptolemy (1541), 434a and 502b.

40. This work was attributed to Nostradamus; for the specific reference, see Nostradamus (1538), 5v.

41. See *De Subtilitate*, 18, in Cardano (1551), 602, as well as in Cardano (2013), 950.

42. In this case, then, the eye's lens would actually act like a glass or crystal lens in bringing incoming light to focus on the retina. However, as is clear from Della Porta's overall analysis of sight and the lens's role in it, such focusing would be unnatural, which is of course why the pupil is designed to prevent it. This is more than a little ironic because within roughly a decade Johannes Kepler would show that it is precisely by bringing light to focus on the retina that the lens fufills its proper function in vision.

43. See *De aspectibus*, I, iv, 6–7 in Alhacen (1572), 4, and in Alhacen (2001), 15–16 (Latin) and 350–51 (English). Alhacen's ocular model is illustrated in the diagram that accompanies note 69 below. The outer circle represents the sclera with the cornea at front. The inner, incomplete circle represents the uvea, the open portion at front being the pupil, which nests right up against the edge of the cornea. Accordingly, if the uveal sphere were completed at front, the anterior portion would extend slightly beyond the cornea, and the resulting common section would be a circle.

44. For book 2 of the *Sphaerica*, see Theodosius (1558), 18–46. Nowhere in book 2, or in books 1 and 3 of the *Sphaerica*, does Theodosius discuss the intersection of two spheres, although one might infer from book 2 that this intersection would form a circle.

45. See *Perspectiva*, III, 8–9, in Witelo (1572), 89–90; also in Witelo (1991), 115–17 (English) and 303-5 (Latin).

46. As is clear from the diagram of the eye Della Porta provides in the first proposition of this book, his argument is based on the incorrect assumption that the cornea forms a perfect continuation with the sphere of the eye rather than bulging outward, as it actually does.

47. Galen, *On the Usefulness of the Parts*, X, 3, in Galen (1541), 176v

48. *Problems*, XXXIII, 4, in Aristotle (1552), 51r.

49. Galen, *On the Usefulness of the Parts*, X, 3, in Galen (1541), 176v. This episode of snow blindness is described in *Anabasis* IV, 5 12–13.

50. Solinus, *Collectanea rerum mirabilium*, LII, 25, in Solinus (1520 [as LXV]), 82v-83r.

51. Ibid., XV, 23-24, in Solinus (1520 [as XXV]), 40v–41r.

52. See *On Sense*, 3, 339b20–340a3.

53. The cubital measure (*cubitalis*) to which Della Porta refers was approximately 1.5 "feet" in length, which translates to roughly 18 inches or 46 cm.

54. See Pseudo-Plutarch, *On the Opinions of the Philosophers*, IV, 15, in Plutarch (1530), 151v. Assumed to be genuinely Plutarchan until the late nineteenth century, *On the Opinions of the Philosophers* (*Placita philosophorum*) was shown by Hermann Diels in his *Doxographi Graeci* to have been composed by someone else on the basis of a source attributed to a certain Aëtius (not Aëtius of Amida); for a brief discussion of this attribution, see Richard D. McKirahan, *Philosophy Before Socrates* (2nd ed.; Indianapolis, IN: Hackett, 2010), 4.

55. See *Problems*, XXXI, 20, in Aristotle (1552), 49v (*Problems*, XXXI, 19, in Aristotle [1984], 1510).

56. Ibid. In chapter 5 of *De coloribus*, which was attributed to Aristotle during the Middle Ages and Renaissance but is now known to be spurious, the color green, as applied to vegetation, is explained in terms of the moisture contained by the given plant and its relative clarity and warmth; as the moisture grows colder, the green tends increasingly toward black; see Aristotle (1552), 77vb–78ra. Presumably, the clear, warm moisture contained by green plant would allow the visual flux to pass through relatively unobstructed, thus keeping the eyesight from being overstrained.

57. See *De Humana Physiognomia*, III, 4, in Della Porta (1586), 195.

58. *Problems*, I, 68, in Alexander of Aphrodisias (1552), f. 95v.

59. See *Perspectiva*, II, I, 2, in Bacon (1996), 171.

60. This is a gross overgeneralization; such nocturnal animals as wolves and owls have round rather than slitted pupils, but as Della Porta correctly observes, their pupils are capable of dilating to a greater extent than those of diurnal animals.

61. See *On the Parts of Animals*, III, 4, 665b10-25, in Aristotle (1550c), 81r.

62. This notion that all the other components of the eye exist for the sake of the lens is fully consistent with the way the eye was understood by medieval physicians,

primarily on the basis of the optical writings of Ḥunayn ibn 'Isḥāq (809–873), known as Johannitius in the Latin West. For an extensive discussion of Ḥunayn's account of the anatomy and function of the eye, see Bruce S. Eastwood, *The Elements of Vision: The Micro-Cosmology of Galenic Visual Theory according to Ḥunayn ibn 'Isḥāq* (Philadelphia, PA: American Philosophical Society, 1982). Whereas Ḥunayn took the centrality of the lens literally, placing it dead center in the eye, Della Porta took it figuratively insofar as he located the lens toward the front of the eye. Thus, it is in the middle of the eye according to a straight on view through the pupil but not from the side, as is eminently clear from the proposition immediately following this one.

63. As we saw in note 37 above, Della Porta is referring here to cataracts, which he supposes are caused by congealed moisture between the cornea and the lens that interferes with the passage of light and images into the eye. Once the congealed moisture is pushed out of the way so the lens is no longer occluded, vision is restored. As we also saw in that note, it was not this imaginary congealed moisture but the lens itself that was actually couched.

64. In short, peripheral vision is due to refracted rays. For the paradigmatic "mathematical" account of such refracted vision, which is based on virtual rays, see Alhacen, *De aspectibus*, VII, vi, 34–37, in Alhacen (1572), 266–70; also, in Alhacen (2010), 100–114 (Latin) and 296–307 (English). As will become clear later on in book 6, Della Porta rejects this account.

65. In other words, any ray that passes to point B on the pupil through any point between H and G will be refracted less than ray HB, so it will end up bypassing point F at the edge of the lens. Accordingly, no point between H and G will be seen.

66. The thrust of Della Porta's argument here is not that the crossed rays would reach the back of the eye in reversed order so as to be seen that way by the retina. Rather, his argument seems to be based on the supposition that, in reaching the back of the eye, the crossed rays would enter the hollow optic nerve in reverse order, the result being that we should see everything in reverse order. This seems to be inconsistent, however, with his later account of vision, which precludes images entering the optic nerves because those nerves are solid rather than hollow.

67. See Vesalius, *De Fabrica*, III, 14, in Vesalius (1568), 497, line 1. It is certainly the case that Alhacen and his Perspectivist disciples, who belong to Della Porta's category of mathematicians, supposed perforce that the lens lies quite close to the cornea at the front of the eye. Consequently, in that regard, as in several others, they were at odds with the vast majority of medieval and early Renaissance physicians.

68. I have chosen to translate *pyramis* here and elsewhere as "cone" even though its literal meaning is "pyramid." The reason for my choice is that in its most general sense this figure is composed of all rays passing into the eye through the circular pupil, so it is actually a cone. Consequently, even when the figure is defined by a particular base in a particular object rather than by the pupil, I will still refer to it as a cone. In addition, although sometimes referring to this cone as a *cone of vision* (*visus*) or a *visual* (*visualis*) *cone*, Della Porta does not mean to imply that it is formed of visual rays emanating outward from the center of the eye. All he means is that it is the cone of intromitted radiation according to which external objects

are seen directly along the perpendiculars rather than along the refracted rays of peripheral vision.

69. See Alhacen, *De aspectibus*, I, iv, 6, in Alhacen (1572), 4, and in Alhacen (2001), 15 (Latin) and 350 (English); Witelo, *Perspectiva*, III, 8, in Witelo (1572), 89, and in Witelo (1991), 303–5 (Latin) and 115–17 (English); Bacon, *Perspectiva*, I, iii, 1, in Bacon (1996), 32–37; Pecham, *Perspectiva*, I, 33, in Pecham (1542), 31–32 (unnumbered), and in Pecham (1970) 118–19. Thus, in the accompanying

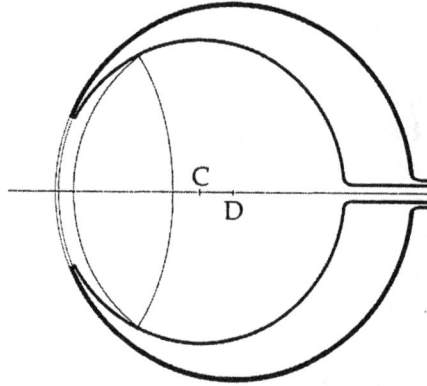

figure, which represents the eye as understood by all these thinkers, the visual axis passes through the center of the cornea, then through the pupil nested up against it, then through the lens and the vitreous humor behind it, both contained by the uveal sphere inside the eyeball, and finally through the hollow optic nerve. Accordingly, point D on the visual axis is the center point of the eyeball, and point C ahead of it is the center point of the uveal sphere inside the eyeball.

LIBER QUARTUS

PROOEMIUM

Postquam oculum construximus singulares eius partes, et partium munia exequuti sumus, ingerit se nunc in hoc ordine demonstratio an fiat visus per intromissionem. Et optime quidem fecerimus si una visionem partim oblique partim refracte fieri declaramus et vera loca in quibus res videantur demonstremus, quae maioribus intentata et obscura fuere. Sic passim ad alia progrediendo, universam videndi formam absolvemus.

Visionem fieri per imaginem receptionem. PROP. 1

Inter vetustiores et recentiores philosophorum et medicorum proceres ad sacietatem usque de videndi ratione exagitata quaestio est num radiorum ex oculis emissione vel illatione fieri contingeret. Res adeo ambagiosae difficultatis, ut plerique rerum scientia non incelebres sigillatim rationes reddere non sunt ausi. Nos ab aliis praetermissas mathematicas [88] rationes afferre, et alia ad veritatem indagandam quantulacumque fuerint, referre iam tentabimus.

Democritus et Epicurus (ex Macrobii recensione) ex corpore opinati sunt "simulachra quaepiam iugi fluore manare . . . quae quasi corporum exuviae ad nostros oculos veluti receptacula . . . et deputatas a natura sedes remeare." Argumento, ut Aristoteles refert, quia rei visae simulachra in oculo apparent, nam quum pupilla sit corpus laeve, tersum, terminatumque ab aqueo profundo, potest idolum repraesentare. Sed hi falsi sunt, ut posteriores probavere. Putabant enim sicut vox ad aures ultro venit, odor in nares influit, et sapor palato ingeritur, sic imagines rerum in oculos penetrare.

Sed res aliter se habet, nam facies nostra oculo obiecta recto meatu profluit, unde posteriorem sui partem respicere deberemus. "Histrio perso-

BOOK FOUR

INTRODUCTION

N ow that we have put the eye and its individual components together and have followed out the functions of its parts, it is time, according to this procedural order, to dedicate ourselves to showing whether sight occurs through intromission. And we will certainly have done our best if we show that vision occurs partly at a slant and partly by means of refraction and if we reveal the precise locations where objects appear, an issue that was broached by our predecessors but remained obscure. Thus, by proceeding here and there, we will unravel the entire process of seeing.

PROP. 1: *Vision is accomplished through the reception of an image.*

Among the chief ancient and modern natural philosophers and physicians, the question of what causes seeing, whether it might depend on the emanation of rays out from the eyes or on the entry of rays into the eye, has already been exhaustively tackled. This issue is therefore of such tortuous difficulty that several thinkers, not lacking renown for a rigorous intellectual grasp of things, did not dare to propose individual explanations. We will now try to adduce mathematical [88] explanations ignored by others and to propose other explanations, however trifling they might be, in order to tease out the truth.

According to Macrobius's account, Democritus and Epicurus assumed that "simulacra continually flow from an object like skins from bodies to our eyes, which are like receptacles, and those simulacra pass on to the seats allotted by nature."[1] In defense of the idea that images of a visible object appear in the eye, Aristotle offers the argument that, since the pupil is a smooth and polished body bounded by water to its depth, it can manifest an image.[2] But these assumptions are false, as later thinkers have demonstrated. For these earlier thinkers assumed that just as the voice comes from outside to the ears, or odor flows into the nose, or flavor is presented to the tongue, so the images of things penetrate the eyes.[3]

The matter is otherwise, though, for if the image of our face flows straight out, away from our eye, then we ought to face the back side of it

nam sibi detractam ea parte videt qua induit, scilicet non faciem, sed posteriorem cavernam," sed nos adversam videmus in speculo imaginem. Alia ratio.
Si imagines, quum inspicimus, ex rebus evolant vel undique nullo inspiciente
simulachra emicant? Si quis teneat, quum inspicit, "quaero cuius imperio
simulachra praesto sunt intuenti?" Si perpetuo fluore rerum manant sim-
ulchra, quaero quo vinculo ad permanendum iuncta[1] permanent? Et quomodo spectra colorem retinent, si colorem non nisi corpora retinere possunt.
Tertia ratio. Quomodo tot rerum imagines, coeli, montium, maris, pecundum,
quas uno oculorum iactu oppidoquam parvo videre possumus?

Alii sunt qui dixerunt per radios ab oculis egressos res contueri, sed
posteriores tanquam falsissimum redarguerunt. Nam vel illi radii corporei,
vel corporis expertes sunt. Si corporei sunt, si lumen ex oculo ad rem obiectam
egreditur, homo ad astra aspiciens, lumen ad astra usque pertingeret; si
corporei essent, universa oculi quantitas quantacunque exinaniretur, nec ad
ea pertingere posset. Praeterea, si per aera transeundum esset, daretur corporis penetratio. Si absque corpore sunt, radii non sentirent rem visam, quia
non sunt sensus nisi in corpore, neque corporeus oculus incorporeas res
contuetur.

Subsequentes aetates philosophi utraque opinione coacti sunt usi,
putantes has ambages evadere. Democritus et Epicurus, Plutarcho referente,
"per quosdam radios ex oculis exertos subiectam rem ferientes, rursusque
redeuntes, et in eosdem resilientes," per hos incursus visualem [89]
potentiam fieri arbitrabantur; "Empedocles imaginibus radios admiscuit et
id coniunctum radios vocavit." Addit Aristoteles ignem in oculis adnatum
esse ac foris emicare ut lumen de laterna, quod si tunicis non clauderetur,
fortius extramitti. "Hipparchus ab oculis ad corporum lineamenta radios
exporrectos perinde ac manuum admotu summa corpora pertrectare[2] eorum
apprehensionem ad eosdem referre."

"Stoici per radiorum ex oculis in ea quae videri queant[3] emissionem
aerisque simul intentionem." Addit Galenus, "per circumfusum aerem, quasi
per baculum," neque visile ad spectatricem facultatem traduci. Plato in *Timeo*
"per συναγχέιαν vel corradiantiam fieri existimavit. Lumine quidem ex
oculis exerto ad aliquantulum intervalli per cognatum sibi aerem effluere; a
corporibus vero illato occursante illi aeremque intermedium diffuso amoli-
tuque facilem cum visuali igne intendente."

Galenus Platonis opinionem sequitur roboratque, ut libro *De placitis
Hippocratis et Platonis*. Ait enim ignis qui mundum suaviter illuminat similis

---

[1] "[V]vincta" in original text.
[2] "[P]errectare" in original text.
[3] "[Q]uerunt" in original text.

[as it propagates outward]. "When he takes off his mask, an actor sees the side that he removed, that is, not the face of the mask but the rear opening,"[4] yet we do not see an image of the back of our head in a mirror. Here is another consideration. Do images fly off from things only when we see them, or do simulacra emanate in all direction where there is no viewer? If someone maintains that they emanate only when the viewer sees, then "I ask by whose command the simulacra are presented to the viewer?"[5] If simulacra of things continually flow out, I ask by what chain they remain bound together in order to keep their unity? And how do the images retain their color if only bodies can retain color? Here is a third consideration. How can we see the images of so many things, of the heavens, of mountains, of the sea, of cattle, that are projected into the tiny compass of one of our eyes?[6]

There are others who claimed that things are seen by means of rays emanating from the eyes, but later thinkers refuted this as utterly false. For either those rays are corporeal, or they are devoid of body. If they are corporeal, and if light emanates from the eye to an external object, then when someone looks at the stars, the ocular light should reach the stars. But if the rays of that light were corporeal, the entire amount would be emptied from the eye, and the light still could not reach the stars. Besides, if the radiation were transmitted through the air, penetration of a body would be involved. If the rays are incorporeal, they would not sense a visible object because the senses exist only in a body, and the corporeal eye does not perceive incorporeal things.

In later times, thinking they would avoid these ambiguities, philosophers made use of both opinions together. According to Plutarch, Democritus and Epicurus opined that the visual [89] power was activated by a return "along rays that are emitted from the eyes, strike the object, and then reenter the eyes; Empedocles mixed rays with images and called the resulting combination rays."[7] Aristotle adds that [according to Empedocles] fire is innate to the eyes and that it emanates outward like light from a lantern, and if it were not enclosed by the tunics, it would be emitted more strongly.[8] "Hipparchus says that rays extend from the eyes to the outer edges of bodies, much as by the motion of our hands we touch the outer boundaries of objects, and they convey back the apprehension of those objects to those same eyes."[9]

"The Stoics argue that vision is due to the emission of rays from the eyes to what can be seen coupled with a simultaneous stretching of the air."[10] Galen adds [that the Stoics claim that vision occurs] "through the surrounding air as if by means of a cane,"[11] but the visible object is not brought to the visual faculty. In the *Timaeus*, Plato supposed that vision "occurs by means of συναγχέιαν, or mutual radiation. Indeed, by reaching out from the eyes, light flows to a certain extent through the cognate air, and when the light from external objects encounters the visual fire emanating from the eye, it diffuses through the intermediate air and moves it aside easily."[12]

Galen follows and confirms Plato's opinion, as, for example, in the book *On the Doctrines of Plato and Hippocrates*. For he says there that the fire that

et congenitus oculorum nostrorum igni. Quum diurnum lumen se visus radio applicat, duo lumina inter se similia et convenientia, quo oculorum acies dirigatur, ibi unius[4] externi internique luminis fit concursus. Id totum igitur eandem passionem sortitum; quum aliquid tangit vel tangitur, sensum efficit videndi.

Et his rationibus persuadet. Corpus quod videmus duorum alterum erit: vel ex se ad nos aliquid mittit, vel vis aliqua sensitiva a nobis ad se usque pervenire expectat. Si aliquid mittit, per pupillae exiguum foramen pertranseundum erit; quomodo igitur alicuius montis, vel vastae rei spectrum in oculos ingreditur? Quod est absurdum. Et si innumerabiles sunt, qui id vident, quomodo ad singulos eaedem feruntur. Et si spiritus intuitivus exit, quomodo in tanta diffusione dilatari potest omnes visui res expositas circumfusus complectatur?

Euclides philosophus Platonicus per radios ex oculis egressos visum fieri dixit, probavitque his rationibus. Sensuum receptacula alia concava, alia non concava constructa sunt. Auditus, olfactos, et gustus concava sunt, et extrinsecus ultro procidentia corpora locum aptum invenirent, et permanerent nec ex vestigio transilirent. Visus contrariam habet formam, nempe convexam, unde in eo radii recipi non possunt. Praeterea, falsum esse ex corpore continuo simulachra [90] profluere et nostros sensus lacessere, nam si id verum esset, quaerentes acum illico nostro obtutui ultro se offerrent, et librum legenti omnes literae occurrerent.

Heliodorus Larisseus quod eiecto lumine emissoque fieret visio hoc argumento innititur. Quod oculis insit fulgor splendorque ut eum noctu ad cernendum nullo extrinsecus lumine indigent, et in tenebris eorum coruscant oculi, exemplum bubo, felis, et alia noctivaga animalia. Et Tiberius ille Romanorum Imperator "paulisper haud alio modo quam luce clara contueretur omnia, paulatim tenebris se offundentibus." Addunt ad id Platonici animalium quorumdam vires quae cum radiis ex visu ipso diffunduntur.

"Sunt in Africa familae quaedam effascinantium, Isigono et Nimphrodoro testibus, quorum laudatione intereant probata, arescantque arbores, moriantur infantes. Esse eiusdem generis in Triballis et Illyriis, adiicit Isigonus, qui visu quoque effascinent interimantque quos diutius intueantur iratis praecipue oculis; quod eorum malum facilius sentiant puberes," aliisque locis alias esse gentes eiusmodi virtutis tradit Plinius. Aphrodiseus esse homines

qui summa animi nequitia perpolluti, liberalem aliquam ... formam conspicati enecant; radios sui conspectus vitiatos transmittunt, quasi veneno quodam vitio

---

[4]"[M]inus" in original text.

gently illuminates the world is similar to, and produced along with, the fire of our eyes. When daylight conjoins with the visual radiation, both lights being similar and suited to one another, that is when the external and internal light conjoin to form a single entity that the gaze of the eyes is directed. This aggregate thus shares the same effect; anything it touches or by which it is touched arouses the sensation of sight.[13]

And he argues persuasively with the following reasoning. Our seeing an object will be accomplished in one of two ways: either the object sends something from itself to us, or some sensitive power needs to reach from us to it. If the object sends something to us, then it will have to be transmitted through the narrow opening of the pupil, so how does the image of some mountain, or of some massive object enter into the eyes? This is absurd. And if such objects are innumerable, then how are their images conveyed individually to those who see them? But if the sensing spirit goes out, how can it be expanded in such a wide dispersion as to encompass all the things exposed to sight?[14]

The Platonic philosopher Euclid claimed that vision occurs by means of rays emanating from the eye, and he supported it by the following reasoning. Some sense-receptacles are concave in structure; others are not. The organs of hearing, smell, and taste are concave, and bodies proceeding from outside can find in them an appropriate lodging where they can remain without failing to leave an impression. The visual organ has a contrary shape, convex in fact, so rays cannot be taken in by it. Furthermore, it is false to say that simulacra [90] flow continually from an object and stimulate our sense, for if that were true, then when we looked for a needle, its images would reveal themselves to our sight immediately, and all the letters would occur at once to anyone reading a book.[15]

Heliodorus of Larissa relies on the following argument that vision must occur through the projection and emission of light. The horned owl, cat, and other nocturnal animals whose eyes glitter in darkness provide evidence that gleaming and brightness are inherent in the eyes so that at night they need no external light for seeing.[16] Moreover, Tiberius, the emperor of the Romans, "could for a short time see everything just as in bright daylight, until darkness gradually closed over him."[17] The Platonists add to this the powers of certain animals, which are propagated along with the rays emanating from the eye.

"According to the testimony of Isigonus and Nymphodorus," Pliny tells us, "there are certain tribes of witches in Africa who can kill meadows, dry up trees, and kill infants by praising them. There are the same sort among the inhabitants of lower Moesia and Illyria, Isigonus adds, who bewitch and destroy those whom they look at for a while, especially with angry eyes; and adults feel their evil most easily," and there are other people in other places with this sort of power.[18] Alexander of Aphrodisias [says that] there are men

whose souls are so thoroughly vitiated by evil that they kill off any noble sort of thing they look at; they transmit corrupted rays with their glance as if by spreading

sui animi radios veluti telum illinant, iaciantque in rem cuius invidia tenentur et desiderio. Prodit igitur de pupilla radius exitialis qui per oculos obducti hominis subdit vertitque quam naturam aggressus est maloque habitu depravat . . . Quod vero aditus per oculos pateat contagioni; constat etiam ex iis, qui mortua hominum corpora contueri recusant, ne radii sui conspectus refracti aliquam sibi lucem pestemque referant.

"Lupi visus noxius est, vocemque homini quem priores contemplentur adimere ad praesens . . . Basiliscus . . . hominem vel si aspiciat tantum dicitur interimere"; sic et menstruatae mulieres obtutu speculum inficiunt obnubilantque.

Ex Peripateticorum placitis sancitum est, quorum princeps Aristoteles, rerum simulachra ad sensum vehi, etsi ipse in *Meteorologicis* et *Problematis*[5] extraferri dixerit. Illuminato enim aere qui inter visum et obiectam rem interpatet, ex corpore species quaedam per aeris medium ad oculum usque pertensa aspectum vulnerat, oculusque per virtutem animae insitam illius colores intuetur et iudicat. Nos dilucidius rem [91] aperiemus et perficiemus. Forma rei videndae in ipsa re materialem habet naturam, sed lumine susceptam materiam subito eximere et spirituali quadam ratione emicare. Et quo magis ab obiecto veniens appropinquat oculo, eo magis ex latiori base accommodare se ipsam ad pupillam et oculum, sicut et corpora liquida per se locis insinuant. Unde per radiorum lineas pyramidales ex se ad oculum figuram portendit, et cono crystallinum verberat.

Quod fiat per simulachrorum in oculos illapsum ita probamus. Si vehemens lumen Solis per se vel a corpore polito reflexum pupillam invaserit, adeo laedit, ut lucis reliquiae diu in oculis residentes post clarioris luminis prospectum, clausis oculis etiam in obscuro remanet quousque a sua natura fuerit dissitum. Lumen igitur pupillam subit. Praeterea, si cubiculi fenestrae claudantur et per angustum foramen ingrediatur secunda lux, albaque papyrus ei obex opponatur quae lumen illud excipiat, quod in platea et oppositis parietibus illustrabit, suis formis et coloribus vestita in albo papyro repraesentabit.

Dico quod sicut corpora a Sole illustrata per arctum fenestrae foramen in subiectum papyrum lumen repraesentat, sic idem visarum rerum spectra

---

[5]"Prolematis" in original text.

their rays like a net poisoned with a certain imperfection from their soul, and they cast them on a thing for which they maintain hatred or desire. Thus, a toxic ray propagates from the pupil, and it is applied and directed through the eyes of a man whose nature it has attacked and perverts by evil possession. That rays are emitted by the eyes is evident in disease, however, a point that is also established among those who are reluctant to look at the dead bodies of men so that the reflected rays of their glance will not bring back any light or disease with them.[19]

"The stare of a wolf is harmful, and if wolves look at a man before he sees them, they take away his voice for the duration"; [and in addition] "the basilisk . . . is said to kill a man just by looking at him,"[20] and likewise, with a glance, menstruating women stain and cloud over a mirror.[21]

According to the opinions of the Peripatetics, among whom Aristotle stands foremost, it is well established that the images of things are conveyed to the sense, even though Aristotle himself claimed in the *Meteorology* and *Problems* that visual rays are conveyed out from the eyes.[22] For when the air that lies between the eye and an external object is illuminated, a certain likeness proceeding from the object to the eye through the aerial medium affects the visual power, and the eye perceives and judges its colors by means of a power instilled in it from the soul. We will expound on this point more clearly [91] and deal with it more completely. The form of a visible object as it inheres in the object itself has a material nature, but when exposed to light, the material nature suddenly disappears, and it becomes manifest through a sort of spiritual cause.[23] Also, the closer the spiritualized form coming from the object gets to the eye, the more it accommodates itself to the pupil and the eye as it becomes increasingly distant from the broader base of the radiant cone on the object itself, just as liquids automatically infiltrate areas through narrow openings. Consequently, the form indicates the shape of any visible object through lines of rays forming a cone that extends from the object to the eye, and it strikes the crystalline humor at the vertex.[24]

That this occurs by means of images entering into the eye we can prove as follows. If strong, direct sunlight or sunlight reflected from a polished body enters the pupil, it offends the eye such that, after a glance at the extremely bright light, vestiges of the light persist for a while in the eyes when the eyes are closed or else remain in a dark place, until the effect is naturally dissipated. Light therefore affects the pupil. Furthermore, if the windows of a room are closed and secondary light[25] is allowed to enter through a narrow opening, and if a piece of paper poses a white barrier toward it that captures this light, the window will render what is in the streets and facing walls, and it will depict them clothed with their shapes and colors on the white paper.[26]

I say that just as the light shining through a narrow window onto paper held near it represents objects illuminated by the Sun, so it depicts the images of visible things by shining through the opening of the pupil and falling

per pupillae foramen subiens in crystallinum depingit. Nemo igitur tam durae cervicis insanaeque mentis ab oculo ad Solem per papyrum reflectionem fieri argutabit, tum quod papyrus non reflectit, tum quod reflexio ad pares angulos fieri iam sancitum est, sed quoquo versus paginam inspexeris, semper easdem figuras inspicies. Tertium addemus argumentum. Ex sui regione quis aliquam figuram intentissime aspiciat; alter homo ex laeva vel dextra parte digitum ostendat. Si aspiciens manum iniecerit ut digitum tangat, profecto longe a digito aerem verberabit, quod erit signum visionem fieri per intromissionem. Rem demonstrare clarius iudicamus. [92]

　　Sit oculus DC, aspiciat rem sibi oppositam E,[6] alter homo ex dextra sui parte digitum oblique in A ponat. Aspiciens si voluerit tangere digitum A, manum iniiciet, ubi non A est, sed in B, et ibi non inveniet. Ratio est quia A ingrediens per pupillam C refrangitur in D; relabitur in B. Tunc ex centro circuli F, per A punctum rei visae trahatur cathetus FA. Coit cum DC in B, et ibi apparebit.

　　Si radius ex oculo egredetur contrarium evenerit. Sit res visa B, pupillae punctus D, quaerens videre B per DB. Ubi ad C perventum est, cadet a perpendiculari FE et veniet ad A. Producta catheto ex B ad F, occurret lineae CA in A. Ibi igitur videbitur B, sed contrarium experimur.

　　Nunc argumenta solvemus. Dicimus quod omnia fere argumenta utrinque adducta ἀντιστρέφοντα sunt (a nobis reciproca dicta) quod propositum argumentum referri convertique in eum possit a quo deductum est, et utrumque pariter valet, ut ex Gellio videre est. Qui fieri dicunt visionem

_____

[6]"FE" in original text.

onto the crystalline humor.[27] No one, therefore, is so hardheaded and mentally deficient that he will argue that it occurs by means of reflection from the eye to the Sun via the paper, not only because paper does not reflect, but also because it has been firmly established by now that reflection occurs at equal angles, whereas from whatever side you look at the paper, you will always see the same shapes projected on it. We will add a third argument. Let someone stare intently at some shape facing him directly, and let another person show his finger on the left or right side. If the viewer raises his hand to touch the finger, he will actually strike the air far away from the finger, which will be an indication that vision occurs by intromission. We think we can demonstrate this point quite clearly. [92]

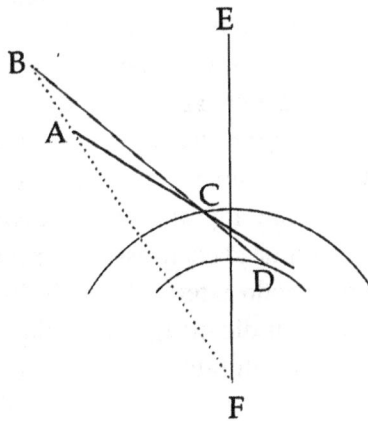

Figure 4.1

Let D and C [in Figure 4.1] be [on] the [lens and cornea of the] eye, let the eye look at facing object E, and let someone else on the viewer's right side place a finger off to the side at A. If the viewer wants to touch the finger at A, he will put his hand not where A is but at B, and he will not encounter it here. The reason is that the image of A entering through the pupil at C is refracted to D, and [the extension of refracted ray DC] inclines toward B. From center point F of the circle, then, draw cathetus FA through point A of the visible object. It intersects DC at B, and here is where it will appear.

If the ray were to exit from the pupil, the opposite would happen. Let B be the visible object and point D on the pupil of someone trying to see B along ray DB. When that ray has reached C, it will recede away from normal FE and will reach A. If the cathetus is extended from B to F, it will intersect line CA at A. Here, then, is where B will be seen, yet we experience the opposite.[28]

We will now refute the previously raised arguments. We say that virtually all the arguments adduced on both sides are ἀντιστρέφοντα (called *reciprocal* by us) because the proposed argument can be turned back and redirected toward what it was deduced from, and both are equally valid, as is apparent from Aulus Gellius.[29] There are those who say that vision occurs by extramis-

extramittendo, quia idolum montis aut vastae alicuius quantitatis ingredi per angustum pupillae foramen non potest. Respondet alter quia neque egredi[7] ab eodem foramine[8] radii possunt ut tantam molam intercipiant, si eadem lex erit exeundi introeundique. Si in oculum penetrans radius acum affert, cur egrediens eam non offendit? Sic et reliqua.

Quod autem una cum Empedocle tradit Plato effici visum per luminis ab oculo emissionem, proinde ac ex laterna. Id esse omnino vanum autumat Aristoteles libro *De sensu et sensili*, quando si ita foret, deberemus etiam in tenebris uti oculis radio inde promicante ad medium illustrandum idoneo. Verum quod infert Plato restingui radium offensantibus tenebris vanius multo est. "Quae enim luminis extinctio esse potest? Extinguitur siquidem humido aut frigidio, calidum et siccum," ut in prunis videmus, sed neutrum in lumine recognoscitur.[9] Quod sic est, ac ob debilitatem nos latet, interdiu claritatis excellentia perire debuerat aut aqua aut glacie. Quippe ignita corpora haec omnino patiuntur, sed in oculorum lumine nil prorsus eiusmodi sit. Unde certa coniectatione Platonis opinionem ut minus veritati subnixam ruere. Praeterea, vel lumen lumini coniungitur quoniam coniunctio corporum est, sed qualitates intenduntur. Neque id etiam assequerentur, nam et intenduntur quae eiusmodi sunt rationis, modo exterius lumen et interius [93] oculorum non sunt eiusdem rationis. Qui dicunt ignem oculo contineri vana dicunt, quoniam intra aquam, cuius naturam sapit oculus, ignis conservari non potest.

Euclides dixit naturam fabricasse instrumenta sensus concava quia receptura erant sonus, odores, et sapores, sed oculum convexum construxerat quia foris radios emissurus erat. Dico in illis bene dixisse, sed oculus quia spirituales radios excepturus erat, non nisi convexus futurus erat, eiusmodi enim figura ad id opus aptissima erat, nec[10] aliter fieri oportuerat, ut suo loco declarabitur. Quin immo nec necessario lumen ab oculo recipi volunt, ut videmus quum densioribus tenebris obvallatos, nec luce prorsus ulla oculum contingente, ex longinquo multa visu apprehendamus ubi comperiatur lumen retegens obiecta.

Praeterea verum est oculum obiectae rei imaginem in pupilla ostendere, nam extima corneae superficies mire pellucet, intra pupillam vero nigricans videtur, sicut speculis retrorsum allinitur plumbum ut obscurior illa densitas sit imaginum sufflamen, quod per vitri delapsae lucidum laevorem densiore obiecto suppressae non evanescerent. Sed falsum est quod inde causetur

---

[7]"ingredi" in original text.
[8]"foramene" in original text.
[9]"recognoscit" in original text.
[10]"nel" in original text.

sion because the image of a mountain or a vast object of any size cannot enter through the narrow opening of the pupil. Someone else responds that rays cannot exit from that same opening so as to encompass such a huge mass, if the same rule applies to exiting and entering. If a ray, in penetrating the eye, pushes through with a sharp point, why does it not harm the eye in exiting? And the same holds for the rest of the arguments [in favor of corporeal extramission or intromission]

Plato, however, joins with Empedocles in proposing that vision is accomplished by the emission of light from the eye, much like light shining from a lantern. In his book *On Sense and the Sensible*, Aristotle asserts that this is utterly groundless; if light were to pass out through the pupil in that way, we should also use radiation from the eye in darkness so that it should sufficiently illuminate the intermediate space as it diffuses outward. But Plato's conclusion that the radiation is extinguished by encountering darkness is even more groundless.[30] "For what can extinguish light? Something hot or dry is certainly extinguished by something moist or cold," as we see in the case of burning coals, but neither of these qualities is recognized in the case of light. Granted that this is so and escapes us because of its weakness, the intense brightness during daylight ought to be extinguished either in water or ice. Indeed, burning bodies suffer such extinction completely, but surely nothing of this sort is the case for the light of the eyes.[31] According to clear thinking, then, Plato's opinion collapses insofar as its basis is so far from the truth. Besides, even if [such] light is conjoined with [equivalent] light, it is as a conjunction of bodies, but qualities are intensified.[32] Nor would they also agree that the intensified qualities are of the same kind insofar as external light and the interior light [93] of the eyes are not of the same kind. Those who say that fire is contained by the eye speak groundlessly because in water, of whose nature the eye partakes, fire cannot be maintained.

Euclid maintained that nature made some sense organs concave because they were to be receptive of sound, odors, and flavors, but she made the eye convex because it was to be open to the emission of rays.[33] I say that in the former cases he spoke well, but because it had to receive spiritual rays, the eye could only be convex, for a shape of that sort was most appropriate for this process, and it could not have been done otherwise, as will be shown in its proper place. Nor indeed do those who espouse extramission agree that light must be received by the eye, since we see when we are surrounded by deep shadow; and when no light reaches our eye, we can visually apprehend many objects from afar, where [visual] light can be understood to touch them.

In addition, it is true that the eye reveals images of objects in the pupil, for the outer surface of the cornea is wonderfully transparent, whereas within the pupil it appears to blacken, just as lead is applied to the back of mirrors so that its darker opacity forms an impediment to images in order that, when they pass through the clarity and smoothness of glass and are restrained by a denser body, they would not disappear. But it is not true that this is the

visio, nam res obliqua oculo opposita etiam videtur, cuius tamen imago in oculo non videtur sed ad pares angulos in alteram partem delabitur, necessaria enim foret rerum oppositio ut videretur. Si animalia vident noctu, id evenit ex visus praestantia, adeo enim visu vigent, ut splendore in tenebris videant peculiari natura a Creatore donata. Nobis autem secus evenit, nam ea bonitate visu deficimus. Et animalia quae visu affascinant vel interimunt, non radiorum extramissione id evenit, sed aerem eis contiguum inficiendo et paulatim ad suum adhaerentem transiens ad hominem usque pertingit.

Lucem secundam idola ad nos deferre. PROP. 2

Lux reflexa ex speculis, vel ex marmore polito, aut metallo eandem lucem generat, ut in speculis cernimus; immo si in speculo fuerint literae exaratae aut alia signa impressa, in opposito pariete lumen reflectit idem, et cum eisdem caracteribus impressis, ut latius alibi docuimus. At si haec lux non ex [94] corpore polito sed ex terra vel lapide reflectitur, generat lumen, sive secundam lucem. Et si prima lux supra virides arbores aut rubras res inciderit, reflectit secundam lucem iisdem coloribus defaedatam.

Haec reflexio non ad pares angulos, ut prima lux, sed ex omni parte, ut ex centro ad circumferentiam; nec colores solum, sed formas, effigies, et omne quod in luce illuminatur reflectit, unde lux secunda secum rerum imagines suis coloribus et figuris insignitas adfert. Hae nempe imagines non semper in opposito videri possunt, nam vehemens lux Solis debiles illas reflexas obscurat, sed si clauso et obscuro loco resilierint, videntur personae et motus, at si per arcta foramina immittantur in obscurum cubiculum clariora rerum simulachra conspicientur, ut diximus.

Lux Solis invisibilis est nec videtur nisi reflexa, nam si per obscurum antrum a foramine in parietem ingreditur et per alterum egreditur, nec aliquid obex ei fuerit, invisa per illud transibit. Si manum opposueris, videbitur manus iluminata, et reflexa tunc antrum illuminabit. Solem in die videmus, quia eius radii a Terra repercutiuntur; nocte non videmus, quia non habent quo percutiant. Solum astra lucida conspiciuntur, quia solida et Solis lumen transverberant. In crepusculis videmus, quia radii vapores circa Terram exis-tentes offendunt. Et matutini clariores serotinis, quia matutini vapores crassi-ores ex anteactae noctis frigiditate.

Feriens Sol oculos visionem impedit; si ferit fenestrae foramen, secum imagines non affert, sic ignium et candelarum. Sed secunda lux imagines et

cause of vision, for an object facing the eye from the side is also seen, even though its image does not appear in the eye but inclines toward the other side at equal angles, for objects must be in a facing position in order to be seen. If animals see at night, it follows that they do so from the superiority of their sight, for on that basis they have powerful sight in order that, being granted a particular nature by the Creator, they can see brightness in darkness. With us, however, the opposite happens, for by the Creator's beneficence we are deficient in sight. Also, in the case of animals that bewitch or kill by sight, this does not happen by the emission of rays but by infecting the air near them, and having that infection gradually diffuse into the surrounding air until it reaches someone.

PROP. 2: *Secondary light conveys images to us.*

Light reflected from mirrors, polished marble, or polished metal generates the same light, as we see in mirrors; indeed, if there are letters engraved or other symbols inscribed in the mirror, it reflects that same light on a facing wall, and it does so with the same incised characters, as we explained at length elsewhere.[34] If, however, this light is not reflected from [94] a polished body but from earth or stone, it generates light, if only secondary light. And if primary light shines on green trees or red objects, it reflects secondary light adulterated with the same colors.[35]

This reflection occurs not at equal angles, as is the case with primary light, but in all directions, as from a center to a circumference; nor does it reflect colors only, but also the forms, likenesses, and everything that is subject to illumination in the light, so along with itself the secondary light conveys the images of objects marked out with their colors and shapes. These images cannot of course always be seen when they are in a facing position, for strong sunlight obscures those weak, reflected images, but if they are brought back to a closed, dark place, people and movement are seen, and if the images of objects are passed through narrow openings into a dark room, they will be seen quite clearly, as we said.

Sunlight is invisible and can only be seen when reflected, for if it enters a dark space through an opening in the wall and exits by another, and if there is nothing to block it, it will pass unseen through that space. If you place your hand in front of it, your hand will appear illuminated, and the light reflected from your hand will then illuminate the space. We see the Sun during the day because its rays are reflected from the Earth; we do not see it at night because there is nothing to reflect its light. The stars are seen to shine only because they are solid and capture the light of the Sun.[36] We see the Sun at twilight because the rays strike the vapors lying about the Earth.[37] During the later rainy season the morning vapors are even clearer because morning vapors are thicker on account of the coldness of the previous night.

When it strikes the eyes, sunlight impedes vision, and if it strikes the opening of a window, it does not convey images with itself, as is also the

simulachra secum affert. Aristoteles, quod secunda lux et imagines ferat et videre faciat ignorans, ait in *Problematis,*

> Quam ob causam manu prae lucerna aut Sole obiecta melius inspicere possumus? An quia lucernae Solisve lux nostris occurrens luminibus reddit suam ob nimietatem infirmiora, quippe quae rem etiam sibi cognatam interimere prae sua possit exuperantia? At vero manu ita arcente, nec luminum acies offici potest, et res visui obvia nihilominus in luce posita est. Unde fit ut acies plenius agat, et res spectanda nihilominus videatur.

In magnis defectibus stellae ob eandem causam nobis apparent, quod suis temporibus accidisse Thucidides memoriae mandavit, et nos quoque vidimus. Ex profundis puteis stellae interdiu videntur, et si Sol sit in meridie, in lapidicinis et [95] metallicis elaborantes homines quotidie eas aspiciunt. Non est igitur necesse in visione ut oculus illuminetur, sed res visa, et nos in tenebris commorantes venientes imagines non videmus, sed venientes se spectari sinunt.

Lumen per aerem imagines deferre. PROP. 3

Si aer praeterea non esset, nec visui locum omnino foret, "qua in parte operae pretium arbitror Plotini placita inserere, astruentis corpus 'per se minime necessarium esse ad emicantem ex oculo radium sustinendum atque perferendum,' qui nullo fultus tibicine aut substrato, suapte natura fertur, nec metuendum ne radius ipse decidat, qui naturam habet luminis. 'Inter visum et visibile medium per se necessarium est lumen externum vel proprium, vel obiecti valde lucentis.'"

Visionem fieri in instanti. PROP. 4

Antiquos et neotericos exercuit haec questio; utrum in instanti fiat visio. Eadem enim est ratio delationis specierum ad oculum in instanti et delationis radii a Sole ad corpus illuminatum. Alacen Arabs in tempore ferri his rationibus probavit. Non comprehenditur forma a sentiente nisi post eius perventum in corpore sensibili, nec ab ultimo sentiente comprehenditur nisi post perventum ad concavum nervi communis; ut lux a corpore luminoso delata ad oppositum corpus et quum id tempore fiat, igitur visio fit in tempore, etsi sensum lateat. Praeterea, nec sensus sentit venientes formas, nisi postquam passus est ab illis. In passione est alteratio, et nulla alteratio nisi in tempore.

case with fires and candles. Secondary light, however, does convey images and likenesses with itself. Not knowing that secondary light can both convey images and cause seeing, Aristotle asks in his *Problems*:

> Why can we see objects better if we place our hand in front of a lamp or the Sun? Is it because the light of the lamp or the Sun striking our eyes makes them weaker because of its excess, since it can destroy anything cognate to it because of its superiority? But if it is shaded in this way by the hand, the gaze cannot be offended by the light, while the object revealed to sight is nonetheless located in the light. Hence, it follows that the sight operates more fully, and the visible object can be seen despite [the strong sunlight].[38]

During total solar eclipses the stars appear to us for the same reason, a phenomenon that Thucydides remembered happening in his time and one that we have also seen.[39] From deep wells the stars are seen during daylight, and if the Sun is at the meridian, men working in stone quarries and in [95] mines see them daily.[40] In the exercise of sight, therefore, it is necessary for the visible object, not the eye, to be illuminated, and when we are lingering in dark places, we do not see the images arrive, but in arriving they let themselves be seen.

PROP. 3: *Light conveys images through the air.*

Moreover, if there were no air, no pathway whatever would exist for sight, "for which I deem it worthwhile in part to introduce the opinion of Plotinus, who affirms that 'a body by itself is the least that is necessary for supporting and sustaining the radiation emitted from the eye.' Supported by no underpinning or substrate, this radiation is sustained by its own nature, and it should not be feared that the ray itself, which has the nature of light, might collapse. 'Between the eye and the visible object external light by itself is a necessary intermediate, whether on its own or in an object that is very brightly lit.'"[41]

PROP. 4: *Vision occurs in an instant.*

The question of whether vision takes place in an instant has exercised ancient and modern thinkers. Indeed, the reason that images are conveyed to the eye in an instant is the same as the reason that rays are conveyed from the Sun to an illuminated body in an instant. The Arab Alhacen showed that it is conveyed in time according to the following reasoning. The form is perceived by the viewer only after it has reached the sensitive body, and it is perceived by the final sensor only after it has reached the hollow of the common nerve[42]; just as in the process of reaching from a luminous body to a facing body light does so over time, so vision occurs over time, even though it escapes sensible notice. Besides, the sense does not perceive the arriving forms until after it has been affected by them. There is change involved in being affected, and change occurs only in time. Therefore, the

Ergo forma aut lux non nisi in tempore defertur, et alia etiam eiusdem rationis adducit argumenta. Alchindus contraria his affert ex experientia quoque in contrarium est. Si igitur visio fit in tempore, etsi tantillum illud erit temporis, erit tamen parva magni temporis pars. [96]

Esto punctus A visus, res visa B perveniat ad A tempore imperceptibili. Sit CA[11] duplo maius. Ergo tempus quo deferretur C ad A duplo temporis maius, etsi tempus ad huc imperceptibile. Sit res visa D feratur ad A. Veniet ergo in tempore triplicato quam B[12] ad A. Sed quia eodem tempore videtur B[13] et D, idest et quae prope nos sunt et stellae in subitanea oculi apertura, impossibile est igitur ut tempore fiat. Aliam addimus. Lux incorporea est, sic et radii deferentes imagines, moventur ergo sine tempore.

Sed qui dicunt ad probandam repentinam lucis celeritatem ab oriente ad occidentem deferri in instanti, improprie loqui videntur, quum fieri non possit ut ab oriente in occidentem lux mittatur; illuminat semper Sol hemispherium, immo maius, quo illustrato adeo sensim radius repit, et sibi succedit, ut non percipies quomodo ferri videatur.

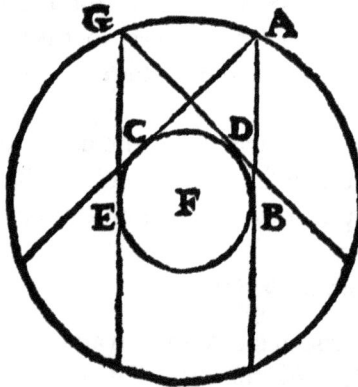

Esto A Sol in nostro orbe, Tellus F. Sol illuminet hemisphaerium BDC[14]; moveatur in G. Radius qui prius erat BC fiat DE insensibiliter. Quando igitur ex B in C momentum fuit, hac motionem impossibile est deprehendi. Visio

---

[11]"sic C" in original text.
[12]"C" in original text.
[13]"A" in original text.
[14]"BDE" in original text.

form or the light is transmitted only over time, and Alhacen also adduces other arguments based on the same sort of reasoning.[43] Al-Kindī offers arguments opposed to these, and the temporality of light transmission and vision is also contravened by experience.[44] Therefore, if vision occurs in time, even though it will be a tiny amount of time, it will still be a small portion of a larger amount of time. [96]

Figure 4.4a

Let A [in Figure 4.4a] be the viewpoint, and let the image of visible object B reach A in an imperceptible amount of time. Let distance CA be twice as great as BA. Consequently, the time in which the image of C is transmitted to A is twice as great, even though that time is still imperceptible. Let the image of visible object D be transmitted to A. It will therefore reach A in three times the amount of time as the image of B reached A. But since B and D are seen at the same time, that is, things that are near us as well as the stars are seen as soon as the eyes are opened, it follows that it is impossible for it to take time.[45] We add another rationale. Light is incorporeal, as are the rays transmitting images, so images move without taking time.

But those who say that the sudden spread of light from east to west proves that it is transmitted in an instant seem to speak improperly, since it cannot happen that light is transmitted from east to west; the Sun always illuminates a hemisphere, even more so, in fact,[46] by the way it illuminates, the radiation creeps along bit by bit and succeeds itself so that you will not perceive how it may seem to move.

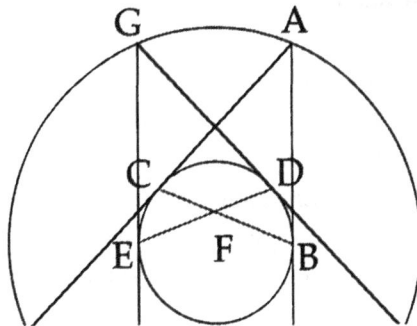

Figure 4.4b

Let A [in Figure 4.4b] be the Sun in our orb, and let F be the Earth.[47] Let the Sun illuminate hemisphere BDC, and let it move to G.[48] Let the radiation that shone on BC earlier be generated imperceptibly on [arc] DE. Therefore, when the light has moved from B to C, this motion cannot possibly

igitur in instanti[15] perficitur. Lux educens visibiles species ipsa sese effundit; easque simul per aerem transuectas in eius extremam superficiem a qua palpebra coniungitur, statim aperta palpebra, pupillae species offertur eique coniungitur minimo temporis momento. Neque est verum quod inquit Ptolemaeus,[16] apertis oculis, eorum radios [97] temporis momento ad caelum ferri; sed aer pingitur omnibus speciebus, quare aer ubicunque aliqua specie terminatur, illico species praesto est quamobrem et oculus.

Visionem fieri per rectas lineas. PROP. 5

Ut Solis radii per foramen angustum excidentes et recto prorsus tendentes nusquam deflectentes sed inflexibili via vadentes, talem intellige et visus viam. Et obiectis etiam corporibus emicant radii, qui ad oculos usque protensi suapte natura referentur, quod ratione et experientia perspicuum est. Praeterea, si consulueris eos qui dioptrae radiis, chorobatis, et eiusmodi instrumentis utuntur, videbis lineas ad visum venientes per pinnacidia et foramina invicem opposita recta ferri, et si cui eorum quid obieceris, visum impedies, et sic clare perspicies non in obliquum sed in rectum radiorum fieri porrectionem. Species ex corpore fluunt et per lucem ex corpore elicitae, et in pupillam delatae. Si inter oculum et rem tabula ponatur, tollitur visio, idest si obvium sese efferat opacum corpus. Praeterea, visionem in instanti fieri iam sancitum est, et quum lineae rectae sint omnium brevissimae, ut forma igitur quam citissime ferretur, rectis lineis ferri necessarium erat.

Visionem fieri per pyramidem. PROP. 6

Quum nos oculos ad rem aliquam contuendam convertimus, ex rei visae oris radii emanant, qui ad oculos usque porriguntur, et pupillam pertranseuntes crystallinum verberant. Imaginari itaque debemus radios ex oppositae superficiei fimbriis fluentes, quasi tenuissima fila in oculos illabi et in oculi centro, velut in manipulum, unum colligit.

Hoc modo visionis pyramidem contemplari licebit; cuius basis in re obiecta, vertex in crystallino erit. Si obiectae rei basis quadrata vel multilatera, sic quadrata vel multilatera pyramis erit; si circularis, conus. Tandem ex basi pyramis nominabitur. Nam ut oculus perfectissime aspiceret non nisi per pyramidem opportunum fuerat, nam si [98] aliter vidisset, fluxa et imbecillis

---

[15]"istanti" in original text.
[16]"Ptolaemeus" in original text.

be perceived.[49] Vision is thus completed in an instant. The light that elicits visible images from objects is itself emanated, both the light and the images are transmitted together through the air to the eye's outer surface where the eyelids are joined, the eyelids are immediately opened, and the image is presented to the pupil, with which it is conjoined in a minimal extent of time. Nor is what Ptolemy says true, that is, that as soon as the eyes are opened, their rays [97] are transmitted in an instant of time to the heavens[50]; rather, the air is painted with all the images, so the air is bounded everywhere by some image, and for that reason, wherever an eye is present an image is also present.

PROP. 5: *Vision occurs by means of straight lines.*

Just as the rays of the Sun shine through a narrow opening and continue straight on to follow a path that never deviates but remains unbending, so you must imagine the path for vision. Rays also shine forth from external objects, and by their very nature they will be extended to the eyes, which is obvious on the basis of both reason and experience. Furthermore, if you consult those who make use of radiation through a dioptra, a chorobates, or instruments of the same sort, you will see that the lines reaching through the facing sighting holes and openings in line with each other arrive directly to the eye, and if you block any of those openings with something, you will prevent sight [through them], and so you will perceive clearly that the propagation of rays occurs not at a slant but along a direct line. Images flow from a body and are elicited from the body by light, and they proceed into the pupil. If a plank is placed between the eye and an object, sight is eliminated, that is, if an opaque body gets itself in the way. In addition, it has already been established that vision occurs in an instant, and since straight lines are the shortest of all, it was therefore necessary that the form be conveyed along straight lines so that it would be conveyed as quickly as possible.[51]

PROP. 6: *Vision is accomplished by means of a cone.*

When we turn our eyes toward some object to look at it, rays emanate from the bounding surfaces of the visible object, and these rays are propagated to the eye and, after passing through the pupil, strike the crystalline humor. We ought, therefore, to imagine that the rays flow from the facing surface in fibers that are like the thinnest threads that reach the eye and are gathered together, as if into a single handful, at the center of the eye.[52]

   This is how the cone of vision should be thought of; its base will be on the visible object and its vertex in the crystalline humor. If the base of the visible object is square or polygonal, the cone will be a square or polygonal pyramid; if the base is circular, it will be a cone. Ultimately, it will be denominated according to the base of the cone.[53] In order for the eye to see perfectly, it was appropriate that it be done only by means of a cone, for if [98] it saw by any other means, vision would be changeable and weak: that

fuisset visio: videlicet, si per parallelas lineas, et oculus latior esset ut parallelas recipere posset, quum parallelae in pyramide ex una parte contrahantur.

Exemplum erit in Sole, qui a centro ad suam circumferentiam lumen diffundit ex centro in plures partes, et lumen in pyramides divisum, nam in circumferentia laxum et diffusum est, in centro vero in cono coarctatur basis in angustum collecta; fortior ibi erit. Igitur ex illuminato corpore lumen cum coloribus ad oculos veniens in pupillae angustiis se accommodat; in conum se colligit. Et quod in longa progressione debilitatum et diffusum est, se contrahit, unde in arctum compressa conus, omnium vigor ibi aderit.

Idem evenit in specillis, nam pyramidem diffusam magis contrahunt et coarctant, ob id cum illis acutius videmus. Nos ad hanc rem melius contemplandam artificio contenti erimus quo Durerius libro suo de pictura utitur. Figit enim claviculum capite per foramen in pariete, ex quo trahit filum ex una parte pondusculum habens. Deinde aliquantisper longe a clavo mensam sternit. Supra impagem figit, habeatque valvam quae apte claudi et referari possit. Habeatque impago duo alia fila cera affixa. Mox stylum capit in cuius acie filum ex clavo propendentem colligat, et socio praebet retinendum. Demum locata super mensam figura polygona vel, si mavis, solida, iubeat socio ut stylum cum filo ad angulos figurae imponat. Tunc fila in impage locata donec se intersecent in loco ubi filum pertransit, affigit. Hoc facto, socius filum remittat, et clausa valva, notet punctum intercisorum filorum. Postea aperi valvam et idem faciat ut prius donec omnia figurae puncta in sua valva signaverit. Et lineis continuabis puncta, et figuram habebis.

Euclides a visibus figuram comprehensam conum verticem habentem vocat. Sed vere conus non est, sed multilatera pyramis. Idem et Larisseus ait, nisi conum intellexerit omne quod ab oculo comprehenditur, quum in visionem fertur; quia pupilla rotunda est, universae visionis terminus circulus erit semper, et ob id conus.

Visionem partim recte, partim refracte fieri. PROP. 7

Visio nostra partim recta, partim refracta fit. Recta visio veritatem et perfectionem; obliqua vel refracta res aliter [99] quam sint ostendit et imperfectionem.

is, if it occurred by means of parallel lines and the eye were wider so that it could receive parallels, since parallels can be gathered together on one side into a cone.

An example will be found in the Sun, which emanates light from its center to its circumference in several directions, and the light is divisible into cones, for at its circumference the light is spread out and diffuse, but in the center, at the vertex, the base is gathered into a narrow compass, and the light will be stronger at this point. Hence, in reaching from the illuminated body to the eyes and entering the narrow spaces of the pupil, the light, along with the colors accompanying it, accommodates and bundles itself into a cone. And because it is weakened and diffused during its long propagation, it contracts so that when it is compressed within the narrow confines of the vertex, the power of all the rays is present at that point.

The same thing happens in spectacle lenses, for they increasingly draw the propagated cone together and contract it, and for that reason we see things more sharply with them.[54] In order to make this point better understood, we will be content with describing a device that Albrecht Dürer uses in his book on painting. He fixes a small nail into a hole in a wall, and from it he extends a thread that has a small weight on the other end. Then, at a fair distance from the nail he sets up a table. On it he attaches a framework, and it should have a door that can be readily opened and closed. The framework should also have two other threads attached with wax. Next he takes a stylus to whose point he attaches the thread hanging from the nail and gives it to an associate to hold. Then, having placed a polygonal figure or, if you prefer, a solid shape on the table, he should direct his associate to place the stylus with its thread at the vertices of the figure. With the threads positioned in the framework until they intersect at the place where the thread passes through, he fixes them in place. When this is done, the associate should give back the thread, and with the door closed, the artist should mark the point where the threads intersect. Afterwards, he may have the door opened and do the same thing as before until he has marked all the points of the figure on the door. Then you will connect the points, and you will have the figure.[55]

Euclid calls the encompassing figure extending from the eyes a cone that has a vertex.[56] However, it is not actually a cone but a many-sided pyramid. Heliodorus of Larissa says the same thing, except that by cone he understood everything that is apprehended by the eye when it is produced in vision; because the pupil is round, the boundary of the entire field of view will always be a circle, from which it follows that the figure is a cone.[57]

PROP. 7: *Vision is due in part to direct radiation and in part to refracted radiation.*

Our vision is in part direct and in part refracted. Direct vision shows everything perfectly, as it actually is; oblique or refracted vision shows things imperfectly, in a way other [99] than they actually are. And since our

Et quia de hac refracta visione maiores nostri nil aut parum meminere, ob id in multos errores inciderunt, ideo de ea nos pro viribus conscribemus.

Nos ea perfecte videmus quae intra pyramidem conspicimus, scilicet quae ab obiecta re proficiscens, et per pupillae hiatum pertransiens perpendiculariter crystallinum ad recta ferit, et haec intra radiosam pyramidem concludi dicitur. Radiosa pyramis est figura corporis oblongi, ab cuius basi brevissimae rectae in opposito tractae ad unicam cuspidem terminantur; cuius basis visa superficies est, latera pyramidis visivi radii, qui externi sunt, cuspis intra oculi centrum consideret, si eo usque progrederetur.

In pyramidis medio axis est, quae sola obiectum perpendiculariter ferit, ut pares utrinque angulos et sibi cohaerentes faciat. Haec axis, vel hic radius, omnium radiorum acerrimus et vivacissimus est, nec usquam res clarior perfectiorque videri potest, hic enim unica, quasi quadam congressione,[17] a caeteris radiis constipatus fovetur, ut merito radiorum princeps ac rex dici mereatur. Radii vero medii in pyramide sunt multivida illa quae ab extrinsecis radiis septa intra pyramidem continentur, atque hi radii, velut chamaeleon, polypus, caeteraque animalia facile pavitantia proprietate quadam rerum colores suscipiunt, ne a venatoribus capiantur, ut et ipsi faciunt, nam a contactu supeficiei usque ad pyramidis cuspidem ita colorum et luminum reperta varietate inficiuntur, ut quovis loco intercideretur, eodem inhaustum lumen acceptumque colorem expromant; inveniendo colorum proportio et stylus servatur et custoditur.

Medii hi vero radii visionem quasi minus perfectam faciunt. Est et refracta vel obliqua visio rerum quae ad pyramidis latera consistit, quae quum per arctum pupillae foramen ingredi nequeunt, illico corneam vel aqueum humorem contingentes refranguntur et ad crystallinum refracti veniunt. Haec visio omnibus imperfectior est, et quanto a recta pyramide longius recesserit, tanto imperfectissima; quippe nec rerum certos colores nec locum, sed lucem et confusam quandam visionem affert. Exemplum apponemus.
[100]
Sit oculi pupilla EF, centrum oculi K, oculus EFIO, crystallinus MGHN. A centro K pyramidales lineae trahantur in oppositum planum CD. Totum id igitur intra pyramidem contentum CD perfecte videtur, omnium perfectissime axis PK. Sit quoque pars plani CAB. Veniat B ad O punctum; refrangetur in H; relabitur in Y. Cathetus ex K et B occurrit HY in Y, et in Y punctus rei

---

[17]"congessione" in original text.

predecessors have said nothing or nearly nothing worth remembering about this refracted vision, they have fallen into many errors on that account, so we will write about it to the best of our abilities.[58]

We see perfectly what we view within the cone of radiation, that is, the cone that proceeds from the visible object and passes through the opening of the pupil to reach the crystalline humor directly along perpendicular lines, and this is said to be encompassed within the cone of radiation. The radiant cone takes the shape of an oblong body from whose base straight lines extend the shortest distance to end at a single facing point. Its base is the surface of the thing seen, and the sides of the visual cone are formed by the rays that lie on the edges, whereas the vertex would come to ground in the eye, at the center, if the cone were to reach all the way to that point.[59]

The axis lies in the middle of the cone, and it is the only ray that conveys the form of the object orthogonally so that it forms both angles equal and uniform with one another. This axis, or ray, is the strongest and liveliest of all the rays, and nothing can be seen more clearly and perfectly along any other ray, for this ray alone is nurtured by the other rays pressing on it, as if by a certain association, so that it deserves to be called the noblest of rays and their king. The rays within the cone, however, are the ones that see a great deal and are fenced within the cone by the outer rays, and like a chameleon, octopus, or other animals that are easily frightened and have the ability to take on the colors of things in order not to be captured by predators, these rays do the same, for from contact with the surface of a visible object to the vertex of the cone, they are tinged with the variety of colors and light they encounter, so that wherever they may be, they reveal the light and color they have taken on and acquired, and having been encountered by these rays, the proportion and tone of the colors is maintained and preserved.[60]

These intermediate rays, however, yield less perfect vision. Also, the vision of things that lie to the sides of the cone is refracted or oblique because the images of these things cannot enter the narrow opening of the pupil directly but on striking the cornea or the aqueous humor they are refracted and reach the crystalline humor refracted. This kind of vision is the most imperfect of all, and the farther from a direct line in the cone it gets, the more imperfect it is; in fact, it yields clear impressions of neither colors nor location, but instead it conveys a somewhat confused light and visual apprehension. We will append an illustration. [100]

Let EF [in Figure 4.7a] be the pupil of the eye, K the eye's center, EFIO the outer surface of the eye, and MGHN the crystalline lens. Draw lines in the radiant cone from center point K to facing plane CD. Everything within the cone subtended by CD is therefore seen clearly, and what lies on axis PK will be seen most perfectly of all. Now let CAB be part of a plane. Let the image of B reach point O; it will be refracted to H, and [the extension of refracted ray OH] inclines toward Y. The cathetus from K through B intersects HY at Y, and point B on the object will appear at Y. Likewise, the

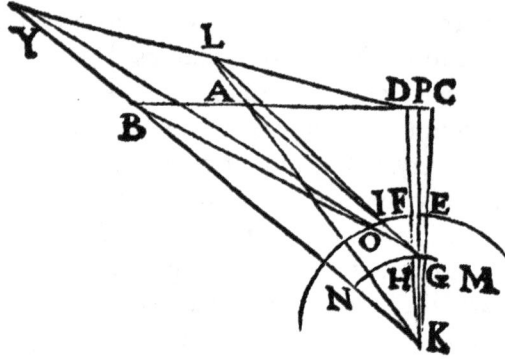

B videbitur.[18] Eodem modo punctus plani A veniet in I, refrangetur in G. Relabitur in L, et occurret catheto in L extensae ex KA, et in L videbitur A punctus, non suo loco ubi est. Nec eodem colore colorata, ut diximus, sed claritatem et indistinctam rerum molem induet, et quanto ab P[19] longius recesserit, imperfectius videbitur. Sed minus confusa visio in L quam in Y et minus in D quam in L. Quod punctus B videatur in Y hoc pacto demonstrabimus, ne idem toto hoc opere sit repetendum.

Diximus alibi omnis aquae corpus rotundum esse, et si cyathum aquae plenum in lunae globo locetur, etiam suprema eius aquae superficies illi rotunda est et mundi centro concentrica. Et res in aquis videbitur ubi linea visus distentia cathetum a re visa ad centrum mundi tracta secatur. [101]

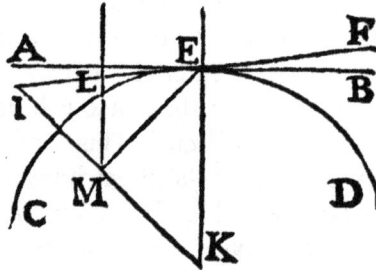

Sit aquae superficies AB, ut nobis videtur plana sed re vera circularis est, sicut DEC. Sit oculus F; sit res visa M. Veniat ad aquae superficiem E;

---

[18]"videbatur" in original text.
[19]"F" in original text.

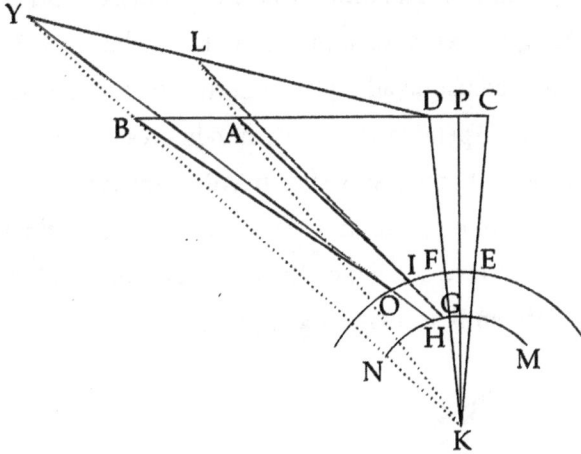

Figure 4.7a

image of point A on the plane will reach I and will be refracted to G. [Refracted ray GI] inclines toward L, and it will intersect the cathetus extended along KA at L, and point A will appear at L, not where it actually is. Nor, as we said, will it appear colored with its true hue, but it will assume the clarity and indistinct shape of things seen off to the side, and the farther away it moves from P [on the axial ray], the more imperfectly it will be seen. Still, vision is less confused at L than at Y and less at D than at L.[61] In this way we will demonstrate that point B should be seen at Y so as not to repeat the same thing throughout this work.

We said elsewhere that the entire body of water [in the universe] is round, and if a ladle full of water were situated on the globe of the Moon, the top surface of the water would also be round and share the same center point as the world. And so an object will appear in water where the extension of the line of sight intersects the cathetus drawn from the visible object to the center of the world. [101]

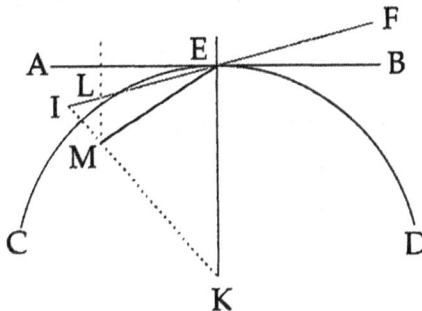

Figure 4.7b

Let AB [in Figure 4.7b] be the surface of water, which appears plane to us but in reality is circular, as represented by DEC. Let F be the eye, and

refrangitur ad oculum F. Trahatur cathetus a mundi centro K per M, et occurret FE in longum extensae in puncto I; ibi videbitur M. Sed quia lineae a centro Terrae ductae usque ad aquae superficiem ob nimiam distantiam a mundi centro, quasi perpendiculares sunt, non in I res videbitur. Sed producta ab M perpendiculari ad aquae superficiem AB, occurret FE in L; in L igitur videbitur. Eadem est in oculo ratio, nam quando aliquid refracte vel oblique conspiciatur, non suo loco sed ubi cathetus a re visa per oculi centrum ducta occurret lineae deferentis formam. Exemplum.

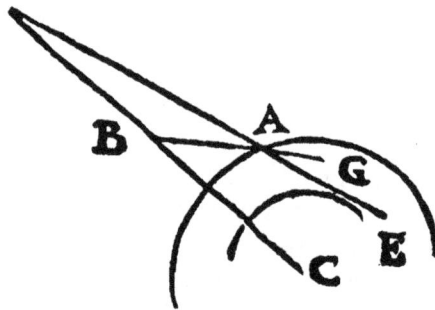

Esto DF oculus—densius scilicet medium—nam aer tenuior est aqueo humore in oculo concluso. Centrum oculi C, res conspecta B. Veniat oblique ad F; non in G decurrit, sed refrangitur in E; extenditur EF[20] in A. Extendatur cathetus ex C per B quousque occurret EF in A. Ibi igitur videbitur B. Noctu [102] cubili stratus lumen matutini crespusculi per inferiorem portae rimam subiens oblique oculo incidebat, ob id refractum in laquearibus cubili respiciebam. Et putans disruptum parietem, pavidus cubili surgens, videbam lumen ad suum locum regredi, nam directe illud oculis tunc occurrebat, et ex eo tempore semper haec sum speculatus.

Non omnis quae oblique oculo obiiciuntur videre posse. PROP. 8

Quando cathetus non occurret lineae extensae formam deferentis, res videri non poterit, ut in speculationibus pilae crystallinae vidimus, nam vel aequidi-

---

[20]"GF" in original text.

let M be the visible object. Let its image reach E on the surface of the water, and it is refracted to the eye at F. Draw the cathetus from center point K of the world through M, and it will intersect FE at point I; and so I is where M will be seen. But since the lines drawn from the center of the Earth to the surface of the water are virtually perpendicular to that surface because of the inordinate distance from the center of the world [which coincides with the Earth's center], the object will not be seen at I. Instead, if the normal is dropped from M to surface AB of the water, it will intersect FE at L, so it will be seen at L. The same reasoning applies to the eye, for when something is seen refracted or at a slant, it is seen not where it actually is but where the cathetus extended from the visible object through the center of the eye will intersect the line conveying the image. Here is an illustration.

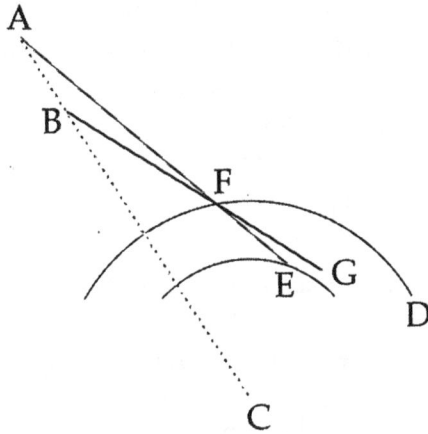

Figure 4.7c

Let DF [in Figure 4.7c] be the eye—that is, consisting of a denser medium—for the air is rarer than the aqueous humor contained by the eye. Let C be the center of the eye and B the object looked at. Let its image reach F obliquely; it does not continue straight to G but is refracted to E [on the crystalline lens], and refracted ray EF is extended to A. Extend the cathetus from C through B until it will intersect EF at A. This is therefore where B will be seen.[62] One night, [102] when I was in bed, the light of morning twilight entered through the lower edge of a door and spread out to reach my eye obliquely, and from the bed I therefore saw it refracted on the paneled ceiling. Thinking that the wall was cracked apart and leaping in fright from the bed, I saw the light return to its proper place, for it reached my eyes directly then, and from that time I have constantly watched out for this phenomenon.

PROP. 8: *Not everything that is exposed to the eye at a slant can be seen.*

When the cathetus will not intersect the extension of the line conveying the form, the object cannot be seen, just as we saw in our discussions of the

stans erit, vel retro oculum fieri deberet visio, ut vidimus.

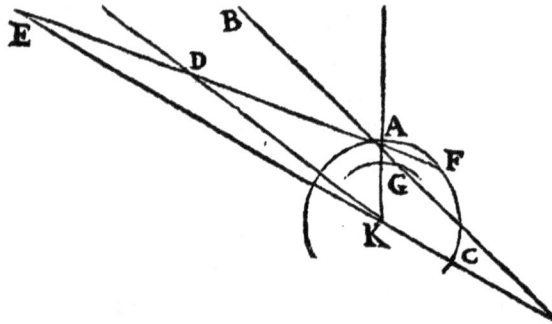

Esto oculus FA, esto res visa D veniat ad pupillam A. Non venit ad F sed refrangitur ad G; extenditurque ad B. Trahatur cathetus ex K per D, et sit KD. Aequidistabit BA; non ergo uspiam concurret. Ergo res videri non poterit, per decimam secundi. Si vero fuerit in E, trahaturque cathetus per EK; concurret retro oculum in C. Non igitur in hoc casu videri poterit, per undecimam eiusdem. [103]

Visionem fieri veluti per infundibulum. PROP. 9

Etsi vulgo dici soleat visionem per pyramidem fieri, non tamen per integram sed curtam et detruncatam, vel ad modum infundibuli. Nam radiosa pyramis sola quae perpendicularis ad crystallinum producitur, si crystallinus non obiiceretur procul dubio ad conum cuspidatum procederet. Sed crystallino obice; eam decurtat. Praeterea, obliqua pyramis plus deformatur, nam lineae ad oculum se frangentes coincidunt quasi per infundibulum.

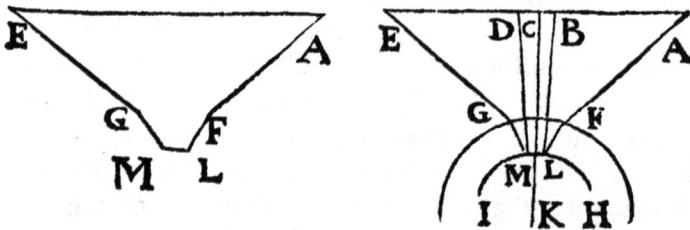

Sit visibile obiectum ABCDE,[21] oculi pupilla FG, crystallinus HI, Occurrit recta pyramis BDLM,[22] quae si usque ad K descenderet, procul dubio perfecta

---

[21]"ABCD" in original text.
[22]"BCLM" in original text.

glass sphere, for either it will be parallel, or the visual apprehension should take place behind the eye, as we saw [in propositions 10 and 11 of book 2].

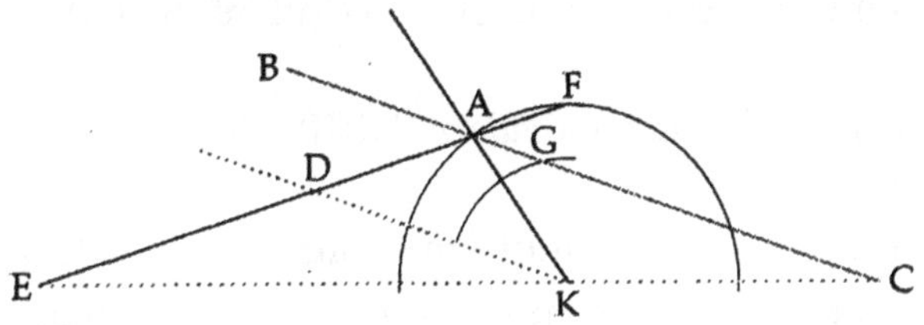

Figure 4.8

Let arc FA [in Figure 4.8] be the outer surface of the eye, and let the image of visible object D reach A on the pupil. It does not continue to F but is refracted to G, and [refracted ray AG] is extended to B. Draw the cathetus from K through D, and let it be KD. It will be parallel to BA, so it will never intersect it. Consequently, the object could not be seen, according to the tenth proposition of the second book. If the object were at E, however, draw the cathetus through EK, and it will intersect refracted ray AG at C, behind the eye. In this case, then, it cannot be seen, according to the eleventh proposition of the same second book. [103]

PROP. 9: *Vision takes place by means of a sort of funnel.*

Although vision is commonly said to take place according to a cone, it does not do so by means of a complete cone but rather a shortened and truncated one in the form of a funnel. For without doubt only the radiant cone that is generated perpendicular to the crystalline humor would continue undeviated to its pointed vertex, if the crystalline humor did not intervene. But the crystalline humor does intervene, and it cuts the cone short. Furthermore, an oblique cone is even more skewed, for the lines, being refracted at the eye, converge as if through a funnel.

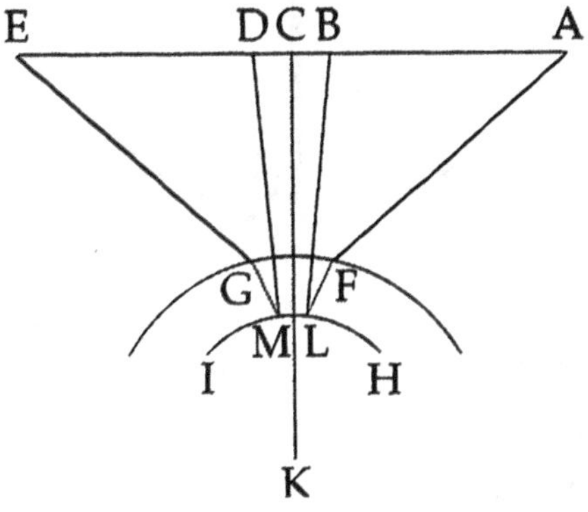

Figure 4.9

Let ABCDE [in Figure 4.9] be the visible object, FG the pupil of the eye, and HI the crystalline humor. Cone BDLM is incident upon the eye

pyramis esset. Sed crystallini occursu scinditur ab ipsius superficie HI. Esto et refracta pyramis AE. Veniat E ad oculum G; refrangitur in M. Sic A ad L occurret crystallino HI; decurtabitur, et erit veluti infundibulum AEFGLM.[23]

Oculum plus quam coeli quartam videre. PROP. 10

Maxima diligentia ab opticis perquisitum est utrum pyramidis angulus quem facit in pupilla fuerit rectus necne—id est si oculus quartam coeli partem uno intuitu videre possit—qua in re turpiter lapsi sunt.

Larisseus ait visionis angulum esse rectum a natura in oculo conformatum, quia anguli acuti obtusique incerti sunt, nam augendo et minuendo infiniti [104] sunt specie, rectus vero finitus. Natura quod est infinitum veluti indeterminatum horret, finitum vero tanquam praestantius amplectitur, unde rationi consentaneum retur visionis angulum rectum esse, idque etiam phaenomenis ostendit. Si quis caelum aspiciat, eius tetrartemorion uno obtutu conspicit, et si ad horizontis circulum oculum convertamus, eandem ambitus partem conspicit. Constat igitur mathematica ratione rectum angulum tetrartemonem continere. Praeterea, in circuli circumferentia locatus oculus totam diametrum conspicabitur, quod non eveniet nisi angulus rectus esset; quod ut clarius eius sententia perspiciatur, addemus exemplum.

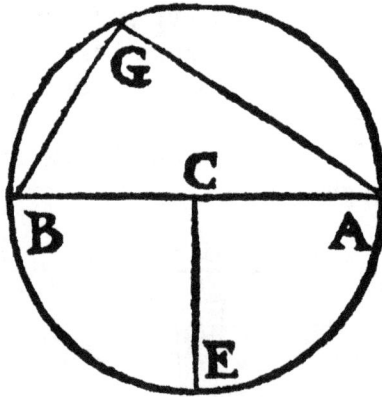

Si oculus positus fuerit in C, videbit coeli quartam AE. Idemque videbit horizontis quartam EB, et ACE et ECB recti anguli sunt. Praeterea, si oculus in G[24] positus fuerit, scilicet in circumferentia, totam diametrum videbitur, ergo AGB[25] rectus est angulus.

---

[23]"ABFGLM" in original text.
[24]"D" in original text.
[25]"ADB" in original text.

along the perpendicular, and if it were to continue down to [the eye's center point] K, it would certainly form a perfect cone. But when it strikes the crystalline humor it is cut by its surface HI. Now let the cone based on AE be refracted. Let the form of E reach G on the eye, and it is refracted to M. Likewise, the form of A will meet the crystalline humor HI at L, and the cone on base AE will be cut short, so it will be like funnel AEFGLM.

PROP. 10: *The eye sees more than a quarter of the sky.*

Whether or not the angle of the radiant cone formed in the pupil is right has been examined with utmost diligence by optical theorists—that is, the question of whether the eye can see a quarter of the sky with one glance—but they have been badly mistaken in the matter.

Heliodorus of Larissa says that the visual angle in the eye is right in conformance with nature, for acute and obtuse angles are indefinite, since they are infinite [104] in kind according to augmentation or diminution, but the right angle is definite. Nature abhors what is infinite just as she abhors what is indefinite, but she embraces the definite as so much the more preferable, so it is assumed that the visual angle's being right conforms with reason, and it is also manifest in the phenomena. If someone looks at the sky, he sees a quarter of it with one glance, and if we turn our eye to the circle of the horizon, it sees the same portion of the circumference.[63] It is therefore established by mathematical reasoning that the right angle subtends a quarter of a circle. Moreover, when the eye is situated on the circumference of a circle, the entire diameter will be seen, which would not happen unless the visual angle were right; and we will add an illustration in order to make Heliodorus's opinion clearer.

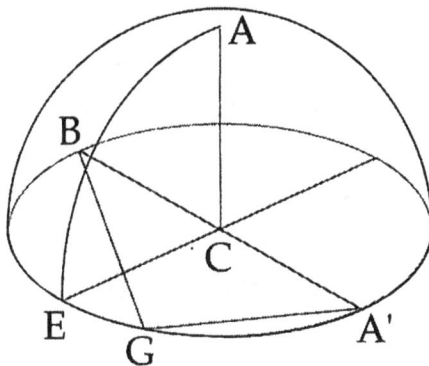

Figure 4.10a

If the eye were located at C [in Figure 4.10a], it would see quarter AE of the sky. The same eye would also see quarter EB of the horizon, and ACE and ECB are right angles. Furthermore, if the eye were located at G, that is, on the circumference, it would see the entire diameter [BA'], so A'GB is a right angle.

Vitellius dicit quod maximus angulorum sub quo fit visio est quasi rectus, quod diameter foraminis uveae quae subtenditur illi angulo in centro visus est quasi aequalis lateri cubi inscriptibilis sphaerae uveae vel lateri quadrati inscriptibilis circulo magno illius sphaerae.

Rogerius Baccon arduae dicit id esse determinationis, tamen foraminis aperturam; etsi sit latus quadrati inscriptibilis intra totam sphaeram, non videre coeli quadrantem. Ratio est, quia in centro Terrae oculus non moratur, commorans igitur super superficiem non videt coeli quadrantem. Oportet quod maiorem recto angulum contineret. Sit caeli quadrans AB, Terra EF,

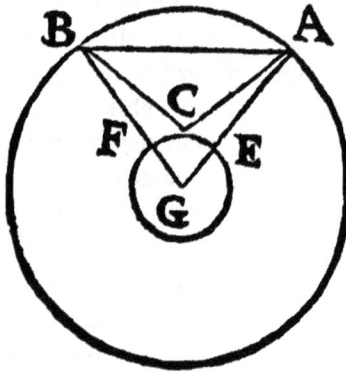

et a centro D lineae DA et DB trahantur ad angulum rectum in D. Et a superficie Terrae C lineae CA et CB protrahantur. Clarum est angulum ACB maiorem esse recto ADB. Igitur [105] ut oculus existens in C, quartam coeli videre posset AB, maiorem recto ACB foraminis aperturam habere deberet.

Praeterea, nec sphaera uveae concentrica est glaciali, unde etsi glacialis et mundi concentricae essent, non ob id uvea concentrica est. Concludit ergo non posse videre coeli medietatem. Causa universalis vero ait diametrum uveae esse latus quadrati quod describitur intra sphaeram uveae, et si sphaera uveae et glacialis concentricae essent, sub latere quadrati videretur, et esset angulus ad centrum rectus. Sed quia centrum glacialis interius est, remanet latus aperturae uveae minus lateris suis quadrati inscriptibilis.

Sed hi omnes sine experientia et ratione locuti sunt, sed quid de caetero eis credendum sit, aliis iudicandum reliquo. Primo in oculi dissectione pupillae foramen usque ad duodecimam—ne dicam vigesimam—partem suae sphaerae aperiri, et in luce usque ad acus tenuitatem constringi diximus.

Witelo says that the largest angle under which vision occurs is virtually right because the diameter of the opening in the uvea that subtends this angle at the center of sight is virtually equal to the side of a cube that can be inscribed in the sphere of the uvea or the side of a square that can be inscribed in a great circle of that sphere.[64]

Roger Bacon claims that it is difficult to determine this, notwithstanding how open the aperture of the pupil is, and even if the side of a square could be inscribed within the entire sphere of the uvea, the eye would not see a quarter of the sky. His explanation is that the eye does not occupy the center of the Earth, so in lying on its surface it does not see a quarter of the sky. In order to do so it must subtend more than a right angle.[65] Let AB [in Figure

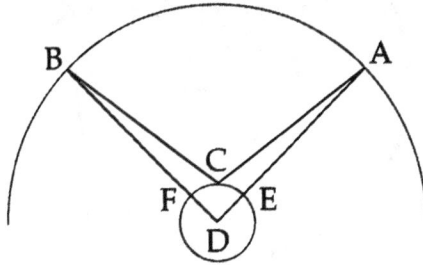

Figure 4.10b

4.10b] be a quarter of the sky, EF the Earth, and from center point D draw lines DA and DB at a right angle at D. Then extend lines CA and CB from C on the Earth's surface. It is obvious that angle ACB > right angle ADB. Consequently, [105] with the eye lying at C, it could see a quarter AB of the sky but would have to have a pupillary opening that accommodates angle ACB, which is greater than a right angle.

In addition, the uveal sphere is not concentric with the glacial humor [according to Bacon], so even if the glacial humor and the universe were concentric, it follows that the uvea is not concentric with it. He therefore concludes that the eye cannot see half the sky. But he claims that the overarching explanation is that the diameter of the opening in the uvea constitutes the side of the square that is inscribed in the uveal sphere, and if the uveal sphere and the glacial humor were concentric, the sky would be seen according to the side of a square, and the angle at the center of sight would be right. But since the center of the glacial humor is further inside the eye [than the center of the uveal sphere], the side of the uveal opening remains less than the sides of a square that can be inscribed in it.[66]

All of these authorities have spoken without experience or reason,. however, and I leave it to others to judge which among the rest ought to be believed. For a start, according to the dissection of the eye, the opening of the pupil extends up to a twelfth—indeed, I should say a twentieth—of its uveal sphere, and we have said that it is narrowed to the width of a needle

Praeterea, proprium pupillae munus esse constringi et aperiri, ut vidimus; neque usquam in eodem statu esse. Quando igitur determinanda erit suae aperturae qualitas?

Sed re vera oculus maiorem quam coeli quartam videt, et praecipue sero, quando Solis abest praesentia, sed id alio modo quam ipsi putant, sed partim directe aspiciendo, partim refracte. Et id sensui manifestum est nobis ambulando et sedendo quod a nostri dextra parte usque ad sinistram videmus, et si id non esset, satis homo infeliciter viveret, ut demonstrabimus. [106]

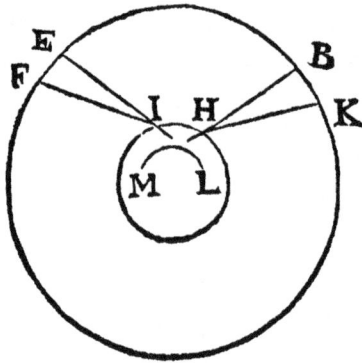

Sit oculus HI, et eius pupilla accedat radius F a dextris. Refrangatur in M; extendatur in E. Sic K veniet ad H; refrangitur ad L. Extenditur ad B. Videbitur tota coeli pars BE, quae est ultra quartam.

point in strong light. Besides, as we have seen, it is an intrinsic property of the pupil to be narrowed and to be opened and never to be in the same state. At what point, therefore, will the quality of its opening be determined?

In reality, though, the eye sees more than a quarter of the sky, especially in the late hours when the Sun has set, and this happens in a way different from the one those authorities suppose, partly from looking directly along unrefracted radiation and partly from refracted radiation. And it is empirically obvious to us when we walk around or stay seated that we see from our right side all the way to our left, and if that were not the case, a person would live fairly unhappily, as we will demonstrate. [106]

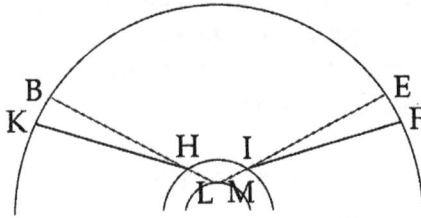

Figure 4.10c

Let HI [in Figure 4.10c] be the eye, and let a ray from F reach it from the right. Let it be refracted to M, and let [refracted ray MI] be extended to E. Likewise, a ray from K will reach H, and it is refracted to L. [Refracted ray HL] is extended to B. The entire segment BE of the sky, which is more than a quarter, will be seen.[67]

NOTES

1. *Saturnalia*, VII, xiv, 4, in Macrobius (1535), 326, lines 30–34. In this case the term *simulacrum* reflects Lucretius's usage, according to which it denotes not an image but rather an atom-thick replica of the object's surface that is transmitted outward.

2. This summarizes Aristotle's brief discussion in *On Sense*, 2, 438a5–12 of Democritus's claim that the eye is composed of water.

3. For this argument, see *Saturnalia*, VII, xiv, 5, in Macrobius (1535), 326–27. The "images" at issue in this passage are physical replicas rather than the immaterial images seen in a mirror or that pass through transparent media.

4. *Saturnalia*, VII, xiv, 7, in Macrobius (1535), 327, lines 4–6.

5. Ibid., lines 8–9.

6. This entire paragraph is a close paraphrase, with directly quoted snippets, from ibid., lines 1–17.

7. Pseudo-Plutarch, *On the Opinions of the Philosophers*, IV, 13, in Plutarch (1530), 151v.

8. This summarizes Aristotle's account of Empedocles in *On Sense*, 2, 437b24–438b2.

9. Pseudo-Plutarch, *On the Opinions of the Philosophers*, IV, 13, in Plutarch (1530), 151v.

10. Aulus Gellius, *Attic Nights*, V, 16, in Aulus Gellius (1560), 147.

11. Galen, *On the Doctrines of Plato and Hippocrates*, VII, in Galen (1535), 107, lines 25–26.

12. Pseudo-Plutarch, *On the Opinions of the Philosophers*, IV, 13, in Plutarch (1530), 151v.

13. See Galen, *On the Doctrines of Plato and Hippocrates*, VII, in Galen (1535), 104, lines 9–23.

14. This entire passage is an extremely close paraphrase of ibid., 101, lines 17–28.

15. See the prologue to the Theonine recension of Euclid's *Optics* in Euclid (1895), 146–151, and in Euclid (1537), 517. For an explicit characterization of Euclid as a Platonist, see part two of the prologue to Proclus's commentary on book 1 of Euclid's *Elements*, in Proclus (1560), 39, and in Proclus (1992), 57.

16. See *Optical Hypotheses*, 2, in Heliodorus (1573), 118–19 (unnumbered: Italian), and 132 (unnumbered: Greek and Latin); see also Heliodorus (1897), 4–5.

17. Pliny, *Natural History*, XI, 54, in Pliny (1559), 293, lines 25–27.

18. Ibid., VII, 2, in Pliny (1559), 153, lines 48–51.

19. *Problems*, II, 54, in Alexander of Aphrodisias (1552), f. 105ra.

20. For the first quoted sentence, see Pliny, *Natural History*, VIII, 34, in Pliny (1559 [as VIII, 22]), 195; for the second, see *Natural History*, XXIX, 19, in Pliny (1559 [as XXIX, 4]), 770.

21. See Aristotle, *On Dreams*, 2, 459b28–460a11.

22. See, for example, *Meteorology*, III, 2, 372a29-30, and *Problems*, XXXI, 15.

23. The idea that sensible forms, including visible ones, undergo a spiritual transformation when passing from the physical object into the media that convey them to the senses traces back to Averroes; see, for example, *Averroes: Epitome of Parva Naturalia*, trans. Harry Blumberg (Cambridge, MA: Mediaeval Academy of America, 1961), 15–16.

24. Actually, as Della Porta argues in proposition 15 of the previous book, the vertex of the radiant cone lies behind the crystalline humor, or lens.

25. According to Alhacen and his Perspectivist followers, secondary (or accidental) light (*lux secundaria/accidentalis* or *secunda lux* in Della Porta's words) takes the form of *lumen*. This, in turn, is the species or similitude of *lux*, or primary light, which constitutes the inherent quality of luminosity in such objects as the Sun. Thus, objects are rendered visible by incorporating secondary illumination from luminous sources. For a clear articulation of the *lux/lumen* distinction, see Roger Bacon, *De multiplication specierum*, I, 1, in Bacon (1983), 2–5.

26. Della Porta is of course describing a camera obscura here, a device he had already described in great detail in *Magiae Naturalis*, XVII, 6, in Della Porta (1589), 266–67. In that description, Della Porta suggests putting a convex lens in front of the aperture to augment the clarity and brightness of the inverted image produced on a white screen by the light projected through the aperture.

27. In retrospect, this claim seems remarkably prescient insofar as it appears to anticipate Kepler's account of how the pupil and the lens of the eye function in producing clear, inverted retinal images of the kind Della Porta describes in *Magiae Naturalis*, XVII, 6, when a convex lens is placed in front of the small opening in the camera obscura. In that same chapter, moreover, Della Porta explicitly likens the eye to a camera obscura. In fairly short order, however, it will become clear that Della Porta failed to follow up on this insight.

28. Of course, in order for this to be true, AC, when extended, would have to touch the lens, contrary to the way it is represented in the actual diagram provided in the text.

29. See Aulus Gellius, *Attic Nights*, V, 10, in Aulus Gellius (1560), 138–40.

30. For Plato's account of visual fire and its emission, see *Timaeus*, 45b–c; on Empedocles and the lantern analogy, see Aristotle, *On Sense*, 2, 437b24–438a2; for Aristotle's argument against the emission of visual fire, see ibid., 437b11–23; and on the extinction of visual fire by darkness, see Plato, *Timaeus* 45d.

31. Quotation from *On Sense*, 2, 437b16-17, in Aristotle (1550c), 187va; this entire passage from the end of the direct quotation is a close paraphrase of ibid., 437b18-23.

32. In other words, if light is a material efflux, then adding one such light to another will not result in increased intensity but, rather, increased volume.

33. See the prologue to the Theonine recension of Euclid's *Optics* in Euclid (1895), 150–51, and in Euclid (1537), 517.

34. See *Magiae Naturalis*, XVII, 1, in Della Porta (1589), 261.

35. This distinction between primary and secondary light, the latter being produced by opaque but nonreflective surfaces upon which primary light shines, was commonplace in medieval optics and remained so throughout the Renaissance; see,

for example, Alhacen, *De aspectibus* IV, ii, 2–3, in Alhacen (1572), 102–3, and in Alhacen (2006), 5 (Latin) and 297 (English).

36. The idea that starlight is borrowed rather than intrinsic was by no means unique to Della Porta, who alludes to this idea in the "Preface to the Readers," where he mentions the passage in Job 37:18 that refers to the heavens as a mirror. We also saw in book 1, proposition 21 above, that John Pecham supposed starlight to be reflected sunlight.

37. Della Porta is clearly referring here to Ibn Mu'ādh's *On Twilight and the Rising of Clouds*, which was attributed to Alhacen in Risner's edition of the *De aspectibus*; see Alhacen (1572), 283–8; see also A. Mark Smith, "The Latin Version of Ibn Mu'ādh's Treatise "On Twilight and the Rising of Clouds,'" *Arabic Sciences and Philosophy* 2 (1992): 38–88.

38. *Problems*, XXXI, 28, in Aristotle (1552 [as XXXI, 29)], 50ra.

39. See Thucydides, *History of the Peloponnesian War* II, 28.

40. That the stars can be seen from wells and mine shafts during daylight had been a common, albeit false, assumption since at least the time of Aristotle, and it remained so until relatively recent times; see David W. Hughes "On Seeing Stars (especially up chimneys)," *Quarterly Journal of the Royal Astronomical Society* 24 (1983): 246–57.

41. Lodovico Ricchieri, *Thirty Books of Readings from the Ancients*, I, 7, in Ricchieri (1566), 24; internal quotations from Marsilio Ficino's prologue to Plotinus, *Ennead*, IV, book 5, ch. 2, in Plotinus (1580), 442.

42. The point here is that visual perception occurs in stages, starting with a brute, sensitive impression in the lens, and culminating in a judgment-based apprehension of what that impression represents, which is carried out by the *ultimus sentiens* (i.e., the final sensor) at the front of the brain, just beyond the optic chiasma, which constitutes the so-called "common nerve." Accordingly, the process has to take time, since the stages cannot occur simultaneously, especially since judgment is a step-by-step process.

43. For the full argument, see Alhacen, *De aspectibus*, II, ii, 21, in Alhacen (1572), 37–38, and in Alhacen (2001), 121–25 (Latin) and 445-8 (English).

44. See al-Kindī, *De aspectibus*, proposition 15, in al-Kindī (1912), 25–27.

45. For the source of this argument, see Witelo, *Perspectiva*, III, 55, in Witelo (1572), 110, and in Witelo (1991), 354–55 (Latin) and 171–72 (English). The flaw in this reasoning, of course, is that, if we assume the points are radiating light continuously, then radiated light from all three will be present from beforehand when the eyes are opened.

46. In other words, since the Sun is significantly larger than the Earth, rays from its edges tangent to the Earth will encompass more than half the Earth.

47. Why Della Porta locates the Sun in "our orb" is unclear; perhaps he simply means within the sphere of our universe. There is no doubt, however, that circle AG is meant to represent the Sun's orbit (or orb).

48. Obviously, since the Sun is being taken as a point in comparison to the Earth, the portion BDC of the Earth illuminated by the Sun will be less than a hemisphere, although it will expand continually as A withdraws ever further away until it becomes virtually hemispherical.

49. I take Della Porta's point here to be that, if one assumes that the light from A spreads from B to C according to a succession of rays, the motion according to which that spread occurs is so swift as to be absolutely imperceptible. Consequently, it would be virtually instantaneous.

50. Della Porta is actually referring here to Hero of Alexandria's *Catoptrica*, which was translated in 1269 by William of Moerbeke under the title *De speculis* and misascribed to Ptolemy, who was presumed to be the author until fairly recent times; for the appropriate location in that work, see Hero (1518), 250vb, lines 58–64.

51. The argument that light propagates in a straight line because of its unimaginable speed, which forces it to follow the shortest path, is to be found in Hero of Alexandria's *Catoptrica*; see Hero (1518), 250vb, lines 50–58.

52. Galen employs the same thread analogy in *On the Usefulness of the Parts*, X, 12, although he uses it to explain visual rays emanating outward in a cone from the center of the eye rather than light rays propagating inward toward that point; see Galen (1541), 179v.

53. Della Porta follows convention in referring to the radiant cone as a pyramid (*pyramis*) insofar as its shape is determined by whatever figure lies at its base. I, however, have translated the term as "cone" throughout because, taken in its entirety, the actual figure is bounded by the pupil, so the field of view at the base is a circle.

54. As Della Porta will show in book 8, where he deals specifically with lenses, this bundling process, which narrows the rays and thus brings them toward convergence, applies to convex lenses but not to concave ones.

55. For Dürer's description of this device in the Latin version of his *Unterweysung der Messung*, see Dürer (1532), 184–85. This is Dürer's illustration of the method,

which entails the helper holding the string taut against a point on the object (a lute, in this case) while the artist moves two threads attached to the opening, one vertically and the other horizontally, until they intersect where the string passes through the opening. Then the artist closes the flap with the paper on which the lute is to be drawn and marks the point of intersection on that paper. Repeating this process, he eventually gets a succession of marks that outline the object.

56. See Optics, postulate 1, in Euclid (1537), 518, and in Euclid (1895; as postulate 2), 2–3.

57. See Optical Hypotheses, 3, in Heliodorus (1573), 116–18 (unnumbered: Italian) and 132–33 (unnumbered: Greek and Latin); see also Heliodorus (1897), 4–7.

58. Actually, Alhacen (and Witelo following faithfully in his footsteps) has a considerable amount to say about refracted vision in De aspectibus, VII, vi, 34–37, in Alhacen (1572), 266–70, and in Alhacen (2010), 100–14 (Latin) and 296–307 (English).

59. According to Alhacen and his Perspectivist followers, although all the rays entering the front surface of the lens along the perpendicular tend toward the center of the eye, only the axial ray ever reaches that point. The rest are refracted away from it when they enter the vitreous humor at the posterior surface of the lens. See, for example, Alhacen, De aspectibus, II, i, 2, in Alhacen (1572), 24–25, and in Alhacen (2001), 82–84 (Latin) and 419–20 (English); see also Witelo, Perspectiva, III, 21 in Witelo (1572), 94, and in Witelo (1991), 317 (Latin) and 128–29 (English).

60. The entire description of the radiant cone in this and the preceding paragraph is a close paraphrase, along with several more-or-less direct quotations, from book one of Leon Battista Alberti's On Painting in Alberti (1540), 14–17. It should be noted, however, that unlike Della Porta, Alberti bases his account on visual rays rather than light-rays.

61. Della Porta's diagram, which I have reproduced fairly faithfully, is misleading, insofar as it has the pupil limited by FE while allowing radiation to reach the crystalline lens through points O and I to the left of F. Properly rendered, the diagram ought to show the pupillary opening located between the corneal surface and the anterior surface of the lens, in which case the rays entering the cornea at O and I could be still funneled through the pupil in order to reach the crystalline lens.

62. This is simply a repetition of the example given in proposition 1, pp. 194–95 above.

63. See Optical Hypotheses, 5, in Heliodorus (1573), 119–20 (unnumbered: Italian), 134 (unnumbered: Greek and Latin); see also Heliodorus (1897), 6–9.

64. See Witelo, Perspectiva, IV, 3, in Witelo (1572), 119.

65. See Bacon, Perspectiva, I, viii, 3, in Bacon (1996), 116–23.

66. For Bacon's explanation of this point, see ibid., 119–21.

67. In fact, according to this demonstration, it is not segment BE of the sky that is seen, but rather, segment FK, whose image is BE. According to modern measurements, the maximum horizontal field of view for a single, healthy eye is around 155°, although the field of relatively clear vision within that range is considerably smaller.

LIBER QUINTUS

PROOEMIUM

Postquam affatim visionem per intromissionem et partim recte partim refracte fieri recensuimus, mox ubi fractionis linea praecisius incumbit, ne de iis amplius posterius verendum sit, nunc operae pretium duximus problemata quaedam attexere secundum Euclidis et maiorum placita. Sed aliter quam ipsis visa sunt solvenda, et fortasse notioribus principiis et experientiis.

Proximiora maiora, remotiora minora cerni. PROP. 1

Quia universa fere quae hoc libro clauduntur de visione huic innituntur propositioni ut quae propriora sint auctiora, remotiora quidem defectiora videantur, ob id quae a nobis exigenda sunt, ut solidiori fundamento nitantur, solertius hoc problema trutinandum censemus.

Prius audiatur Euclides quid sic probat id praemittens axioma: ubi maiori angulo spectata maiora videri. Sic demonstrat. [108] Sit oculus B,

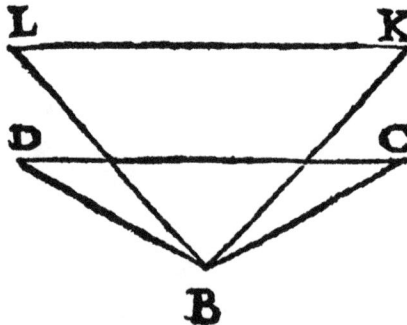

linea propius[1] admota CD, longius vero KL priori parallela et aequalis. Perveniat linea KL ad oculum B per ductus KB, et linea CD per CB, DB.[2] Quia

---

[1]"proprius" in original text.
[2]"CD" in original text.

BOOK FIVE

INTRODUCTION

Now that we have accounted sufficiently for the fact that vision is accomplished by means of intromission and that it is due partly to direct and partly to refracted radiation, and having subsequently shown exactly where the line of refraction falls, we deem it worthwhile to add some problems according to the opinions of Euclid and other predecessors, so that their opinions will no longer be accorded blind deference. But these problems will be resolved in ways other than foreseen by them, and perhaps on better-known principles and empirical grounds.

PROP. 1: *Nearer things are perceived as larger, and farther ones as smaller.*

Because absolutely everything about vision that is included in this book depends on the supposition that things that are more immediate are more accessible to us, whereas things that are at a greater remove are grasped more imperfectly, we propose to resolve this problem more fruitfully according to things that are requisite for us in order that they be placed on a more solid footing.[1]

Attend first to what Euclid demonstrates, proposing as an initial axiom that things viewed under a larger angle appear larger.[2] He proves the proposi-

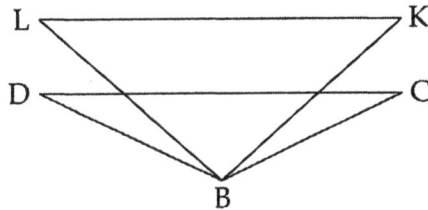

Figure 5.1a

tion as follows. [108] Let B [in Figure 5.1a] be the eye, CD a line brought up near to it, and KL a line farther away and parallel and equal to the former. Let the image of line KL reach the eye at B along lines KB [and LB], and let

angulus CBD[3] maior est KBL, igitur maior videbitur CD quam ipsa KL.

Sed Plotinus Platonicus inficiatur hoc ab Euclide depromptum axioma verum esse, eiusque rationes redarguit quas, quia alibi attulimus, huc afferre non necessarium visum est. Sed uterque lapsus est. Suppositionem cui Euclidis demonstratio innititur non omnino veram esse censemus, nam obiectum vastae magnitudinis, si propius[4] oculo adhaeserit, maxima illius pars eum latebit, at longiuscule remotum, latiorem partem videbit, nam quum species in conum desinat quo aciei cuspis, longius protenditur ea laterales axes latiorem intercipiunt magnitudinem. Exemplo res dilucidior fiet. [109]

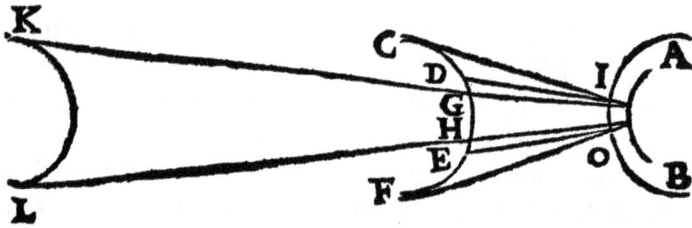

Sit visibile obiectum vastae magnitudinis CF; oculo AB valde proprius, deferatur ad eum pyramis. Quia cuspis longe pupillam superat, totum conspicari non potest, sed ex pupillae latitudine non videbitur nisi GH, quia hos ambos lateralium linearum ductus pyramidis linea circularis CF proscindit in G, H. At si longius idem obiectum producatur et eo usque processum sit ut angulus coni maximus pupillam impleat, ut in LK, totum videbitur, vel ipsius pars maior.

At si maius videri putabitur proprius admotum, ex refractione evenit, nam FO ad E, et CI ad D decurrit, ut posterius videbimus. Praeterea nec Euclides memorat visionem in crystallino fieri, nec cognovit partim recte, partim refracte videri.

Sed ad nostra transeamus re solertius perpensa. Quae aequalis magnitudinis lineae sinistrorum et destrorum ad oculum incumbunt, proximiores proclivibus lineis et reclinantibus accedunt, et remotiores rectius super oculum oriuntur. Quae reclinant franguntur, et fracta longius amandantur et quousque cum suo catheto decussentur, remotius producuntur quo loco

---

[3]"SBD" in original text.
[4]"proprius" in original text.

the image of line CD reach B along CB and DB. Therefore, since angle CBD > angle KBL, CD will look larger than KL.

However, the Platonist Plotinus denies that this axiom proposed by Euclid is true, and he refutes his arguments, but since we have dealt with them elsewhere, it seemed unnecessary to provide them here.[3] On the contrary, both sides have been passed over. We judge the supposition on which Euclid's demonstration rests to be not entirely true, for if a huge object lies close to the eye, most of that object will be out of its sight, but when the object lies somewhat farther away, the eye will see the previously unseen portion, for since the image terminates at the vertex of the radiant cone, which constitutes the point of sight, the farther away it extends, the wider the magnitude its lateral rays intercept. This point will become clearer through an illustration. [109]

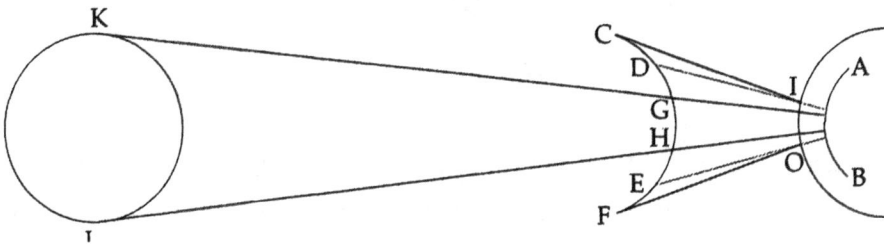

Figure 5.1b

Let CF [in Figure 5.1b] be a huge visible object, and with the eye AB quite close, let the radiant cone reach it. Because the cone's vertex lies relatively far from the pupil, the entire object cannot be seen; instead, only GH will be seen because of the width of the pupil, since the circular line CF cuts both of the leading edges of the extended cone at G and H. If, however, the same object is drawn farther away and is moved until the largest angle at the vertex fills the pupil, as is the case with object LK, the entire object will be seen, or at least most of it.

If the object will be judged to look larger when it is drawn nearer, however, this is due to refraction, for [after refraction at the surface of the eye, the refracted line of sight resulting from] ray FO extends straight to E, and [the refracted line of sight resulting from] ray CI extends straight to D, as we will see later.[4] Furthermore, Euclid does not mention that vision occurs in the crystalline humor, nor did he know that seeing is due partly to direct and partly to refracted radiation.

But let us approach our point in a more considered way. Of the right and left lines that reach the eye from equal magnitudes, those that come from nearer objects reach along slanting, inclined lines, whereas those coming from farther objects strike the eye more directly. Those that are inclined are refracted, and the farther the resulting refracted rays extend until they intersect their catheti, the farther they are extended out to where the images are

videntur. Et quia flexu hebescit visus, incondita et obscura videt eo loco. Contra vero in remotis, quia rectius aciei cuspis progreditur, minusque frangitur, minus igitur a vero formae puncto reclinabit. Rem subsignabimus exemplo.

Esto magnitudo AB aequalis et utrinque parallela, et punctus B oculo occurset in puncto E, ex doctrina tertii correlarii primae [110] propositioni libri nostri vestigetur diameter IF. Et quia circuli pars IE sexta est totius circumferentiae, linea BE ad diametri finem declinabit F. Extendatur EF, et occurset catheto ex oculi centro H per B punctum, extendaturque cathetus quousque occurrat lineae FE in G puncto. Igitur punctus B refractus in G videbitur satis remote.

Secus autem eveniet remotiori puncto D, nam et ipse oculo in O occurret. Inventa diameter erit QR, linea ergo DO[5] prope R refrangetur, et erit OL. Longius prorogetur in M, et centro H per D trahatur cathetus usque ad eum locum ubi decussabit refractionis lineam in M. Punctus tandem D in M videbitur propinquior lineae parallelae BD[6] quam G. Unde quanto quacunque ex parte procliviores oculo occursent radii, tanto refractionis linea, quousque cum catheto coeat, longius egredietur, et visa res[7] longius emicabit; quare si idem ex altera parte feceris, A in P et C in N videbitur; maior igitur PG[8] ipsa MN.

---

[5]"DG" in original text.
[6]"BM" in original text.
[7]"res" repeated in original text.
[8]"PQ" in original text.

seen. And as sight is weakened by the bending of rays, the eye sees any object as confused and obscure at that point. The opposite, however, happens when the objects are distant because the more directly the ray proceeds to the point of sight at the vertex of the cone, the less it is refracted, so the less it will incline away from the actual point of the object producing the image. We will address this point with an illustration.

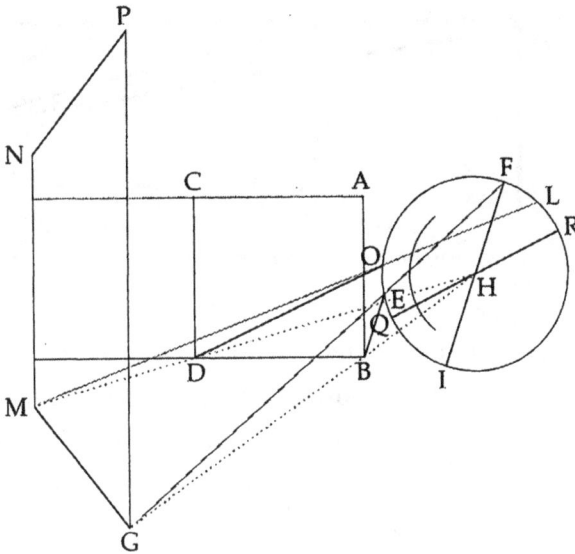

Figure 5.1c

Let magnitude AB [in Figure 5.1c] be equal and parallel to CD, let the image of point B intersect the eye at point E, and let it reach diameter IF [along EF after refraction], according to the third corollary [110] of the first proposition of our book.[5] And since segment IE of the circle is one-sixth of the entire circumference of the eye, line BE will incline toward endpoint F of diameter IF [after refraction]. Extend EF, let it intersect the cathetus passing from center point H of the eye through point B, and extend the cathetus until it intersects line EF at point G. Thus, after the refracted ray is extended to G, the image of point B will lie fairly far away from point B itself.[6]

Something different will happen with farther point D, though, for the image of this will fall on the eye at O. Diameter QR will be found, so line DO will be refracted toward R, and the refracted ray will be OL.[7] Let it be extended to M, and draw the cathetus from center point H through D to where it will intersect the line of refraction at M. Ultimately, point D will appear at M, closer to parallel line BD than G is. Hence, the more the rays strike the eye at a slant from the side, the longer the line of refraction will continue out until it intersects the cathetus, and the [relatively] farther away the visible object will appear to lie. Therefore, if you do the same thing on the other side, A will appear at P and C at N, so PG will look larger than MN.[8]

*Parallelae in plano existentes visui obiectae concurrere videntur, nunquam
tamen concurrunt. PROP. 2*

Audiamus prius maiorum placita. Euclides, quem sequuti sunt omnes, id ita
probat. Sint parallelae lineae utrinque productae AB, CD, oculus E, et sint

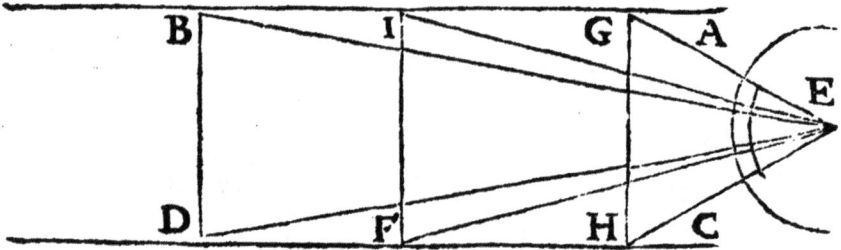

tria intervalla etiam parallela inter eas constituta GH, IF, BD. Protrahantur
[111] ex intervallorum extremitatibus lineae ad oculum GE, HE, IE, FE,
BE, DE. Et oculus angulo GEH videat intervallum GH, et angulo[9] IEF intervallum IF, et angulo BED intervallum DB. Quoniam angulus GEH maior est
angulo IEF, quia intra se continet, tanquam sua parte totum, videbitur GH
ipsa IF maior, et eodem modo IF ipsa BD.

Sed hoc falsum esse monet experientia, nam quanto res longius producetur, angulus acutior fiet. Mox acutissimus tandem longiuscule; nullus
erit angulus, quia lineae constituentes quodammodo naturales sunt. Ergo
spatium intra duarum parallelarum tractus constitutum, adeo augustari videtur, ut parallelae concurrere videantur, quod nunquam eveniet.

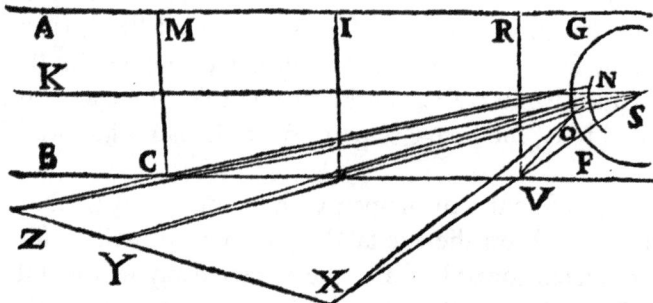

Sed nos ita solvimus. Sunt parallelae utrinque ductae GA, FB, oculus
ON, quo K punctum recta acie respiciat. Veniet ad oculum oblique linea RV

---

[9]"angulus" in original text.

PROP. 2: *Parallel lines that lie in the same plane as the eye appear to converge, yet they never do meet.*

Let us listen first to the opinions of our predecessors. Euclid, whom all the rest followed, proves it in this way.[9] Let the two parallel lines AB and CD

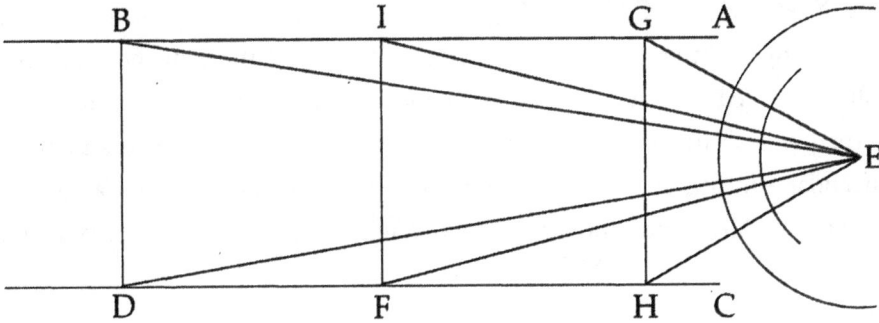

Figure 5.2a

[in Figure 5.2a] be drawn, let E be the eye, and also let the three intermediate lines GH, IF, and BD be set up parallel to one another. From the endpoints of those intermediate lines draw [111] lines GE, HE, IE, FE, BE, and DE to the eye. Let the eye see intermediate line GH under angle GEH, intermediate line IF under angle IEF, and intermediate line DB under angle BED. Since angle GEH > angle IEF because it contains IEF, as the whole contains its part, GH will appear larger than IF, and, by the same token, IF will appear larger than BD.

But experience shows this to be false, for the farther away the object is drawn, the more acute the angle will become. Finally, at the farthest distance the angle is most acute, and so there will be no angle because the established lines are somehow natural. Accordingly, the space between the extension of the two parallels is set, so it appears to contract in the way that parallels seem to converge, but that will never happen.[10]

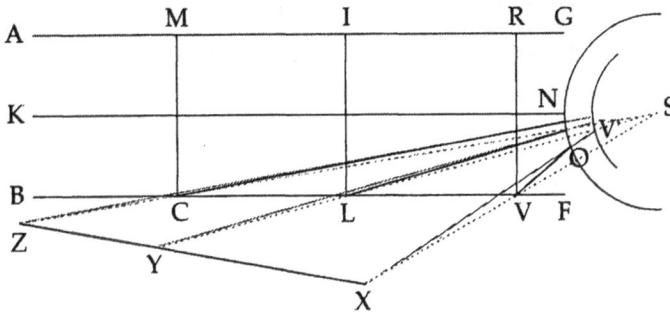

Figure 5.2b

We resolve the problem as follows. GA and FB [in Figure 5.2b] are two extended parallels and ON the eye, from which point K lies on the direct

proxime, minus vero propiusque.[10] IL, remotissima MC. Sit obliquus lineae tractus VO pupillae occurrens; refrangetur ad V. Producatur in X, et excitetur cathetus per oculi centrum S et V[11] punctum rei visae. Et conveniet cum visuali linea OX in X; igitur in X punctus V videbitur. Eodem modo punctus L in Y, et punctus C[12] in Z. Et producta exili linea per puncta X, Y, Z, semper angustabitur dum unica fiet cum linea FB, et tunc visus non amplius obliquatur, sed verum pupilla cernit quicquid ex regione opponitur. Unde lineae pares ab utrinque concurrere videbuntur. Eodem modo ad sinistrum ducendae sunt lineae. Sic latissima prope oculum platea videbitur, angustissima in fine NK[13]; spatium vero AB, quod aciei mucroni incumbit, nunquam concurrere videbitur. [112]

Parallelae lineae supra et infra oculum existentes concurrere videntur, nunquam tamen concurrunt. PROP. 3

Octava suppositione hoc theorema solvit Euclides: Altiora videri quae sublimioribus radiis cernerentur, depressiora vero quae humilioribus. Sed quum in hoc nil alienum a praedicto sit, nulla aliorum theorematum additione opus erat, frustra enim per plura fieri dicuntur quae per pauciora fieri possunt. Nos eodem modo solvemus quem supra memoravimus. Est enim globatae figurae oculus, et pupilla orbiculari meatu fenestrata qua se visibilium species insinuant. Nil itaque impedit quin undique linearum concursus in crystallino fiat, et undique transitus inhibeatur spectrorum. Visiva enim acies ubi in medio parallelarum incedit aequaliter a lateribus franguntur obliquae lineae. Rem exemplo apertiorem reddemus.

Sit altitudo RIM planitiei parallela FB, oculus SVO, parallelae ut prius aequis intervallis distinctae. Et puncto R superiori deorsum educito radium ad pupillam O; refrangetur ad V. Prolongetur; pervenit ad C.[14] Extrahatur cathetus ex S centro oculi transiens per R quousque concurrat cum radio visus in C. Punctus ergo R sublimius videbitur in C.[15] Sic punctus I in D, et [M] in P, et protracta linea per puncta, erit CDP ex alto deprimi videbitur. Idem, ut [113] superne monuimus, inferne facias, et planitiem elevari videbis visus acie in medio respiciente.

---

[10]"propriusque" in original text.
[11]"O" in original text.
[12]"X" in original text.
[13]"MN" in original text.
[14]"E" in original text.
[15]"E" in original text.

axial line of sight. The form of line RV, nearest to the eye, will reach it along oblique lines, whereas that of IL, which is not as near, will arrive along less oblique lines, and that of the farthest object MC will arrive [along the least oblique lines]. Let the oblique continuation of line VO reach the pupil, and it will be refracted to V'. Extend [refracted ray V'O from O] to X, and drop the cathetus through center point S of the eye and point V on the visible object. It will intersect line of sight OX at X, so point V will appear at X. Likewise, point L will appear at Y, and point C at Z. When a thin line is drawn through points X, Y, and Z, it will continually approach line FB until it coincides with it, and at that point the line of sight is no longer oblique, but instead the pupil properly discerns whatever faces it in a direct line. Consequently, equal lines will appear to converge from both sides. Likewise, lines can be drawn to the left [above AR]. A street will therefore appear widest near the eye and narrowest at the end of NK, but the space AB, which is compressed toward point K of direct sight, will never appear to narrow completely. [112]

PROP. 3: *Parallel lines lying above and below the eye appear to converge, yet they never do meet.*

Euclid resolves this theorem on the basis of the eighth supposition, which states that things discerned by higher rays appear higher, whereas things discerned by lower rays appear lower.[11] Because there is nothing different in this theorem from what was said before in the previous one, however, it was unnecessary to include any other theorems, for they say that it is vain to do with more what can be done with fewer.[12] We will therefore prove this in the same way that we laid out above. For the eye is of a spherical shape, and the pupil is perforated by a round opening through which the images of visible objects wend their way. Therefore, nothing whatever prevents lines coming from all directions from converging on the crystalline humor, and nothing inhibits the passage of images from all directions. For where the line of sight reaches the middle between the parallels, the oblique lines of sight to the sides are equivalently refracted. We will make this point clearer by an illustration.

Let RIM on top [in Figure 5.3] be a plane parallel to FB, let the eye be SV'O, and as before, let equal parallel intervals RV, IE, and MB be marked out. From point R above, a ray is extended down to O on the pupil, and it will be refracted to V'. Let [refracted ray OV'] be extended to reach C. From center point S of the eye, drop the cathetus that passes through R until it intersects the refracted line of sight at C. Point R will therefore appear elevated at C. So, too, point I will appear at D and M at P, and if a line is drawn through those points, it will be CDP and will appear to slope downward from above. As [113] we showed above, you can do the same thing with the lower line, and you will see the plane slope upward with respect to the line of sight in the middle.[13]

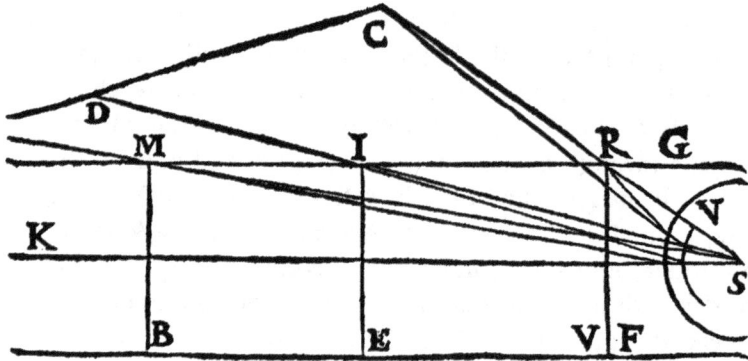

Oculus alterutri parallelo plus inhaerens minus remotior, vero magis ea concurrere videbit. PROP. 4

Etsi hoc theorema minus ex dictis necessarium videretur, quo tamen non mirentur ambulantes per caenacula aut dormitoria caenobiorum aut per regias aulas quum videant summa laquearia magis descendere quam pavimenta ascendere apponere constituimus. Quod non alia evenit ratione nisi quia aspicientes oculorum aciem figant in opposito pariete ad sui corporis altitudinem, et maior erit verticis distantia ad rectum quam ad planum, ob id minor erit refractio ex inferiori parte. Etsi minus id egere exemplo videremus, ut rudioribus consuleremus apponere contentis erimus.

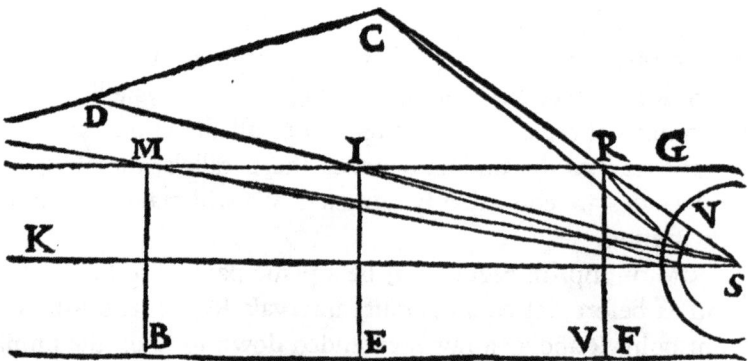

Sit oculus S, et ubi aciei mucronem[16] diriget in opposito pariete ibi pone lineam A, et ab oculo S transversam lineam ducito SA. Parallelus longe a

---

[16]"mueronem" in original text.

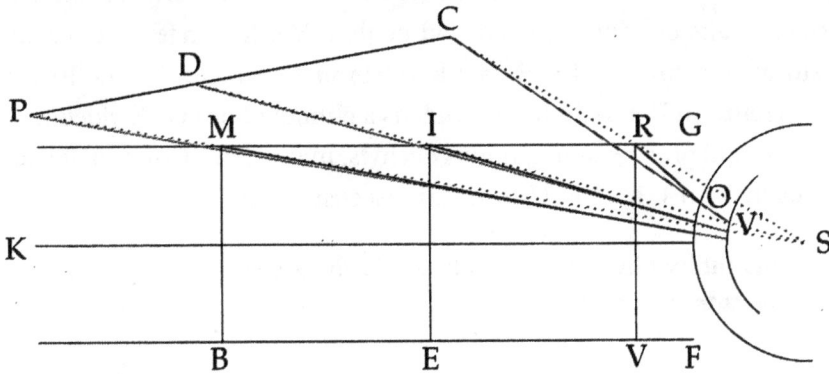

Figure 5.3

PROP. 4: *The less distant the eye is from each of the parallel lines, the more sharply it will see them converge.*

Although this theorem may not seem to be particularly necessary according to what has been said, we have decided to add it so that those who walk through refectories or the dormitories of monks or through royal palaces should no longer marvel when they see that the ceiling slopes downward more sharply than the tiled floor slants upward. The only reason this happens is that viewers fix the gaze of their eyes on the facing wall at the height of their body, and the distance from their head straight up to the ceiling is greater than that to the ground, so the refraction will be less on the lower side. And even though it would seem to us that this does not really need an illustration, we will be happy to append one in order to provide for those who are not well informed.

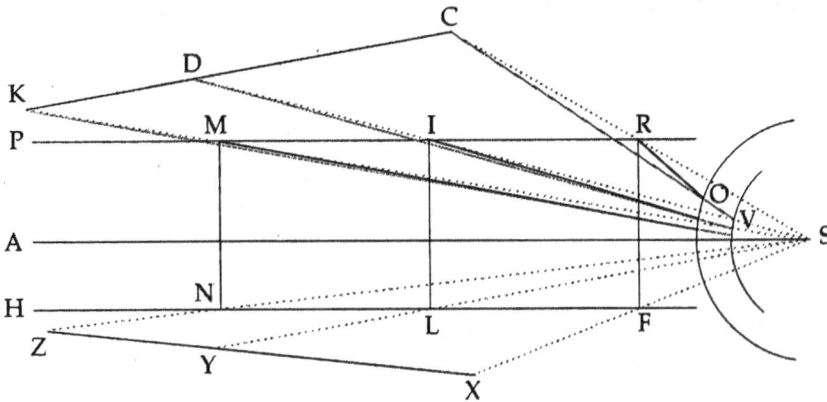

Figure 5.4

Let S [in Figure 5.4] be the eye, place point A where that eye will direct its axial line of sight on the facing wall, and draw transverse line SA from

vertice superius distans sive rectum sit RP; propinquior sive planities FH. Descendat satis oblique R punctus ad oculum VS. Refrangetur, et occurret catheto in puncto C, et [114] sic R videbitur in puncto C remotissime. I vero videbitur in D et M in K; descendens a culmine linea CDK videbitur. At F punctus, qui prope oculum, minus declivis; minus refrangitur. Et videbitur punctus in X, et L in Y, et N in Z. Erit ascensus linea XYZ.

Aequalibus intervallis in plano oculis subiecto existentibus, quae remotiora, minora spectantur. PROP. 5

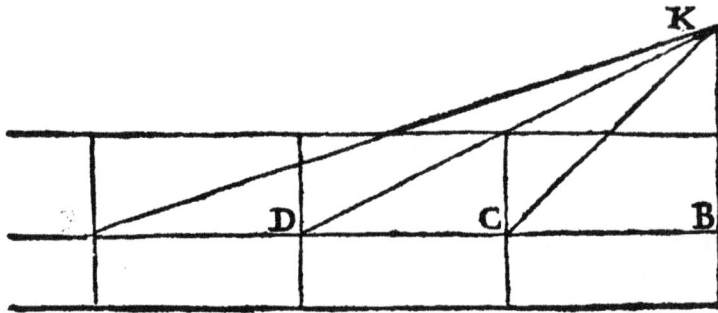

Ita probat Euclides. Sit oculus K,[17] aequalia quidem intervalla BC, CD, DF. Ex oculo[18] prodeant linearum ductus KB, KC, KD, KF. Sed BK ad rectos angulos ipsi plano FB substrato superstet, aut subsistat. Quia maior est angulus BKC ipsa CKD,[19] et CKD ipso DKF, maior igitur videbitur BC ipsa CD, et CD ipsa DF.

    Plotinus Platonicus inquit plura oppida longius remota minus inter se distare videntur quam sint, rationemque affert quod tunc rei dimensionem plane assequimur "quando singulas eius partes gradatim intuemur, at quum res longius se posita est, particulae plurimae latent, merito igitur totius mensura nos fugit."

    Secus autem ex nostris principiis. [115] Sit planum substratum FHBC, aequalia intervalla SD, DE, et sublatus oculus GIO. Veniat punctus D ad O. Refrangitur, et progreditur ad X. Cathetus ex G oculi centro per D extensa occurret praedictae in X. Punctus igitur D in X videbitur, sic E in M, et T in A. Maior igitur est distantia XM[20] quam MA.

---

[17]"A" in original text.
[18]"oculis" in original text.
[19]"CFD" in original text
[20]"XH" in original text.

the eye at S. Let RP be a parallel line or plane directly and high above the head, and FH below and nearer the head. Let the image of point R fall at a significant slant to VS, the eye. It will be refracted, and [when extended, the resulting refracted ray OV] will intersect the cathetus at point C, and [114] so R will be seen quite far away at point C. Point I, however, will be seen at D and M at K, and so line CDK will appear to slope downward from its apex [at C]. The ray from point F, which is near the eye, is less sharply slanted, so it is less refracted. Point F will be seen at X, L at Y, and N at Z. XYZ will be the upwardly sloping line.[14]

PROP. 5: *When equal spaces lie on a plane below the eyes, the ones that lie farther away appear smaller.*

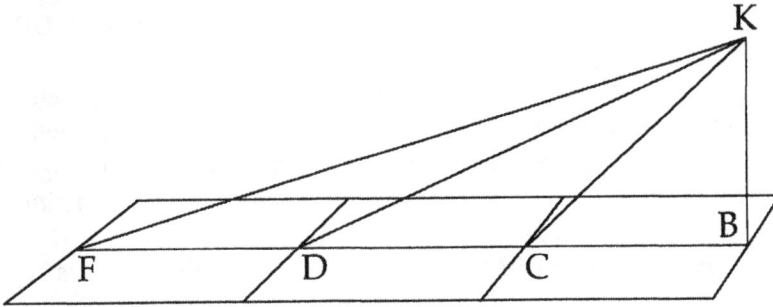

Figure 5.5a

Euclid proves this as follows. Let K [in Figure 5.5a] be the eye, and BC, CD, and DF the equal spaces. Let lines KB, KC, KD, and KF be extended from the eye. Now let BK stand or be posed upright at right angles to plane FB below. Since angle BKC > angle CKD, and angle CKD > angle DKF, BC will therefore look longer than CD, and CD will look longer than DF.[15]

The Platonist Plotinus claims that several forts lying at a great distance appear to lie at a smaller distance among themselves than they do, and the explanation he gives is that we arrive at the magnitude of a thing clearly "when we see its individual segments successively, but when the thing is removed to a great distance, several such segments cannot be seen, so it follows appropriately that the measure of the whole escapes us."[16]

This, however, is otherwise according to our principles. [115] Let plane FHBC [in Figure 5.5b] lie below the eye, let SD, DE [and ET] be equal intervals on it, and let GIO be the eye raised above the plane. Let the image of point D reach O. It is refracted, and [the extension of the refracted ray] continues to X. The cathetus extending from center point G of the eye through D will intersect the aforesaid refracted ray at X. Point D will therefore be seen at X, as will point E at M and point T at A. Distance XM is therefore greater than distance MA.

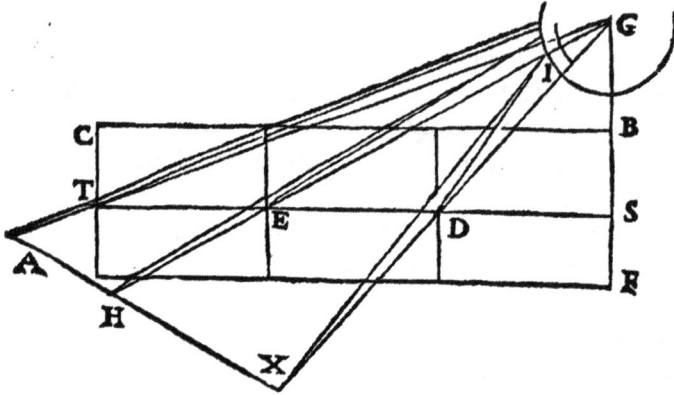

Parallelae in imum aut summum extensae concurrere videntur. PROP. 6

Eadem quoque demonstratione ostendere possumus cur intra puteum con-
spicientes paralleli parietes in summo latiores in immo concurrere videntur,
et ex imo cocleas dum aspicimus in imo latiores, in summo angustiores
videantur. Quod quia ex eadem dependent demonstratione, sola delineatione
rei contenti esse poterimus.

  Sit aspicientis oculus in summo putei E, et paralleli parietes AB, CD,
intervalla eorum aequalia F, G, H. Accedat proximior punctus F ad oculum

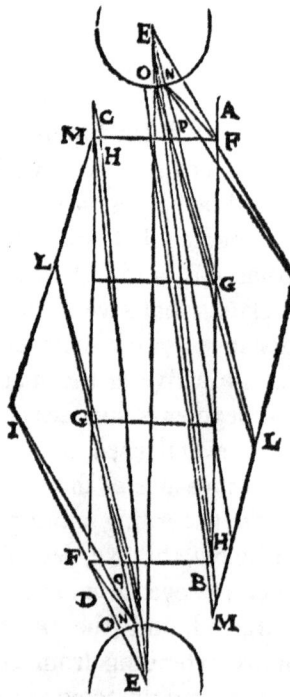

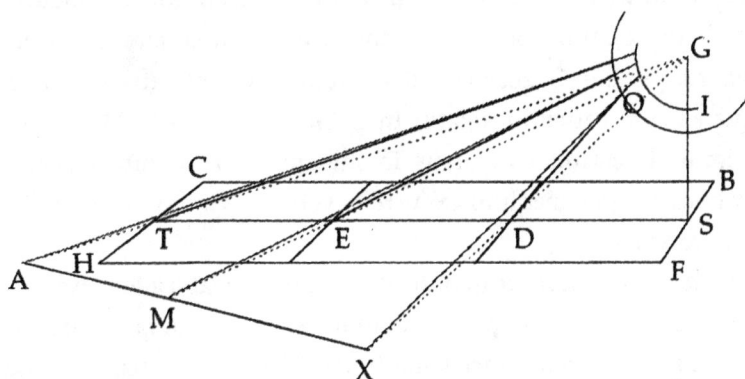

Figure 5.5b

PROP. 6: *Parallels extending downward or upward appear to converge.*

On the basis of the same demonstration we can show why parallel sides viewed in a well appear wider at the top while converging at the bottom and also why, when we look from the bottom, spiral staircases appear wider at the bottom and narrower at the top. But as both phenomena depend on the same demonstration, we can be satisfied with one illustration.

Let E [in Figure 5.6] be the eye of a viewer at the top of a well, let AB and CD be parallel walls, and let F, G, and H mark off equal intervals on

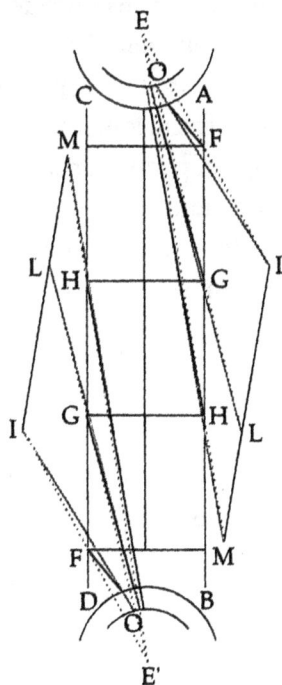

Figure 5.6

O. Relabitur ad E, inde refractam lineam protraham donec concurrat cum futura catheto ex centro oculi E et puncto F, et erit concursus in puncto I. Sic enim ex puncto G accedat ad oculum linea, et refrangatur, et fluat quousque ea cum catheto iungatur in L. Sic punctus H in M. Angustabitur [116] igitur linearum concursus in imo putei. Hoc autem evenit, quia centrica linea EE per medium excurrens veritatis vindex recta, obliquas ad se convenire videbit.

Si oculus mox statuatur in imo, et sursum aspiciat, eadem evenient quae supra evenerant; stans ad perpendiculum erectus, latera in arctius coire videbit. Nec Euclidis in hoc loco superfluitas [117] insimulanda, duodecimo enim theoremate probat quae obiiciuntur longitudinem habentium, quae sunt a destris ad sinistra procedere videntur, et quae a sinistris ad dextera, quod ex nona suppositione probat. Dixerat enim ibi, quae dexterioribus spectantur radiis dexteriora apparere, quod ex eodem principio demonstrari poterant, nam lineae quae ab oculis longe concurrunt dexterae sinistrorsum, et sinistrae destrorsum accedere videntur.

Aequalium altitudinum supra oculum positarum, quae propinquiora altiora videbuntur. PROP. 7

Per octavam suppositionem hoc theorema Euclides solvit: "Quae sub humilioribus radiis videntur humiliora apparent." Et quae prope oculum sunt altioribus radiis conspiciuntur. Nos autem eodem principio demonstramus. Quae enim prope oculum feriunt, et refracta longius fluunt, quousque cum catheto iungantur; sic a suo loco longe superius conspiciuntur, si magnitudo supra fuerit. Eadem res exemplo dilucidior.

Sint tres aequales altitudines AB, CD, EF, et sit AB prope oculum, et accedat A[21] punctus ad pupillam M. Frangatur, et fluat, et deinde cathetus

---

[21]"a" in original text.

them. Let the image of nearer point F reach the eye at O. [After refraction] it inclines toward E, so let me extend the refracted line until it intersects the cathetus to be drawn from center point E of the eye and through point F, and the intersection will be at point I. Likewise, let the line of incidence from point G reach the eye, let it be refracted, and let [the resulting refracted ray] extend out until it intersects the cathetus at L. The same with point H, which is seen at M. Thus, [116] the lines will narrow toward convergence at the bottom of the well. This happens, moreover, because the centric line EE' passing through the middle is the standard of absolute straightness toward which the viewer will see the oblique lines converge.

Now, if the eye [centered on E'] lies at the bottom and looks up, the same things will happen that happened above, and the sides standing upright along the perpendicular will appear to come together in a narrower compass. And Euclid should not be accused of redundancy in this regard, [117] for in the twelfth proposition of his *Optics* he demonstrates that of things having length, those that lie to the right appear to slope toward the left, whereas those to the left appear to slope to the right, which he demonstrates on the basis of the ninth supposition. For he said there that whatever is viewed with rays more to the right will appear more to the right, which could be demonstrated according to the same principle, for of lines that converge at a great distance from the eye, the more rightward of the left-hand ones and the more leftward of the right-hand ones appear to converge.[17]

PROP. 7: *Of equal heights placed above the eye, those that lie closer appear higher.*

Euclid resolves this theorem on the basis of the eighth supposition: "Things that are seen with lower rays appear lower."[18] Also, things that are near the eye are seen with higher rays. We demonstrate this, moreover, on the basis of the same principle. For when they are refracted, the rays from things that lie near the eye extend out quite far until they meet with the cathetus, and so they are seen far higher than they actually are if the magnitude lies above. This very point can be made clearer with an illustration.

Let AB, CD, and EF [in Figure 5.7] be three equal heights, let AB be near the eye, and let the image of point A reach the pupil at M. Let it be

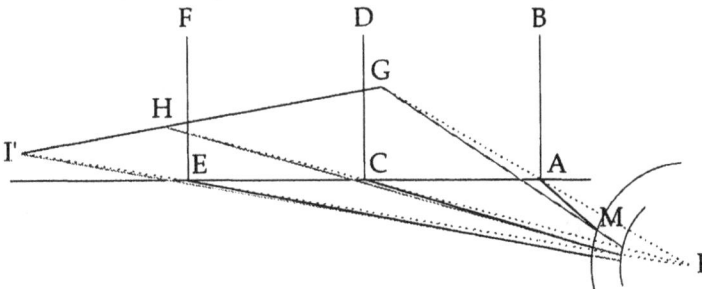

Figure 5.7

ex oculi centro I per idem punctum protrahatur donec incurrat priorem lineam in G. Sic punctus A sublimius [118] in G apparebit. Idem eveniet de puncto C, qui in H humiliori loco videbitur, et E in I. Et hoc modo in plateis domorum aequalium altitudines in plano erectarum, quae prope oculum ad astra tolli videntur, quae vero longe absunt ad solum descendere.

Aequalium altitudinum sub oculo positarum remotiores sublimiores videri. PROP. 8

Per septimam suppositionem probat Euclides quod "quae sublimioribus radiis videntur sublimiora apparent." Nos autem ita probamus. Ubi visionis mucro oppositum locum feriet, quae proximiora sunt obliquius feriunt oculum et longius in imo refranguntur; minus autem quae longe; ob id remotiores exaltari videbuntur, mergi vero propinquiora. Exemplum.

Sint tres aequales altitudines sub oculus substitutae AB, CD, EF, et sit AB prope oculum L. Et veniat punctus [A] ad pupillam M; frangitur. Et decurrat linea. Mox cathetus trahatur ex L centro oculi et puncto A donec occurret[22] priori lineae MG,[23] et sic videbitur in G,[24] et E in I.[25] Trahatur exilis linea per puncta concursus, et erit GHI; remotior pars I[26] sublimior est, depressior vero G, quae propinquior fuerat. [119]

Si erectae perpendicularis magnitudinis summitatem inspexeris procumbere videbis. PROP. 9

Quamplurima theoremata quae sequuntur ab Euclide, Ptolemaeo, et Alhazeno praetermittuntur, et ex iam dictis sequuntur quae pictoribus, perspectivis,

---

[22]"accurret" in original text.
[23]"G" in original text.
[24]"H" in original text.
[25]"L" in original text.
[26]"L" in original text.

refracted, let [the refracted ray] be extended, and then extend the cathetus from center point I of the eye through the same point [A] until it intersects the previous line of refraction at G. Thus, point A will appear elevated [118] to G. The same will happen with point C, which will be seen at a lower spot H, and E at I'. This is how, in streets with houses built at equal heights on a plane, those houses near the eye appear to reach to the stars, whereas those that lie far away appear to sink to the ground.

PROP. 8: *The farther of equal heights placed below the eye appear higher.*

By the seventh supposition [of his *Optics*] Euclid argues that "things seen with higher rays appear higher."[19] We, however argue, as follows. When the direct line of sight strikes a facing location, the images of things that are nearer strike the eye more obliquely and are refracted farther toward the bottom, whereas the images of things farther away are refracted less far downward so that things that are farther away will appear to rise up, and things that are nearer will appear to sink down. Here is an illustration.

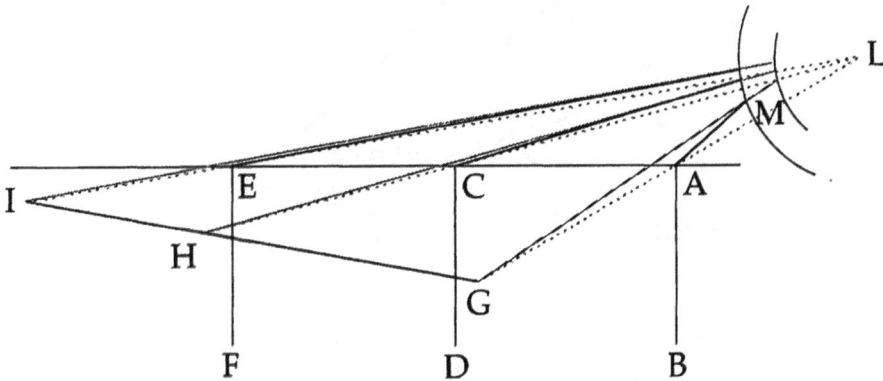

Figure 5.8

Let AB, CD, and EF [in Figure 5.8] be three equal heights situated below the eye, and let AB be near the eye centered on L. Let the image of point A reach M on the pupil, where it is refracted. Extend line [of refraction MG]. Then draw the cathetus from center point L of the eye through point A until it intersects the foregoing line of refraction MG, and so A will be seen at G, [C at H, and] E at I. Draw a thin line through the points of intersection, and it will be GHI, whose farther point I is higher, but [whose nearer point] G is lower, and that point is where the nearer point A was seen. [119]

PROP. 9: *If you look at the top of a height that stands perfectly upright, you will see it lean forward.*

Many of the propositions that follow are left out by Euclid, Ptolemy, and Alhacen, and among these, which we just mentioned, there follow some that

lapicidis, architectoribus, et caeteris, qui circino et visu sua opera examinant, quam maxime futura necessaria iudicamus. Nam nisi quis aciei cuspidis lineam quam centricam dicere possumus, quum per oculi centrum et circuli basis visivae pyramidis pertranseat, consideret, in maximos et fere infinitos errores prolabetur. Et maiorum dicta inter se adversantia pernoscet.

Ad rem igitur nostram redeuntes, si quis erectam ad solum magnitudinem inspexerit et superne in summo aciem exaltaverit, ad oculum procumbentem videbit, nam ex visivis lineis solum centrica veritatem assequitur. Videbit igitur aedificii apicem in suo loco. At imum oblique oculis feriet, frangetur linea, et longe a suo loco retrocedere facit. Unde retrocedens imum, videbis procumbere summum. Hoc exemplo fiet res clarior.

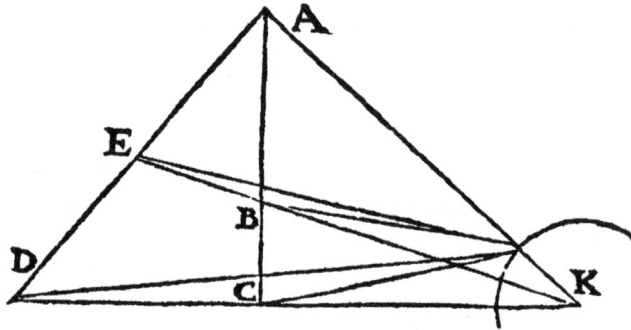

Sit erecta perpendicularis altitudo ABC. Oculus K in imo existens, A[27] summum punctum feriat. Videbit aedificii apicem in suo loco. Veniet ad oculum imi punctum C in O; reclinabit[28] ad centrum. Et producta excurrat tantum quantum cum futura catheto [120] concurrat, et erit in D, retrocedens ab ipso imo. Sic punctus B in E videbitur; igitur rectam ABC magnitudinem procumbentem videbis, ut AED.

Erectae magnitudinis si imum inspexeris, semper summum retrocedere videbitur. PROP. 10

Contra autem, si imam aedificii partem quis conspexerit etiam ad perpendiculum erectam, vertex resupinari videbitur, nam quum centrica verum in imo

---

[27]"a" in original text.
[28]"relinabit" in original text.

we judge to be absolutely necessary for painters, practitioners of perspective,[20] stone masons, architects, and so forth, who operate with due diligence using compass and eye. For unless one takes into full account the line of sight from the viewpoint that we can call the centric line, since it passes through the center of the eye and the center of the base circle of the radiant cone, he will fall into the worst of errors and an almost infinite number of them.[21] He will also learn how the opinions of our predecessors are in conflict.

To return to our subject, then, if someone looks at a height standing upright on the ground and lifts his gaze up to the top, he will see it lean forward toward his eye, for among all the lines of sight, only the centric one apprehends the object as it is in actuality. Therefore, the viewer will see the top of the height where it actually is. But the image of the bottom will strike the eyes at a slant, and so the line [along which it is conveyed] will be refracted, and that causes the image to lie far behind the bottom's actual location. Consequently, since the bottom appears to draw away, you will see the top lean forward. This point will become clearer with an illustration.

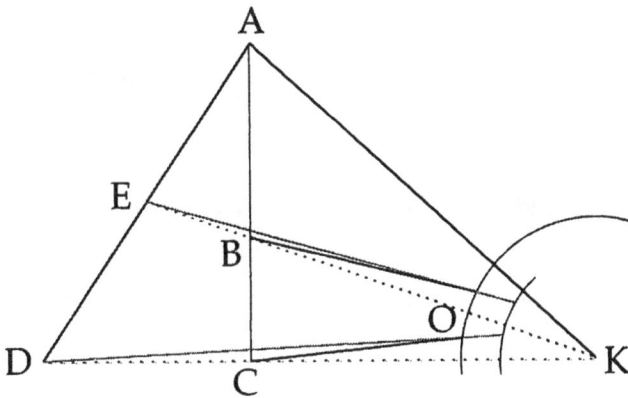

Figure 5.9

Let ABC [in Figure 5.9] be the upright height. With K the eye lying at the level of the height's bottom, let the image of point A at the top reach the eye along the centric line. The eye will see the top of the building in its actual location. The image of point C at the bottom will reach the eye at O, and it will bend toward center point K. Let [the resulting line of refraction] be extended far enough to intersect the cathetus, [120] which is to be extended, and it will do so at D, which draws away from the bottom itself. Similarly, point B will be seen at E, so you will see upright height ABC leaning forward, as along AED.

PROP. 10: *If you stare at the bottom of an upright height, the top will always appear to lean back.*

On the other hand, if one looks at the bottom part of a building, which still stands perfectly upright, its top will appear to lean backward, for since the

consequatur, obliqua ex summo veniens ad oculum linea; prae acie refracta fallitur, et falsa oculo exhibet resupinam magnitudinem. Exemplo fiet clarior.

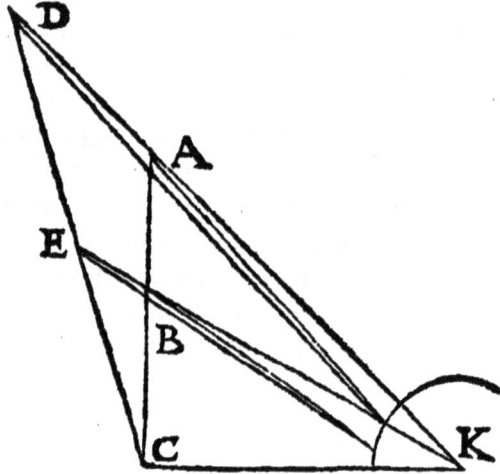

Sit eadem altitudo ABC; oculus in imo existens centrica sua punctum C inferiorem feriat. Punctus A ad oculum veniens obliquabitur, et refracta excurret ad D, longius. Sic B ad E, minus longe. Unde perpendicularis altitudo ABC transversa retro CED ab summitate resupinata apparebit.

Vitruvius huius animadversionis meminit, inquiens:

Membra omnia, quae supra capitula columnarum sunt futura, idest epistylia, zophori, coronae, tympana, fastigia, acroteria inclinanda sunt in frontis suae cuiusque altitudinis parte duodecima, ideo quod quum steterimus [121] contra frontem, ab oculo lineae duae si extensae fuerint, et una tetigerit imam operis partem, altera summam, quae summam tetigerit, longior fiet, ita quo longior visus lineae in superiorem partem procedit, resupinatam facit eius speciem. Cum autem (uti suprascriptum est) in fronte inclinata fuerint, tunc in aspectu videbuntur esse ad perpendiculum, et normam.

Sed Vitruvii opinio quoquomodo vera, falsave videtur, et dum optices gnarum architectorem optat, se penitus illius ignarum ostendit. Sed ut quae ipse dicit et nos refellimus plenius intelligantur, eius rei descriptionem apponemus.

Sint membra AB supra columnas collocanda BC, oculus D, radii extensi DA, DB. Quia AD longior est BD, ergo AB supinata videbitur. Sed propositio

centric line is actually focused on the bottom, the line reaching from the top to the eye is slanted; being refracted ahead of the gaze, [the resulting line of sight] is in error and gives a false impression to the eye that the height leans backward. This will become clearer with an illustration.

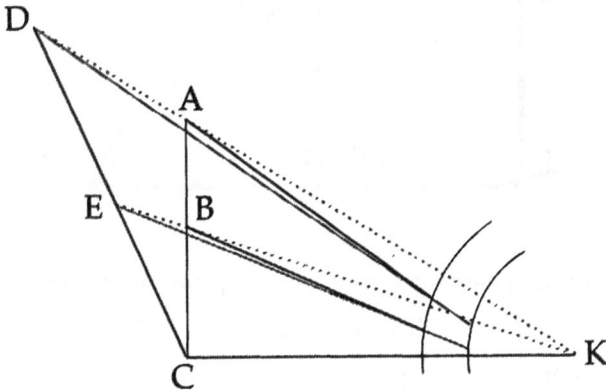

Figure 5.10a

Let ABC [in Figure 5.10a] be the same height [as in the previous proposition], and with the eye at the bottom, let its centric line fall on lower point C. On reaching the eye, the image of point A will arrive at a slant, and its refracted image will be displaced to D, which is farther away than A. So, too, B will be seen at E, less far away. Consequently, upright height ABC will appear turned away and slanted backward from its top along CED.

Vitruvius mentions this observation, saying that:

> All the elements to be placed above the capitals of the column, that is, the epistyles, friezes, cornices, tympana, raking cornices, and acroteria, should have a front surface that inclines outward to one-twelfth its own height. This is why, when we stand [121] opposite any façade, and two lines might be extended from our eye so that one would touch the lower margin of any part of the building, and the other touch the very top, that line which reaches the upper margin will be the longer of the two. Inasmuch as a longer line of sight extends to the upper part of the building, it will make it seem to tilt backward. But if, as we stated earlier, the elements of the façade are made to incline, they will seem perfectly vertical to the viewer.[22]

But however true Vitruvius's opinion seems to be, it also seems false, and while he aims his discussion at an architect who is versed in optics, he shows himself to be profoundly ignorant of that discipline. In order, however, to make what he says and our refutation of it more understandable, we will provide an illustration of this point.

Let AB [in both diagrams of Figure 5.10b] be members assembled on columns BC, let the eye be D, and let DA and DB be rays extending [to the eye]. Since AD > BD, then AB will appear to lean backward.[23] But, as far as

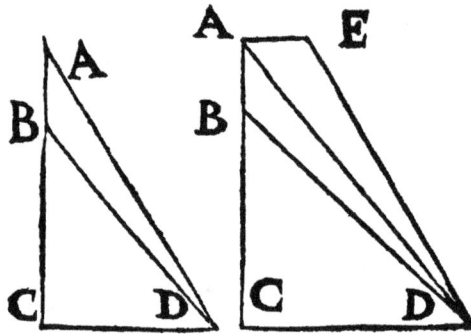

haec a nullo, quod sciam, perspectivae scriptore adhuc demonstrata est, nec scio unde excerpserit. Multasque habet in se difficultates et nodos, et multorum opiniones afferre oporteret, quas omittimus ne taedium lectoribus ingeramus.

Praeterea duodecima eius parte inclinata, non erunt aequales lineae. Nam si oculus ab aedificii imo abfuerit per suam altitudinem, tunc lineae DB et DA aequales fiunt, ut in E, nam quum AB et AE aequales sint et axis ad rectum illum angulum secat, erit uterque triangulus aequalis AED, et ABD, et sic latera ED, DB aequalia erunt. At solum duodecima pars inclinabitur; longe ab imo C seducendus est oculus.

Praeterea, ut opticae ignarus, neque locum praescripsit quo sisteret oculus membra illa inspecturus. Nos autem dicimus si oculus verticem obnixe aspexerit, propendere potius quam resupinari videbit, ratione [122] iam dicta in praecedenti propositione et in sexta. Quia semper circa centricam angustantur lineae, si oculus prope imum fuerit, at si oculi cuspidem columnis affixerit inspector, tunc videbit membra illa aliquantisper resupinari, ut diximus.

Erectae magnitudinis perpendicularis si medium inspicietur, illud surgere, extrema resupinari videbuntur. PROP. 11

Si erectae magnitudinis umbilicum centrica inspectoris feriet, illum tuberosum et globosum inspiciet; extrema vero recedere cuiusdam arcus similitudine. Exemplum sic oculis subiiciam. Sit perpendicularis erectus paries ABC, oculus F, cuspis medium feriat B, et sit centrica FB. Summus parietis apex ad oculum veniat F. Frangitur et relabitur extra, et ubi attigerit cathetum,

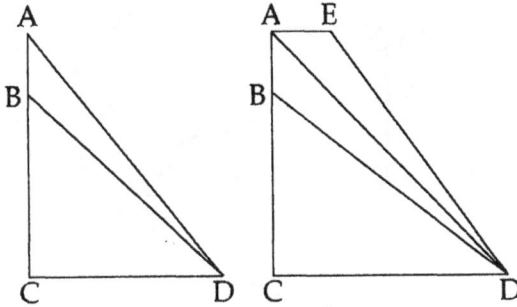

Figure 5.10b

I know, this proposal [to lean the members forward by a twelfth part] has not yet been demonstrated by any writer on optics,[24] nor do I know where Vitruvius got it. It has many intrinsic difficulties and knots, and it would require applying the opinions of many thinkers, which we omit in order not to bore our readers.

For one thing, if it is inclined by a twelfth part, the lines will not be equal. For if the eye were to lie away from the bottom of the building at a distance of its height, then lines DB and DA become equal, as in E, for since AB and AE are equal and the [centric ray along the] axis intersects the height at a right angle, both triangles AED and ABD will be equal, and so sides ED and DB will be equal. But it will be inclined only one-twelfth part, and the eye is removed a long way from the bottom.[25]

In addition, as someone ignorant of optics, Vitruvius did not specify where the eye that was to look at that member should have been situated. We, on the other hand, say emphatically that if the eye looks at the top, it will see it lean forward rather than backward, according to the explanation [122] already given in the previous proposition, as well as in the sixth proposition.[26] Because lines of sight are always narrowed around the centric line, if the eye lies near the bottom and if the viewer fixes the line of sight directly ahead at columns, then he will see those members lean back to some extent, as we have said.

PROP. 11: *If the midpoint of a perfectly upright magnitude is looked at, the magnitude will appear to bulge out, whereas the endpoints will appear to incline backward.*

If the centric line of a viewer reaches the middle of an upright magnitude, he will see it as bulging and curved, and he will see the ends recede in something like an arc. Accordingly, I will subjoin an illustration as a visual aid. Let ABC [in Figure 5.11] be a perfectly upright wall, let F be the eye, let the axial line of sight reach midpoint B, and let FB be the centric line. Let the image of [point A at] the very top of the wall reach the eye at F. It is refracted and inclines away toward the normal, and where [the refracted

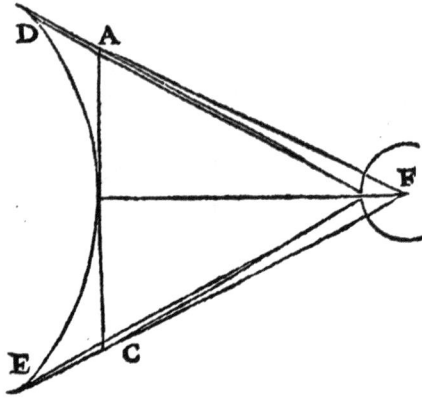

ibi fac punctum D. Idem facies de imo puncto, et ubi C videbitur in E, ibi aliam notam ascribe. Deinde prolonga lineam per puncta quae super annotasti, et singulari quodam modo se flecti videbitur, ut in medio tuberosa ventricosaque appareat, quia summum et imum recedunt. [123]

Recta linea oculis exposita, cuius medium centrica feriat, convexa videbitur.
PROP. 12

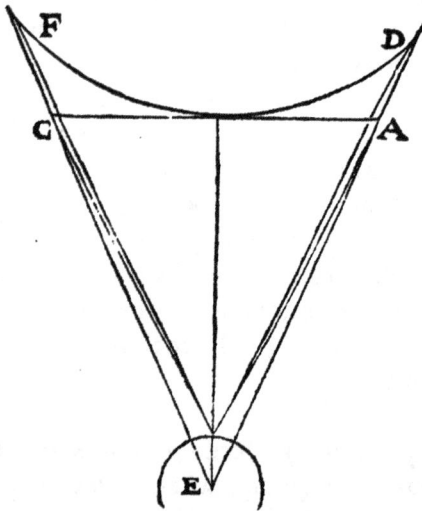

Idem eveniet de muris in plano iacentibus erectorum aedificiorum, nam si recta fuerit, oculis in medio respicientibus, convexa repraesentabitur, nam extrema retrocedere videbuntur. Exemplum ante oculos ponimus.

  Sit recta linea ABC describens ima erectorum aedificiorum fundamenta. Oculus sistatur in medio F, ut cuspide percurrat medium ipsius B. Veniet

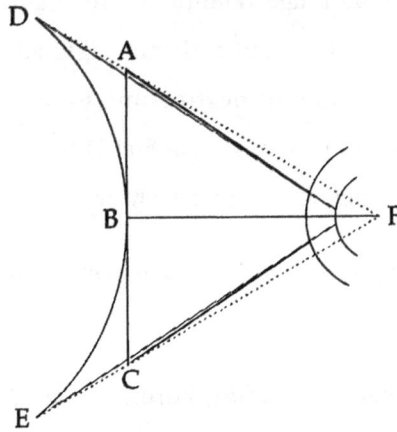

Figure 5.11

ray's extension] reaches the cathetus, put point D. You will do the same with the bottom point C, and make another mark where C will be seen at E. Then pass a line through the points you marked down, and every point on it will appear to deviate in some way, so that in the middle the wall appears bulging and distended because the top and bottom appear to recede. [123]

PROP. 12: *If a straight line is exposed to the eye, and if the centric ray reaches its midpoint, it will appear convex.*

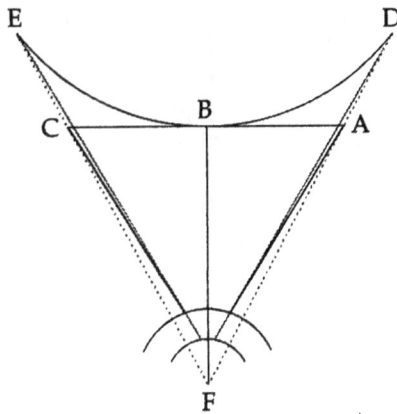

Figure 5.12

The same thing will happen with walls of upright buildings lying in a plane, for if they are straight and the eyes face their middle, they will show themselves as convex because the ends will appear to recede. I place an illustration before your eyes.

Let ABC [in Figure 5.12] be a straight line representing the bottom foundations of upright buildings. Let the eye at F be situated in the middle

ad eum extremus punctus longe remotus C; reclinabit ex fractione. Tunc longius producatur, et ubi convenit cathetam, illic adiice notam E. Eodem modo facito per aliam partem, ut dextrorsum, ut sinistrorsum, procurrent lineae. Postremo de puncto E ad punctum B et D trahe lineam suam curvam extremitates omnes adiungendo, et globosam te signare comperies. [124]

Concava linea oculis exposita, cuius medium centrica feriat, recta videbitur. PROP. 13

Contra vero, si muri concavi fuerint, eorumque medio visivae pyramidis cuspis insinuetur, non amplius concavi sed recti videbuntur, nam eorum extremae partes ad oculos oblique venientes et ultra suos limites retro apparentes; concava superficies recta demonstrabitur. Verum haec exactius explicantur subiecto exemplo.

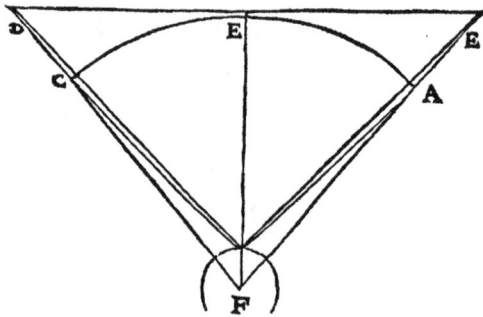

Esto concava superficies ABC; oculus F, cuius centrica FB medio dirigatur. Concavae lineae extremum C ad F oculum veniat. Refrangetur et retro iaculabitur in D. Ex alia parte idem eveniet ex A; videbitur in E. Protracta linea erit recta EBD. Euclides hoc contra quam nos statuimus docet convexum inspiciendo planum videri; idque quia paulatim inspicitur, quod ad rem facere nullo modo videtur. [125]

Rectorum aedificiorum toruli ubi eos centrica percusserit, ibi tumidiores videbuntur. PROP. 14

so that its axial line of sight reaches B at the line's midpoint. The image of endpoint C, far away, will reach it and will be bent by refraction. Extend [its line of refraction] farther, and where it intersects the cathetus, add mark E at that point. Do the same thing on the other side, so that on the right, as on the left, the lines will extend symmetrically. Finally, draw the curved line from point E to point B and point D by joining all the endpoints, and you will verify that it designates a curved line to you. [124]

PROP. 13: *If a concave line is exposed to the eye, and if the centric ray reaches its midpoint, the line will appear straight.*

On the other hand, if the walls are concave, and if the axis of the cone of vision reaches their middle, they will no longer appear concave but straight because the images of the parts at their ends reach the eye obliquely and appear beyond their boundaries, so a concave surface will show itself as straight. But these points are explained more precisely according to the illustration added below.

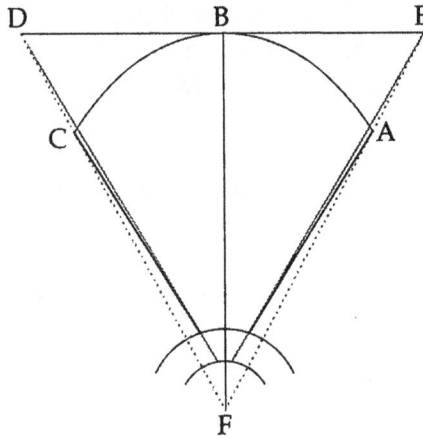

Figure 5.13

Let ABC [in Figure 5.13] be a concave surface, and let F be an eye whose centric line FB is directed at the middle. Let the image of endpoint C of the concave line reach the eye at F. It will be refracted and [the resulting line of refraction] will ultimately be projected beyond at D. The same will happen on the other side at A; it will be seen at E. The line extended [through these points and midpoint B] will be straight line EBD. Contrary to what we have established, Euclid teaches that when a convex surface is looked at, it appears flat; but when this is considered briefly, it in no way seems to fit the phenomenon.[27] [125]

PROP. 14: *At the point where the centric ray strikes convex moldings of upright buildings, the moldings appear more swollen.*

Rectorum ante oculus positorum aedificiorum toruli, zophori, et alia promi-
nentia ornamenta, etsi ad normam et ad rectum ordinata consummataque
fuerint, ubi oculi cuspidem direxeris, ibi prominentem partem inspexeris,
et praecipue si ex obliquo steterit, nam gibositas conspectius oculis se insinuabitur. Exemplum.

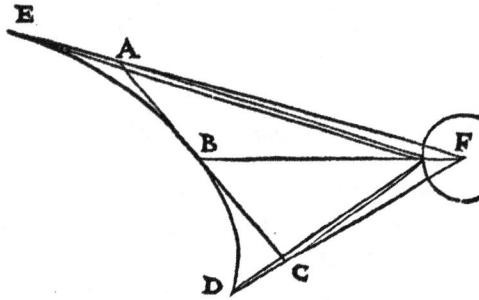

    Esto torulus ABC, ex obliquo oculis obiectus FB, cuius mucro FB. Sed
punctus A ad F veniens refractus convenit cathetum in puncto E, et C
in puncto D. Videbitur igitur obliquus totus EBD in medio B gibberosum
et convexum.

Ex alto planum inspicientibus globosum videri. PROP. 15

Si oculum supra planum sublimius steterit, illudque inspexerit, ubi centricam
direxerit, illic turgidum videbit; extrema depressius recedent. Causam pluries

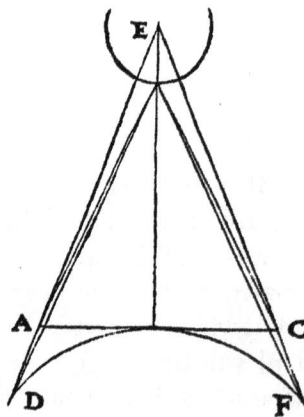

assignavimus, exemplumque parum ab iis variabitur quae supra proposui-
mus. [126]

Even when the convex moldings, friezes, and other projecting ornaments of upright buildings situated in front of the eyes are arranged and finished by carpenter's square and ruler, you will observe swelling at the part to which you direct your line of sight, especially if your eye is situated off to the side, for the swelling will be presented more conspicuously to the eyes. Here is an illustration.

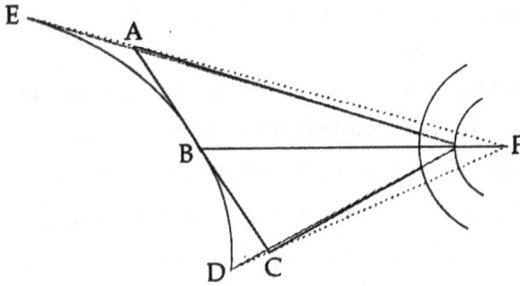

Figure 5.14

Let ABC [in Figure 5.14] be a convex molding aslant to the eyes according to FB, and let FB be the axial line of sight. Once refracted, the ray from point A reaching the eye at F intersects the cathetus at point E, and the refracted ray from C intersects the cathetus at point D. Consequently, the entire slanted surface EBD will appear protuberant and convex at midpoint B.

PROP. 15: *From high above, a flat surface appears rounded to viewers.*

If an eye is situated high above a flat surface and looks at it, it will see the plane as swollen at the point where it directs its centric ray, and the outer

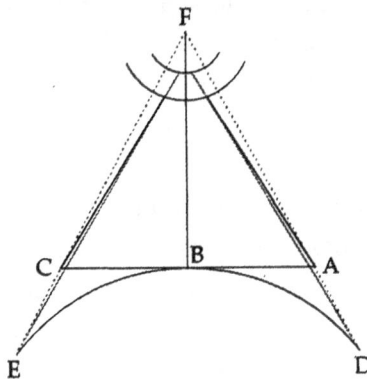

Figure 5.15

edges will appear to sink down. We have determined the cause of this phenomenon several times, and its illustration will be a variant of the ones we proposed above. [126]

Sit planum ABC, oculus in medio inspector F, ubi visus acies dirigetur B. Et extrema A, C veniant ad oculum F. Ex refractione retrocedunt, unde videbuntur retrocedere. Hac ratione quum mare ex alto inspexerimus, in medio convexum videmus, et praecipue ubi centrica fixa erit, et extrema compressiora. Et ibi visus debilissimus, quia vastae magnitudinis mare.

Visionem terminatam esse. PROP. 16

Veteres, Macrobio referente, humanum radium per planitiem extentum, sine impedimento non progredi alterius 180 stadiis (stadium continet octavam partem mille passuum, ut sint millia passuum 22 et semis). Aelianus vero docet quod humanus visus in mari usque ad 30 [millia] passuum[29] potest. Sed hoc simpliciter prolatum utrumque falsum est, nam sunt homines qui acutissime vident, et sunt lusciosi. Strabo vocatus a M. Varrone 135 millia passuum videbat: "Solitum Punico[30] bello a Lylibeo Siciliae promontorio exeunte classe Carthaginis portu omnem numerum dicere." Et [127] sunt alii tam obtusi visus ut quae ante pedes habeant non cernant. Oportet quoque distinguere visibilia, velut splendores magnitudines vastas plusquam 50 millia passuum videri. Stellas etiam in firmamento videmus ob earum magnitudinem et splendorem.

Coelum, ubicunque homo fuerit, rotundum fere semper videri. PROP. 17

Maximus est ineptorum numerus qui putat cœlum globosum videri quod vere globosum fuerit, quasi eorum visus ad extremum usque cœli ambitum protendantur. Sed ubicunque homo steterit, si oculos sursum sustulerit se circumvolvendo, rotunda quadam cœli conclusione contineri videbitur. Ad oculum enim, tanquam ad centrum venientes radii undique aequales— longitudinis puta 22 millia passuum—nec ultra progredientes; oportet rotundum videri.

Idemque in horizonte evenire fatemur, nam ubicunque homo fuerit, se convertendo rotundum horizonta cernet. Eadem ratione sed re vera non rotundum sed fere rotundum aut ovalis figurae dicas, nam intercapedo a summo cœlo ad capitis verticem brevior est quam a vertice ad horizonta. Sunt qui dicant id evenire quia a capitis vertice ad caelum minor est distantia quam a vertice ad horizonta, hoc modo. Linea enim AB minima est, quia minor ipsa BD et BD ipsa BE, ex tertio Euclidis, ob id ovalis figurae videri.

---

[29]"passus" in original text.
[30]"Pontico" in original text.

Let ABC [in Figure 5.15] be the plane, F the eye viewing it in the middle, and B where the centric line of sight will be directed. Let the images of endpoints A and C reach the eye at F. After refraction those images recede toward the back, so they will appear to fall behind. According to this reasoning, when we look at the sea from on high, we see it as convex in the middle, especially where the centric ray is fixed, and the edges appear pressed down. And it is here at the far edges that sight is weakest because the sea is huge.

PROP. 16: *Vision is limited.*

According to Macrobius, the ancients claimed that unimpeded human visual radiation extending along a plane propagates outward no more than 180 stades (one stade contains an eighth of a mile, so there are 22.5 miles in 180 stades).[28] Aelian, however, instructs us that at sea human sight can extend to 30 miles. But the opinion put forth by each is simply false, for there are men who see most acutely, and there are weak-sighted men. A man called Strabo by Marcus Varro saw 135 miles: "During the Punic Wars he was in the habit of telling from the promontory of Lilybaeum in Sicily the entire number of ships in a fleet leaving the port of Carthage."[29] [127] There are others whose vision is so dim that they cannot see what they have at their feet. It is also necessary to distinguish among visible objects, such as huge, bright magnitudes that can be seen more than 50 miles away. We see even the stars in the firmament because of their size and brightness.

PROP. 17: *Wherever a person is, the heavens always appear nearly spherical.*

A great number of foolish people believe that the heavens look spherical because they actually are spherical, as if their sight extended to the farthest edges of the heavens. But wherever a man stands, if he maintains his eyes upward while turning around, it will appear by a certain deduction that the heaves are enclosed by a sphere. In fact, just like rays reaching a center, rays coming to the eye from every direction appear equal in length—a length, say, of 22 miles—and do not propagate any farther; so the heavens must be perceived as round.

We grant that the same thing happens at the horizon because, wherever a person is, he will see the horizon as circular when he turns around. But using the same reasoning, you might say that it is actually not spherical but almost spherical or of an oval shape, for the interval from the zenith of the sky to the top of the head is shorter than from the top of the head to the horizon. There are those who might argue in the following way that this happens because the distance from the top of the head to the heavens is less than the distance from the head to the horizon. Indeed, line AB [in Figure 5.17a] is the shortest possible because it is shorter than BD, and BD is shorter than BE, according to [proposition 7 of] the third book of Euclid,[30] so arc

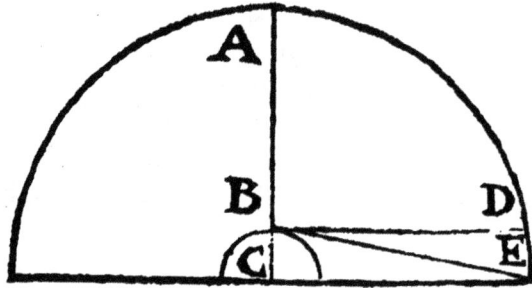

Sed hoc falsum est, nam in tam longa distantia differentia fere nulla est.

Sunt non parvae authoritatis viri philosophi qui id evenire dicunt quod inter nos et infimum horizonta multa sint interiacentia corpora, sed inter verticem et cœli summum nullum intermissum corpus, et haec intersita corpora faciunt ut distantia [128] productior videatur. Sed hoc quoque falsum, nam vera causa est quod, dum sursum cœlum aspicimus, aciei cuspis ad zenith dirigitur; horizon oblique oculum feriet, et ratione refractionis longius conspicari. Haec clara sunt ut exemplo non indigant, sed ut huius disciplinae initiatis demereamur non gravabimur apponere.

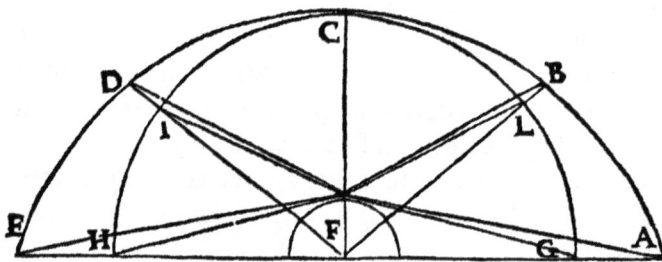

Caelum erit GLCIH, zenith capiti imminens C quod centrica ferit. Punctus H ad oculum F venit, frangitur, et relabitur in E, unde punctum H longe in E apparebit. Eodem modo I punctus in D, et C suo loco, et G, L in A, B. Si haec puncta exili circumscribes linea, ovatam figuram te fecisse cumperies circa horizonta. Ob id non pauci fuere ex antiquorum philosophorum scholis

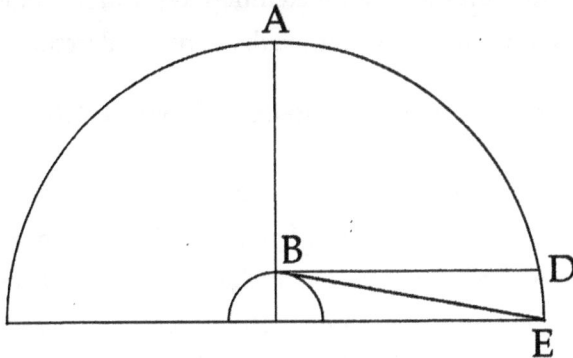

Figure 5.17a

DA will appear oval in shape. But this is false, for at such a great distance there is virtually no difference.

There are natural philosophers of no small authority who claim that the sky's appearing oval happens because between us and the lowest horizon there are many intervening objects, but between the top of the head and the zenith of the sky there is no intervening body, and these intervening objects cause the [horizontal] distance **[128]** to look more stretched out.[31] But this, too, is false, for the true cause is that, when we look upward at the sky, our centric line of sight is directed toward the zenith, whereas the [images of all the points on the] horizon reach the eye obliquely, and because of refraction they appear farther away. These points are clear enough that there should be no need for an illustration, but in order to oblige initiates of this discipline, we will not be unwilling to supply one.

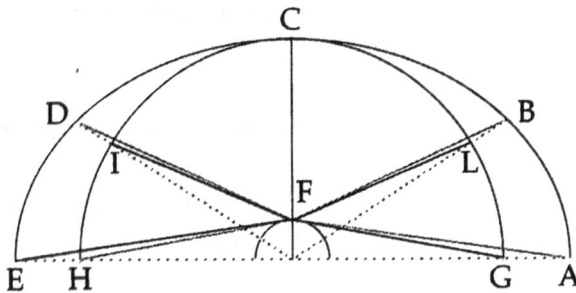

Figure 5.17b

GLCIH [in Figure 5.17b] will be the heavens, and C the zenith overhead to which the centric ray projects. The image of point H reaches the eye at F, is refracted, and [the refracted ray] inclines toward E, so point H will appear farther away at E.[32] Likewise, point I will appear at D, C will appear in its actual location, and G and L will appear at A and B, respectively. If you pass a thin line around these points, you will affirm that they form an oval shape around the horizon. This is why there were not a few ancient

non minime authoritatis qui ovatum cœlum putaverunt et confessi sunt hac, ut spero, ratione fulti, quod ovatum oculis eorum videretur.

Oculo in theatri rotundi medio constituto, ovale videri. PROP. 18

Ex his etiam pendet ratio cur, oculo in areae medio existente vel theatri rotundi, spatio ovale videatur. Circumferentiae enim obviam partem centrica ferit, quae vero destrorsum et sinistrorsum fuerint oblique ad oculum venient, et ex refractione longe a suis punctis unde accesserant relabuntur; ob id circularis figura qualis apparebit. Exemplum effingere nitar, etsi ex superiore minore necessarium videatur. [129]

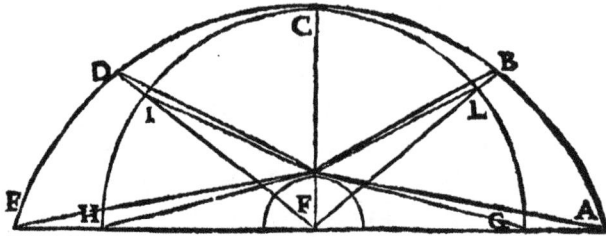

Sit areae[31] vel theatri medium F, ubi oculus statuatur, theatrum GLCIH, ex oculi regione punctus C ubi centrica feriet. Venit H ad F oculum, veniens oblique, frangitur, et longius emittitur in E; punctus I in D. Ex alia parte idem facies, si G et L punctus in A et B videbitur. Exili linea circumscribendo puncta, videbitur ABCDE.

Est locus in quo aequales magnitudines inaequales videantur. PROP. 19

Haec quae sequuntur problemata ad multiplices perspectivae usus, quam vulgus practicam vocat, necessaria sunt, quum res aliter quam sunt oculis repraesentandae sint per linearum flexus, circuitus, et obliquitates. Possunt enim res aequales inaequales videri et inaequales aequales. Hoc problema Euclides ita docet.

Sint magnitudines aequales BC, CD, et circa BC magnum circuli [130] segmentum circinabimus BFC, supra vero CD semicirculum. Et a puncto B transversam BF ducemus, ubi se duo circuli se mutuo abscindunt ad F, et

---

[31]"area" in original text.

natural philosophers in schools of no little authority who believed that the heavens are oval and said so on the basis, so I hope, of this reasoning because they appeared oval to their eyes.

PROP. 18: *If the eye is situated in the middle of an amphitheater, the amphitheater looks oval.*

The reason why, when the eye is the middle of an arena or an amphitheater, it appears to be oval throughout its extent depends on these conclusions as well. For the centric ray reaches the directly exposed part of the circumference, whereas the images of the parts that are to the right and left will reach the eye obliquely, and on account of refraction they end up far from the points they came from, which is why a circular shape will appear oval. I will try to fashion an illustration, even though it may hardly seem necessary, given the previous proposition. [129]

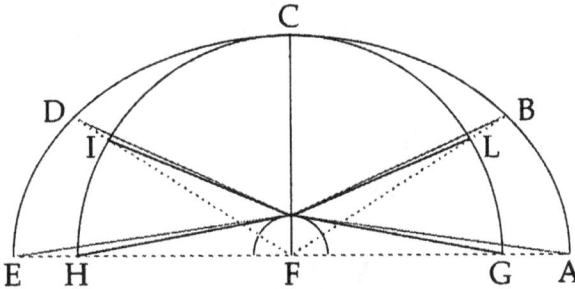

Figure 5.18

Let F, where the eye is situated [in Figure 5.18], be the center of the arena or the theater, let GLCIH be the theater, and let point C be where the centric ray will extend in a direct line from the eye. The image of H reaches the eye at F and, arriving at a slant, is refracted, and it is displaced far out to E, and the image of point I is displaced to D. You will do the same on the other side, insofar as points G and L will be seen at A and B, respectively. When a thin line is passed over the points, it will appear as [oval] ABCDE.[33]

PROP. 19: *There is a place from which equal magnitudes may appear unequal.*

The problems that follow are necessary for many applications in perspective, which the vulgar call practical, because things are presented other than they actually are to the eyes by the bending, curving, or slanting of lines. In fact, equal objects can appear unequal and unequal ones equal. Euclid instructs us about this problem as follows.[34]

Let BC and CD [in Figure 5.19a] be equal magnitudes, and around BC we will draw large segment BFC of a circle, [130] but we will draw a semicircle on CD. Then from point B we will draw transverse line BF to F, where the two segments of the circles cut one another, and we will do the

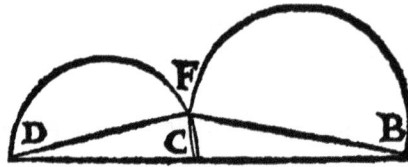

similem ex F ad D. Erigimus etiam lineam ex C ad F. Quoniam angulus CFD est in semicirculo, rectus est, et angulus BFC in maiori segmento minor recto. Apposito oculo in angulo BFD, videbit BC minorem ipsa CD, quia quae sub minori angulo minora videntur. Sed hoc falsum est, quia impossibile est ut oculus F per obtusissimum angulum videat BFD, ut monuimus. Sed quomodo fieri possit nostro more exemplum oculis spectandum subiecimus.

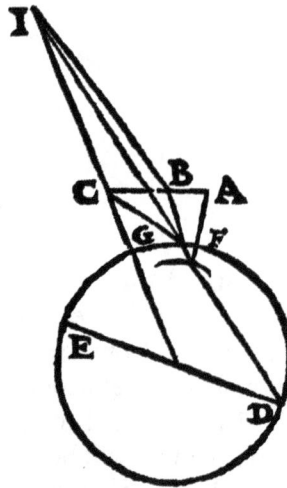

Apponatur ita oculus ut visivo angulo partem unam recta aspiciat, et alteram obliquam, ut longe refracta aequalis ei videatur. Sint AB et BC aequales magnitudines. Veniat ad oculum F, G, A, B recte, et punctus C declinans ad G. Quia GC declinat, vestigetur diametri parallela DE. Et quia GE sexta est circuli pars, punctus G[32] declinabit ad D, extendeturque in I; unde punctus

---

[32]"H" in original text.

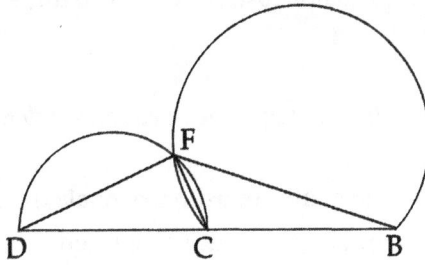

Figure 5.19a

same from F to D. We also drop a line from C to F. Since angle CFD lies in a semicircle, it is right, whereas angle BFC in the larger segment is less than right. If the eye is placed on [vertex point F of] angle BFD, it will see BC as smaller than CD because what is seen under a smaller angle [i.e., CFB < CFD] appears smaller. But this is false because, as we have shown, it is impossible for the eye at F to see by means of angle BFD, which is extremely obtuse.[35] According to our fashion, however, we will place an illustration of how this can be done before your eyes.

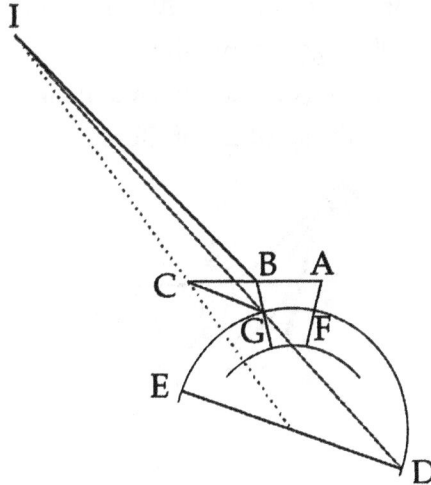

Figure 5.19b

Place an eye [whose outer surface is DFGE in Figure 5.19b] so that it sees one part [of line ABC] straight on according to the visual angle, whereas it sees another part, which is equal to it, obliquely so that it sees its image refracted to a distance. Let AB and BC be equal magnitudes. Let the images from A and B reach F and G on the eye directly along the perpendicular, and let the ray from point C be inclined to G. Because GC is at a slant, find parallel diameter DE. And since GE is one-sixth of the circumference of the circle,[36] the ray from point G will incline toward D, and [refracted ray GD]

C videbitur in I. Igitur BI maior ipsa BA conspicietur, sic BC maiorem ipsa
BA conspicietur. [131]

Est locus in quo inaequales magnitudines aequales videntur. PROP. 20

Contra vero locus est in quo oculo apposito inaequales magnitudines aequales
videantur. Euclides ita probat. Sint duae magnitudines quarum maior CB,

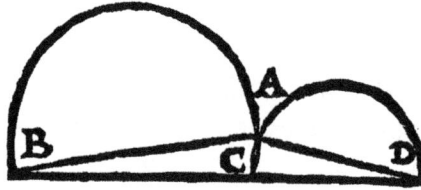

minor CD. Super utramque describantur duo maiora circuli segmenta sed
similia, scilicet quae angulum aequalem suscipiant, ut aequalis sit DAC eo
qui sub CAB. Connexis lineis DA, AC, AB, quod per 33, 3 in similibus
segmentis aequales anguli, ergo magnitudines aequales videbuntur.

Ex superiori ratione falsam dicimus. Exemplum nostrum in medio apponemus. Sint duae magnitudines, AB maior, BF minor. Veniat directe magni-

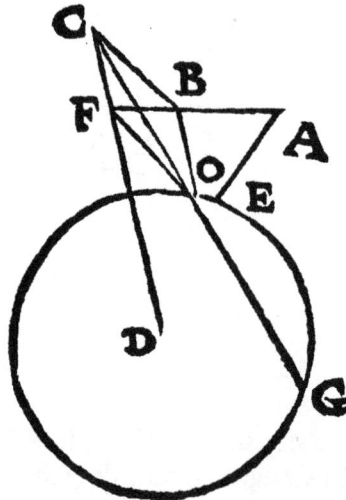

tudo AB ad pupillam EO, quae ut est videbitur. Veniat punctus [132] F ad
O oblique. Relabitur ad G, extensaque linea ad C, occurret catheto ex D in

will extend to I, so point C will be seen at I. Image BI will thus appear larger than BA, so BC will appear larger than BA. [131]

PROP. 20: *There is a place from which unequal magnitudes appear equal.*

On the other hand, there is a place where the eye can be situated to see unequal magnitudes as equal. Euclid proves it this way.[37] Let there be two

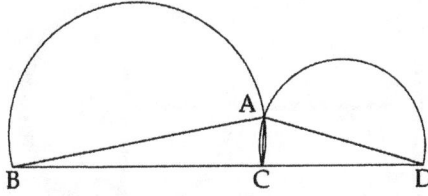

Figure 5.20a

magnitudes of which CB [in Figure 5.20a] is the larger and CD the smaller. On each of them describe two similar major segments of a circle, that is, segments that contain an equal angle, so that [the angle contained by arc] DAC is equal to [the angle contained by arc] CAB. When lines DA, AC, and AB are joined, since equal angles are in equal segments, according to book 3, proposition 33 [of Euclid's *Elements*], the magnitudes will appear equal.[38]

According to previous reasoning, we say that this is false. We will lay our Illustration out in the open. Let there be two magnitudes, AB [in Figure

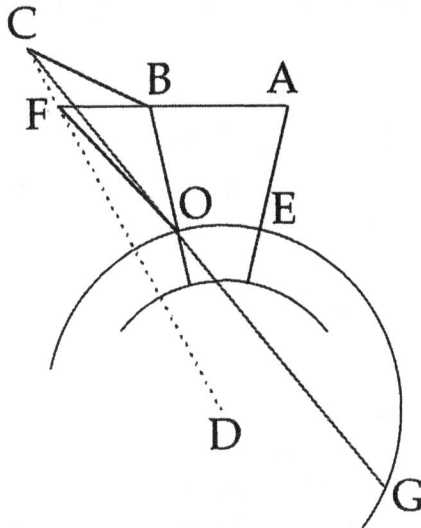

Figure 5.20b

5.20b] the larger and BF the smaller. Let the image of magnitude AB reach the pupil EO directly, so it will appear as it actually is. Let the image of point [132] F reach O obliquely. It inclines toward G, and when the

C. Connexa BC aequalis est AB; videbitur ergo aequalis BF[33] ipsi AB.

Sphaerae hemisphaerio semper minus[34] *spectabitur. PROP. 21*

Euclides contrarium sentit. Ait enim sphaera binis spectata oculis si dimetiens sphaerae aequa intervallo oculorum fuerit, ipsiusque hemisphaerium spec-

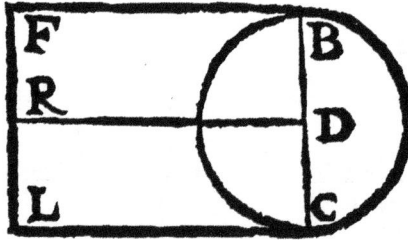

tabit. Sphaerae dimetiens BC, centrum D; et supra tria puncta B, D. C oriantur tres perpendiculares, diriganturque in rectum, quarum terminis designetur linea FRL,[35] ipso dimetienti aeque remota. Loceturque alter oculus in F, alter in L. Quia diametro eorum distantia par est, si manente axe RD,[36] circumferatur parallelogramum BFCL; circunscripta figura circulus erit, quare hemisphaerium tantum ipsius spectabitur sphaerae.

At si oculorum intervallum illud sphaerae superabit, plus hemisphaerio [133] conspicietur. Esto oculorum intervallum CB, centrum sphaerae K. A

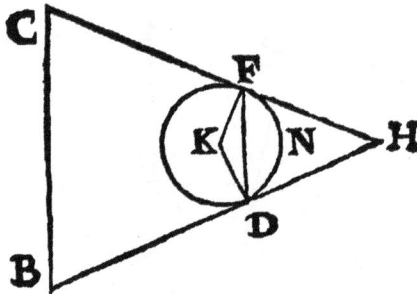

punctis C, B procidant radii tangentes sphaeram in punctis D, F; productique mutuo congrediuntur in H. Pars igitur FND minor est hemisphaerio, ex iis quae probaverat in 23 theoremate; remanens pars FKD visa maior erit.

---

[33]"BC" in original text.
[34]"mius" in original text.
[35]"FKL" in original text.
[36]"D" in original text.

line of refraction GO is extended to C, it will intersect the cathetus dropped
from D at C. When it is joined, BC = AB, so BF will appear equal to AB.

PROP. 21: *Less than a hemisphere of a sphere will invariably be seen.*

Euclid is of the contrary opinion. For he says that, if the sphere is viewed
with both eyes, and if the diameter of the sphere is equal to the distance

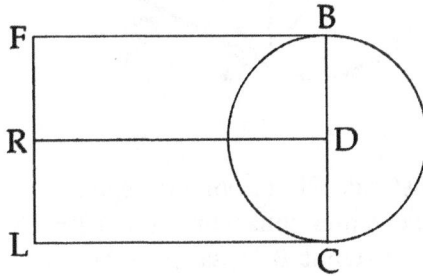

Figure 5.21a

between the eyes, the viewer will see its hemisphere. BC [in Figure 5.21a]
is the diameter of the sphere and D its center. Drop three perpendiculars
from the three points B, D, and C, extend them straight out, and let line
FRL, which is parallel to the diameter, define their endpoints. Place one eye
at F, the other at L. Since the distance between them is equal to the diameter,
if axis RD remains fixed, rectangle BFCL can be rotated around it, and the
circumscribed figure will be a circle, so only a hemisphere of the sphere will
be seen.[39]

If, however, the space between the eyes is greater than [the diameter
of] the sphere, more than a hemisphere [133] will be seen. Let CB [in

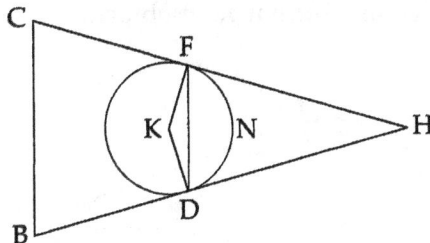

Figure 5.21b

Figure 5.21b] be the space between the eyes and K the center of the sphere.
From points C and B, let rays extend tangent to the sphere at points D and
F, and when they are extended they meet each other at H. Portion FND is
therefore less than a hemisphere, according to what [Euclid] demonstrated
in proposition 23 [of his *Optics*], and so remaining portion FKD, which is
seen, will be larger.[40]

Contra vero si oculorum intervallum minus sphaerae dimetiente fuerit, minor illius pars spectabitur. Probatque hoc modo. Esto sphaera cuius cen-

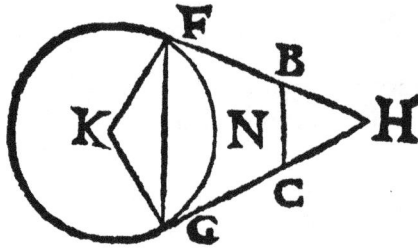

trum K; oculorum distantia BC minor ipsa sphaerae diametro. Procedant radii ab oculis B, C, tangentes sphaeram, et sint BF, CG, congredianturque mutuo in H. Quoniam distantia BC ipsa sphaerae diametro minor est, igitur ab H procidentes radii minorem sphaerae partem[37] intercipient, per ea quae supra probaverat.

  Sed haec falsa sunt, supponit enim eodem tempore utrumque oculum videre posse, quod suo loco falsum demonstrabimus. Semper enim uno oculo aspicimus, et si binis inspectare necessum fuerit, velocissimo ictu visiva virtus huc illuc discurrit, quod idem esset ac si oculo huc illuc moveretur. Semper igitur cuiusque sphaerae oculus minus hemisphaerio spectabit, idemque sensit vigesimo tertio problemate.

Quanto oculus sphaerae propinquior fuerit, tanto minorem eius partem videbit.
PROP. 22

Probat Euclides oculum plus haerens sphaerae vel columnae quod idem existimat, minus videt, et plus videre existimabit, idque hoc modo probat. [134] Esto sphaera cuius centrum K, et ab oculo D ad centrum procedat

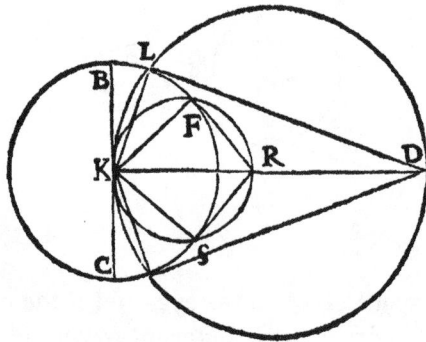

---

[37]"patrem" in original text.

On the other hand, if the distance between the eyes is less than the diameter of the sphere, a smaller portion of it will be seen. Euclid proves it

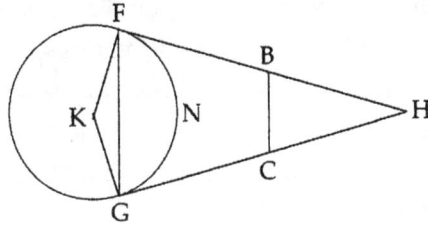

Figure 5.21c

this way. Let there be a sphere with center K [in Figure 5.21c], and let the distance BC between the eyes be less than the diameter of that sphere. Let rays from eyes B and C extend tangent to the sphere, let them be BF and CG, and let them intersect one another at H. Since distance BC is less than the diameter of the sphere, the rays extending from H will intercept a smaller portion of the sphere, according to what Euclid proved above.[41]

But these conclusions are false, for he supposes that both eyes can see at the same time, which we will show is false at the appropriate place [in book 6]. In fact, we always see with one eye, and if it is necessary to look with both, the visual faculty shifts back and forth with extremely rapid jerks, which is tantamount to having one eye move back and forth. Consequently, an eye will always see less than a hemisphere of any sphere, and Euclid is of the same opinion in the twenty-third proposition [of his *Optics*].[42]

PROP. 22: *The closer the eye gets to a sphere, the smaller the portion of it the eye will see.*

Euclid demonstrates that the closer the eye judging it gets to a sphere or a cylinder, the less of the sphere or cylinder it actually sees but the more it will judge that it sees, which he proves as follows. [134] Let there be a sphere with center K [in Figure 5.22], and let line DK reach from the eye

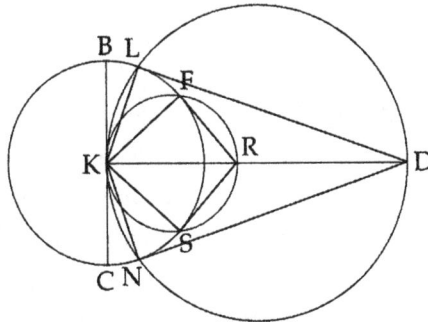

Figure 5.22

linea DK. Et ex puncto K oriatur recta BC ad perpendiculum, circa DK circulus describatur, diriganturque tangentes DL et DN, connectanturque KL, KN. Igitur anguli ad L, N, recti sunt, quia ad semicirculum ipsamque sphaeram inibi attingunt. Mox haereat magis oculus sphaerae in R, et secundum RK describatur circulus, et ab R dirigantur tangentes circulum iam descriptum RF, RS, et connectantur FK, SK. Unde sub angulo D spectatur NL; sub R, SF. Sed maius est LFSN ipso SRF; apparet enim minus, quia angulus ad R maior est eo qui ad D.

Sed hoc falsum, quia eorum quae prope sunt minor pars videtur directe, maior refracte, quae vero longe, maior pars directe, minor vero refracte. Et quia ex fractione hebescit visus, languide, nec suo loco, et maiora quam fuerint videt; non tamen ratio ex angulo provenit.

Si[38] in linea supra circuli centrum orientem stet oculus, circulus rotundus videbitur. PROP. 23

In hoc problemate cum Euclide convenimus. Ipse ita probat. Esto circulus cuius centrum K, ab eo oriatur perpendicularis linea KB. Oculus sit in B,

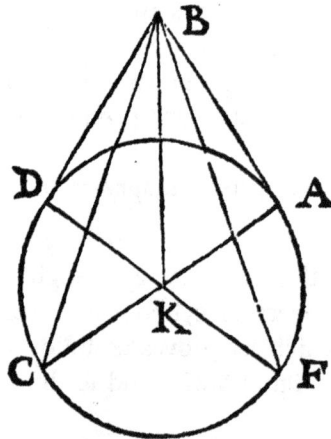

trahantur per centrum diametri AC, FD. Dico has binas diametros aequales videri, et inde circulum [135] aeque rotundum. Connectantur BA, BF, BC, BD. Quoniam BK et KF binis BK, KC alterae alteri aequales sunt, et aequales anguli, quia recti, ergo et bases BC, FB sunt aequales. Eadem quoque ratione BA, BD aequales sunt, sunt et diametri aequales; angulus igitur qui DBF angulo CBA aequalis. Videbuntur ergo aequales diametri.

---

[38]"se" in original text.

at D to the center. Then from point K drop straight line BC orthogonal to DK, draw a circle around [diameter] DK, extend tangents DL and DN [to the sphere], and join KL and KN. Therefore, the angles {KLD and KND} at L and N are right because they intercept a semicircle within the same sphere.[43] Now let the eye get closer to the sphere at R, draw a circle around [diameter] RK, drop tangents RF and RS from R to the circle [BLNC] already drawn, and join FK and SK. Consequently, arc NL is seen under the angle [LDN at] D, whereas arc SF is seen under the angle [FRS at] R. But arc LFSN > arc SF, and of course it appears smaller because the angle [FRS at] at R is larger than the one at D.[44]

This is false, however, because a smaller portion of things that lie close is seen directly, while a greater portion is seen by means of refraction; but a greater portion of things lying far away is seen directly, whereas a smaller portion is seen by means of refraction. And since sight is weakened by refraction, it will see things less clearly, displaced from their actual location, and larger than they are; and the explanation is in no way based on the visual angle.[45]

PROP. 23: *If the eye stands on a line dropped directly to the center of a circle, the circle will appear uniformly round.*

In this proposition we agree with Euclid. He proves it thus. Let there be a circle with center K [in Figure 5.23], and on it erect perpendicular line KB.

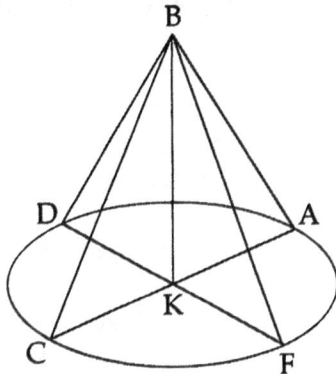

Figure 5.23

Let the eye be at B, and draw diameters AC and FD through the center. I say that both diameters will appear equal, and therefore the circle will appear [135] uniformly round. Join BA, BF, BC, and BD. Since BK and KF [in triangle BKF] are equal to both BK and KC [in triangle BKC], one to the other, and since both pairs form equal angles because they are right, bases BC and FB [of the respective triangles] are equal. Also, by the same reasoning, BA and BD are equal, and the diameters are also equal, so angle DBF = angle CBA. Consequently, the diameters will appear equal.[46]

Si in linea supra circuli centrum obliquam sed semidiametro aequalem oculus steterit, circulus non videbitur rotundus. PROP. 24[39]

Haec contra Euclidis sententiam est, inquit enim ipse, si quae ex centro excitatur non fuerit ad angulos rectos ipsi [136] plano, aequalis autem fuerit quae ex centro, circulus aequalis apparebit.

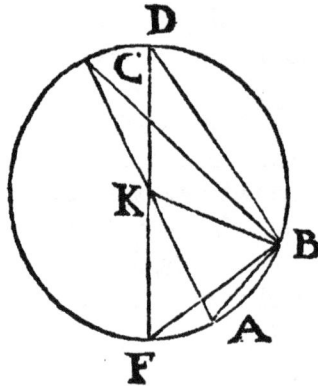

Sit circulus cuius[40] centrum K, et ab ipso K excitetur non ad angulos rectos ipsi plano KB sed aequalis ei quae ex centro circuli, connectaturque BD,[41] BC, BA, BF. Quoniam DK,[42] KB, KF sunt aequales, et rectus est angulus qui sub DBF, quia in semicirculo, eademque ratione qui sub ABC, et quae sub aequalibus angulis aequalia spectantur, unde diametri, et ex inde circulus aequalis videbitur.

Hoc falsum probabimus, nam si quae ex centro obliqua fuerit, alia pars propius altera longius erit. Ea quae prope recte, quae longius vero refracte videntur, et refracta longe a loco in quo sunt videntur. Impossibile est ergo ut oblique iacens oculus plano aequalem circulum videat.

Ex pupillae dilatatione omnia se ipsa minora[43] *videri. PROP. 25*[44]

Mydriasis, sive platicoria, sive pupillae dilatatio est quum maior fit pupilla morbo quam iustum poscat, ut iridis circulo proxima fiat. Interdum ex toto visum impedit, interdum permultum, ut omnia quae conspiciantur ab iis minora[45] quam sint videantur esse. Causam dicunt excrementitium "humorem illapsum qui aut acervatim aut palatim fertur," unde visionis tunica distenditur et foramen dilatatur.

---

[39]"23" in original text.
[40]"cuus" in original text.
[41]"CD" between BD and BC in original text.
[42]"Dk" in original text.
[43]"maiora" in original text.
[44]"24" in original text.
[45]"maiora" in original text.

PROP. 24: *If the eye stands on a line dropped obliquely to the center of a circle and equal to a radius, the circle will not appear uniformly round.*

This is contrary to Euclid's opinion, for he says that if the line erected at the center is not at right angles with the circle's [136] surface but is equal to the radius, the circle will look uniform throughout.

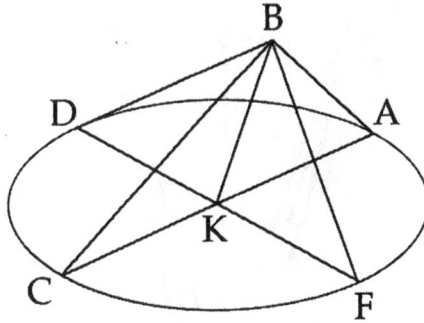

Figure 5.24

Let there be a circle with center K [in Figure 5.24], and from K erect a line KB that is not at right angles to the circle's surface but is equal to the radius that extends from the circle's center, and join BD, BC, BA, and BF. Since DK, KB, and KF are equal, since angle DBF is right because it is contained in a semicircle, since angle ABC is right by the same reasoning, and since things that are viewed under equal angles look equal, it follows that the diameters and, by extension, the circle will look equal throughout.[47]

We will prove that this is false, for if the line from the center is oblique, one side of the circle will lie nearer to the eye and the other farther away. Things that are nearer are seen [relatively] directly, whereas things that lie farther off are seen by means of refraction, and things seen by means of refraction appear to lie farther away than they actually are. It is therefore impossible for an eye positioned at a slant with respect to its surface to see a circle as uniform throughout.[48]

PROP. 25: *Everything appears smaller than it is because of a dilation of the pupil.*

Mydriasis, platicoria, or dilation of the pupil occurs when the pupil is made so much larger than it ought to be by disease that it can become nearly the size of the circle of the iris. Sometimes it interferes with vision to a great extent and sometimes totally, so that everything seen by those suffering from this condition looks smaller than it is. They say the cause is "an irruption of excremental humor produced either by accumulation or from here and there" so that the uveal tunic is distended and the pupillary opening is dilated.[49]

Avicenna et medici omnes inquiunt ab iis omnia minora videri quam sint. Aetius causam affert ex diffusione visivi spiritus. Sed investiganda est mathematica causa. Nos dicimus quod cui pupilla stricta est, omnia fere obliqua veniunt et refranguntur; cui lata recte accedunt, ob id minus refranguntur. Exemplum, sed haec distanti variari dicemus.

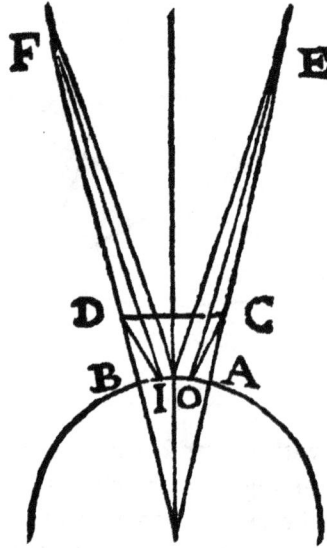

Sit oculus cuius pupilla AB lata, sit magnitudo CD. Ad eam veniat; videbit eam directe, ut est. Sit autem stricta OI; veniat C ad O. Refrangetur, et relabitur ad E, cui catheto occurret. Ex altera parte veniet ad F. Maior est igitur EF ipsa CD. [137]

In pupillae angustia maiora omnia videri. PROP. 26[46]

Phthisis seu tabes pupillae morbus est, quando ipsa fit angustior rugosiorque, ut omnia quae conspiciantur ab iis maiora quam sint videantur esse. Causa morbi est condensatio quam ex siccitate acceptam ferunt. Aegineta id accidere dixit quando res lumini obiectae sunt. Exemplum in superiori apposuimus. Sed ne putet quis haec contraria esse illis quae iam diximus—nam in serotino et matutino crepusculo res maiores videntur ob pupillae laxiorem obtutum quam meridie in pupillae angustiis—hoc animadvertendum in naturali pupillae dilatatione et constrictione omnia opportuniora videri secundum naturae necessitatem. Nam a natura factum est ne plus dilatetur angustie-

---

[46]"25" in original text.

Avicenna and all the physicians say that everything seen by those suffering from this condition appears smaller than it is.[50] Aëtius ascribes the cause to the emission of visual spirit.[51] But the mathematical cause has yet to be examined. We say that for anyone whose pupil is narrow, nearly all incoming images reach the eye obliquely and are refracted, whereas for anyone with a wide pupil, the images reach more directly, so they are less refracted. Here is an illustration, but we will admit that the effect varies with distance.

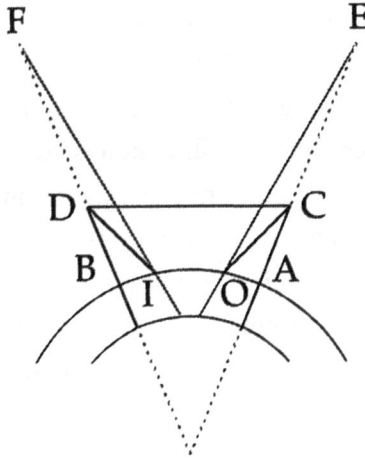

Figure 5.25

Let there be an eye whose pupil AB [in Figure 5.25] is wide, and let CD be a magnitude. Let its image reach the pupil, and the eye will see that object directly, as it actually is. On the other hand, let OI be a narrow pupil, and let the image of C reach O. It will be refracted, and [its refracted ray] inclines toward E, where it will intersect the cathetus. On the other side, [the ray from D refracted in the eye] will reach [the cathetus at] F. The resulting image EF is therefore larger than CD. [137]

PROP. 26: *In the case of a narrowed pupil everything appears magnified.*

Phthisis, or wasting of the pupil, is a disease in which the pupil becomes narrower and more shriveled so that everything looked at by such pupils appears larger than it is. The cause of this disease is compression that the eyes suffer on account of dryness. Paul of Aegina said that this happens when visible objects are in light.[52] We provided an illustration above.[53] But in order that no one think that this is contrary to what we already claimed—for things appear larger at evening and morning twilight because of a wider view through the pupil than they do at midday, when the pupils are narrowed—it should be taken into account that everything is seen more suitably in the natural dilation and constriction of the pupil according to the necessity of nature. For by nature the pupil is made to dilate and contract no more

turque quam deceat, quae vero a morbo fiunt praeter naturam accidunt et naturae opera pervertunt.

Senes propinqua minus videre. PROP. 27[47]

Ex supradictis iam pendet ratio cur senes, ut acutius videant res ex oculis longius divertunt, velut legere aut filum in acus foramen immittere. Aristoteles senes procul videre dixit, quia procul radii non coeunt, abducunt rem ubi coire maxime [138] radii solent. Sed id falsum est, putat enim ipse hominem duobus oculis simul videre et esse locum visionis ubi radii illi simul coeant. Sed cur senibus radii illi remotius coeunt et non propinquius adhuc sub iudice iis[48] est. Et quod hoc quoque falsum sit, monoculi senes rem etiam longius amandant ut perfectius videant.

Sed vera ratio est quod senibus pupilla diducitur reseraturque, ut caetera quoque membra non recte suo funguntur officio. Humor quoque incrassatur, unde maiori luce ad videndum indigent, et fere iis similes sunt qui mydriasi laborant, necesse enim habent ut quae videre velint lucidiora sint magisque coacta; quod utrumque crystallinis specillis emendatur, haec enim refractione radios uniunt, et lux multiplicatur in eis, humor enim quem natura pellucidissimum oculis indidit, sordescit senio et fit tenebrosior. Interrogati a me senes cur res longius abstrahant ut perfectius videant, respondent, prope confusa omnia videri, et hebescere visum, quod non nisi ex pupillae latitudine evenire contingit.

---

[47]"26" in original text.
[48]"lis" in original text.

than necessary, whereas things are forced by disease to occur unnaturally and to pervert the operations of nature.

PROP. 27: *Old people see close things poorly.*

The reason that old people move things away from their eyes in order to see them more clearly, for example, to read or to thread a needle, depends on what has already been said. Aristotle claimed that old people see at a distance because their visual rays do not converge except at a distance, so they move the object to where the rays usually converge [138] to the greatest extent.[54] But this is false, for he thinks that a person sees with both eyes at the same time and that the visual focus is where the rays from the two eyes come together. But why the rays in older people come together far away and not nearer has yet to be determined by those who follow Aristotle. And this is also false because old people with one eye move an object away in order to see it more clearly.

   The true explanation is that in older people the pupil is relaxed and open so that, just like the rest of the bodily members, they do not fulfill their function properly. Also, the [aqueous] humor thickens, so older people need more light to see, and they are just like those who labor under mydriasis, for by necessity they want the things they see to be brighter and more concentrated, both of which conditions are improved with crystalline lenses because they bring the rays together through refraction, and the light is augmented in them, since the humor that nature endowed the eyes with, initially quite clear, degenerates and becomes darker in an old person. Asked by me why they move things farther away to see them more clearly, old people respond that everything that lies near seems blurry, and their vision is weakened, which happens to arise only from the width of the pupil.

NOTES

1. The epistemological passage that Della Porta describes, from what is closer and better known to what lies farther away and more dimly grasped, harks back to the prologue to Aristotle's *Physics*, I, 1, 184a10–184b14.

2. See Euclid, *Optics*, postulate 4, in Euclid (1895) 2–3 and 154–55, and in Euclid (1537), 518; the theorem at issue is proposition 5, in Euclid (1895) 8–9 and 158–61, and in Euclid (1537), 518.

3. See Plotinus, *Enneads*, II, viii, 2, in Plotinus (1580), 190–91.

4. As it stands, this demonstration seems to show that object CF will look smaller than it should because, when CI and FO from the object's endpoints are refracted, the resulting refracted rays ID and OF, which form lines of sight, will intersect the object at D and E, and DE is obviously smaller than CF, not larger. In the subsequent portion of this theorem, however, Della Porta will show that the image of CF seen along lines ID and OE lies far beyond DE and will therefore appear much larger than CF itself.

5. This citation is difficult to fathom because there is no such corollary in this treatise. The relevant locus would be book 2, proposition 2, where Della Porta shows that, when it refracts into a glass sphere, a ray, such as BE, will intersect the diameter at some given point depending on the size of arc IE. In this particular case, moreover, IE is 60° (i.e., the arc subtended by one side of an inscribed hexagon), so, as Della Porta shows in book 2, proposition 2, the refracted ray EF will intersect the diameter where the diameter itself intersects the sphere, just as Della Porta observes in the next sentence. Of course this assumes that refraction in water (i.e., the aqueous humor) is identical to refraction in glass, an assumption that Della Porta should have known full well to be false.

6. Stipulating that arc IE be 60° is not only arbitrary and unnecessary, but also problematic, given the difference in density and refractive power between glass and water. All that matters is that ray BE reaches the eye at a greater slant than ray DO and therefore that it be more sharply refracted to the anterior surface of the crystalline lens.

7. In other words, since arc QO < 60°, refracted ray OL will intersect diameter QR beyond R and thus outside the circle.

8. Why Della Porta included such complicating factors as the insistence that arc EI = 60° and the tacit assumption that glass and aqueous humor are identically refractive is a mystery. Nonetheless, the basic logic of his analysis in this theorem is clear: namely, that when an object is near the eye, it will appear relatively farther away and larger than it would if viewed from a greater distance. This conclusion contravenes the size–distance invariance hypothesis, according to which the same object appears to be the same size no matter the visual angle under which it is viewed, that angle varying according to distance.

9. See *Optics*, prop. 6, in Euclid (1895) 8–11 and 160–63, and in Euclid (1537), 519–20.

10. As Della Porta points out here, if we follow Euclid's reasoning, the two parallels should eventually appear to meet perfectly when the center of sight is far enough from their distant endpoints. In the next paragraph, however, he will attempt to show that this perfect convergence will never occur.

11. In the Heiberg critical edition, this is postulate 5 of Euclid's *Optics* and of the Theonine recension; in the 1537 edition of Euclid's works it consists of postulates 7 and 8. See Euclid (1895) 2–3 and 154–55, and Euclid (1537), 518.

12. This is a standard scholastic articulation of what came to be known as "Ockham's Razor" in the nineteenth century.

13. This demonstration represents Della Porta's alternative to propositions 10 and 11 of both versions of Euclid's *Optics*, in Euclid (1895), 16–20 and 166–69, and in Euclid (1537), 521.

14. Implicit in this illustration, therefore, is that lower line XYZ will slant upward more gently than upper line CDK slants downward, thus explaining why "the ceiling slopes downward more sharply than the tiled floor slants upward" in rooms with high ceilings.

15. See Euclid, *Optics*, proposition 4, in Euclid (1895), 6–9 and 158–59, and in Euclid (1537), 519.

16. Marsilio Ficino, "Commentary" to *Enneads* II, viii, 2, in Plotinus (1580), 187, lines 37–38.

17. For Euclid's demonstration of this phenomenon, see *Optics*, proposition 12, in (1895), 20–21 and 170–71, and in Euclid (1537), 521.

18. Euclid (1537), 518. The theorem at issue is proposition 14, in Euclid (1895), 22–23 and 170–73, and in Euclid (1537), 521.

19. Ibid., 518. Euclid's actual proof is in proposition 13, in Euclid (1895), 22–23 and 170–71, and in Euclid (1537), 521.

20. Although it is extremely doubtful that Della Porta uses the term *perspectivist* here to designate those who deal with optical theory (*perspectiva*), it is not clear precisely what he does mean by it. Within the context, however, it seems probable that he meant to refer to draughtsmen, who need to know how to represent things as they actually appear in three dimensions.

21. The reference to the "centric line" (*linea centrica*) clearly harks back to Alberti's *De pictura*; see note 60 to book 4. As Della Porta describes it, of course, this line forms the so-called visual axis, which was understood by medieval and early Renaissance optical thinkers to run through the center of the pupil, the center of the eye, and the center of the optic nerve.

22. Vitruvius, *De architectura* III, 5, 13, in Vitruvius (1583, as III, 3, 3), 113. Translation from Vitruvius (1999), 53.

23. According to Della Porta's reasoning in proposition 10, however, AB will appear to lean backward only if the axial line of sight is focused on point B.

24. Within this context it is clear that *perspectiva* refers to the science of optics, not to the practice of draughtsmanship.

25. What Della Porta is getting at here is far from clear. The point seems to be that, in this particular case, when the eye lies at D such that DC = height AC, as in the right-hand diagram of figure 5.10b, and when the cornice is slanted outward

along line BE according to the equality of AB and AE, lines of sight DB and DE reaching the bottom and top respectively of the cornice will be equal. Presumably, then, if one stares along centric ray CD at the bottom of the wall, slanted cornice BE will look upright because the two rays ED and BD will be refracted in such a way that E's backward displacement counterbalances B's backward displacement by just the right amount to make EB look perfectly vertical. As Della Porta points out, however, in this case the cornice is not inclined at a distance of one-twelfth the altitude, and, moreover, the eye stands inordinately far away. By implication, then, Vitruvius's prescription does not apply universally, if at all.

26. In other words, according to the same reasoning in proposition 6 that explains why the sides of a well appear to converge to someone looking upward from the bottom, a tall wall will appear to lean forward when the viewer looks toward its top.

27. See Euclid, *Optics*, proposition 22, in Euclid (1895), 32–37 and 180–81, and in Euclid (1537), 523.

28. See Macrobius, *Saturnalia* VII, 14.15, in Macrobius (1535), 327. Della Porta's "mile" is one thousand paces (*millia passuum*), so the "stade" is one eighth that, or 125 paces. The Roman pace was around 5 feet, so the resulting mile is around 5,000 feet, which is quite close to the modern English mile of 5,280 feet.

29. Pliny, *Natural History* VII, 21, in Pliny (1559), 164. Despite having set a limit of 30 miles for the distance of human vision at sea, Aelian, in *Various Histories* XI, 13, recounts the same story of the sharp-eyed Sicilian who could see all the way from Lilybaeum to Carthage, roughly 140 miles; see Aelian (1548), 176.

30. See Euclid (1589), 334–37. Since the small semicircle is supposed to represent the Earth, line BD represents the horizon line, so nothing below point D would be seen from B.

31. See, for example, Bacon, *Perspectiva*, II, iii, 6, in Bacon (1996), 226–29; Pecham, *Perspectiva communis*, I, 65, in Pecham (1970), 142–43; and Cardano, *De Subtilitate*, III, in Cardano (1551), 148, and in Cardano (2013), 209.

32. According to the figure, points H and G lie below the horizon, so they cannot be seen. Nonetheless, the basic point holds according to Della Porta's logic: namely, that the rays reaching the eyes from the sides will be refracted into the eye, and the more oblique the rays, the greater the refraction and consequent image displacement.

33. Although this is the same figure as in proposition 17, the small semicircle now represents not the Earth but the eye, whose center point is F. Consequently, all the incoming rays except for the axial ray from C reach the eye's surface obliquely and are refracted into the eye. Suffice it to say that this figure is a problematic representation of the situation insofar as all the rays enter the eye at the same point—that is, where the axial ray enters the eye—in which case the rays from H and G would strike the eye before reaching that point and could therefore not possibly be seen.

34. See *Optics*, proposition 44, corollary, in Euclid (1895), 98–99; proposition 45 in Euclid (1895) and 232–33; and proposition 49, in Euclid (1537), 534.

35. Della Porta seems to be referring to book 4, prop. 10, where he shows that one eye can see more than a quarter of the sky at a single glance. Nowhere in that

proposition, however, does he specify how much more than a quarter of the sky it can see, so book 4, prop. 10 provides no grounds for his claim that it would be impossible for the eye at F to see all of BD because of the size of angle BFD.

36. This is a gratuitous stipulation on Della Porta's part and was presumably meant to give the proposition a patina of mathematical precision; see note 6 to proposition 1 above.

37. See *Optics*, propositions 45 and 46, respectively, in Euclid (1895), 96–99 and 232–35, and proposition 49, in Euclid (1537), 534.

38. See Euclid (1589), 436–37. Proposition III, 21 is more relevant here than III, 33; see ibid., 399–400.

39. See *Optics*, proposition 25, in Euclid (1895), 40–43 and 184–87, and Euclid (1537), 524–25.

40. For proposition 23, see Euclid (1895), 36–39 and 180-3, and Euclid (1537), 523–34. In that proposition Euclid proves that, if the eye is placed at H, it will see less than half the sphere. For Euclid's proof that more than half the sphere will be seen if the distance between the eyes is greater than the sphere's diameter, see *Optics*, proposition 26, in Euclid (1895), 42–45 and 186–87, and Euclid (1537), 525.

41. See *Optics*, proposition 27, in Euclid (1895), 44–47 and 186–89, and Euclid (1537), 525.

42. See note 40 above for the full citation.

43. This is a bit misleading, since angles KLD and KND lie within a circle, that is, KLDN, not within the sphere BLRC.

44. For the sphere, see, Euclid, *Optics*, proposition 24, in Euclid (1895), 38–41 and 182–85, and in Euclid (1537), 524. For the cylinder, see proposition 29, in Euclid (1895), 46–49 and 190–91, and in Euclid (1537), 525–26.

45. Della Porta's explanation here traces back to proposition 1 above, where he establishes that things looked at from close up look larger than they should because of the refractive displacement of the endpoints.

46. This proposition is based on the first part of proposition 34 of Euclid's *Optics*, in Euclid (1895), 60–63 and 198–201, and in Euclid (1537), 527–28.

47. This proposition is based on the second part of proposition 34 of Euclid's *Optics*, in Euclid (1895), 64–65; it is given as a separate proposition (i.e., 35) in the Theonine recension, in Euclid (1895), 200–3, and in Euclid (1537), 528.

48. The claim that "things that are nearer are seen [relatively] directly, whereas things that lie farther off are seen by means of refraction" would seem to contradict Della Porta's argument in proposition 22 above that the closer the eye gets to a sphere, the larger the portion it sees appears to be because of refraction (see note 44 above). Implicit in this demonstration, however, is that, being nearer to edge A of the circle, the eye at B sees that portion of the circle more directly, whereas it sees the farther edge at C at a greater slant, so B's view of that edge is more refracted. As a result, the image of the side of the circle at C is more displaced than that of the side at A, which means that the side toward C will look longer than the side toward A.

49. Quotation from Aëtius, *Tetrabiblos*, II, 3, 52, in Aëtius (1542), 357.

50. See Avicenna, *Canon*, III, iii, 4, 12, in Avicenna (1556), 429.

51. See Aëtius, *Tetrabiblos*, II, 3, 52, in Aëtius (1542), 357.

52. See *Medical Compendium*, III, in Paul of Aegina (1551), 156.

53. Accordingly, if AB in Figure 5.25 represents the normal pupil and OI the abnormally constricted pupil, then image EF seen abnormally through OI will be larger than image CD seen normally through AB.

54. See Aristotle, *Problems*, XXXI, 25; in Aristotle (1552; as XXXI, 26), 50ra.

LIBER SEXTUS

PROOEMIUM

Postquam huc provecti sumus ut de visione tractatum sicut decuit absolverimus, videbatur ultima his manus imponenda, nisi problemata quaedam superesse viderentur, quae aliter accepta, nostram institutionem infirmarent. Eaque expendisse valde praestiterit, eoque libentius ad ea enodanda accedimus: cur scilicet binis oculis donatus homo et unusquisque per se satis suo fungitur officio solus, simplicem rem non duplicatam aspiciat? Sed videamus ecquid maiores senserint.

Binis oculis obiectam rem simplicem conspici. PROP. 1

Aegerrime nostrorum maiorum animos inquirentium contorsit disquisitio cur binis oculis simplicem rem contuentibus simplex visio occurrat, iidemque simptomate quodam correpti duplicatas res videant. Principio ab Aristotele incipiemus, qui quaerit in *Problematis*, "Cur vinolentis quod unum est interdum multa appareant?" Aitque conspectus initia [140] intus esse et a vino moveri, et quum initia moventur conspectus, vis utraque obiectae rei partem attingit, ob id res una geminata videtur. Idem quoque oculo subter oppresso evenit, et quum initia conspectus moventur, alterum cum altero coire nequit. Idem etiam evenit ex motu intrinseco et extrinseco vini. Et alibi: "Cur distracta conspectuum societate res una geminari videtur?" Rationem dixit, "quod utriusque oculi radius non ad idem punctum devenit, et quod bis noster animus inspexit inde geminatum se inspexisse existimat."

Aphrodiseus eadem quae Aristoteles habet, quod "oculorum musculi varie vino completi factique languidi depravantur vertunturque alter sursum, alter deorsum; unde radii oculorum non eundem in locum concursare pos-

BOOK SIX

INTRODUCTION

N ow that we have advanced to the point of completing our treatment of vision, as was appropriate, this treatment seemed to be the last thing to which these hands should be applied, except that certain problems seem to remain, which, unless resolved, would weaken our instruction. We therefore broach those problems that might be especially worth taking into account and by so doing unravel them in a pleasing way by explaining why, being endowed with two eyes, each of which is adequate on its own to fulfill its function, man does not see a single object double. But let us see what our predecessors might have thought.

PROP. 1: *A visible object appears single to both eyes.*

The question of why a single object looks single to those looking with two eyes has tied the minds of our predecessors who looked into it into painful knots, as has the question of why those looking with two eyes see things double when suffering a particular pathological disorder. We will begin with Aristotle, who asks in the *Problems*, "Why does a single thing sometimes appear multiple to people who are drunk?"[1] He responds that the wellsprings of vision [140] are internal and are moved by the wine, and when the wellsprings of vision are moved, the power[2] of each eye reaches a different part of the visible object, so a single object appears doubled. The same thing also happens when the eye is pressed from below, and since the wellsprings of vision are moved, the one cannot converge with the other.[3] The same thing arises from external motion and the internal motion of the wine. Elsewhere he asks, "Why does a single object appear doubled to those with distorted vision?" The reason, he claimed, is "that the radiation from each eye does not reach the same point, and so our mind judges that what the two eyes looked at were looked at double."[4]

Alexander of Aphrodisias maintains the same thing as Aristotle, arguing that "the muscles of the eyes, being variously filled with wine and rendered sluggish, are distorted, one turning upward, the other downward. Consequently, the radiation from the eyes cannot converge at the same spot, so

sunt, ob id conspectum rei geminari necesse sit. Constat hoc etiam eo quod qui alterum oculum premit sursumque impellit gemina omnia videt." Galenus dicit quod nervi optici in uno loco circa cerebrum coeunt ex quo egressae per utramque pupillam, et revertentes eaedem rectae lineae ad eundem locum non coeunt.

Avicenna medicinalem affert rationem ex oculorum manu contactu, ex ebrietate, veloci motu, et pupillae morbo simplicem rem geminatam videri. Idque ex quadruplici causa evenire. Putat enim primum visionis instrumentum esse crystallinum humorem sed penes illum non perfici visionem, nam si in eo visio fieret, res geminata semper videretur. Sed simulachrum a spiritibus, qui videndi ratione praediti sunt, per opticos nervos concavos deferri ad eorum occursum, et spiritibus aliis communicant qui communis sensus virtutem retinent, et hic videndi virtus perficitur. Hi communicant imaginativae, quae velut penu conservat et in usu promit. Prima igitur causa cur res una geminata videatur est instrumenti dimotio a suo loco, nam imagines in crystallinum venientes, quum a spiritibus diversimode in opticorum nervorum coitu deferantur, rem singularem duplicem vident.

Secunda causa ex spirituum perturbatione evenit, ut in ebrietate, nam etsi imago in crystallinum fuerit, spiritus qui illam laturi sunt ad nervorum decussationem opticorum si conturbentur, visionem conturbant et ad sinistrum, dextrumque videtur.

Tertia, quod post nervorum decussationes sunt spiritus communi sensui famulantes qui imagines illi diiudicandas ferunt, qui si conturbentur velociter [141] ante et retro, simplicis rei duplicem imaginem exhibent; ut qui ignem circumvertit, nam priusquam una pars iudicetur, sequitur velociter altera, et sic altera post alteram, ut igneus circulus conspiciatur. Sic spiritus in anteriore cerebri parte circumvoluti, imagines plures, et se circumvolventes ostendunt.

Postrema ex pupillae morbo accidit, videlicet constrictione, dilatatione, et conversione a suo situ, ut rerum simulachra contorqueantur. Albertus ab his eadem. Ex opticorum nervorum dimotione, ut quae ab oculis adferantur spiritibus communis sensus non se contingant in uno puncto; idemque evenire in digiti suppositione. Vel dum moventur visivi spiritus ex cerebro ad oculos propter inclusum ventum extra metas a natura praescriptas moventur dextrorsum et sinistrorsum, apprehendunt rem unam quasi duplicem. Vel redundantia eiusdem spiritus in communi sensu intra tunc reddunt accipiuntque, et priusquam una compleatur inchoat altera, ut unum stet, alterum

the view of an object must be doubled. This is also confirmed by the fact that anyone who presses one eye and pushes it upward sees everything double."[5] Galen says that the optic nerves meet at a single place near the brain from which they go out to both pupils, and when the same straight lines return to the same place, they do not meet.[6]

Avicenna offers a medical explanation for why a single object can appear double through contact of the eyes by a hand, intoxication, swift motion, and disease of the pupil.[7] This occurs according to a fourfold cause. For he believes that the primary organ of vision is the crystalline humor but that vision is not completed there, for if vision were to be achieved in that humor, an object would always appear double [because each eye has a crystalline lens]. But the image is conveyed through the hollow optic nerves to their juncture [at the optic chiasma] by the [visual] spirits, which are endowed with the faculty of seeing, and there they unite with the other spirits that the faculty of common sense contains, and here the power of seeing is fulfilled.[8] These unite with the imagination, which retains the images as if in a storehouse and proffers them when needed. Thus, the first reason that a single object might appear double is a displacement of the organ, for since the images reaching the crystalline humor are conveyed by the spirits in different ways to the conjunction of the optic nerves, they see a single object double.

The second reason arises from a disturbance of the spirits, as in the case of intoxication, for although the image is in the crystalline humor, if the spirits that are to convey them to the intersection of the optic nerves are confused, they confuse the visual apprehension and the object appears both to the left and to the right.

The third reason is that the spirits that serve the common sense after the intersection of the optic nerves and that convey the images to be judged by it, display the image of a single object as double if they are swiftly agitated [141] to and fro, an example being someone who swings a torch around, for before one portion of the swing can be determined, another quickly follows, and thus one after another, so that a fiery circle can be seen. Thus, when the spirits in the anterior portion of the brain are whirled about, they exhibit a host of images also whirling about.[9]

The last reason arises from a disease of the pupil, that is, constriction, dilation, or displacement, so that the images of things are distorted. Albertus Magnus offers the same reasons drawn from these.[10] Because of a displacement of the optic nerves, the images that are conveyed by the spirits from the eyes to the common sense cannot come into contact at one point, and the same happens with a finger pressing upward against the eye. Or else, while the visual spirits move from the brain to the eyes, if they move toward the left or right in a way constricted beyond the norms prescribed by nature, they apprehend a single object as double. Or else an excess of such spirits in the common sense rejects and receives images, and before one is fully formed, another begins to form, so that one stands still while the other

moveatur. Sic gutta cadens videtur linea et mota virga circulus. Vel ex uvae morbo, ut diximus.

Inter recentiores perspectivos, Alhazenus rem exactius docet, dicit enim quod duae rerum formae ad crystallinum pervenientes, si ibi visio perficeretur, duae unius rei formae viderentur, sed forma recepta in superficie glacialis pertransit suum corpus et per foramen opticorum nervorum ad communem nervum pervenit, et ibi altera alteri supponitur, ut ex illis duabus formis unica efficiatur. Et tunc ultimum sentiens unam rei imaginem sentit, et dum ordinantur non comprehendit illas, nam si non dum ordinatas comprehenderet, duas sentiret.

Ratio quae eos ad id affirmandum duxit est quod, quando oculus pervertitur a suo situ, res una geminata videtur; quum uterque oculus suam recipiens imaginem, alterum alteri supponere non possit ultimo sentienti, unde et dexter oculus unam, sinister alteram sentit. Argumentum afferunt quod fiat per opticos nervos, quod quando aliquo morbo humoribus infarciuntur, perit visus, et remediis exinaniti, redit visus.

Vitellius rem lineis demonstrat quae Alhazenus longissimis verbis. Dicit enim quod in medio cuiusque visivae pyramidis axis est qui transit per foramen opticorum nervorum, et quando duo axes in puncto uno coeunt nervi oppositi, puncti circundantes axem alter alteri etiam supponitur, et simul concordant in ultimo sentienti. [142]

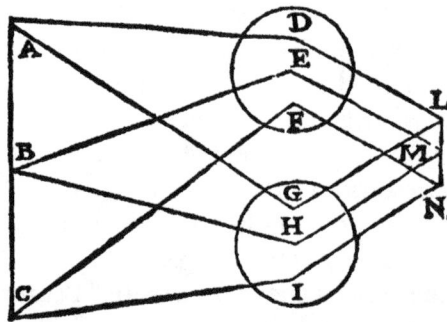

Esto visa res ABC, dexter oculus GHI, sinister DEF, pyramis dextra cuius basis ABC, conus GI. Veniat ad dextrum oculum AC per nervum opticum concavum HM usque ad sensum communem LMN. Sic punctus basis A per G veniet in L, et punctus alter basis C per I veniet in N, axis

moves. Thus, a falling drop of water appears as a line and a twirling rod as a circle. Or else diplopia occurs because of a disease of the uvea, as we have said.

Among more recent optical theorists, Alhacen shows this very thing, for he says that since two forms of things reach the crystalline humor, if vision were completed there, then two forms of a single thing would be seen, but the form impressed on the surface of the glacial humor passes through its body and reaches the common nerve [at the optic chiasma] through the hollow of the optic nerves, and at this point one form is superimposed on the other so that from these two forms a single one is formed. The final sensor then senses one image of the thing, and until they are properly arranged, it does not perceive them [as single], for if it were to perceive them while they were not so arranged, it would sense them as double.[11]

The line of reasoning that led these thinkers to affirm this is that, when the eye is displaced from its position, a single object appears doubled; since each eye takes on its own image, one image cannot be superimposed on the other at the final sensor, so the right eye senses one, and the left another. These thinkers offer an argument that is based on the optic nerves because when they are clogged with humors from some disease, vision ceases, but when those humors are dispelled by remedies, sight returns.

Witelo shows with a diagram what Alhacen demonstrates at great length with words. For he says that the axis, which is in the middle of any visual cone, passes through the hollow of the optic nerves, and when the two axes converge on a single point of the opposing nerve, the points surrounding the axis are also superimposed one on the other, and they correspond perfectly in the final sensor.[12] [142]

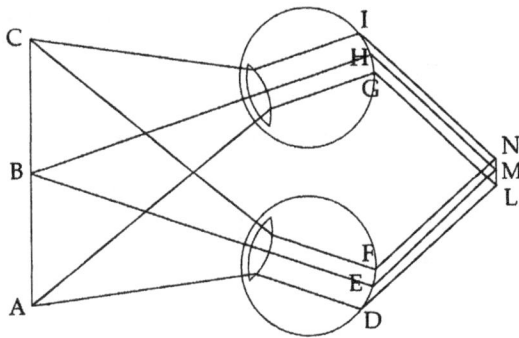

Figure 6.1a

Let ABC [in Figure 6.1a] be the visible object, GHI the right eye, DEF the left one, ABC the base of the right-hand cone, and let GI [lie on the far side of] its vertex. Let the image of AC reach the right eye and pass through the hollow optic nerve HM to the common sense at LMN. Thus, the image of point A on the base will reach L via G, and the image of the other point

vero pyramidis BH veniet in M. Sic sinistrae pyramidis punctus basis C venit
per F in N, et punctus A per D in L, et axis pyramidis BE veniet in M. Sic
axis punctus M dextrae pyramidis erit idem cum puncto basis sinistrae,
et L punctus basis dextrae et sinistrae pyramidis, sic N utriusque quoque
pyramidis, et hoc modo congruentes puncti eiusdem rei per utrunque oculum
ultimo deferenti coaptantur.

Hae sunt antiquorum et recentiorum opiniones; nos autem aliter sen-
timus. Virtus enim sentiens vidensque una est, et si duo sint oculi, quae
utrique in parvo temporis momento, ubi necesse erit praesto est, et tam
facile, ut nosmet ipsi quomodo id fiat minime sentiamus; famulantia instrumenta tam citam eiusmodi virtuti obedientiam praestant, ut prius opus exe-
quantur quam imperatum sit. Oculos binos natura largita est nobis, a dextris
unum, a sinistris alterum, ut si a dextris aliquid visuri sumus, dextro utamur,
at si a sinistris sinistro. Unde semper uno oculo videmus, etsi omnes apertos
et omnibus videri existimemus.

Idque his argumentis probamus. Inter utrumque oculum mediet aliqua
tabula vel paries ut alterum ab altero disterminet, ac dextro oculo librum
opponamus et legamus. Si quis sinistro alterum librum ostendet, non solum
non legere sed [143] neque paginas videre valuerit, nisi brevi temporis
momento videndi virtutem a dextro oculo subtrahat, mutuetque sinistro.
Idem in aliis sensibus evenire videmus. Si dextra aure aliquem loquentem
audiverimus, non poterimus alterum sinistra admittere, et si utrumque audire
velimus, neutrum audiemus; vel si aliquod dextra audiemus, tantundem
sinistra deperdemus. Sic etiam si una manu scribemus altera lyram tangere
aut aureos numerare nequimus.

Alterum argumentum erit, si quis baculum ante se locaverit, et illum
parietis alicui rimae ex regione existenti obiecerit, notaveritque locum,
quando sinistrum oculum clauserit, non videbit baculum ab opposita rima
dimotum. Ratio est quia unusquisque dextro oculo cernit, ut dextra manu
et pede utitur, et qui sinistro oculo, manu, aut pede pro monstro habetur.
At si dextrum oculum clauserit, illico baculus ad dextram partem accedit.

Tertium erit argumentum quod natura duos oculos, alterum alteri fecit,
ut alter a dextris alter a sinistris hominem ab incursantibus defenderet. Et
id in animalibus clarius apparet, nam per semipedem inter se distant, ut in
bubus, equis, et leonibus videre est. In avibus alter alteri oppositus est ut si
a dextris[1] vel a sinistris videndum sit virtus visiva velocissime accedens,

---

[1] "destris" in original text.

C on the base will reach N via I, whereas the ray along axis BH of the cone will reach M [via H]. Likewise, the image of point C on the base of the left-hand cone arrives at N via F, and the image of point A arrives at L via D, while axial ray BE of the cone will reach M [via E]. Accordingly, point M on the axis of the right-hand cone will be the same for the corresponding point on the left-hand base, and point L will correspond to the same point on the base of the right-hand and left-hand cones, so N will also correspond for both cones, and in this way corresponding points of the same object conveyed through both eyes to the final sensor are superimposed.[13]

These are the opinions of ancient and modern thinkers, but we think otherwise. For the power of sensing or of seeing is unique, and although there are two eyes, this power is ready for each at an instant's notice, when necessary, and with such ease that we may scarcely notice how it happens; and these subservient instruments lend such swift obedience to a power of this kind that they fulfill their function as soon as ordered. Nature has bestowed two eyes on us, one on the right, the other on the left, so that if we need to see something on the right, we may use the right one, and if we need to see something on the left, we may use the left one. Consequently, we always see with one eye, even though we may judge that both are open and think we are seeing with both.

We prove this according to the following arguments. Let someone put a panel of some kind or a screen between the two eyes so as to separate one from the other, and let us hold a book to the right eye and read it. If someone displays another book to the left one, the reader will be unable not only to read it but even [143] to see the pages, unless for a brief moment of time he takes the power of seeing from his right eye and lends it to the left. We see the same thing occur in the other senses. If we are listening with our right ear to someone talking, we cannot attend to someone else with the left one, and if we try to listen to both, we will hear neither; or if we hear something with the right ear, we will suffer a commensurate loss of aural acuity in the left. So, too, if we write with one hand, we cannot play a lute or count gold coins with the other.

Another argument will be that, if someone places a staff in front of himself and blocks some hole in the wall directly in line with it, and if he takes note of the place, then when he closes his left eye, he will not see the staff move from the facing hole. The reason is that, just as one uses the right hand or foot, he looks with the right eye, and anyone who uses the left eye, or hand, or foot is taken to be a monster. But if the one looking at the staff closes his right eye, the rod immediately shifts to the right side.[14]

A third argument for why nature made two eyes side by side will be that one eye can guard a person against attackers on the right and the other against attackers on the left. This appears more clearly in animals, for their eyes are half a foot apart, as is seen in cattle, horses, and lions. In birds, one eye is on the opposite side from the other so that, if the visual power is to deal quickly with what needs to be seen on the right or on the left, the

animal sua munia obire possit. Unde non simul videre possunt rem eandem. Et si quis situm oculorum in homine inspiciat, non in recta linea sunt, ut sine oculorum contorsione rem eandem inspiciant, etsi in recta linea videntur in radice tamen cohaerent.

Sed quaeso quae tam dira hominum insania est imaginari spiritus deferre spectra per foramina nervorum opticorum et naturales spiritus intellectualia simulachra deferri. Quomodo per foramina nervorum opticorum[2] transeunt, si nobis diligentissime intuentibus vix occurrerunt et adeo tenuissimi ut obtutum effugiant, si vere foramina dici possunt (et sunt medicorum plerique, qui perforatos esse negent)? Quomodo se simul circa cerebrum se decussant, si se vix tangunt, immo alter ab altero distat? Sed quae larvae eorum intellectum praestringunt, ut putent per tam longum iter deferri imagines per lineas quasdam[3] ad libitum concinnatas, ut eundem semper situm conservent?

"Strabones ab ortu naturae gemina non conspiciunt," Aphrodiseus dicit, "quod oculos non sursum deorsumque depravatos habent sed dextrorsum et sinistrorsum; quod enim oculi suum obtinent situm, idest directum, radii eundem in locum profluant, atque identidem [144] referant necesse est," quod contrarium est antedictis. Quod linearum illa falsa sit imaginatio, ponamus ut strabo, qui pupillas contra se positas habet. Quomodo eveniet illa linearum positio, quin rem duplicem non videant? His enim A veniens ad dextrum oculum transiret per G in R mox in O, at ad sinistrum A per D,

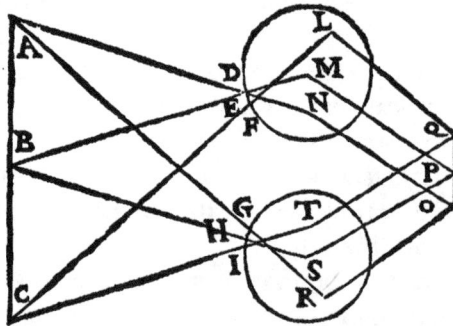

per N ad O itidem. Sed R in dextra dextri oculi, N vero in dextra sinistrum oculi, et sic ordo visionis perverteretur, ut dextra sinistra, et sinistra dextra viderentur.

---

[2]"opicorum" in original text.
[3]"quasdas" in original text.

animal can fulfill its functions. Hence, both eyes cannot see the same thing at the same time. And if one looks at the situation of the eyes in humans, they are not in a straight line, so that humans can look at the same thing without contorting their eyes, even though the eyes seem to come together in a straight line at the source.

But I marvel at how dreadfully unsound it is of people to imagine that spirits convey phantasms through the hollows of the optic nerves and that natural spirits convey intellectual images.[15] How do they pass through the hollows of the optic nerves, if those hollows barely show up when examined most diligently by us and are therefore so tenuous as to escape notice, if they can actually be called hollows (and there are many physicians who deny that they are perforated)?[16] How do they intersect one another near the brain, if they barely touch each other and are actually separated one from the other?[17] But what specters dull their intellects so that they believe that images are conveyed through such a long pathway by certain lines produced at will so as to maintain the same situation throughout?

"Those who are cross-eyed at birth do not see double by nature," Alexander of Aphrodisias says, "because they have eyes that are not distorted up and down, but rather to the left and right; since their eyes maintain this orientation, that is, straight ahead, the rays flow out to the same location, and it necessarily follows that they bring back the same image,"[18] [144] which is contrary to what has been said before. Let us imagine a cross-eyed person who has pupils situated against one another in order to show that the idea of such lines is false. How will it happen that according to this situation of lines, those pupils do not see a doubled object? For by such

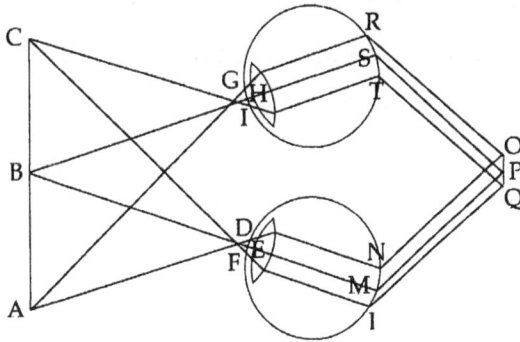

Figure 6.1b

lines the image of A [in Figure 6.1b], which reaches the right eye, will pass through G to R and then to O, whereas at the left eye the image of A will pass through D and through N to the same point O. But R is to the right in the right-hand eye, whereas N is to the right in the left-hand eye, and so the arrangement of sight would be skewed so that what is to the right would be seen to the left, and what is to the left would be seen to the right.[19]

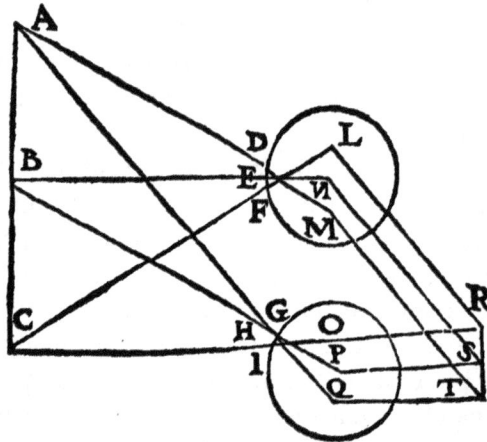

Praeterea, oblique aspicientibus omnia eo more perverterentur hoc modo. [145] Esto conspicienda magnitudo ABC. Veniat[4] C ad dextrum oculum per pupillam I ad Q, mox ad commune sentiens T,[5] sed C idem ad sinistrum oculum per F pupillam ad L, mox ad idem sentiens R. Non coincidit igitur T, sed labitur in R. Sic A dextro oculo in R, in sinistro vero in T, conversa et contorta omnia viderentur. Vana igitur et puerilis eorum imaginatio, sed ne illis vaniores videamur, omittendos putamus.

Galenus quod utroque oculo videamus ita demonstrat, sed profecto si ita nos medicinalem artem docuisset, de nobis actum esset. Inquit enim; quae oculo dextro cernimus alibi apparent quam quae sinistro; et quae sinistro videmus alibi quam quae dextro, at quae utroque oculo in horum medio; idque exemplo demonstrat.

Esto dextra pupilla A, sinistra B, magnitudo visenda DC; ex utraque pupilla A, B visione illapsa videamus DC. Sed a pupilla dextra EF, a sinistra GH. At si uno oculo videatur aliter eveniet, nam sinistro oculo clauso B, magnitudo CD in sinistram partem videbitur FE,[6] clauso vero dextro, magnitudo CD videbitur in HG[7]; si ambo simul referentur, in CD videbitur.

At si qui demonstrationem lineis non assequatur, hoc utatur experimento. Stans iuxta columnam utrumque oculum vicissim claudat. Et quum

---

[4]"veniant" in original text.
[5]"I" in original text.
[6]"HG" in original text.
[7]"FE" in original text.

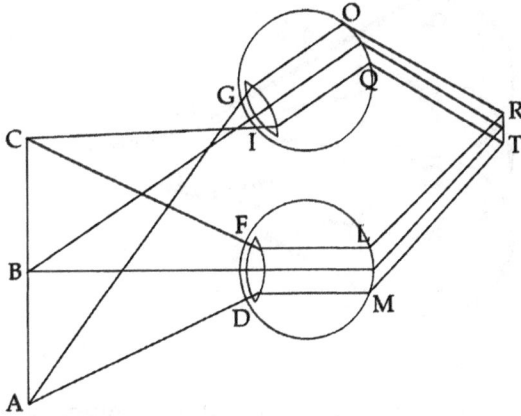

Figure 6.1c

Furthermore, everything seen off to the side by viewers should be skewed in that same way, as follows. [145] Let ABC [in Figure 6.1c] be the magnitude to be viewed. Let the image of C reach the right-hand eye through I on the pupil to Q and then to the common sense at T, but let the image of that same point C reach the left-hand eye through F on the pupil to L, then to R in the same common sense. It does not coincide with T, therefore, but is diverted to R. Thus, with the image of A in the right eye at R [via G and O] but at T in the left one [via D and M], everything would appear inverted and distorted. Therefore, the idea of those who believe in such visual lines is vain and puerile, but in order that we not appear more vain than they, let us take into account some things we have omitted.[20]

Galen shows that we see with both eyes in this way, and of course, if he has taught us the art of medicine accordingly, it should be pursued by us. For he says that what we see with the right-hand eye appears different from what we see with the left-hand one, whereas what we see with the left-hand eye appears different from what we see with the right-hand one, and what we see with both eyes is a mean between the two. He demonstrates this point with the following illustration.[21]

Let A [in Figure 6.1d] be the right-hand pupil, B the left-hand one, DC the magnitude to be seen, and let us see DC by means of visual radiation emitted from both pupils A and B. But DC will be seen as EF by the right-hand pupil and as GH by the left-hand one. If it is seen with one eye, the outcome will be different, for if the left-hand eye at B is closed, magnitude CD will appear as FE on the left, whereas if the right-hand eye is closed, magnitude CD will be seen at HG, but if both are brought to bear together, it will be seen at CD.

If, however, one does not follow the demonstration by means of a diagram, let him try this experiment. Standing near a column, let him close each eye in turn. When he closes the right-hand one, he will not see some

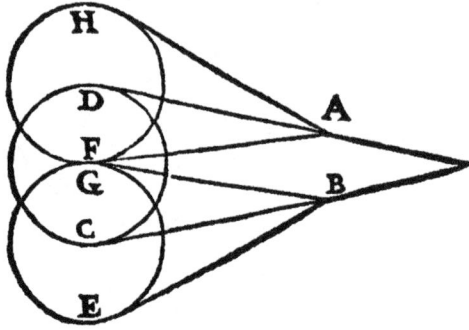

dextrum clauserit, quaedam eorum quae in dextra columnae parte videbantur non videbit, et quum sinistrum clauserit, quaedam quae prius viderat sinistro nunc dextro non videbit, et [146] si utrosque aperuerit, utrasque partes videbit. At si a columna paululum digressus utrumque oculum aperire ac vicissim claudere incipiet et columnam intuebitur, illam repente transilire videbit. Si dextrum clauserit, versus dextram partem veniet, si sinistrum, sinistram versus, et simul utrisque aspicienti medium locum videre putabitur eius, qui seorsum utrique abscondebatur.

Sed horum causa alia est quam ipse putat. Dum oculo A circulum DC inspicio, quia dextro oculo inspicio, semper eodem loco erit, at si dextrum clausero, dextrum eundem circulum in FE videbo, ut diximus, et si dextrum et sinistrum claudemus mox, binis inspiciemus Si alibi quam suo loco videbuntur, hoc ex refractione eveniet, ut diximus, quia uni recte, alteri oblique cadit.

Quod visio per spiritus non fiat qui referant imagines primo sensitivo clarum est, argumentaque quae afferunt ridicula, et prorsus exibilanda, frustra enim fiunt per plura, quae per pauciora fieri possunt; nos apparentias absque eorum ope servabimus. Dicunt quod clauso uno oculo spiritus ad alterum accedunt, ob id surgere, et pupillam dilatari. Hoc falsum est, nam si lente clauseris non eveniet, si vero digito premes, turgidulum fieri alia ratione contingit, nam ubi presseris, ibi acurrit sanguis. Cur autem compresso digito oculo simplex res geminata videatur ratio est quod ubi compressio, ibi virtutis concursus; oculus delicatissima partium pars est ibi virtus visiva occurrens.

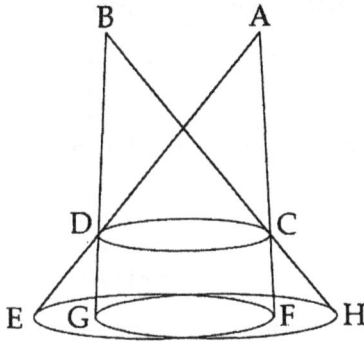

Figure 6.1d

of the things that he saw on the right-hand side of the column, and when he closes the left-hand eye, he will not see with the right-hand eye some of the things he had previously seen on the left-hand side, and if [146] he opens both eyes, he will see the things on both sides. If, moreover, while moving back a bit from the column he will begin to open and close each eye in turn as he looks at the column, he will see it suddenly jump back and forth. If he closes the right-hand eye, the column will shift to the right-hand side, if he closes the left-hand eye, the column will shift to the left-hand side, and if he looks with both eyes, he will judge its location to lie in between, which had been concealed from each separately.[22]

But the reason for these things is different from what Galen thinks. When I look at circle DC with the eye at A, since I am looking with the right-hand eye, it will always be at the same spot, but if I close the right-hand eye, I will see the same right-hand circle at FE, as we said, and then if we close the right-hand and left-hand eyes in turn, we will see with both. If the images appear to lie elsewhere than in their proper place, this will arise from refraction, as we said, because the image of one is directly incident, whereas that of the other is obliquely incident.[23]

That vision is not accomplished by means of spirits that convey images to the initial sensitive faculty is obvious, and the arguments that those who propose in favor of this theory are ludicrous and should be hissed off the stage forthwith, for things that are accomplished by more that can be accomplished by fewer are done in vain.[24] We will save the appearances without the help of their arguments. They say that when one eye is closed the spirit flows to the other, so it surges, and the pupil is dilated.[25] This is false, for if you close your eye slowly, it will not happen, but if you press with your finger, the pupil happens to dilate for another reason, for where you press is where blood rushes in. The reason a single thing appears doubled when the eye is pressed by a finger is that where the pressure is applied is where visual power coalesces. The eye is the most sensitive of all the organs, which is where the visual power flows. When it is divided between both eyes, the

Binis divisa oculis geminatam rem conspicabimur. Idem autem eveniet, si conatu et vi binis oculis rem unam conspicari velimus.

Refractionem obliquam duplici modo accidere posse. PROP. 2

Diximus superioribus libris obliquam refractionem maxime adiuvare visionem, nam ita insensibiliter visivae rectae pyramidi adnectitur et ordinate, ut refracte videns homo se recte videre putabit, et haec a dextris et a sinistris rectam pyramidem ambit. Et exemplum iam adduximus. Secus autem in altera refractione, nam quum homo visivam virtutem utroque oculo dividit, et cum conamine aspicit, oculos pervertit, nec situm servant a natura ordinatum, et irregulariter cadunt, unde [147] apparitiones rerum mirabiles eveniunt et longe a suis locis res videntur, et ex oculorum duplicitate, duplicata singularis imago rei conspicabitur. Exemplum.

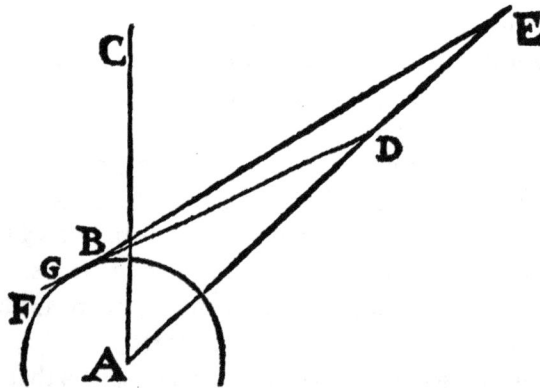

Sit oculi centrum A, centrica vero AC. Veniat rei videndae spectrum a sinistris in dextrum citra centricam; et sit E, pupilla B. Erit linea deferens formam EF; refrangetur in G. In contrarium extensa, erit in D. Trahatur a centro A cathetus per E, et occurret GD in D. Ibi erit imago rei.

Simplicem magnitudinem utrique oculo citra centricam oblique incidentem geminatam videri. PROP. 3

object will appear double. The same thing will happen, moreover, if we try to see an object as single with both eyes by exerting effort and force.

PROP. 2: *Refraction at a slant can happen in two ways.*

We said in previous books that refraction of oblique rays is a crucial aid in vision, for on that account it is bound by and arranged insensibly within the cone of direct vision so that a person seeing by means of refraction will think he is seeing directly, and this [additional range of oblique rays] surrounds the cone of direct vision on the left and right sides. We have already adduced an example of this [in book 4, proposition 9]. But it is otherwise in the case of another kind of refraction, for when a person divides the visual power between both eyes, and when he looks with effort, he subverts the eyes, and they do not maintain their proper disposition, as arranged by nature, but are irregularly inclined, [147] so the images of things become extraordinary and appear to lie far from their actual locations, and because of the doubling of the eyes, the image of a single thing will be seen double. Here is an illustration.

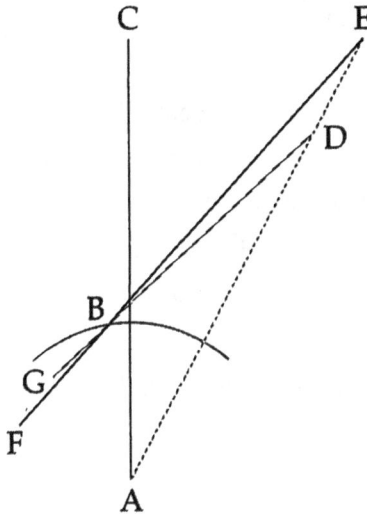

Figure 6.2

   Let A [in Figure 6.2] be the center of the eye and AC the centric ray. Let the image of a visible object reach from the left to the right beyond the centric ray; let the object be E and the pupil B. The line conveying the image will be EF, and it will be refracted to G. When the line of refraction is extended out, it will reach D. Draw the cathetus from center point A through E, and it will intersect the extended refracted ray GD at D. This is where the image of the object will lie.[26]

PROP. 3: *The image of a single magnitude that is obliquely incident to both eyes in front of the centric ray appears doubled.*

Diximus simplicem esse naturalem videndi virtutem, et simpliciter intuendo simplicem[8] rem binis oculis conspectam simplicem videri. At si naturalem hanc virtutem dividere velimus, quod non eveniet sine conamine, distorquendo situm oculorum, ut unicuique oculo suae virtutis pars ascribatur, contingit rem unicam duplicatam a duobus oculis videri, cuius rei exemplum.

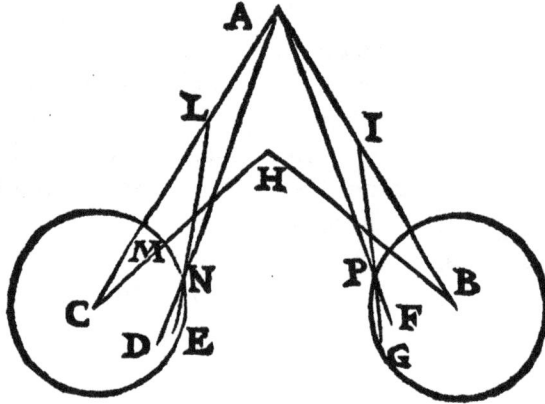

Sit visa magnitudo A, oculorum centra B, C, et inspiciant firmiter punctum H iuncti. Oculos contorquendo, conantes utroque oculo H videre. Tunc punctus A cadit dextro oculo in ND. Refrangitur in contrariam partem a centrica HC[9]; pervenit in E. **[148]** Extenditur; erit EL. Trahatur cathetus ex C ad A; occurret EL in L. Igitur A a dextro oculo videbitur in L.

Eodem modo cadat A sinistro oculo citra centricam in P. Descenderet in F, sed refrangitur in G. Extenditur in I.[10] Deducta catheto ex B et A, occurret extensae prius lineae in I. Sic punctus A videbitur a sinistro oculo in I. Sic clauso oculo dextro C, non videbitur L, et clauso B, abscondetur I. Sed oportet ut centrica firmiter aspiciat H, et ambobus reseratis, videbuntur I, L, et A peribit. At si oculi in suum situm redeunt, et oculi consueto more aspiciant, peribunt I, L, et solum A videbitur. Et si denuo virtus dividatur, descendet A ad I et L.

Simplicem magnitudinem utrique oculo cis centricam oblique incidentem geminatam videri. PROP. 4

---

[8]"semplicem" in original text.
[9]"HE" in original text.
[10]"L" in original text.

We have argued that the faculty of sight is naturally undivided, and when it looks in an undivided way, an individual object viewed by both eyes in a single glance appear single. But if we want to divide this natural power by contorting the position of the eyes, which will not happen without effort, and if we therefore assign part of its power to each eye, it happens that a single object appears double to the two eyes, an illustration of this point being as follows.

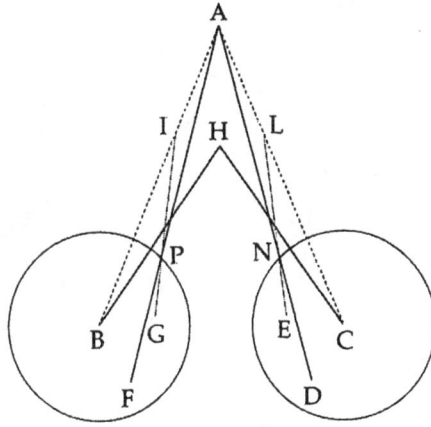

Figure 6.3

Let A [in Figure 6.3] be the visible magnitude, B and C the centers of the eyes, and let them join together in looking steadily at point H [along centric rays CH and BH]. With their position distorted, both eyes strain to see H. In that case the image of point A falls along ND to the right-hand eye. It is refracted away from the centric ray HC, and it reaches E. [148] When [the refracted ray from E] is extended, it will form EL. Draw the cathetus from C to A and it will intersect EL at L. Therefore, A will be seen at L by the right-hand eye.

Likewise, let the image of A reach the left-hand eye at P in front of the centric ray. It should continue to F but is refracted to G. [The resulting refracted ray GP] is extended to I. Thus, when the cathetus is drawn from B to A, it will intersect the previously extended line at I. Point A will therefore be seen at I by the left-hand eye. Hence, when the right-hand eye at C is closed, L will not be seen, and when the left-hand eye at B is closed, I will disappear. But the centric ray has to look steadily at H, and when both eyes are open, I and L will be seen, but A will disappear. If the eyes return to their natural situation, however, and if the eyes look in their usual way, I and L will disappear, and A alone will be seen. And if the power is divided again, A will drop down to I and L.[27]

PROP. 4: *The image of a single magnitude appears double to both eyes when it strikes the eye obliquely outside the centric ray.*

Possumus et alio modo simplicem magnitudinem oblique oculis incidentem geminatam videre, differtque a superiori, quod hic clauso oculo dextro peribit sinistra imago, et clauso sinistro peribit dextra. Idque eveniet quod ibi citra centricam, hic cis oblique oculis incidit magnitudo. Exemplum. [149]

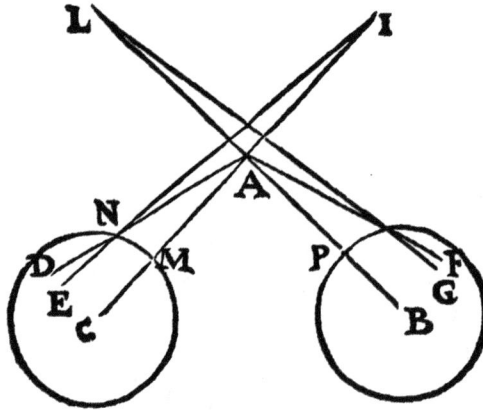

Sit oculus dexter C, sinister B, res videnda A; cadat cis centricam CA in pupillam N, et sit AN. Non venit in D, sed refrangitur in E; extenditur in I. Trahatur cathetus ex CA, occurret extensae in I. Igitur in I videbitur magnitudo A. Occurrat eodem modo A sinistro oculo per O pupillam. Venit non in F, sed refrangitur in G; extenditur in L. Ducta catheto BA, occurret extensae in L; sic A videbitur in L. Sic clauso oculo B, aboletur L, et clauso C, peribit I. Sic virtutem dividendo vi, videbitur A dividi, et ad duplicatum iter accingitur, nam ab oculo C videbitur egredi ad I,[11] et ab oculo B egredi ad L; et virtutem uniendo et oculos ad usum convertendo, I et L ad A redeunt.

Utrique oculo duplicem magnitudinem oblique incidentem simplicem videri.
PROP. 5

Haec conversa est praecedentis, ibi enim simplex magnitudo duplex videbatur; hic duplex simplex videbitur. Causa est quod in contrarias partes axis cadunt lineae. Exemplum. [150]

---

[11]"L" in original text.

We can also see the image of a single magnitude double in another way when it strikes the eyes obliquely, and it differs from the previous one because in this case, when the right-hand eye is shut, the left-hand image will disappear, whereas when the left-hand eye is closed, the right-hand image will disappear. This will happen because in the previous case the image of the magnitude reaches the eyes obliquely inside the centric ray, whereas in this case it reaches the eyes obliquely on the other side. Here is an illustration. [149]

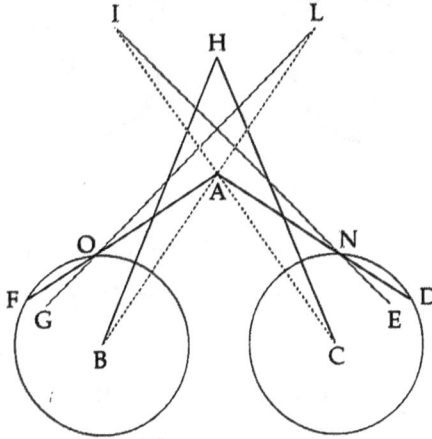

Figure 6.4

Let C [in Figure 6.4] be the right-hand eye, B the left-hand one, and A the visible object. Let a ray from A reach N on the pupil outside of centric ray CA,[28] and let it be AN. It does not reach D, but is refracted to E, and [the refracted ray EN] is extended to I. Draw the cathetus along CA, and it will intersect the extended refracted ray at I. Magnitude A will therefore be seen at I. Likewise, let the image of A reach the left eye through O on the pupil. It does not reach F but is refracted to G, and [the refracted ray] is extended to L. When cathetus BA is drawn, it will intersect the extended refracted ray at L, so A will be seen at L. Accordingly, when the eye at B is closed, L will disappear, and when the eye at C is closed, I will disappear. Thus, by dint of a divided power, A will appear divided, and it is reached by two paths, for it will appear to shift to I by the eye at C and to shift to L by the eye at B, and by unifying the power and turning the eyes to their normal function, I and L return back to A.[29]

PROP. 5: *A doubled magnitude lying at a slant to both eyes can appear single.*

This is the converse of the preceding theorem, for in that case a single object appeared double, whereas here a double object will appear single. The reason is that the lines of incidence fall on opposite sides of the axis. Here is an illustration. [150]

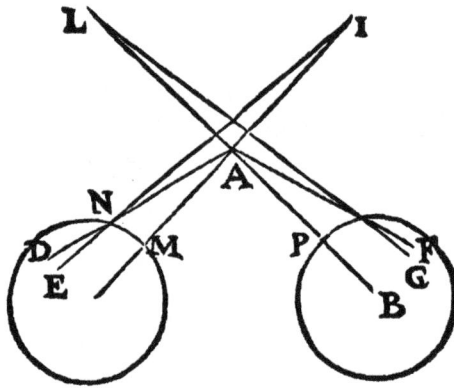

Sit una magnitudo visa I, cadat ad oculum dextrum C citra centricam in N, et sit IN. Veniet in E, sed refrangitur in D; extenditur in A. Cathetus vero IC occurret DN in A; punctus igitur I videbitur in A. Alter punctus L eodem modo cadat in O pupillam, et sit LOG. Refrangitur in F; extenditur in A. Et a puncto B et L extendatur cathetus BL. Occurret OA in A; punctus igitur L videbitur in A. Duplex igitur magnitudo I, L videbitur simplici loco a binis oculis in A. Sic clauso oculo B, aboletur A dimidium, clauso vero C, peribit A dimidium. Et si I magnitudo nigra fuerit, L alba, videbitur magnitudo in A alba, nigra; scilicet alba ex parte B, nigra ex parte C.

Utrique oculo binae magnitudines recte et oblique incidentes, ternae videntur.
PROP. 6

Si recte oculus unam aspiciat magnitudinem refracte vero alteram; alter vero oculus refracte unam aspiciat, duae illae magnitudines tres videntur. [151]

Sit dexter oculus C; aspiciat directe A; et oblique B punctum, qui cadat cis centricam in D pupillam. Veniret in M, sed frangitur in E; extenditur in F. Extensa catheto, occurret EF in F; punctus ergo B videbitur in F. Idem punctus B cadat oblique oculo H per pupillam O; relabitur in I.[12] Extendatur

---

[12]"L" in original text.

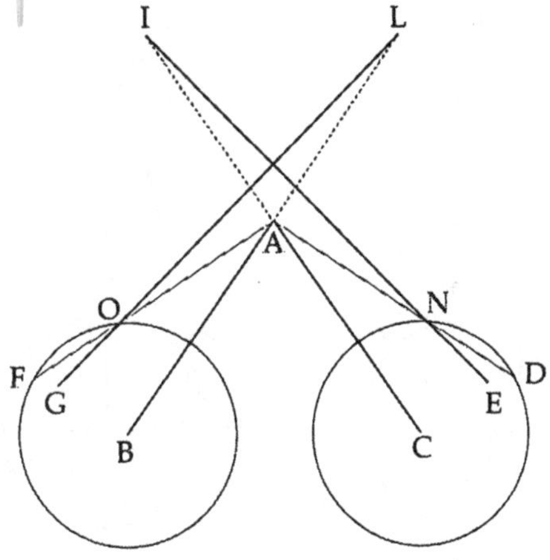

Figure 6.5

Let the image of one magnitude I [in Figure 6.5] reach the right eye C at N, outside the centric ray [CA], and let the ray [conveying the image] be IN. It will continue toward E [if unrefracted], but it is refracted to D, and [the resulting refracted ray DN] is extended to A. Cathetus IC, however, will intersect the extension of DN at A, so point I will be seen at A. Similarly, let the image of the other point L reach the pupil at O, and let the ray [conveying it] be LOG. It is refracted to F, and [the resulting refracted ray FO] is extended to A. From point B and through L let cathetus BL be extended. It will intersect OA at A, so point L will be seen at A. Consequently, the two magnitudes I and L will be seen by both eyes at a single location A. Hence, when eye B is closed, half of the doubled image A is done away with, and when eye C is closed, half of the doubled image A will disappear. Moreover, if magnitude I is black and L white, the magnitude at A will appear both black and white, that is, white on the side of B and black on the side of C.[30]

PROP. 6: *When the images of two magnitudes reach both eyes directly and obliquely, they appear tripled.*

If one eye sees one magnitude directly but another one by means of refraction, and if the other eye sees one of the magnitudes by means of refraction, those two magnitudes appear tripled. [151]

Let C [in Figure 6.6] be the right-hand eye, and let it see magnitude A directly [along centric ray CA]; let it see point B obliquely, and let B's image reach point D on the pupil outside the centric ray. It would continue to M, but it is refracted to E, and [the resulting refracted ray ED] is extended to F. When the cathetus [CB] is drawn, it will intersect EF at F, so point B will be seen at F. Let the image of the same point B reach the eye at H obliquely through point O on the pupil, and it is bent toward I. Extend

IG, occurret catheto exeunti ex H ad B in puncto G; punctus ergo B videbitur in G. Sic A, B magnitudines videbuntur G, F, A ter. Sic clauso oculo H, occultabitur G, et videbitur A, F a dextro. Clauso vero C ascondetur A, F, et videbitur G, A.

Ut binae magnitudines duplicatae videantur. PROP. 7

Utrique oculo potest magnitudo altera recte, altera oblique occurrere, sic binae magnitudines quatuor videbuntur. Exemplum.

Sit dexter oculus H, sinister C, et magnitudines binae A et B. Uterque oculus directe et aequidistanter suam magnitudinem in opposito existentem

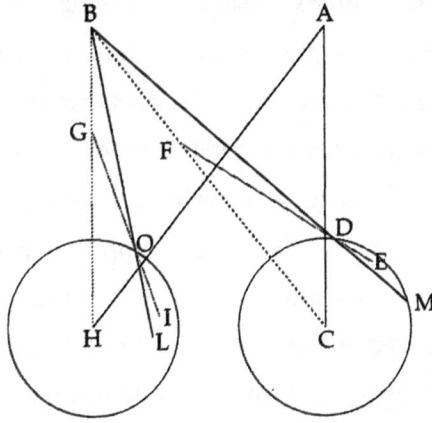

Figure 6.6

[refracted ray] IG, and it will intersect the cathetus dropped from H to B at point G, so point B will be seen at G. Accordingly, magnitudes A and B will appear tripled at G, F, and A. When the eye at H is closed, therefore, image G will be hidden, and images A and F will be seen by the right-hand eye. On the other hand, when the eye C is closed, A and F will disappear, and G and A will be seen.[31]

PROP. 7: *How each of two magnitudes might appear doubled.*

The image of one magnitude can reach each eye directly and that of another obliquely, so two magnitudes will appear quadruple. Here is an illustration.

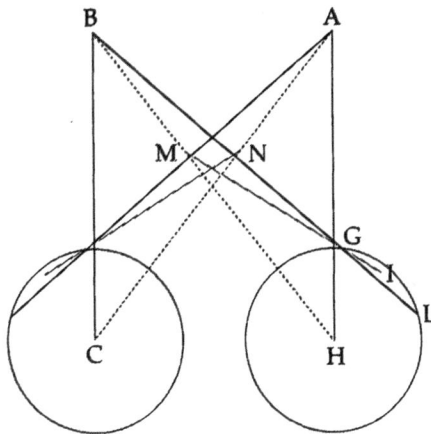

Figure 6.7

Let H [in Figure 6.7] be the right-hand eye, C the left-hand one, and A and B the two magnitudes. Let each eye see the magnitude that faces it

aspiciat, oculus H, A, oculus C ipsam B. Oblique vero B occurrat oculo dextro H cis centricam in pupilla G; descenderet in L, sed lex refractionis contorquet in I. Dirigitur in rectum in M; ducta catheto ex H,[13] B, occurrit iam directae in M. Sic punctus B videbitur in M. Eodem modo A videbitur [152] in N ab oculo C. Et quanto oculi vehementius contorquentur virtusque visiva intenditur, eo puncta M, N se magis appropinquabunt. Sed maxime animadvertendum quod magnitudines M, N videntur minus coloratae et languidae, quia per refractionem videntur. Et si A[14] magnitudo nigra erit et B[15] alba, videbitur N nigra, et M alba. Et si H oculum clauseris, peribit A, M[16]; si C, ipsa N, B.[17]

Binae magnitudines utrique oculo oblique cadentes quatuor videntur. PROP. 8

Possumus et alio modo demonstrare ut binae magnitudines oculis nostris quaternae videantur, aliter quam superiori sermone retulimus, nam ibi altera recta, altera obliqua cernabatur, hic vero utraque magnitudo utrique oculo oblique cadit, et si altera alba, altera nigra sit, cohaerentes alba nigra media videbitur, cuius exemplum. [153]

Esto dexter oculus G, sinister H; magnitudo obiecta ipsi G sit A alba, et magnitudo obiecta ipsi H nigra sit B. Oculus igitur G aspicit B; cadit in pupillam P oblique. Veniret in N, sed labitur in M ex refractione. Extendatur

---

[13]"N" in original text.
[14]"B" in original text.
[15]"A" in original text.
[16]"N" in original text.
[17]"M, A" in original text.

directly and along parallel lines, eye H seeing A directly [along centric ray HA] and eye C seeing B directly [along centric ray CB]. On the other hand, let the image of B reach the right-hand eye H obliquely at G on the pupil to the outside of the centric ray; it would continue to L except that the law of refraction bends it to I. [The resulting refracted ray IG] is directed straight to M, and when the cathetus is drawn from H to B, it intersects the straight line just drawn at M. Thus, point B will be seen at M. Likewise, A will be seen [152] at N by the eye at C. In addition, the more radically the eyes are turned inward toward M and N and the visual power is strained, the more points M and N will approach one another. It must be strongly emphasized, however, that magnitudes M and N appear less vividly colored and weaker [than A and B] because they are seen by means of refraction. If, moreover, magnitude A is black and B white, N will appear black and M white. Also, if you close the eye at H, A and M will disappear, whereas if you close the eye at C, N and B will disappear.[32]

PROP. 8: *The images of two magnitudes reaching both eyes obliquely appear quadrupled.*

We can show in a way other than the one we set forth above how two magnitudes can appear quadrupled to our eyes, for in the previous case one magnitude was seen directly and the other obliquely, but in this case the images of both magnitudes reach both eyes obliquely, and if one is white and the other black, they will appear as an intermediate black-and-white combination, an illustration of which is as follows. [153]

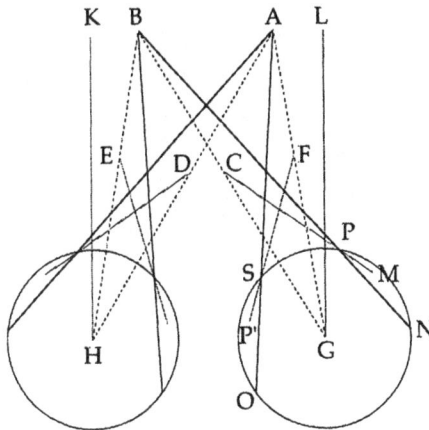

Figure 6.8

Let G [in Figure 6.8][33] be the right-hand eye and H the left-hand one, let visible magnitude A [closer to] G be white, and let the visible magnitude B [closer to] H be black. Eye G therefore sees B, but its image falls obliquely to P on the pupil. It would continue to N, but it is diverted to M by refraction.

ex P in C, et cathetus ex G, B occurrat ei in C. Videbitur igitur nigra magnitudo B in C nigra. Eidem oculo G etiam oblique cadat magnitudo alba A in pupillam S. Veniret in O, sed frangitur in P. Extendatur PS in F, trahaturque cathetus ex A, G. Occurret PF in F; igitur magnitudo alba A videbitur in F. Eodem modo ab oculo H magnitudo alba A videtur in D alba, et magnitudo nigra B videtur in E nigra. Unde si claudatur oculus G, aboletur alba F et C nigra, et si claudetur oculus H, peribit D alba et E nigra. At si intentius spectabuntur A, B, et virtu[18] conetur dividi, magnitudines apparentes in C, D coibunt, et videbitur alba. nigra, scilicet alba ex parte D, nigra ex parte C.

Si tres magnitudines in longitudine lineae ponantur in oculorum prospectu, apparebunt quinque. PROP. 9

Si in longitudine regulae quae oculis perpendiculariter opponatur tres chylindri opponantur, et medius directe [154] oculis cadat, extremi autem oblique, chylindri quinque conspicientur.

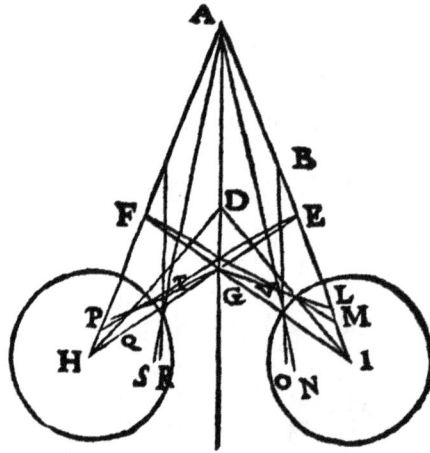

Sit regula ADG, tres chylindri in ea siti A, D, G, sinister oculus I, dexter H. Inspiciat H dexter oculus directe D, et cadat a puncto A ad pupillam T linea AT. Veniet in S; refrangitur in R. Extendatur TR in C; ducatur cathetus HA. Occurrit ei in C; oculus ergo H videbit A in C. Sic postremus chylinder G occurrat eidem oculo H, per pupillam T. Veniet in P; refractio contorquet

---

[18]"virsu" in original text.

Extend [the resulting refracted ray MP] from P to C, and the cathetus drawn from G to B should intersect it at C. Consequently, black magnitude B will appear black at C. Let the image of white magnitude A also fall obliquely to the same eye G at point S on its pupil. It would continue to O, but it is refracted to P'. Let [refracted ray] P'S be extended to F, and draw the cathetus from A to G. It will intersect P'F at F, so white magnitude A will be seen at F. Likewise, from eye H white magnitude A appears white at D, and black magnitude B appears black at E. Therefore, if the eye at G is closed, white image F and black image C disappear, and if the eye at H is closed, white image D and black image E will disappear. Moreover, if A and B are looked at intently, and if an effort is made to divide the visual power, the magnitudes appearing at C and D will combine, and they will appear white and black, that is, white on the side of D and black on the side of C.[34]

PROP. 9: *If three magnitudes are placed along a line within the field of view of the eyes, they will appear as five.*

If three cylindrical pegs are placed along the length of a ruler that stands perpendicular to the [line joining the centers of the] eyes, and if the image of the middle one reaches [154] the eyes directly while the ones at the ends reach them obliquely, five cylinders will be seen.

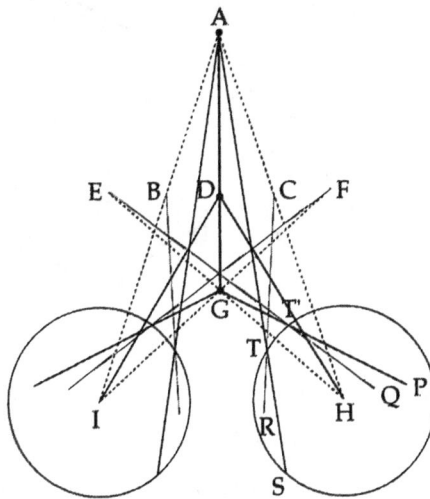

Figure 6.9

Let ADG [in Figure 6.9] be the ruler, A, D, and G the three cylindrical pegs located on it, I the left-hand eye, and H the right-hand one. Let the right-hand eye H view D directly [along centric ray HD], and let radial line AT fall from point A to point T on the pupil. It would continue toward S but is refracted toward R. Extend TR to C, and draw cathetus HA. It meets TR at C, so the eye at H will see A at C. Likewise, let the image of the last cylindrical peg G reach the eye at H through T' on the pupil. It will continue

ad Q. Extendatur QT ad E; cathetus ducatur ex H, G. Occurret prius extensae in E; igitur G in E videbitur. Eodem modo oculus I videbit A in B et G in F. Clauso oculo H, peribit chylinder C, E, et clauso I, peribit B, F.

Si bini oculi obiectae regulae remotiorem partem recte aspexerint, oblique proximiorem ex interiori parte, bifida videbitur. PROP. 10

Si regula oculis recte opponatur, at pars una naso, altera in rectum porrigatur mensura unius cubiti, et oculi recte remotiorem partem aspexerint, oblique inferiorem, dico inferiorem [155] partem naso propinquam bifidam videri, cuius causa est quod ea pars quae oculis directe aspicitur, suo loco videtur, quae autem obliqua non suo loco.

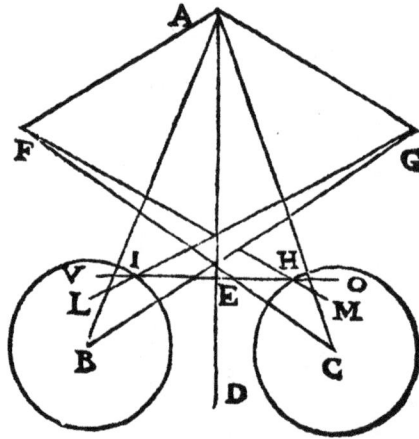

Esto regula AED naso D opposita perpendiculariter; oculus dexter B, sinister C. Inspiciat uterque directe A remotiorem partem; quippe suo loco videbitur radio centrico CA, BA. Sed pars E venit oblique ad oculum B per pupillam I.[19] Veniret in N, sed frangitur in L. Extenditur in G; occurret catheto EB in G. Ergo punctus E videbitur in G, et linea EA videbitur AG. Eodem modo punctus E oculo C videbitur in F, et AE videbitur AF, unde AG et AF bifida apparet. Clauso oculo B, abscondetur AG, et clauso C, abscondetur AF.

---

[19]"L" in original text.

toward P, but refraction bends it toward Q. Extend QT' to E, and draw the cathetus from H to G. It will intersect the previously extended ray QT' at E, so G will be seen at E. In the same vein, the eye at I will see A at B and G at F [and D at D along centric ray ID]. When the eye at H is closed, cylindrical pegs C and E will disappear, and when the eye at I is closed, cylindrical pegs B and F will disappear.[35]

PROP. 10: *If both eyes look at the farther end of a ruler directly and the nearer end on the inner side obliquely, the ruler will appear split.*

If a ruler is posed straight toward the eyes and extends from the nose on one side to a distance of one cubit on the other, and if the eyes look straight at the farther endpoint and obliquely at the nearer one, I say that the lower [155] part that is nearer the nose appears split, the reason being that the end the eyes view directly is seen in its actual location, whereas the end viewed obliquely is not seen in its actual location.

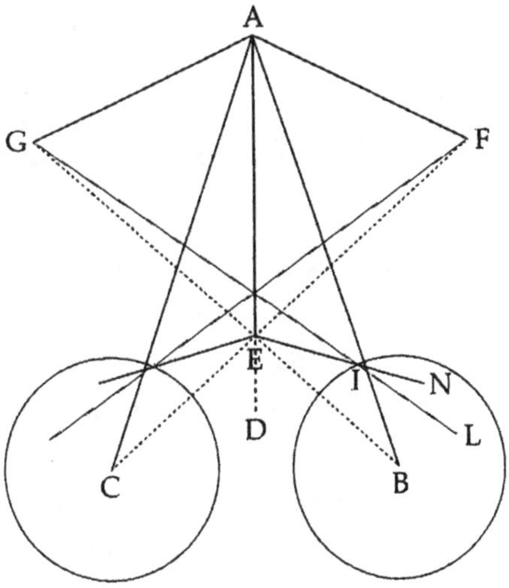

Figure 6.10

Let AED [in Figure 6.10] be a ruler posed orthogonally to the nose at D, and let B be the right-hand eye and C the left-hand one. Let both eyes view the farther end A directly; it will of course be seen in its actual location by centric rays CA and BA. But the image of the segment at E reaches the eye at B obliquely through point I on the pupil. It would reach N but is refracted to L. [The resulting refracted ray LI] is extended to G and will intersect [extended] cathetus EB at G. Point E will therefore be seen at G, and line EA will be seen as AG. Likewise, point E will be seen at F by the eye at C, and AE will be seen as AF, so AG and AF appear split. When the eye at B is closed, AG will disappear, and when the eye at C is closed, AF will disappear.

Si bini oculi obiectae regulae proximiorem partem aspexerint directe, oblique vero remotiorem, remotior pars videbitur bifida. PROP. 11

Regula oculis opposita ex regione nasi ut una pars naso alia in rectum porrigatur naso perpendiculariter, et oculi inferiorem naso partem proximiorem aspexerint, superior lineae pars oblique inspecta, bifida videbitur. Ratio est quae oculus recte suo loco; quae oblique non suo aspicit loco. Exemplum. [156]

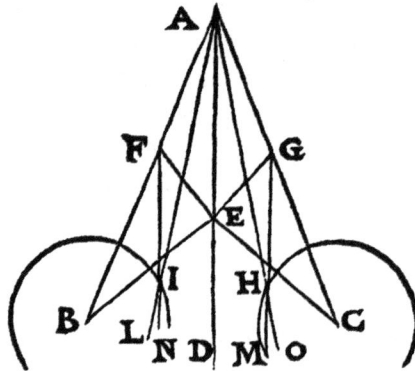

Esto regula oculis opposita AED, dexter oculus B, sinister C. Uterque directe aspiciat punctum E; suo loco E videbit. Oblique vero aspiciat A. Cadat linea incidentiae dextro oculo AI. Veniret in L; labitur in N. Extendatur NI in F[20]; trahatur cathetus ex BA, occurret lineae NF[21] in F. Punctus ergo A videbitur[22] in F, et iniunctis EF, linea AE videbitur EF. Eodem modo cadat oculo dextro C punctus A per H. Veniret in O; refrangitur in M. Extenditur, et occurret catheto CA in G. Punctus A videbitur in G, et ducta linea GE, videbitur AE EG. Sic linea EA ab utroque oculo videbitur EF,[23] GE,[24] et clauso oculo C, aboletur linea GE; clauso vero B, peribit FE.[25]

Si bini oculi obiectae regulae mediam partem aspexerint recte, at inferiorem superiorem quam oblique, decussata in medio videbitur. PROP. 12

Haec propositio ex utraque composita est, nam prius in remotiori [157] sui parte recte inspecta, in inferiori bifida videbatur, et post inferiore recte

---

[20]"G" in original text.
[21]"BA" in original text.
[22]"videbirur" in original text.
[23]"EA" in original text.
[24]"GA" in original text.
[25]"FA" in original text.

PROP. 11: *If both eyes look at the nearer end of the ruler directly and the farther end obliquely, the farther end will appear split.*

If the ruler is posed in line with the nose so that one end is at the nose and the other extends orthogonally from the nose, and if the eyes look [directly] at the lower end by the nose, whereas the upper end of the line is viewed obliquely, it will appear split. The reason is that the end the eye looks at directly is seen in its actual location, whereas it does not see the end it views obliquely in its actual location. Here is an illustration. [156]

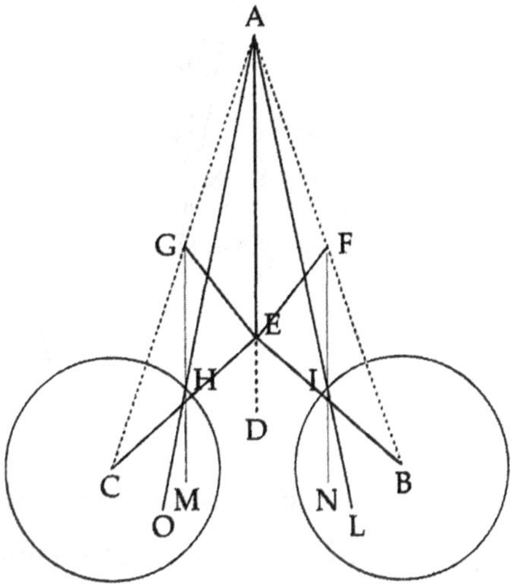

Figure 6.11

Let AED [in Figure 6.11] be the ruler facing the eyes, B the right-hand eye, and C the left-hand one. Let both eyes view point E directly [along centric rays CE and BE], and each will see E in its actual location. On the other hand, let each of them view A obliquely. Let line of incidence AI reach the right-hand eye. It would continue to L, but is diverted to N. Extend [the resulting refracted ray] NI to F. Draw the cathetus along BA, and it will intersect line of refraction NF at F. Point A will therefore be seen at F, and with EF joined, line AE will appear as EF. Likewise, let the image of point A reach the left-hand eye at C through H. It would continue to O, but it is refracted to M. [The resulting refracted ray HM] is extended, and it will intersect cathetus CA at G. Point A will be seen at G, and when line GE is drawn, AE will appear as EG. Hence, line EA will appear as EF and GE with both eyes, and when the eye at C is closed, line GE disappears, whereas when the eye at B is closed, FE will disappear.

PROP. 12: *If the two eyes look directly at the middle of a ruler and obliquely at the lower and upper ends, it will appear broken in the middle.*

This proposition is combined from both the previous ones, for earlier when the farther end [157] of the ruler was viewed directly, it appeared split at

inspecta superior divisa. Nunc inspecta media sui parte, utraque pars superior
et inferior bifida videbitur, veluti in medio decussata.

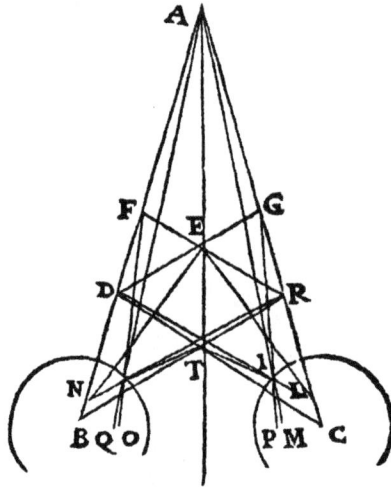

Esto recta linea visa AET, sinister oculus C, dexter B. Aspiciat sinister
C directe punctum E; videbit illud suo loco E. Pars longinquior A, ut diximus,
cadat ultra centricam in dextram oculi partem I. Veniet in M; reflectitur in
P. Extenditur in G ipsa PI[26]; cathetus erigatur ex re inspecta A et centro C.
Occurret PG in G; ergo pars A videbitur in G. Pars regulae inferior T veniet
in pupillam I; procederet in L, sed refractione labitur in S. Extendatur IS;
venit in D. Mox ex re visa T et oculi centro C[27] extendatur cathetus CT
quousque occurret lineae SD. Occurret in D; ergo punctus T videbitur in D.
Ergo linea GED videbitur linea AET ab oculo C. Eodem modo a dextro oculo
B punctus A videbitur in F et T in R, et linea AET videbitur FER, decussantes
se in E. Sic clauso oculo sinistro C, peribit linea GED, et clauso dextro B,
peribit linea FER.

---

[26]"PH" in original text.
[27]"T" in original text.

the bottom, and subsequently, when the bottom end was looked at, it appeared broken at the top. In this case, when the middle is viewed directly, both the upper and lower ends will appear split, as if the ruler were broken in the middle.

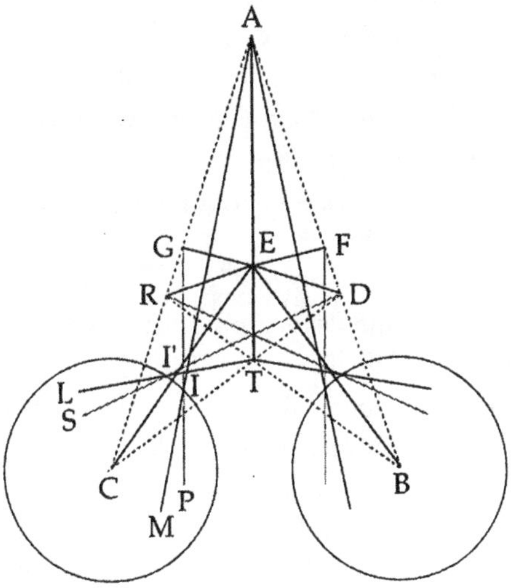

Figure 6.12

Let AET [in Figure 6.12] be the straight line viewed, C the left-hand eye, and B the right-hand one. Let the left-hand eye C look directly at point E, and it will see it at its actual location E. As we said before, the image of the farther point A may fall beyond the centric ray to spot I on the right side of the eye. It will continue toward M but is refracted toward P. Refracted ray PI is extended to G; drop the cathetus from the visible object A and center point C of the eye. It will intersect PG at G, so endpoint A will appear at G. The image of lower end T of the ruler will reach I' on the pupil, and it should continue toward L but is diverted by refraction to S. Extend refracted ray I'S, and it reaches D. Then extend cathetus CT from visible object T and center point C of the eye until it intersects line SD. It will intersect at D, so point T will be seen at D. Line AET will therefore appear as line GED to the eye at C. Likewise, point A will be seen at F and point T at R by the right-hand eye at B, and line AET will be seen as FER, the two cutting one another at E. Consequently, when the left-hand eye at C is closed, line GED will disappear, whereas when the right-hand one at B is closed, line FER will disappear.

NOTES

1. *Problems*, III, 30, in Aristotle (1552 [as III, 29]), 8va.

2. The"power" (*vis*) referred to here is the visual power physically projected out from the eye to the object—in the form of visual rays ('ὀψειξ) according to the original Greek text in 875b13; see Robert S. Mayhew, trans., *Aristotle's* Problems, II (Cambridge, MA: Harvard University Press, 2011), 130.

3. This entire section after the direct quotation is a close paraphrase of *Problems*, III, 30, in Aristotle (1552), 8va–b

4. *Problems*, XXXI, 11, in ibid., 49rb

5. *Problems*, I, 123, in Alexander of Aphrodisias (1552), 99vb.

6. See *On the Usefulness of the Parts*, X, 12, in Galen (1541), 179v.

7. See Avicenna, *Liber de anima*, III, 8, in Avicenna (1508), 16rb–17rb.

8. A faculty of the soul according to Aristotle and his followers, the common sense is supposed to lie either in the heart or at the forefront of the brain, where all the individual sense data are combined into a sensible representation of what has been seen, heard, felt, etc. It is here that the act of sensation is complete in preparation for the subsequent process of perceptual judgment. As Della Porta makes clear in proposition 7 of the following book, he follows Aristotle literally in locating the common sense in the heart rather than in the brain.

9. Avicenna is adverting here to the oculogyral illusion, which arises when a viewer spins swiftly, then stops suddenly, and sees his surroundings swirl about him.

10. See Albertus Magnus, *Liber de sensu et sensato*, I, 11, in Albertus Magnus (1517b), 7rb–7va, also in Albertus Magnus (1890b), 28.

11. See Alhacen, *De aspectibus*, I, v, 27, in Alhacen (1572), 16–7, and in Alhacen (2001), 53–56 (Latin) and 376–78 (English). For Alhacen, as well as his Latin followers, the final sensor (*ultimus sentiens*) is what completes the act of brute visual sensation before that of perception; although Alhacen is unclear where this faculty actually lies (whether in the common sense at the forefront of the brain or at the optic chiasma), Bacon is quite explicit in locating it at the optic chiasma; see *Perspectiva*, I, ii, 2, in Bacon (1996), 62–65.

12. See Witelo, *Perspectiva*, III, 37, in Witelo (1572), 102–3, and in Witelo (1991), 336–38 (Latin) and 151–53 (English).

13. Consequently, images L and N will correspond with their object points A and C, from which it follows that image LN will have the same left-to-right orientation as its object. In short, it will not be inverted. This analysis is based on the supposition that when an object is seen by"corresponding rays" within the two radiant cones within the eyes, that object will appear single. Hence, in this particular analysis, the axis for the cone within eye GHI is HB, whereas the axis for the cone within eye DEF is EB. Accordingly, since the two axial rays converge on B, that point appears single. Meantime, since DA and GA are corresponding rays within their respective cones, that is, symmetrically to the left of their respective axes, they will yield a single image of point A. Likewise, corresponding rays FC and IC lying symmetrically

to the right of their respective axes, will yield a single image of C. Consequently, the entire line ABC, with all its intermediate points, will appear single. This approach to image fusion traces back to Ptolemy's analysis of diplopia in books 2 and 3 of his *Optics*; see Ptolemy (1996), 83–90 and 141–50. Its proximate source, however, is Alhacen's account of image-fusion and diplopia in Alhacen (2001), 562–86.

14. Della Porta thus recognizes the fact of ocular dominance—that is, that most of us favor one eye over the other—although he overgeneralizes the preference for the right-hand eye.

15. In Della Porta's day, and well before, it was generally assumed that the medium through which phantasms and intellectual images or species were conveyed to and through the brain was animal spirit(s) (*spiritus animalis/animales*), which were distilled from arterial blood charged with vital spirit(s) (*spiritus vitalis/vitales*), the medium that provided heat and life to all animals. Nutrition, in turn, was provided by venous blood, which was charged with natural spirit(s) (*spiritus naturalis/naturales*) and which gave rise to arterial blood after respiration. Each type of spirit was associated with a specific organ: natural spirit with the liver; vital spirit with the heart; and animal spirit with the brain. Within the context of this physiological model, then, Della Porta's appeal to natural spirits for perceptual and intellectual functions is peculiar and peculiarly problematic. For a succinct contemporary account of the three types of spirits, see Jean Riolan the elder, *Abridgements of the Whole of Medicine*, in Riolan (1598), 21r–22r.

16. In *De Fabrica*, IV, 4, for instance, Andreas Vesalius claims that, after careful examination of the optic nerves from several large animals, he had never observed the hollow that Galen insisted was there; see Vesalius (1568), 329.

17. Vesalius makes this point in *De Fabrica*, IV, 4, in ibid, 329, the argument being that, although they intersect at the optic chiasma, the optic nerves do not actually unite there as was supposed by Galen and his various followers, including Alhacen and Witelo.

18. *Problems*, I, 124, in Alexander of Aphrodisias (1552), 99vb.

19. In other words, point A, to the left of midpoint B of the object will be seen at O, to the right of midpoint P of the image. The same holds *mutatis mutandis* for point C, to the left of midpoint B; it will be seen to the right of midpoint P of the image, so the image will be inverted.

20. The point of these two examples illustrated by Figures 6.1b and 6.1c is presumably to show that this method of analyzing strabismus flies in the face of Alexander of Aphrodisias's claim that those who are congenitally cross-eyed see single rather than double.

21. See Galen, *On the Usefulness of the Parts*, X, 12, in Galen (1541), 180r.

22. Della Porta's description of this experiment is drawn directly from ibid., 180r–v.

23. In other words, the proper location of the object is determined according to the dominant eye, which sees the object head-on according to direct radiation, which is not refracted in the eye. The other eye sees it at a slant, however, so the incoming rays, being oblique, are refracted in the eye. It is therefore this refraction that makes the object appear dislocated.

24. Here we have another appeal to Ockham's Razor; see note 12 to book 5.

25. See, for example, Galen, *On the Usefulness of the Parts*, X, 5, in Galen (1541), 177r.

26. Note that in this case Della Porta assumes that in passing from air into the denser aqueous humor of the eye, the light refracts *away* from, rather than toward the normal, as it should; cf., for example, book 3, proposition 14. Note as well that locating the image location in front of its object contradicts what Della Porta attempts to establish in book 4, proposition 1. However, as will become evident in a moment, although it runs counter to the"laws" of refraction he has followed up to now, Della Porta adopts this counterassumption in order to explain various aspects of diplopia, or double vision. Presumably, this refraction away from the normal is the"other kind of refraction" to which Della Porta refers in the first paragraph, and it is evidently due to the unnatural conditions under which the visual power is divided.

27. This entire analysis is based on the assumption just discussed: namely, that when the rays from point A strike the surface of the cornea, they are refracted away from, rather than toward, the normal, as they should be, since they pass from a rarer into a denser medium. Della Porta was forced to make this assumption because of his insistence that all vision outside the cone of direct radiation is due to refracted rays and, moreover, that the resulting line of sight originates at the point on the crystalline lens where the refracted rays strikes. Accordingly, it was only by supposing that rays AN and AP refract away from the normal that he was able to account for the empirical fact that, when the eyes are focused to a point ahead of the visible object, the left-hand eye will see the object's image to the left of the point of focus, whereas the right-hand eye will see its image to the right of the point of focus. It is worth noting that Ptolemy, Alhacen, and Witelo explained this case of diplopia by supposing that the image of A appears to the right because it is seen along ray CLA to the right of centric ray CH, whereas the image of A appears to the left because it is seen by ray BIA to the left of centric ray BH, rays CLA and BIA thus being noncorresponding.

28. CA is in fact a cathetus, not a centric ray. The centric rays, which do not appear in the original, printed diagram, are HB and HC in Figure 6.4, both of them coming to convergence on point H beyond object point A instead of in front of it, as in the previous theorem.

29. Unlike the previous case, this one requires that the rays from A refract toward the normal, as they should. However, as both the diagram and text indicate, Della Porta supposes that the eyes are focused on object A so that CA and BA form centric rays. In fact, this case requires that the eyes be focused to some point H beyond A, CH and BH thus forming the appropriate centric rays. As Ptolemy and Alhacen would explain the phenomenon, then, the left-hand eye at B will see A at L along ray BAL to the right of centric ray BH, whereas the right-hand eye at C will see A at I along noncorresponding ray CAI to the left of centric ray CH. Accordingly, as proposed, the image of A will appear to the right of A from the perspective of left-hand eye B and to the left of A from the perspective of the right-hand eye at C.

30. Again, Della Porta is forced to suppose, against theory, that the rays from each object are refracted away from the normal when they pass from the rarer air

into the denser aqueous humor. Ptolemy and Alhacen would explain this combining of images according to the convergence of centric rays BAL and CAI at A. Accordingly, as they are seen by corresponding rays within their respective cones, I and L will be seen at the same point of fusion in the optic chiasma and will therefore appear superimposed on each other.

31. Unspecified by Della Porta, the key condition for this analysis is that both eyes focus on object A along respective centric rays CA and HA in Figure 6.6. Accordingly, A will appear single to both eyes because it is seen by corresponding rays within the respective cones. Consequently, when the eye at H is closed, its images G and A will disappear while the eye at C will continue to see its images F and A. On the other hand, when the eye at C is closed, its images A and F will disappear while the eye at H will continue to see its images G and A. Suffice it to say that, in order to explain the phenomenon, Della Porta is forced to have rays BD and BO refract away from the normal when they enter the aqueous humor.

32. In this case centric rays HA and CB are parallel, which is to say that both eyes are staring off to infinity. Consequently, eyes H and C see their respective objects A and B directly along the centric rays. Meanwhile, from the perspective of the eye at H the image of B appears at M, whereas from the perspective of the eye at C the image of A appears at N. Consequently, four images will be seen, two directly, and two by refraction, and if B were black and A white, the images would appear alternately white, black, white, and black in sequence from A through N and M to B. Furthermore, contrary to what Della Porta claims in the original, uncorrected Latin text, when eye H is closed, images A and M, not A and N, will disappear, and when eye C is closed, images B and N, not A and M will disappear. According to Ptolemaic and Alhacenian analysis, this quadrupling would be explained by the fact that the eye at H sees B along ray HB, which lies to the right of centric ray HA, whereas the eye at C sees A along ray CA, which lies to the left of centric ray CB. Again, Della Porta assumes that rays BNG and AM will be refracted away from the normal on entering the aqueous humor.

33. Since the original Latin text provides no diagram for this proposition, I have provided Figure 6.8 to reflect the conditions of the theorem.

34. In this case, the two eyes should be directing their centric rays to the sides of objects A and B, as represented by HK and GL in Figure 6.8. The resulting demonstration depends, again, on having the image refract into the eye away from, rather than toward, the normal. Della Porta's suggestion about how to make images C and D coincide requires that the viewer attempt to shift centric ray HK of the left-hand eye to coincide with HA and at the same time to shift centric ray GL of the right-hand eye to coincide with GB, in which case the two images will coincide where the two centric rays meet, as illustrated in proposition 5 above.

35. In both cases, when each eye is closed, the other eye will continue to see D, so its image will not disappear. In essence, then, six images, not five, are seen by both eyes, but since the images of D coincide, they count as a single image. A peculiarity of this theorem is that, whereas the two lines of radiation from A both refract in the wrong direction (i.e., away from the normal), the two lines of radiation from G refract toward the normal, as they should according to theory.

LIBER SEPTIMUS

PROOEMIUM

Supersunt adhuc de visione problemata quaedam, scilicet de his quae intra oculum fiunt et extra videri putantur, curioso discenti valde necessaria, ob id nos duximus explananda ut eiusmodi errores ad posteros in infinitum non propagentur. Multi enim sunt qui disquisitionum taedio praepediti quae ab aliis scripta sunt non considerantes, pro veris assumpta quae falsa erroneaque. Sic falsis innixa fundamentis falsissima sunt quae deducuntur. Quae conscribenda duximus haec sunt.

Colores, culices,[1] et alia quae intus in oculo sunt foris videntur. PROP. 1

In aliis quoque sensibus evenire videmus, ut quae intus sint foris existimemus. In auribus ex sufflato spiritu intus fieri videmus bombos, tonitus, et sibilos quos foris existimamus. [159] Sic et in oculis evenire videmus, affectis enim tunicis vel humoribus, multa foris conspiciuntur quae intus habemus. In hyposphagmate suffunditur ceratois sanguine ut aspicientes cuncta subrubra arbitrentur. Hypochyma, sive glaucosis, est in glaucum mutatio crystalloidis humoris prae humiditate, et in morbo quoque regio ovalis humor croceus evadit, ut idola per eosdem humores pertranseuntia crocea defaedataque conspiciantur omnia. Sic et si alius color eorum perspicuitatem interturbet, utpote plumbeus, fumosus, tenebrosusque, eiusdem coloris spectra refert. In suffusionibus pravi humores descendunt in pupillam, vel biliosi vapores sursum ducti, unde languentes exiguos culices, aranearum texturas, lanarum vellera ob oculos versari perperam vident. Et in cornea, si alicuius morbosae tabis vestigia subsident, vel uvea tunica quae substernitur ceratoidi aliquo foramine pervia erit, quae foris spectantur omnia, et perforata et fenestrata

---

[1]"culiccs" in original text.

BOOK SEVEN

INTRODUCTION

At this point there remain certain problems concerning vision, namely, problems involving things that occur inside the eye but are thought to be seen outside. These problems are crucial for any diligent person to learn about, and for that reason we have determined to explain them so that errors of this sort are not passed on indefinitely to our posterity. For there are many who, having gotten tangled up in tedious disquisitions written by others who lack due care, take things that are false and erroneous as true. Based, therefore, on false foundations, the things deduced from them are absolutely false. These things are what we have determined to write about.

PROP. 1: *Colors, floaters, and other things that are inside the eye appear to lie outside it.*

We see that the same thing also happens with the other senses, so that we judge things that are internal to lie outside us. From a puff of air inside our ears we perceive[1] booming sounds, thunder, and whistles that we judge to exist outside. [159] We see the same thing happen in our eyes, for when the tunics or humors are affected, many things are perceived to be outside that we actually possess internally. In the case of hyposphagma the cornea is suffused with blood so that viewers judge everything to be reddened.[2] Cataract, or glaucosis, is a change of the crystalline humor into a gray color in the presence of moisture, and in the disease of jaundice the aqueous humor[3] becomes yellow so that all the images passing through those humors may appear yellow and degraded. So also, if another color, such as a leaden, smoky, or dark color, muddles their clarity, the images take on the same color. In the case of suffusions, defective humors descend into the pupil, or bilious vapors are carried up into it so that those suffering this malady see faint gnats, spider webs, and tiny tufts of wool falsely coursing about in front of their eyes. And in the case of the cornea, if vestiges of any wasting disease remain, or if the uveal tunic lying beneath the cornea is perforated by some opening, everything seen outside will appear both perforated and pierced or hollow, for whatever we cannot see properly we judge to be

videbuntur, vel concava, nam quae videre nequimus concava putamus. Unde quamplurima intus sunt quae foris opinamur.

Splendorum imagines quaedam ante oculos versari apparent quae intra oculum sunt. PROP. 2

Contingit nobis saepe tussiensibus, aut evomentibus, aut in oculis percussis, ut splendorum imagines quasdam foris videamus quae tamen non foris sed intus habeamus; quorum quidem rationem antequam aperiamus maiorum placita referre non pigebit. Empedocles, ut diximus, ignem in oculis esse dicebat. Plato ignis "participes oculorum orbes" deos fecisse qui non ureret sed illuminaret. "Lumenque aliquod oculis inesse, etiam si exiguum. Animalia multa quae noctu vident argumento esse, quando in tenebris eorum coruscant oculi . . . et contortis quoque oculis nostris circulum quendam intus plane intueri." Sed Aristoteles id refellit: hoc non ignis ratione apparere sed ex politionis,

> nam quicquid laevigatum densumque est sui natura fulgorem habet, sed non eiusmodi ut actu medium lucidum efficere valeat. Unde quum nigricans illud in oculorum meditullio (crystallinum existimo), politum maxime sit, mirum [160] non esse debet si splendorem retineat . . . quod si celerius agatur oculus, fulgere percipiatur, in quiete non ita . . . Hoc ideo evenire, quia celeri et festina agitatione, quod unum simplexque est quodammodo duplum videatur. In eo enim motu nigricans in alium transfertur locum in quem fulgorem prius emiserat, et priusquam is deficiat videt, ut extrinsecus illatum quo fit ut videns et res visa unum existens, duo quodammodo censeantur. Quod si tardior esset motus, prius evanesceret coruscatio quam nigricans illud dilaberetur. Et in tenebris hoc praecipue fieri, quum interdiu luminis copia splendorem illum exiguum absorbeat evanidumque reddat.

Haec Aristotelis opinio aliquibus frivola visa est, quum ea pupillae claritas, neque etiam moto oculo ab opposito vidente unquam visa fuerit, neque simul esse posse ut suo existens peculiari loco sui fulgorem conspici possit tanquam alio loco residentem et unum esse videns et re visa. Quibus Alexander obiiciens respondit pupillam magnitudine non carere, fieri enim posse ut illius quota pars sit videns alia vero quae videatur. Idemque per diversas sui partes videns sit et visum, quum fieri non possit quod unum duo sint. Sed aliter sensit in suis *Problematibus*, existimans et ipse hanc rationem fallacem, et ad visivos spiritus refert:

> Abductio enim ventris et pectoris sursum humores et spiritus compellit, qui cum ad oculos veniunt, visionem conturbant depravantque ut foris illas imagines ementi-

hollow. Consequently, there are many internal things that we judge to be outside us.

PROP. 2: *Bright spectral lights that are actually inside the eye appear to course about in front of the eyes.*

It often happens that when we cough, or vomit, or are struck in the eyes, we see certain bright spectral lights outside us that actually belong not outside but within us. Before we broach the explanation, it will not be amiss to refer back to our predecessors. As we said before, Empedocles claimed that there is fire in the eyes. Plato argued that the gods formed "the orbs of the eyes to share" in a fire that would not burn but would illuminate.[4] "There is a certain light in the eye, even though it is tenuous. Support for this claim is found in many animals that see at night, for when they are in darkness, their eyes glitter, and when we contort our eyes, we clearly see a circle of some kind inside."[5] Aristotle, however, refutes this, claiming that such shining from the eyes is due not to fire but to polish,

> for anything that is smooth and dense has luster by its nature, but not of the kind that can render the medium actually transparent. Thus, since what is in the middle of the eye (I judge this to be the crystalline humor) is as polished as possible by being blackened, it should come as no surprise [160] that it retains brightness, which may be perceived to shine if the eye is agitated rapidly but not when it is still. This therefore happens because what is one and simple is somehow seen double because of a swift and rapid agitation. For by the motion in it, the location of the blackening where the brightness previously shone is shifted elsewhere, and the eye sees it before it disappears so that what is conveyed from outside, by which it happens that the viewer and the thing viewed are one, is somehow judged to be two. But if the motion were slower, the shining would disappear before the blackening would be shifted away. Moreover, this happens above all in darkness because during the daytime the abundance of light swallows up this tenuous brightness and renders it evanescent.[6]

This opinion held by Aristotle has been viewed as frivolous by some because the shining of the pupil has never been seen in a moving eye by a person looking at it face on, nor is it possible that at the same time that, existing in its particular place, it can perceive its own brightness while it lies at another place and what does the viewing and the thing viewed are one. Objecting to these critics, Alexander of Aphrodisias responded that the pupil does not lack size, for it can happen that a certain portion of it is seeing, whereas another is being seen. And it is both seeing and seen according to the differences of its parts because it cannot actually be the case that two things are one. But he is of a different mind in his *Problems*, where he judges this very reasoning to be fallacious as well and reverts back to visual spirits:

> For an abduction of the belly and chest forces humors and spirits upward, and when they reach the eyes, they disturb and pervert vision so that the images from

antur. Unde si vapor oblongiori ductu est, imagines etiam hac forma apparere. Item si bilis vaporem illum excitat, aurum imagines imitantur; si pituita albi suspicio occurrit. Ad summum pro figura coloreque[2] causae visa imaginamur.

Alio etiam problemate causam reddens, "cur qui in fronte percutiuntur ignem videant?" Dicit "spiritus visui accommodatus per ictum extenuatus ardet profectusque in aerem, talis videtur qualis per ictum redditus est. Sic enim est duobus lapidibus aut ferris mutuo attritis, aer interveniens ictu extenuatus, maiorem in modum in ignem convertitur; quod idem et in nubibus evenire certum" putat.

Sed pudet, profecto pudet, insigniores hos naturae interpretes tam puerilia effutire: quod laevigatio in oculo splendorem efficiat; quod res celeriter mote suum splendorem loco quo discesserat[3] derelictum aspiciat; quod pars una lucida sui partem alteram lucidam conspiciat; [161] quod nigricans (sive pupilla) quae non nisi oscitatio et inanitas pelliculae uveae est lucida sit; quod vapores ex pectoris compressione ictuve accendantur, et spiritus aquei ignescant; quod spiritus et humores ascendant ad oculos, qui in suis tunicis conclusi sunt—ex cuius ore, nisi mente capti, haec possunt excidere?

Sed veritatem indagare tentandum nobis est. Dicimus igitur aqueum humorem a natura lucidum constitutum, ut lucem affinem suam extrinsecus insinuantem recipiat, et quasi oculi portitor, speciem ab extimo lumine adductam ad intimum crystallinum traiiciat, in quo videndi vis sita est. Idque hoc manifestario argumento probari potest. Si ad hirquos manu oculos comprimas, in adversam partem orbem igneum luminosum et in medio nigrum intueberis, et si canthos temporibus premes ad hirquos, lumen videbis. In quiete id non conspicimus, nam crystallino directe opponitur pupillae hiatus laxus, praecipue noctu. Quum comprimis oculum, crystallinum humorem illum conspicaberis sub tela uveae,[4] et nigrum in medio existentem rotundum pupillae foramen est. Praeterea, quum lucentes felis oculos noctu conspicerem, vidi totum oculum illuminatum intus, eratque aqueus humor, nam si crystallinus esset, non nisi crystallinum clarum vidissem. Memini etiam quum puer essem et anatomicis administrationibus navarem strenuam operam, quum die humanum oculum secuissem, reliquias noctu aqueo humore madefactas quasi lucere vidisse, et causam ignorasse.

Averroes libro *De sensu et sensili* dicit in oculo esse lumen proprium, sed non declarat quo resideat loco. "Augeae micantes oculos celebrat Apollonii interpres ... Augusto adeo nitescebant oculi, ut si intentius aliquem intuitus esset, cogeretur is, ceu splendoris nimietate caligaret vultum submittere,"

---

[2]"caloreque" in original text.
[3]"discessetat" in original text.
[4]"uvaeae" in original text

outside are rendered deceptive. Accordingly, if the vapor is oblong in extent, the images also appear in that shape. Likewise, if vapors from bile disturb the eye, the images assume a golden hue; if phlegm reaches the eyes, a suggestion of white occurs. In terms of ultimate causation, we imagine what we see according to shape and color.[7]

In another problem he provides an explanation for "why those struck in the forehead see sparks of fire." He says that "the spirit adapted to sight is thinned by the blow and actually ignites into the air, and it appears as such inasmuch as it is rendered thus by the blow. For so it is that when two stones or pieces of iron are struck against each other, the air between them, being thinned by the blow, is patently converted into fire. This same thing," he opines "certainly happens in the clouds."[8]

But is it not shameful, truly shameful, that interpreters of these signs of nature babble such puerilities as: that the polish in the eye causes it to glitter; that a swiftly moving object sees its shining left behind in the place where it had departed; that one lucid part can look at another lucid part; [161] that the blackening (or pupil) is not lucid without an opening or space in the uveal tunic; that vapors from compression of the chest or from a blow are ignited and ignite the aqueous spirit; that spirits and humors ascend to the eyes and are enclosed in its tunics—can these things escape the mouths of those interpreters, unless their minds are possessed?

We, however, must make an effort to investigate the truth. We say, therefore, that the aqueous humor is lucid by nature so that it can accept light from outside that is akin to it, and, like the eye's toll collector, it transports the species abstracted from the outermost light to the innermost crystalline humor, where the faculty of sight is seated. And this can be shown by the following obvious argument. If you press the eyes with your hand at the corners, you will see a luminous fire on the opposite side and black in the middle, and if you press the corners for a while, you will see light. At rest we do not see this, for the crystalline lens faces the relaxed opening of the pupil, particularly at night. When you press the eye, you will see the crystalline humor beneath the web of the uvea, and the black lying at the middle is the round opening of the pupil. Furthermore, when I saw the shining eyes of a cat at night, I saw the entire eye lit up inside, and that was the aqueous humor, for if it were the crystalline humor, I would have seen only the crystalline humor as bright. I also remember that when I was a boy and was diligently engaged in the hard work of anatomical procedures, one day, as I was dissecting a human eye, I saw the moist remainders in the aqueous humor shine at night, but I did not know why.

In his book *On the Sense and Sensible*, Averroes says there is an innate light in the eye, but he does not reveal where it resides.[9] "The translator of Apollonius of Rhodes praises the flashing eyes of Augias. . . . In Augustus the eyes shone so brilliantly that if he were to look intently at someone, that person would be curdled, as if submitting his countenance to such an excess

quod si crystallinus lucidus esset id non accideret, quum in profundo lateat. Unde in oculi percussionibus a suo loco dimotus crystallinus, vel aqueus humor, quasi igneus scintillat. Idque dum experirer ipse ex diutina oculi compressione, laesus oculus, si cito casu aliquo revolvatur oculus, igneum orbem in parietibus oppositis coruscantem video. Sed in sequenti haec latius explicabuntur. [162]

Noctu quaedam foris videri, quae intra oculos adsunt. PROP. 3

Tiberius Caesar praegrandes oculos habuit, "et qui (quod mirum esset) noctu etiam et in tenebris viderent, sed ad breve, et quum primum a somno patuissent; demum rursus hebescebant"; haec ex Suetonio. At Plinius: Tiberio Caesari eam oculorum "naturam fuisse, ut expergefactus noctu paulisper haud alio modo, quam luce clara contueretur omnia paulatim tenebris sese obducentibus: nec alii genitorum mortalium fuisse naturam." Sed quod mirum in Tiberio Caesare dicit Suetonius, nec ulli mortalium Plinius, hanc ipse oculorum naturam sortitus sum, plerosque habui amicos, et complures etiam legi eiusmodi naturae consortes.

Sed in huius disquisitione maiores nostros consulamus. Aphrodisaeus quaerit cur "per noctem inter dormiendum repente oculos aperiamus et lucem percipimus?" Redditque rationem "quia spiritus videndi coactus amplificatusque in dormiendo universus, larga copia erumpit per tenebras atque elucet; itaque speciem quandam lucis merito potest conspicari." Et alibi causam reddens cur animalia quaedam noctu cernant, alia minime, etiam ad visivos spiritus reddit, inquiens

qui spiritum visivum crassiorem habent quam ut aerem permeare usque ad rem videndam valeat, Solis absentia refrigeratur et per noctem crassescit ut interdiu videre non possit, at interdiu aeris beneficio qui tenuis et calidus videt. Feles vero hyaenae, et vespertiliones . . . spiritum quidem illum visorium tenuissimum et lucidissimum habent, ita ut per noctem modice incrassescens, idoneus ad rerum conspectum reddatur; interdiu vero supra modum extenuatus expanditur et evanescit. Leones acutiuscula luce praediti sunt, argumento pupillae, quae praefulget, atque humidae partes oculorum, quae dilucidissima est, (humorem aqueum puto); ob id per Solem ingredi acriorem non patitur, et ignem etiam interdiu eam ob rem fugiunt.

Mihi autem videtur ob pupillae laxitatem et aquei humoris perspicuitatem evenire, videmus enim in noctivagis animalibus, ut fele, noctua, bubone, et

of brightness would darken it,"[10] but this would not happen if the crystalline humor were luminous, since it is tucked away deep in the eye. Therefore, when the eye is struck and the crystalline lens or the aqueous humor is displaced, it scintillates like fire. And while I may have experienced the same thing from pressing the eye for a long time, if my eye is injured from a sudden fall and is rotated, I see a fiery orb shining on the facing walls. But these things will be explained more fully in a subsequent place. [162]

PROP. 3: *At night certain things that exist inside the eyes appear to lie outside.*

Tiberius Caesar had extremely large eyes, "and (what was amazing) they could even see at night and in the dark, but just briefly, when first opened after sleep; soon they became weak again"; this comes from Suetonius.[11] Pliny, as well, says of Tiberius Caesar that "the nature of his eyes was such that when he woke up in the night he could see everything differently for a while, just as he would see them in broad daylight, until his eyes were gradually enveloped in darkness; and there was no one else in the human race who had this nature."[12] But whereas Suetonius calls it marvelous in Tiberius Caesar, and Pliny says it does not pertain to any other mortal, I myself have achieved this capacity with my own eyes, and I have had several friends who have as well, and I have also gathered several associates with the same sort of ability.

But let us consult our predecessors in the discussion of this point. Alexander of Aphrodisias asks why "at night we may suddenly open our eyes when half awake and perceive light." He offers as an explanation "that of all the visual spirit that has accumulated and increased during sleep, a large quantity bursts out through the darkness, so a visible form can be discerned thanks to the emitted light."[13] Turning elsewhere to the question of why certain animals see clearly at night, whereas others see very little, he adverts again to the visual spirit, claiming that

among those who have extremely dense visual spirit that is able to penetrate the air up to the visible object but is cooled when the Sun is absent and thus thickens at night, the spirit cannot for that reason see as it does during daylight, whereas with the benefit of thin, warm, daytime air, it does see. On the other hand, cats, hyenas, and bats undoubtedly have a visual spirit that is extremely attenuated and lucid so that, only slightly thickened during the night, it is rendered capable of seeing things then, whereas during the daytime it is attenuated to a great extent by expansion and dissipates. Lions have been endowed with especially penetrating light, as indicated by their pupil, which shines brightly, as well as by the moist parts of their eyes, so it is extremely clear (I take this to be the aqueous humor), and because of this, when they enter sunlight, a dazzling object is not sensed, and for that reason they even shy away from fire during daylight.[14]

It seems to me, however, that this happens because of a relaxing of the pupil and the extreme clarity of the aqueous humor, for in such nocturnal animals

caeteris pupillas ultra modum dilatari ut sola pupilla sine iride et albugine conspicua oculi pars videatur.

Praeterea, eorum aqueus humor pellucidissimus est, ut praesenti inspectione patet, et qui pupillis latis [163] praediti sunt noctu tuto ambulant, ut fere in tenebris videant, praecipue in iuventute, quando humor pellucidior est. Audivi ab amico qui tribus mensibus obscuris carceribus occlusus fuerat et post videre cœpit in tenebris paulominus ac si lumine esset, quod evenerat ex[5] pupillae dilatatione in tenebris, ac carceribus exclusus per multos dies caecutiebat. Quum ego puer essem nocte intempesta expergefactus, interdum cubiculum totum et quae in eo erant supellectilia clarissime conspiciebam, et id tandiu durabat quandiu immotas palpebras haberem; in iuventute crebrius, in senectute rarius quia aqueus humor crassescens tenebrosior redditur. Sed praeteritis annis, quum id accidisset, animadverti facie cubili impressa recubuisse, et quum pulvinar molle, oculi prominentes, capitis pondere oculi in somno vehementer comprimebantur; idque quum iterum tentarem et oculos vehementius intus comprimerem, apertis oculis, splendorem quendam maximum sed supellectilia non videbam, unde animadverti ex oculorum compressione pupillam dilatari, et pupillam, vel crystallinum a suo loco dimoveri. Hic igitur perspicuus humor vectabulum est specierum rerum in cubiculo existentium ad crystallinum, et hic illuminati aeris vice fungitur, unde quae intra oculum sunt, extra in parietibus mentiuntur.

In cerebri distillationibus, iris circa lucernam apparet, quae intra oculum causatur. PROP. 4

Aristoteles in *Meteorologicis* ait:

> Circa candelas ut plurimum Australibus flantibus hyeme tota iris fit, et maxime manifestam fieri humidos oculos habentibus, nam horum visus propter debilitatem refrangitur. Idque fieri aeris humiditate et ab evaporatione a flamma defluente et mista, tunc enim fit speculum et propter nigredinem, fumosa est enim evaporatio illa. Sed iris lucernae ab ea in hoc differt quod id lumen non album sed purpureum apparet circulariter, et irinum non autem puniceum, est enim et visus paucus, qui refrangitur, et nigrum speculum.

Seneca: "In balneis quoque circa lucernam tale quiddam aspici solet, ob aeris densi obscuritatem; frequentissime autem Austro, quum caelum maxime [164] grave et spissum est." Alexander et reliqua Peripateticorum turba ab eo eadem. Profecto huius seculi pluries calamitatem deploravi, ut si aliquis primo erret, omnes mox suas captivantes opiniones primum errorem sequuntur. Sed quomodo id fieri possit quamplurimis diebus advigilavi; tandem

---

[5]"et" in original text.

as the cat, the night owl, the horned owl, and the rest we see that their pupils dilate to such an exaggerated extent that only their pupil, without the iris, appears clearly along with the white of the eye.

Furthermore, as personal inspection shows, their aqueous humor is exceptionally limpid, and those possessed of wide pupils [163] walk confidently at night, as they see perfectly in darkness, especially during youth, when the humor is extremely transparent. I heard from a friend who passed three months cooped up in dark prisons and subsequently began to see a bit in the darkness as if there were light present, which came about because of the dilation of his pupil in darkness, but when he was let out of prison, he was blind for several days. As a boy, when I was awakened in the dead of night, I saw the entire room and the furniture that was in it quite clearly for a while, and it lasted as long as I kept my eyelids motionless. This happens more frequently in youth and less frequently in old age because the aqueous humor, in thickening, is rendered darker. But in later years, when it happened, I noticed I was lying face down in the room, and since the cushion was soft and my eyes were bulging, they were shut tight in sleep by the weight of my head. And when I tried it again and pressed my eyes inward with great force, I saw a certain bright light but not the furniture when I opened my eyes, so I assumed that my pupil dilated because of the pressure on the eyes and that the pupil or the crystalline lens was displaced. This clear humor, then, is the vehicle through which the forms of the objects in the room are conveyed to the crystalline lens, and this is activated instead of illuminated air, so things that exist inside the eye are judged to lie outside, on the walls.

PROP 4: *In the case of distillations from the brain, a rainbow appears around a candle, but it is created inside the eye.*

In the *Meteorology* Aristotle says that

> when south winds are blowing during winter, a full rainbow is usually formed around candles, and it becomes most obvious to those who have moist eyes, for their sight is reflected because of its weakness. And it arises from the humidity of the air and the vapor emanating from the flame and mixed with it, for in that case it forms a mirror because of the blackness, since the emitted vapor is smoky. But the rainbow about the candle represents a departure insofar as the light that shows in a circle appears not white but purple, yet it lacks the reddish color of rainbows because the sight is weak and reflects, and the mirror is black.[15]

Seneca adds that, "in the baths too, because of the darkness of the dense air, something of this kind can regularly be seen around a lamp, most often in a south wind, when the atmosphere is particularly heavy [164] and dense."[16] Alexander of Aphrodisias and the remaining crowd of Peripatetics say the same thing. Indeed, I have often lamented the misfortune of this age that if someone first errs, everyone after him, captivated by his opinions, follows the initial error. But for several days I have been attending to how this

multa contrariari cognovi.

Et primo, si lumen candelae oblongum vel pyramidale est, cur igitur iris non est oblonga vel pyramidalis, si ab omni luminis parte aequales lineae proiiciuntur? Secundo, si circa candelam est, cur accendendo minuuntur recendendo maiores fiunt? Quod contrarium esse deberet, nam quae longius distant, minora videntur. Tertio, papyrum perforando et opponendo lumini candelae, et papyrus sit media intra oculum et candelam, tunc non circa candelam videtur iris sed supra papyrum. Quarto, si Aristotelica causa eveniret, non videretur iris nisi Austro flante, et aere humido existente, et oculis ophtalmia laborantibus. Sed solum oculis laborantibus, semper videtur iris, vel calido vel sicco aere, flantibus Septentrione, Zephyro, et aliis. Non ergo ex illa evenit causa.

Sed veritas est quod iris non circa candelam sit sed intra oculum, nam si circa candelam, non iam ex dictis experientiis, nec in medio aere fieri potest. Non alibi igitur quam intra oculum, et si intra oculum non alibi quam in humore aqueo. Radii enim lucernae pertranseuntes humorem aqueum, qui non tunc lucidissimus et pellucidissimus, ut in sanitate, sed in ophtalmia vel distillatione laborantibus sit roscidus et quasi guttulis conglaciatus; laterales radii pertranseuntes et se refrangentes intra oculum colores generant, ut culpis illa tricolor efficatur; nam primo lumine flavus; postremo, quod crystallinum tangit, purpureus vel caeruleus; viridis in medio. Qui omnes ad refractionum loca percurrentes, circulum circa candelam rotundum efficiunt, quia rotunda pupilla est, et qui primus crystallinum tangit, superior efficitur halurgus; qui ceratoidem flavus et lumini proximior. Clarioris intelligentiae gratia exemplum apponemus.

Sit candela AB, cuius radii perpendiculares pertransuent irrefracti. Obliquus BF ferit pupillam F; refrangitur a cornea in contrariam partem, quia eis centricam GN, et ascendit[6] ad G crystallinum. Extenditur in D. Cathetus ex luminis centro N et re visa B trahitur, occurritque FD in P. Ibi ergo color proximior [165] crystallino caeruleus, unde circulus PDEQ[7] caeruleus. Si enim ex oculi centro traheretur cathetus, non concurreret cum linea incidentiae. Causa quod quanto te elongaveris a lumine eo maior circulus iridis augescit est quod pyramidis lineae, quanto magis producuntur, latiorem spatium intercipiunt. Quod solum suffusionibus, distillationibus vel ophtal-

---

[6]"accendit" in original text.
[7]"PEDQ" in original text.

occurs, and I have at last come to understand several things that contravene the standard explanations.

First, if the candle's light is oblong or conical, then why is the halo not oblong or conical, assuming that equal lines of radiation are projected from every spot of the light? Second, if the circles of light lie around the candle, why do they get smaller when we approach and larger when we draw away. The opposite ought to obtain, for what lies farther away appears smaller. Third, if a piece of paper is perforated and placed in front of the candle's light, and if the paper lies between the eye and the candle, then the halo is not seen around the candle but upon the paper. Fourth, if the Aristotelian rationale were to hold, the halo would be seen only when the south wind blows, when the air is humid, and when the eyes are suffering from ophthalmia.[17] But only when the eyes are suffering from ophthalmia is the rainbow invariably seen, whether in hot or dry air, whether the wind is blowing from the north or west, and so forth. Thus, it does not occur because of that cause.

The truth, however, is that the halo is not around the candle but inside the eye, for if it is around the candle, then it cannot be produced by the experiments just mentioned or in the middle of the air. Hence, it can be nowhere else than inside the eye, and if it is inside the eye, then it can be nowhere else than in the aqueous humor. For when the rays pass from the candle through the aqueous humor, which in this case is not extremely lucid and transparent, as it should be in someone healthy, but instead is in someone suffering from ophthalmia or distillation in whom the humor is cloudy, as if congealed into droplets, the lateral rays that pass through and are refracted within the eye generate colors such that the incoming light forms three colors on account of the disorders; in fact, yellow is first with respect to the light, purple or blue last because it touches the crystalline lens, and green in between. And all of the rays, in passing to the places where they are refracted, form an evenly rounded circle around the candle because the pupil is round, and what first touches the crystalline lens forms the upper circle of purple, whereas what touches the cornea forms yellow, which is closer to the candlelight. For the sake of clearer understanding, we will append an illustration.

Let AB [in Figure 7.4] be a candle whose perpendicular rays pass through the cornea unrefracted. Oblique ray BF strikes the pupil at F and is refracted by the cornea in the contrary direction, since the eye's centric ray is GN, and it continues to the crystalline lens G. [Refracted ray G'F] is extended to D. The cathetus through center point N of the light source and visible object B is drawn, and it intersects FD at P. Here, then, is where the color blue [165] is nearer the crystalline lens, so circle PDEQ is blue. For if the cathetus were drawn from the center of the eye, it would not intersect the line of incidence.[18] The reason that, as you draw farther away from the light the circle of the rainbow gets larger, is that the more the lines of the cone are extended outward the broader the space they intercept.[19] But the rainbow around a candle may be seen only by those suffering from suffusions, dis-

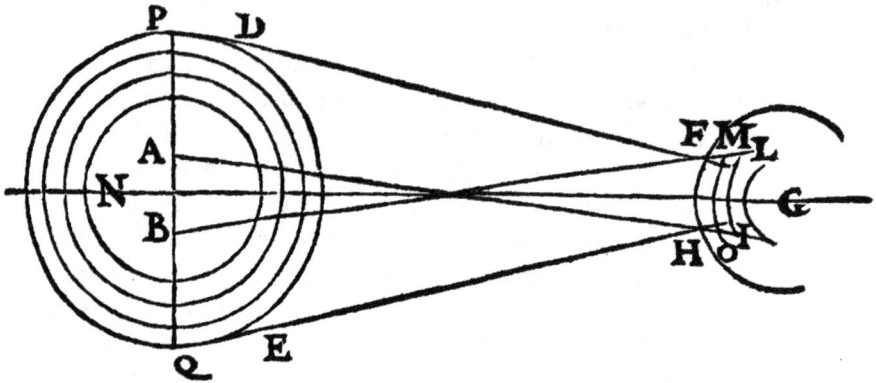

mia laborantibus videatur circa candelam iris, et non Euro, vel Austro flantibus, vel aere humecto existentibus, et candela fuliginosum aerem faciente.
Consulant medicos, et praecipue Demosthenem medicum, dicentem in suffusionibus ubi ad pupillam humores descendunt offundunturque, aliqui circa
lucernam circuli videntur.

*In vertigine omnia circumverti videntur quae firma manent, sed humores intra
oculos moventur. PROP. 5*

Theophrastus vertigines fieri putat libro Περὶ ἰλίγγων[8] vel quum spiritus
alienus circa caput venerit, vel ab humiditate, vel a circulari capitis motu.
Hae causae cerebrum invadentes vel afficientes afficiunt ut omnia circumvolui
videantur, nil enim in visu differt utrum intrinsecus aut extrinsecus movens
sit. Aristoteles in *Problematis* rationem reddens cur ebriis omnia circumferri
videantur dicit quod visus movetur a vini calore, et videntur res moveri,
nihil enim interest an visus moveatur, an res conspectui obvia. Circulariter
moveri videntur, quia res visa conus basis est, et quum visus rem [166]
conspectam non deserat, circumferri videt. Aphrodiseus clarius loquitur:

> Vinum haustum vaporosam spirituum exhalationem ad cerebrum mittit, qui spir-
> itus quum digeri et consumi nequeant, prius quam temporis spatio concoquantur
> per cerebrum volvuntur ... et perturbati ita per conspectivos nervos ad pupillam
> profluunt faciuntque ut obviae res tales videantur quales ipsi sunt, affectum enim
> interiorem foris esse imaginamur, quo sit[9] ut vertigine tantisper tentemur.

---

[8] "ἰλίγγον" in original text.
[9] "fit" in original text.

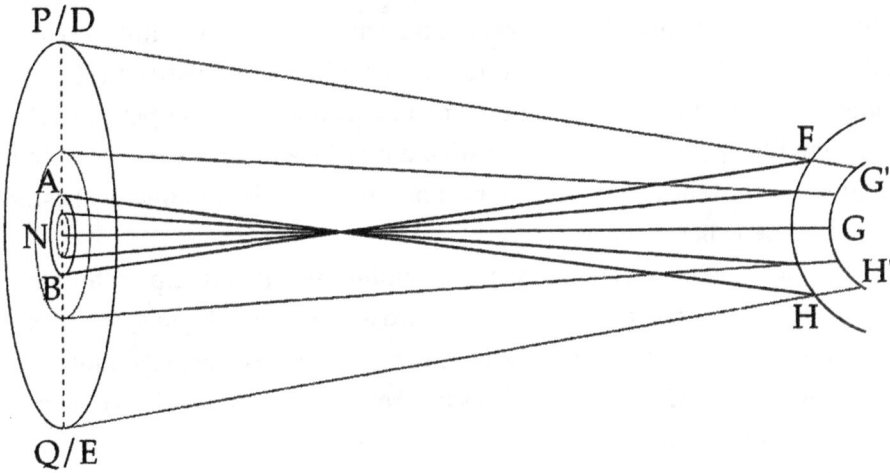

Figure 7.4

tillations, or ophthalmia, not because the east or south winds blow, or because they are in humid air, or because the candle makes the air sooty. Let them consult physicians, especially the physician Demosthenes Philalethes, who says that some circles appear around a candle in the event of suffusions, when humors descend to and pervade the pupil.[20]

PROP. 5: *In vertigo everything that is fixed in place appears to spin around, but it is the humors inside the eyes that move.*

In his book *On Dizziness*, Theophrastus opines that dizziness occurs when a foreign spirit comes to the head, or because of moisture, or because of a spinning motion of the head. When these determinants intrude into or affect the head, they cause everything to appear to spin around, for in the case of vision it makes no difference whether what moves is internal or external.[21] Offering an explanation for why everything appears to spin around to those who are drunk, Aristotle claims in the *Problems* that sight is moved by the heat of the wine, and objects appear to move, for it makes no difference whether the visual radiation or the object exposed to view moves. Things appear to move in circles because the visible object forms the base of the cone, and since the visual radiation does not lose touch with the object [166] viewed, it sees the object spin around.[22] Alexander of Aphrodisias speaks more clearly:

> When it is drunk, wine sends a vaporous emanation of spirits to the brain, and since these spirits cannot be digested and used up, they boil and revolve through the brain before they are digested, and when they are disturbed in this way, they flow through the optic nerves to the pupil and cause objects exposed to sight to appear as if they are disturbed this way, for we imagine the internal effect to lie outside us, and it is in this way, meanwhile, that we suffer vertigo.[23]

Mihi vero aliter videtur, nam fieri non potest ut spiritus ad cerebrum pervenientes per opticum nervum transeant, cuius tam angustus est aditus—si vere
aditus est—ut vix quicquam admittat, et in oculum intromissus a retiformi
tunica, et a crystallino praependitur[10] ne in aqueum humorem pertranseant,
nisi enim illuc pervenerint non perturbabunt visionem. Neque res secundum
circulum se moventur, qui basis est visionis, ut Aristoteles inquit, quia falsum
est, et res se habet aliter in visu.

Nulla res extra videri potest intra oculum, nisi que in aqueo humore,
aut cornea fuerit. Incalescente igitur cerebro aut vino aut aliquo symptomate
facit ut vicina quoque ferveant. Oculi enim cerebro maxime propinqui sunt;
igitur incalescunt. Quae calescunt in vaporem solvuntur. Qui si libere spatiari
datur, in auram abit; si concluditur (nam ceratoide consepitur), circa eam
vertitur,[11] ut accidit in bullientis aquae lebete, incalescens enim uno latere
circumvertitur aqua; sic vapores radialem pyramidem commoventes facit ut
res a suo loco moveri videantur, non circulariter, ut dicunt, sed a dextro ad
sinistrum, ut semper pergat ad sinistrum, et nunquam perveniat. Idem accidit
respiciendo lapides sub aquis fluvii delabentis, nam moveri et excurrere
videntur quum suo loco non redeant, et quanto longius res conspecta erit,
eo magis circumvolvi videtur, propius[12] vero minus.

Res una imo oculo conspecta geminata videri potest. PROP. 6

Aristoteles in *Problematis* rem unam plures videri posse probat. Idque multoties vinolentis accidere atque evenire ut quum visus commoveatur, nunquam
quiescat in re conspecta, et quod eodem tempore varie cernitur; aliqua prius
aliqua serius occurrunt. Visus enim visibilia percipit tempore [167] interpos-
ito; id ergo quum saepissime contineat fieri visum, videbitur res repetita ut
duo sint. Sed obscure admodum Aristoteles loquutus est; veritatem nacti
lucidioribus verbis rem aperiemus.

Diximus in praecedenti quod res circumferri videbatur quae suo loco
manebat quia vapores ab humore aqueo elevati intra oculum circumfere-
bantur. Nunc eadem de causa, si vapores non circumvolvuntur sed recta
moventur vel tumultuarie pulsantes et commoventes pyramidum latera, res
una quae hoc modo videbatur eodem quasi temporis momento, alibi conspicitur, et quia visio et discretio, ut Aristoteles dixit, fit in tempore, quia in ea
visione citissimus est motus, res una plures videntur.

---

[10]"praepeditur" in original text.
[11]"vertuntur" in original text.
[12]"proprius" in original text.

To me, however, it seems otherwise, for it cannot happen that the spirits reaching the brain pass through the optic nerve, whose opening—if there actually is an opening—is so narrow that it scarcely allows anything to enter, then pass into the eye from the retiform tunic and stay suspended in front of the crystalline lens so that they do not pass into the aqueous humor, for unless they reach that point, they will not disturb vision. Nor do objects, which form the base of the visual cone, move in a circle, as Aristotle claims, because it is false, and the matter obtains otherwise in vision.

Nothing outside can be seen inside the eye unless it is in the aqueous humor or the cornea. Consequently, heating the brain either by wine or by some other symptomatic agent causes neighboring things to heat up. Indeed, the eyes are as close as possible to the brain, so they heat up. Things that heat up dissolve into vapor. If it is allowed to spread out freely, this vapor transforms into air; if it is enclosed (in fact, it is hemmed in by the cornea), it circulates around inside it, as happens in a kettle of boiling water, for when one side of the water is heated it circulates, so the vapors that disturb the radial cone cause an object to appear to move in place, not in circles, as they say, but from right to left, such that it continues to move to the left but never arrives. The same thing happens in regard to stones sunk under flowing water, for they appear to move and rush forward when they do not actually move in their place, and the greater the distance at which the object is viewed, the more it appears to rotate, whereas the nearer the distance, the less it appears to rotate.

PROP. 6: *A single object viewed in the depth of the eye can appear double.*

In the *Problems* Aristotle shows that a single object can appear to be multiple. It often happens with drunk people that, since their sight quivers, it never comes to rest on the object that is looked at, and the object is discerned differently at the same time, so that the image of one part impinges [on the cornea and crystalline lens] earlier while another impinges later. For the visual faculty perceives visible objects in an interval [167] of time; consequently, since most of the time what is seen remains steady, an object seen repeatedly will appear as if it were double.[24] But Aristotle has expressed himself extremely obscurely, so we will broach the matter in clearer language so as to get at the truth.

We claimed earlier that an object that remained stationary appeared to revolve because the vapors arising within the eye from the aqueous humor were stirred around. Now, for the same reason, we argue that if the vapors do not swirl around but move in a straight line or batter the edges of the radiant cone to move it rapidly, a single object that has been seen in this way is perceived elsewhere at virtually the same moment of time, and since, as Aristotle said, vision and discrimination occur in time, a single object appears as more than one because the motion is extremely rapid in this sort of vision.

Exemplum habemus in igne, quem si aliquis citissime circumvolvet, circulus igneus apparebit. Et in trochis, qui citissime circumvertuntur, punctus in basi circulus videbitur, sic et in mola pistrini. Unde ex vehementissima motione vaporum in aqueuo humore, duae res videntur una. Averroes in *Sensu et in sensili* dicit visum non videre nisi habuerit integrum organum et non perturbatum. Iracundus quoque in hora suae irae ex ferventi corporis et sanguinis calore visus intercipitur et simplicem rem duplicem videbitur, nam ut ebrietas, ita ex irae excandescentia vapores ex aqueo humore suscitantur. Avicenna plures causas affert, praecipua innititur his quod ubi coeunt optici nervi ibi duae imagines duorum oculorum videntur una, sed vapores vel humores ex cerebro descendentes et nervos concutientes faciunt ut res una plures videantur. Pentheus furiis agitatus quae simplicia erant duplicata videbat. Virgilius:

> Eumenidum veluti demens videt agmina Pentheus
> Et Solem geminum, et duplices ostendere Thebas.

Et Aristoteles etiam causam ad opticum nervum refert. Initia conspectus a vino commoventur, et totum caput sollicitatur, et quum initia conspectus moventur, idem se applicare non quit, sed quasi partem rei objectae vis utraque conspiciendi attingit; merito igitur res una suspicionem recipit geminationis; quod idem oculo subter oppresso solet evenire, conspectus enim movetur initium, ut ei coire cum altero non amplius liceat. Sed hic motus extrinsecus est, vini autem intrinsecus est, nec quicquam interest, idem enim sequitur qualitercunque aspectum commoveris. **[168]**

In deliriis multa simulachra videntur quae nusquam sunt. PROP. 7

Ait Aphrodiseus, "Qui per febrem arduam delirant atque insaniunt, iis de cerebro ipso vapores ad conspectorium spiritum funduntur." Averroes in *Colliget*, quod "propter aliquam aegritudinem . . . aut propter timorem, aut mestitiam . . . solvitur quidam vapor" qui cerebrum pervadit, imprimendo in illud optatae aut formidatae rei simulachrum, movetque una spiritum animalem, et motus ad imaginationem usque pervenit, deinde eodem spiritu communem sensum movet, qui particulares unde res quae intus agitur foris existimatur. Haec Averroes.

Sed humor melancholicus vel mania vim habet aliquando, ut idola reservata in memoria in vigilia videantur quae in somno videnda erant. In

We have an illustration in the case of fire; if someone swirls it around swiftly, a fiery circle will appear. In iron hoops, as well, if they are rolled swiftly, a point on the base will appear as a circle, and so too in the case of a grindstone for corn. Hence, on account of the violent motion of vapors in the aqueous humor, two things appear as one. In *On Sense and the Sensible*, Averroes claims that the visual faculty does not see properly unless it has a sound and undisturbed organ. At the moment someone who is enraged is angry his vision is seized on because of the ferment of his body and the heat of his blood, and he will see a single object doubled, for just as in drunkenness, vapors are roused from the aqueous humor because of an outburst of anger.[25] Avicenna offers several reasons, among which the foremost supporting one is that where the optic nerves come together the two images from the two eyes are seen as one, but the vapors or humors descending from the brain and agitating the nerves cause a single object to appear as more than one.[26] Driven by the furies, Pentheus saw things that were single as double. As Virgil says:

> Like mad Pentheus, when he sees the maenads,
> And sees a double Sun and a duplicate Thebes.[27]

Aristotle as well traces the cause to the optic nerve. The beginning stages of sight are disturbed by the wine, and the entire head is agitated, and when the beginning stages of sight are disturbed, the visual faculty cannot direct itself properly, but instead the power of seeing from both eyes reaches a portion of the external object. It follows appropriately, therefore, that a single object takes on the look of doubling. This also happens customarily when the eye is pushed from below, for the line of sight is moved at the source so that the one line of sight cannot intersect the other on the object.[28] This motion is extrinsic, however, whereas that of the wine comes from within, but there is no difference between them, for the same thing follows no matter how you disturb the line of sight. [168]

PROP. 7: *In delirium many apparitions of things that in no way exist are seen.*

Alexander of Aphrodisias says, "Among those who are deranged and driven mad by high fever vapors flow from the brain itself to the visual spirit."[29] In the *Colliget*, Averroes claims that "because of a given disease, fear, or misfortune, a particular vapor is produced" that pervades the brain, impressing on it the likeness of a longed-for or dreaded object, and by itself it moves the animal spirit, and the movement reaches the imagination and then moves the common sense by means of the same spirit, so that for these particular people what is happening within is judged to be happening outside.[30] So much for Averroes.

But the melancholy humor or madness sometimes has the power to make images that are seen and retained in memory during wakefulness be

somno feruntur idola ad communem sensum, qui est in corde. Sic nobis repraesentat praeterita ac si praesentia essent. In furore vero et mania, quum organa intus perturbentur neque suum servant ordinem, ea nobis extrinsecus videntur quae oculis non aspicimus, non enim vere videntur sed imaginatione. Philosophi phantasmata vocant, et exteriori sensorio defferi dicunt. Memini quum ipse ardenti febre laborarem et iam morti propinquus; videbam prope cubile heremitam ex sacello egredientem ac campanas pulsantem, quumque eum rogarem ne importuno sonitu cerebrum excruciaret, mihi in ludibrium linguam exerebat, qua de re maxime angebar; neque ea phantasmata ante oculo mihi videre videbantur, sed veluti somnium vigilantis. Sic fuit et phantasma Cai Cassii, quando sibi in medio pugnae in campis Philippicis, Caesar apparuit minabundus; et cuius vultu ob terrorem aufugiens, dixit "quid aliud superest, si occidere virum parum est?" Orestes furiis agitatus semper videbat matrem ab eo occisam et furias perferre. Unde inquit apud Euripidem:

> O mater, obsecro te ne mihi immiseris
> Sanguinolentas puellas, et serpentibus horridas.
> Ipsae enim, ipsae iam propius[13] me impetunt . . .
> O Phaebe necabunt me, aspectu canino truces
> Terribiles, inferiorum sacerdotes, tremende Deae.
> **[169]**

Virgilius:

> Aut Agamemnonius scaenis agitatus Orestes,
> Armatam facibus matrem, et serpentibus atris,
> Quum fugit ultrices sedent in limine dirae.

Sic et multa fieri accidunt quae a daemonibus putantur.

Phantasmata foris videri possunt, quae sunt intus. PROP. 8

Nos quibusdam propinatis possumus facere ut quis ea foris videat quae intus in cerebro et in imaginatione fiunt, idque facillime ex iam adducta ratione. Ex vino et cibis sursum feruntur vapores secum naturam editarum rerum ferentes, atque cerebrum invadentes imagines et phantasmata imprimunt ex quibus orta sunt, at si venenosa quaedam admiscuerimus et organa et rerum ordines perturbabunt, ut quae intus fiant foris putet, nec affectus discernere poterit.

---

[13]"proprius" in original text.

seen during sleep. During sleep the images are conveyed to the common sense, which is in the heart.[31] Thus, something from the past is represented to us as if it were present. In raving and madness, though, when the internal organs are perturbed and do not maintain their usual disposition, things appear outside to us that we do not see with our eyes, for they are not actually seen by the eyes but by the imagination. Natural philosophers call them phantoms, and they say that they are transmitted through the exterior sense faculties. I remember when I was suffering from an acute fever and was already near death; from my bed I saw a monk emerge from his cell to ring a bell, and when I begged him not to torment my brain with this jarring noise, he stuck his tongue out at me in derision, at which I was extremely vexed. These phantoms did not seem to appear before my eyes, though, but rather like a dream during waking. So it also was with the phantom that came to Gaius Cassius, when in the midst of battle on the fields of Philippi, Caesar appeared in threatening manner to him, who, fleeing in terror from his visage, asked, "What else is left, if killing the man is not enough?"[32] Troubled by the furies, Orestes continually saw his mother, who was killed by him and who brought the furies down on him. Consequently, according to Euripides, Orestes says:

> Oh mother, I beg you not to send me
> These bloody maidens, bristling with snaky hair.
> Here they are, right now, attacking me,
> O Phoebus, they will kill me, these savages with doglike look, These priestesses
> of hell, these dreadful goddesses.[33] [169]

As Virgil puts it:

> Or like Agamemnon's son Orestes driven on an empty stage
> Pursued by his mother with torches and snakes
> While the avenging Fiends lurk in the doorway.[34]

So, too, many things happen to take place that are attributed to demons.

PROP. 8: *Phantasms that are within can appear to lie outside.*

With certain drinks we can cause someone to see things that are created internally in the brain and in the imagination as if they existed outside, and the explanation for this is quite easily adduced from previous reasoning. From wine and food, vapors are produced that carry with them the nature of the things from which they are produced, and when they seep into the brain, they impress images and phantasms of what they originated from, as if we mix certain poisons that will pervert the organs and the arrangement of things so that what is generated internally the viewer believes to be external and cannot perceive his actual state of mind.

Quum puer essem, mandragorae vinum mancipio meo propinavi, et non multo temporis intervallo horribilia phantasmata se videre dicebat, a bubus cornibus impeti, quod non alia ratione contigit nisi quia bovinam carnem paulo ante comederat. Maniaco solano, vulgo bella donna dicto, et stramonio, in vini spiritu propinatis, cum quorundam avium et quadrupedum cerebro, effeci ut qui id hauserant se in eiusmodi animalia conversos dicerent, et admirationem, risum, et cachinnum spectatoribus patebant maximum, de quibus alibi pertractavimus. Et simile videtur id artefactum quod evenit natura in febribus; nec mirum si aliquando foras aliquid putamus fieri quod intus habemus.

Aliqua maiora videri quae vere non sunt. PROP. 9

Quod videndi instrumentum aqueum futurum iam supra demonstravimus. Sed Deus Optimus, Maximus ne per aquam in visu deciperemur, adeo tenuissimam illam creavit ut vix ab aere tenuitate differet.[14] Sed haec aqua in distillationibus cerebri et in suffusionibus crassescit aliquando adeo ut veram visionem conturbet, ut non solum irides videamus sed res aliquanto maiores [170] quam sint. Et in visu facillime decipimur, unde putamus res foris maiores esse quae non sunt, sed humor oculi crassior factus. Et idem est crasso oculi humore videre res in suo esse, quam sano oculo pyra vel cerasa aquis mersa, nam multo maiora quam sint videntur, ut diximus. Valerius Flaccus:

> Qualis in Alcidem et Thesea Rhoetus[15] iniqui
> Nube meri, geminam Pholoen, maioraque cernens
> Astra ruit. . . .

Affectiones in oculis servatas in conspectis rebus videmus. PROP. 10

Decipimur quoque in visu quum affectiones vehementium visibilium adhuc in oculis servamus et alias res inspiciendo eiusmodi effectibus affectas videmus. Alteratis enim instrumentis, non instrumenta sed objecta putamus alterata. Exemplum erit quum Solis fulgorem aut globum aspicimus aut reflexam lucem ex speculis vel politis corporibus, amoto enim Sole remanet adhuc in oculis splendoris affectio, et quaecunque spectamus primo flava rubraque, mox viridia caeruleaque. Et quousque affectio illa oculis asservatur, eousque obiecta alterata conspiciuntur. Qui diu rotas circumvolventes aspexerint, et si alio transferant oculos, etiam rotata vertigine omnia circumvolvi videbunt,[16] quia adhuc in oculis species illa rotationis asservatur. Sic etiam quum rapidos et praeter fluentes fluvios spectaveris, quae deinde videbis etiam aufugere putaveris dum affectionis reliquiae in oculis perdurabunt.

---

[14]"defferret" in original text.
[15]"Rhaesus" in original text.
[16]"videbit" in original text.

When I was a boy, I gave someone mandrake wine with my own hands, and he said that not long afterwards he saw horrible phantasms, that he was being attacked by the horns of cattle, which happens for no reason other than that he had eaten beef a short while before. In spirit of wine, maniacum solanum, called *belladonna* in the vernacular, or stramonium, can cause those who have drunk it while having certain birds or animals in mind to say they were transformed into animals of that kind, and they were exposed to great astonishment, laughter, and jeering among onlookers, concerning which we have dealt elsewhere.[35] The effect that arises by nature in fevers seems similar, so it is not surprising if we sometimes think that something we consider to be happening outside our body is affecting us internally.

PROP. 9: *Some things appear larger than they actually are.*

We already showed above that the organ of sight had to consist of water. But in order that we not be deceived in seeing through water, God, most Good and Great, made it extremely tenuous so that it would scarcely differ from air in rarity. In the case of distillations from the brain or suffusions, however, this water sometimes thickens so that it distorts proper vision in such a way that we not only see halos but also see things somewhat larger [170] than they are. And we are easily deceived in vision, so we judge things outside to be larger than they are, but with an ocular humor rendered thicker. And seeing a thing as it is in the thick humor of the eye is like seeing pears or cherries immersed in water with a sound eye, for they will appear much larger than they are, as we said. Accordingly, Valerius Flaccus wrote:

> Even as Rhoetus, clouded with excess of wine and seeing Pholoe double and the stars larger than their wont, rushed upon Alcides and Theseus.[36]

PROP. 10: *We see effects that persist in the eyes as if they were in things seen.*

We are also deceived in sight when we retain the effects of vivid visible objects in our eyes or when we see other things subject to effects of this kind. For when the organs are altered, we believe that the objects, not the organs, are altered. Looking at light from the Sun or at its orb, or looking at its light reflected from mirrors or polished bodies will serve as an example, for when we turn away from the Sun, the effect of its brightness remains in the eyes, and whatever effect we see is first yellow and ruddy, then green and blue.[37] And as long as this effect persists in the eyes, objects will appear altered by it. If those who look for a while at spinning wheels and shift their eyes elsewhere, they will still see everything revolve because of dizziness, since that type of rotation still persists in the eyes. So, too, if you were to look at rivers flowing swiftly by, you would believe that what you see afterwards also flows by as long as the effect persists in the eyes.[38]

Ascendentum Solem quasi circumvolvi videri. PROP. 11

Remanet hoc absolvamus problema, quod non mediocris literaturae viros distorsit, cur ascendentum Solem inspiciendo, illum quasi in se ipsum circumvolvi videmus. Sic etiam et descendentem in occasu idem etiam evenire videmus in cuiusque fulgoris repentino occursu. Cuius ratio est, quia ex repentino, cum Solis cum cuiusque fulgoris occursu, pupilla, ut diximus, illico constringitur, cuius constrictu humor aqueus commovetur. Et quia circularis pupilla, circularis eius erit commotio. [171] Commotio enim humoris in causa erit, ut quae foris videantur circumvolvi appareant.

Rerum apparentium distantias intra oculum distingui. PROP. 12

Quae hucusque rationes adduximus, non alio nostra intendit intentio, nisi ut demonstraremus quomodo nos distinguimus magnitudinum distantias, scilicet ut res una ab altera distet, res tam arduae difficultatis, quod inquirentium philosophorum animos non parum distorsit. Sed aliorum opiniones repetamus. Aristoteles non alio modo res inter se distare comprehendi posse dixit, nisi inter utramque intermedio cognito, cuius opinionem tum caeteri philosophi, tum perspectivi sequuti sunt. Plotinus Platonicus quaeritur cur remotissimae a nobis magnitudines, et si inter se distantes stadio, videntur minori inter se spatio distare. Aitque quod ubi confusius formarum cernimus varietatem, ibi nec quantitatem intervalli discernimus, ut alibi longius prosequuti sumus.

Nobis vero rem tam arduam perpendentibus, ita diiudicare visum est. Quemadmodum in speculis convexis res oppositae omnes in semidiametro[17] ex centro ad circumferentiam perspiciuntur, ita et in oculo ad convexi speculi [172] formam conformato evenire existimamus, nam res quae speculo propinquior est proximior superficiei apparebit, et quae longius centro vicinior. Sed vere rei visae punctus est ubi linea formam oculo referens cathetum a speculis centro ad visam rem ductam intersecabit. Cuius exemplum apponamus.

Sit convexum speculum DEF, oculus A, res visa altera longior B, altera propinquior L. Per punctum speculi D video rem B, quia angulus ADC aequalis angulo CDB. Extendatur AD in longum quousque occurrat catheto ex H centro speculi ad rem visam B, et erit in G. Sic res visa L per punctum E videtur ab oculo A, quia anguli AEI, IEL[18] pares sunt. Extensa AE occurret eidem catheto in F, propinquior speculi superficiei DE. Sed haec ad oculum traducamus afferamusque exemplum.

---

[17]"semetrdiamo" in original text.
[18]"IED" in original text.

PROP. 11: *The rising Sun appears to spin.*

It remains for us to solve the problem of why, in viewing the rising Sun, we see it as if it were revolving on itself, a problem that has vexed men whose writings are by no means mediocre in quality. Likewise, when the Sun sets in descending, we see the same thing happen in the sudden flash of its light.[39] The reason, as we said before, is that the pupil is constricted from the sudden flash of light with the Sun, and the aqueous humor is agitated by this constriction. And since the pupil is circular, its agitating effect will be circular. [171] The agitation of the humor will of course be why things seen outside appear to spin.

PROP. 12: *The distances of visible objects are determined inside the eye.*

Up to this point we have adduced explanations with no other intention than to demonstrate how we determine the distances of magnitudes, namely, that one thing lies apart from another, an issue of such sheer difficulty that it tormented the minds of inquiring natural philosophers more than a little. But let us revisit the opinions of others. Aristotle said that the separation between two objects can be perceived only by means of some known thing lying between them, an opinion followed at one time by other natural philosophers and at another by the Perspectivists.[40] Plotinus, the Platonist, asks why magnitudes that are extremely far from us and that lie a stade apart appear to lie a distance of less than that apart. He responds that where we discern a variety of forms in an extremely confused way, at that spot we do not discern the size of the interval, as we have described at length elsewhere.[41]

But now that we have considered this extraordinarily difficult issue carefully, it seems to us that it is determined as follows. In the way that everything facing convex [spherical] mirrors is perceived on the radius extending from the center to the circumference, so, too, we judge that this perception arises in the eye according to the shape [172] of the convex [spherical] mirror, for an object that is closer to the mirror will appear closer to its surface, and one farther away will appear closer to the center. But the actual point of the image of the visible object is where the line conveying the form to the eye intersects the cathetus extended from the center of the mirror to the visible object. Let us provide this example.

Let DEF [in Figure 7.12a] be a convex [spherical] mirror, A the eye, B a more distant visible object, and L a nearer one. I see object B through point D of the mirror because angle ADC = angle CDB. Extend AD until it intersects the cathetus dropped from center point H of the mirror to visible object B, and the intersection will be at G. Likewise, visible object L is seen by the eye at A through point E because angles AEI and IEL are equal. When extended, AE will intersect the same cathetus [HB] at F, which is nearer than G to surface DE of the mirror. But let us give an illustration of how we might apply this to the eye.

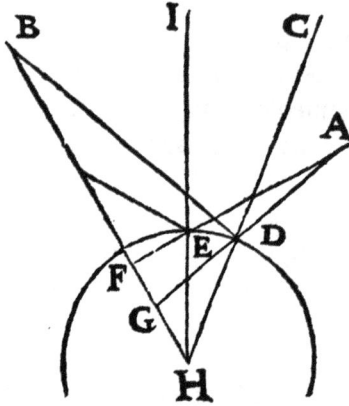

Sit oculus GLIDH, centrum eius F. Sint duae magnitudines, altera remotior A, altera propinquior C; sit centrica FDB.[19] Et veniat A magnitudo ad I[20] pupillam; descendet in D ad centricam in FB. Et C per L pupillam extendatur in E. Sublimior est E ipsa D, quia quanto depressius venerit eo altius feriet centricam, et quanto dexterior punctus erit, dexterior cornea, et crystallinum feriet, et ibi oculus discernit veritatem et aestimabit foris. Sed de his fusius in libris de speculis. Hinc etiam provenit cur natura rotundum oculum affirmarit, quia omnem figuram capere potest, non autem planum.

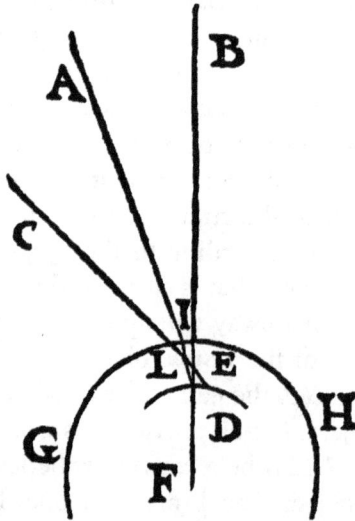

---

[19]"FEDB" in original text.
[20]"L" in original text.

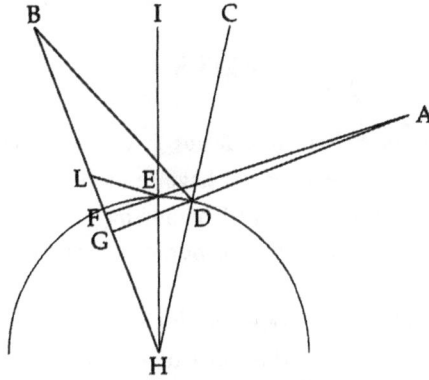

Figure 7.12a

Let GLIDH [in Figure 7.12b] be the eye with F its center [and D on the crystalline lens]. Let there be two magnitudes, A the farther one, C the nearer one, and let FDB be the centric ray. Let the image of magnitude A reach the pupil at I, and it will continue down to D on centric ray FB. Then let the image of C extend through L on the pupil to E. E is lower than D because the lower it reaches the lens the higher it will strike the centric ray, and the more to the right the point will be, the more to the right it will strike the cornea and crystalline lens, and this is where the eye discerns the actual distance and will judge it to be what obtains outside.[42] But these things have been dealt with abundantly in books on mirrors. From this also comes [the explanation for] why nature established a spherical eye because it can take on every figure, which a flat surface cannot.[43]

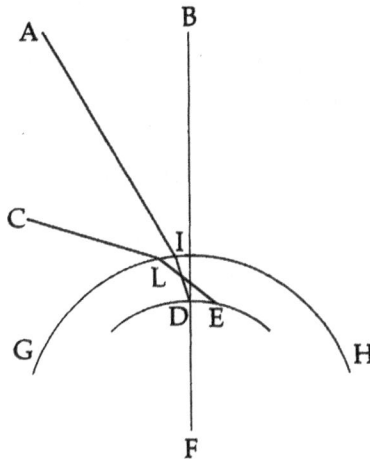

Figure 7.12b

NOTES

1. The actual Latin verb here is *videmus* ("we see"), which clearly does not apply literally to sounds. Consequently, I have rendered it "we perceive."

2. Hyposphagma, or subconjunctival hemorrhage, is a reddening of the eye's surface caused by a broken blood vessel, but it affects the white of the eye, not the cornea.

3. In the Latin text, Della Porta refers to this humor as "ovalis" (egg-like) rather than "aqueus" or "albugineus"; for this particular usage of ovalis, see, for example, Meletius Monachus (1551), 61.

4. *Timaeus*, 45b, in Plato (1551), 712.

5. Ludovico Ricchieri, *Thirty Books of Readings from the Ancients*, XV, 1, in Ricchieri (1566), 547.

6. This entire quotation, which is a pastiche drawn from ibid., 548–49, summarizes Aristotle's account in *On Sense* 2, 437a33–437b10.

7. *Problems*, II, 55, in Alexander of Aphrodisias (1552), 105ra.

8. *Problems*, I, 60, in ibid., 95ra.

9. See Averroes (1550), 192vb.

10. Ludovico Ricchieri, *Thirty Books of Readings from the Ancients*, XV, 2, in Ricchieri (1566), 548; the "translator" to whom Ricchieri refers is most likely Valerius Flaccus, who rendered Apollonius of Rhodes's *Argonautica* (early 3rd century B.C.) from Greek into Latin around 70 A.D. The Augias (or Augeas) of flashing eye is best known for owning the stables that Hercules was commissioned to clean.

11. *Lives of the Caesars*, III, lxviii, 2, in Suetonius (1510), 155r.

12. *Natural History*, XI, 54, in Pliny (1559), 293, lines 25–28.

13. *Problems*, I, 69, in Alexander of Aphrodisias (1552), 95va.

14. *Problems*, I, 68, in ibid.

15. *Meteorology*, III, 4, 374a19-26, in Aristotle (1550b), 206rb, lines 15–25.

16. *Natural Questions*, I, ii, 4, in Seneca (1529), 414, lines 8–10; translation adapted from Seneca (2010), 142.

17. An inflammation of the eyes, often of the conjunctiva, ophthalmia can cause a discharge that interferes with vision.

18. Della Porta seems to be confused here because, by definition, the cathetus dropped from the center of the eye to object point B *will* intersect the line of incidence at B, although that intersection is irrelevant to the determination of image location. Perhaps what Della Porta had in mind is that the cathetus, so defined, will intersect the refracted ray FG' (which is the line of incidence to the lens after refraction) behind the lens, in which case no image will be seen (see book 2, proposition 11). It is presumably for this reason that he referred to the line passing through A, N, and B as the cathetus, since its intersection with the extension of refracted ray FG', and all other equivalent rays, defines where the blue circle of the halo appears to the eye.

19. Presumably what Della Porta means here is that, since the rays from A and B on the lamp cross centric ray GN in order to strike the eye at points F and H, when the eye is removed farther away from the lamp, those rays will strike a wider area on the eye according to the amount they will have spread from the intersection point on the centric ray. The basic point of this proposition is to show that the rays from outer points A and B of the candle flame will be refracted into the eye so as to produce lines of sight FD and HE. Points D and E thus lie on the outer circle PDEQ of the halo around the candle flame, which is blue. Rays from inside the outer circle of the candle flame, as represented by the gray lines, will form a halo.

20. For the reference to Demosthenes Philalethes, see Aëtius, *Tetrabiblos*, II, iii, 51, in Aëtius (1542), 357. The analysis in this proposition is clearly ad hoc. For instance, there is no obvious reason for assuming that the rays from A and B cross the centric line, other, perhaps, than to explain how those rays spread out to strike ever larger areas on the eye as it moves away from the lamp, the result being that the halo around the lamp will appear to increase in size with that movement. The formation of another color band is illustrated by the gray circle inside black circle AB defining the candle flame. Rays from it reach the eye after crossing and are refracted to the lens. The continuations of the refracted rays striking the lens then reach the plane defined by P/D and Q/E, where they form the gray circle between the outer blue circle of the halo and the outer circle of the candle flame.

21. See *On Dizziness*, in Theophrastus (1605), 257.

22. See *Problems*, III, 20, in Aristotle (1552), 8ra,

23. *Problems*, II, 72, in Alexander of Aphrodisias (1552), 107rb.

24. See *Problems*, III, 20 and XXXI, 11, in Aristotle (1552), 8ra and 49rb.

25. See Averroes (1550), 191vb–192ra.

26. See Avicenna, *Liber de anima*, III, 8, in Avicenna (1508), 16rb-17rb.

27. *Aeneid*, IV, lines 469–70, in Virgil (1579), 109r; translation from Virgil (2005), 92.

28. See, for example, Aristotle, *On Dreams*, 3, esp. 462a30–462b7.

29. *Problems*, II, 55, in Alexander of Aphrodisias (1552), 105ra, lines 36–38.

30. *Colliget*, III, 38, in Averroes (1552), 28va.

31. In locating the common sense in the heart rather than the front ventricle of the brain, Della Porta is following Aristotle literally rather than hewing to the more conventional "Aristotelian" model of internal senses introduced by Avicenna and Averroes and adopted by the vast majority of scholastic thinkers from the mid-thirteenth century on; for a discussion of this model and its genesis, see Smith, *From Sight to Light*, 130–66.

32. One source of this story is Valerius Maximus, *Memorable Deeds and Sayings*, I, 8, in Valerius Maximus (1565), 26rb–va; the direct quotation can be found in the commentary by Oliverius Arziganensis on folio 26va.

33. *Orestes*, lines 255–261, in Euripides (1562), 52a.

34. *Aeneid*, IV, lines 471–73, in Virgil (1579), 109r; translation from Virgil (2005), 92.

35. See *Magiae Naturalis*, VIII, 2, in Della Porta (1589), 151–52. See also note 15 to the Introduction, p. lxxxii above.

36. *Argonautica*, III, lines 65–67, in Valerius Flaccus (1548), 67; translation from Valerius Flaccus (1938), 131.

37. Della Porta notes correctly that, immediately after being dazzled by sunlight, we see a positive afterimage with a slightly ruddy golden cast, followed quickly by a negative one that passes from greenish to blue. See also book 9, proposition 6, p. 401.

38. These aftereffects are presumably due to the disturbance of the aqueous humor as it follows the swift rotation of the wheel or the fast flow of the water, the result being that the disturbance persists for a time even after is objective cause is lost sight of.

39. This presumably refers to the "green flash" phenomenon, which occurs as the top edge of the Sun is just at the horizon during rising or setting. I have been unable to find any testimony whatever to the apparent solar spin at sunrise mentioned by Della Porta.

40. See Aristotle, *On Memory*, 2, 452b5-22; see also Alhacen, *De aspectibus*, II, ii, 25, in Alhacen (1572), 39–42, and in Alhacen (2001), 129–39 (Latin) and 451–57 (English)

41. See *Enneads*, II, viii, 1, in Plotinus (1580), 189–90.

42. The radial paths AID and CLE in this illustration are arbitrary insofar as they seem to be determined not by physical rules but by the need to fit the conclusion: That is, the image of C seen along the extension of refracted ray EL will appear below the image of A seen along the extension of refracted ray DI. An unfortunate and presumably unintended consequence of this choice of radial paths, however, is that those images will be reversed in right-to-left orientation because refracted rays LE and ID intersect before reaching lens DE.

43. I take Della Porta to mean by this that, because the lens is curved, it receives images not just according to length and breadth in the plane perpendicular to the centric ray but also in depth. Accordingly, since E in Figure 7.3 lies deeper in the eye than D and thus below it, the image of C appears to lie below that of A.

LIBER OCTAVUS

PROOEMIUM

Superioribus libris de convexa et concava corporum refractione diximus; ex iis enim nec aliunde adductis primordiis specillorum tractationem aggressi sumus; res ardua, mirabilis, utilis, iucunda, nec ab aliquibus adhuc tentata. Utilitatis equidem amplitudinem nequeo satis mirari, quum qui fere lumine orbati sint, eorum ope, etiam ad longissimam distantiam elongent visum, nec eorum causam congoscunt. Nos iis praecognitis, mirabiliores eiusmodi effectus assequuti sumus, ut longa loci intercapedine distracti minutissima quaeque conspicari possimus, ne hoc nostro instituto intactum omnino aliquod relinquamus. Sed lectorum veniam praecamur, si quae minus probata, omissa, et manca in lucem prodeunt, arduum est enim sine duce in tenebris per ambages ambulare. Sed rem aggrediamur. [174]

Lucem ex refractione multiplicari. PROP. 1

Lux diaphanum corpus pervadens multiplicatur in corpore ipso, sed quomodo id eveniat inquirere tentandum. Vitellio id evenire dixit ex radiorum coadunatione et coarctatione, ut videtur in phiala aquae plena, quae non solum maius lumen ostendunt sed urunt. Sed iste homo nescit aliud esse radiorum coadunationem et multiplicationem; neque de omni refractione loquitur, nam est quae non unit, ut in corporibus planis diaphanis. Nos autem dicimus multiplici id ratione evenire. Primo, quia in ipsa refractione est simul reflexio, nam extima aquae superficies laevigata est instar expoliti et praenitentis speculi, unde ex ea reflexi radii circumfusum aerem sublustrant parietesque oppositos lumine vulnerant, immo ex aquae motu moventur et illi. Virgilius:

BOOK EIGHT

INTRODUCTION

In earlier books we have discussed refraction through convex and concave bodies, and in fact we have approached the treatment of lenses according to basic principles adduced from them, not from other sources; this is a subject that is difficult, wonderful, useful, and pleasurable, and one that has not been attempted by others up to now. Indeed, I cannot admire enough how useful lenses are, when those who are virtually deprived of the light of vision extend their sight even to the greatest distance with their aid but do not understand how they work. Knowing these things beforehand and in order not to leave anything entirely untouched in our instruction, we have followed out the most wonderful of these sorts of effects, for instance, that [using lenses] we can discern the tiniest of things from a spot far away. We beg the indulgence of readers, however, if things that are less esteemed, neglected, or defective emerge into the light, for it is difficult to walk through winding ways in darkness without a guide. But let us turn to the subject at hand. [174]

PROP. 1: *Light is multiplied by refraction.*

Light streaming into a transparent body is multiplied in that body by refraction, but we should try to understand how this happens. Witelo claimed that it happens because of aggregation and uniting, as is seen in a vial full of water, for not only do these exhibit more light, but they also burn.[1] This man, however, does not know that uniting and multiplying rays are different things, and he does not talk about every refraction, for there is a kind that does not unite, as for instance in plane, transparent bodies. We, on the other hand, say this happens for a number of reasons. First, it happens because reflection occurs simultaneously with refraction, for the topmost, shiny surface of water is like a polished and glittering mirror, so the rays reflected from it illuminate the surrounding air somewhat and strike facing walls; consequently, from the movement of the water they, too, are moved. As Virgil puts it:

Sicut aquae tremulum labris ubi lumen ahenis[1]
Sole repercussum, aut radiantis imagine Lunae
Omnia pervolitat late loca, iamque sub auras[2]
Erigitur, summaque ferit laquearia tecti.

Et ex superiori radiorium reflexione etiam in corpore diaphono splendor acquiritur. Secundo evenit ex refractorum radiorum reflexione, nam ex offensante ultima vasis vel corporis [175] superficie resiliunt radii, et ad summam superficiem delabuntur, et ex superiori iterum ad infimam. Sic ex multiplici et mutua superficierum reflexione maius lumen ingeritur, ut videmus in speculis Soli vel candelae oppositis, ex mutua enim superficierum reverberatione radiorum ex obliquo quamplurimae conspicientur candelae etsi unica existat. Exemplum.

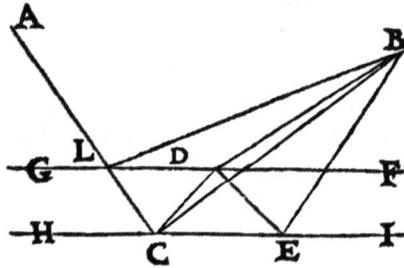

Esto oculus B, candela A, speculum cuius suprema superficies FG, ima IH. Candelae lumen ferit L; resilit ad oculum eius imago per lineam LB. Pervenit ad C; resilit per lineam CB. Mox relabitur ad D; resilit per DE.[3] Sic etiam ex E, unde quatuor locis eius imaginem videmus, et quanto obliquior visus, eius multiplicior figura. Est et alia causa, quod lumen lucidum corpus pervadens eius luciditate induitur augeturque. Lux transiens rubrum vitrum rubra perspicitur, et si vitrum impurum, minuitur lumen, sed per nitidum crystallum clarior splendiorque, ex his per crystallina specilla clariora omnia et nitidiora conspicimus. Per clariora igitur diaphana corpora illabens lumen augetur multiplicaturque.

In specillis convexis inversionis punctum invenire. PROP. 2 [176]

Puncta concursus refractarum linearum tam convexi quam concavi corporis sunt puncta inversionum. Puncta concursus secundo libro prop. 3 determinavimus in crystallina pila, et punctus inversionis non alius quam concursus

---

[1] "alienis" in original text.
[2] "atris" in original text.
[3] "DB" in original text.

Sunlight, or the radiant moon reflected from water
Trembling in a bronze bowl, will glance and flit
All over a room—and then flash suddenly
Onto the coffered ceiling high above.[2]

And so brightness is even obtained through the topmost reflection of rays
in a transparent body. Second, the multiplication of light happens by means
of the reflection of refracted rays, for the rays rebound in striking the inner
surface of a [transparent] vessel or [175] body, and they stream to the top
surface, and likewise from the upper to the lower one. Thus, more light is
generated by multiple and mutual reflections from the surfaces, as we see
in mirrors facing the Sun or candles, for because of a mutual reflection of
obliquely incident rays from the surfaces a number of candles are seen even
though only one exists. Here is an illustration.

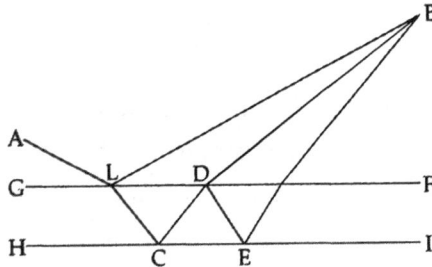

Figure 8.1

Let B [in Figure 8.1] be the eye, A a candle, and let there be a mirror
whose upper surface is FG and whose lower one is IH.[3] The light of the
candle reaches L, and its image rebounds to the eye along line LB. That light
also reaches C [by refraction] and rebounds along line CB. Then it is diverted
to D [by reflection] and rebounds along DE.[4] The same holds at E, so we
see the candle's image at four places, and the farther off to the side the eye
lies, the more the candle's image is multiplied.[5] There is another cause as
well, insofar as the brightness of light streaming into a transparent body is
enhanced and amplified. Light transmitted through red glass is perceived as
red, and, if the glass is unclear, the light is diminished, whereas it appears
clearer and brighter in passing through clear crystal, for which reasons
we perceive everything more clearly through crystal lenses. Hence, light
penetrating very clear, transparent bodies is augmented and multiplied.[6]

PROP. 2: *To find the point of inversion in convex lenses.* [176]

The points of intersection of refracted lines are the points of inversion in a
convex as well as a concave transparent body. In proposition 3 of the second
book we determined the points of intersection in a glass sphere, and the
point of inversion is nothing more than the intersection of the rays on the

radiorum ex oppositis in circulo partibus, uti ex sinistra in dextram, aut ex dextra in sinistram, et sursum deorsumque. Exemplum.

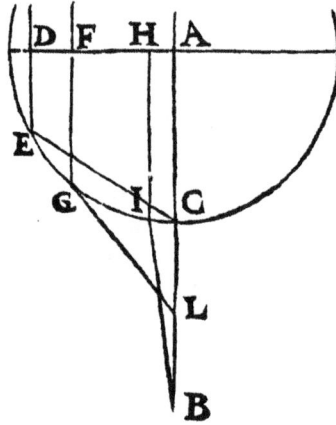

Esto vitrea semisphaera CIGE, et incidat radius DE tangens latus exagoni CE. Quae supra E sunt ad rem nostram non faciunt, nam intra crystalli corpus refranguntur; quae infra I minus, nam languescunt circa diametri lineam AB. Radii ergo incidentes intra sphaerae partem GI et intra perpendicularem lineam LB refranguntur; ibi sunt inversionum puncta, et ibi radiorum coadunatio et concursus.

In convexis specillis, posito oculo in puncto inversionis, magnitudinis imagine specillum complebitur. PROP. 3 [177]

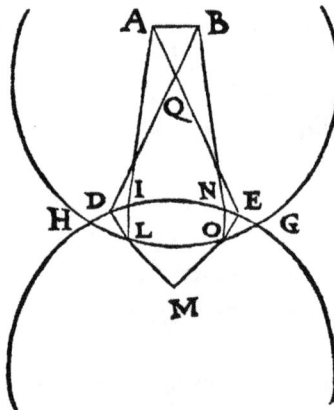

Esto opposita magnitudo AB, convexum specillum GH; veniat B punctus ad I. Per secundam secundi, frangitur in L, et per quintam refrangitur ad M

opposite sides of the circle so that the rays cross from left to right, or from right to left, or from upwards to downwards.[7] Here is an illustration.

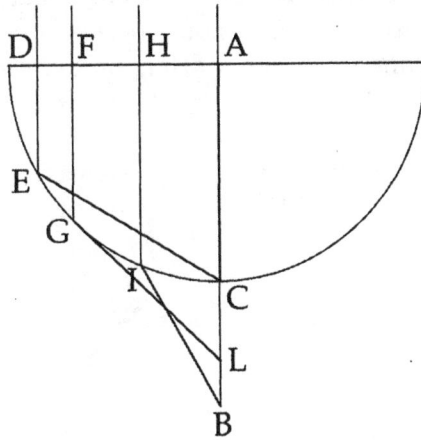

Figure 8.2

Let CIGE [in Figure 8.2] be a glass hemisphere, and let ray DE strike the side CE of an inscribed hexagon. Any rays that strike beyond E do not matter in our account, for they are refracted within the body of the glass, but those that strike inside I are refracted less and indeed descend around diameter AB. Thus, the rays striking segment GI within the sphere are also refracted within perpendicular line LB, and here is where the points of inversion lie and where the rays intersect and unite.[8]

PROP. 3: *In convex lenses, when the eye is placed at the point of inversion, the lens will be filled by the image of a magnitude seen through it.* [177]

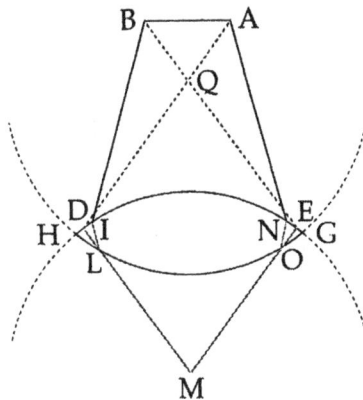

Figure 8.3

Let AB [in Figure 8.3][9] be the facing magnitude and GH the convex lens, and let the image of point B reach I. According to the second proposition

oculum. Eodem modo punctus A veniat ad N. Frangatur ad O; mox refrangatur ad oculum in M. Extendantur catheti ex punctis rei visae A, B et per
centrum inferioris superficiei specilli Q. Sic AQ occurret ML in D, et BQ
occurret MO in E. Totum specilli spatium occupabitur imagine AB, et nil
determinatum sed omne confusum videbit, est enim oculus in M inversionis puncto.

In convexis specillis, magnitudo posita in puncto inversionis, eius imagine
totum specillum occupabitur. PROP. 4

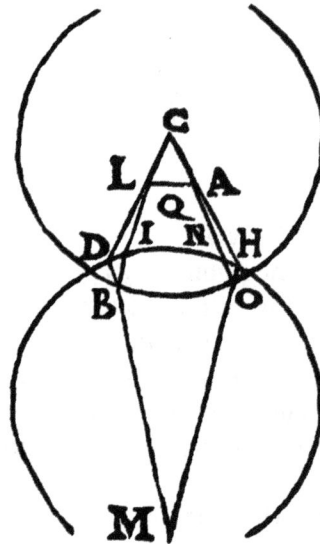

Esto visenda magnitudo AB, centrum inversionis Q prope magnitudinem
AB. Veniat punctus B ad I. Frangitur ad L; ex L frangitur ad oculum M longe
existentem. Eodem modo A punctus venit ad N, ex N ad O, et ex O ad
oculum M. Extendantur MO, ML ad cathetos ex centro C productas et puncta
A, B. Sic occurret CB ipsi ML in D, et CA ipsi MO in H,[4] occupabitur ergo
totum specillum. [178]

In convexis specillis, magnitudine et oculo in puncto inversionis positis, imago
specillum totum occupabit. PROP. 5

Sit magnitudo AB prope punctum inversionis Q. Accedat punctus B ad
specillum GH in puncto I[5]; refrangatur ad L. Mox ad oculum perectabit M.
Extendatur ML quousque occurrat catheto extentae ex centro C et per punctum B, et concurrent in puncto D. Eodem modo A veniet ad N; refrangetur

---

[4]"E" in original text.
[5]"D" in original text.

of the second book, it refracts to L, and according to the fifth proposition of the same book it refracts to the eye at M. Likewise, let the image of point A reach N. Let it refract to O, and then let it refract to the eye at M. Extend the catheti from points A and B of the visible object through center point Q of the lower surface of the lens. AQ will therefore meet ML at D, and BQ will meet MO at E. The entire space of the lens will be occupied by the image of AB, and the eye will in no way see it as definite but as totally confused, for the eye is at point M of inversion.[10]

PROP. 4: *In convex lenses, if a magnitude is placed at the point of inversion, the entire lens will be filled by its image.*

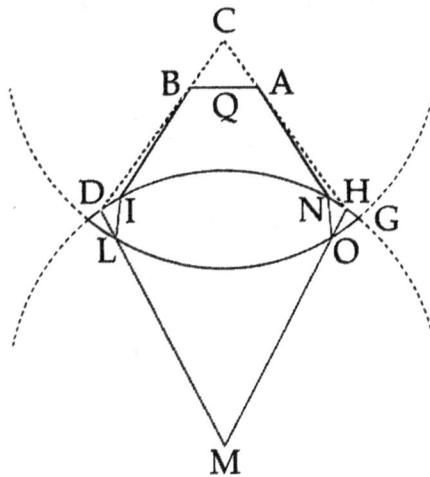

Figure 8.4

Let AB [in Figure 8.4] be the magnitude to be seen and Q the center of inversion on magnitude AB. Let the image of point B reach I. It is refracted to L, and from L it is refracted to the eye far away at M. Likewise, the image of point A reaches N, then goes from N to O, and from O to the eye at M. Extend MO and ML to the catheti dropped from center point C and through points A and B. CB will therefore meet ML at D, and CA will meet MO at H, so [image DH] will occupy the whole lens. [178]

PROP. 5: *In convex lenses, if the magnitude and the eye are each placed at a point of inversion, the image will occupy the entire lens.*

Let AB [in Figure 8.5] be a magnitude at point of inversion Q. Let the image of point B reach lens GH at point I, and let it be refracted to L. Then it will proceed to the eye at M. Extend ML until it meets the cathetus dropped from the center C through point B, and they will intersect at point D. Likewise, the image of A will reach N, will be refracted to O, and will end

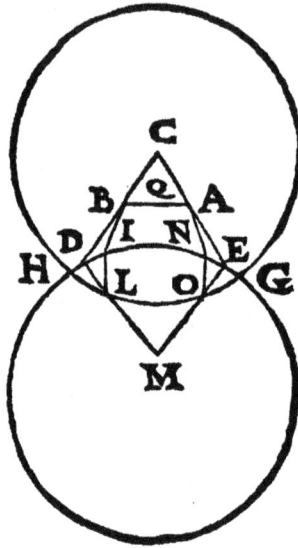

ad O; veniet ad M. Extendatur MO usque ad E, quo loco occurret catheto ex CA directae. Magnitudo igitur AB occupabit totum specillum ED.

In convexis specillis, magnitudine specillo propinqua, oculo prope aut procul posito, semper recta videbitur. PROP. 6

Esto magnitudo AB prope specillum HL; accedat punctus B ad specillum N. Refrangitur ad O; refrangitur ad F oculum. Elongetur FO in E, et trahatur cathetus ex CB. Occurret FO[6] [179] in E; punctus B in E videbitur. Sic A venit in O, et in R, et in F oculum. Extendatur FR; occurret catheto ex CA in D. AB magnitudo recta videbitur in DE. Elongetur oculus in G. Venit

---

[6]"FB" in original text.

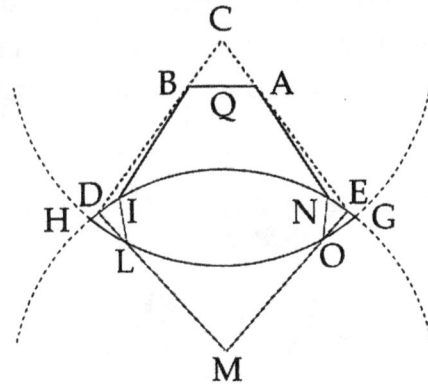

Figure 8.5

up at M. Extend MO to E, where it will intersect the cathetus extending straight from CA. Consequently, the image ED of magnitude AB will occupy the whole lens.

PROP. 6: *In convex lenses, if the magnitude is near the lens, it will always appear upright, whether the eye is placed near or far away.*

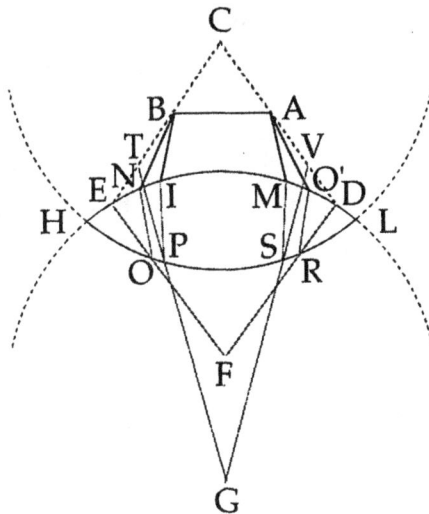

Figure 8.6

Let AB [in Figure 8.6][11] be a magnitude near lens HL, and let the image of point B reach N on the lens. It is refracted to O and then refracted to the eye at F. Extend FO to E, and draw the cathetus along CB. It will intersect FO at E, [179] and so point B will be seen at E. Likewise, the image of A reaches O', then R, and then the eye at F. Extend FR, and it will intersect the cathetus dropped along CA at D. Magnitude AB will be seen upright

punctus B ad I, ad P, ad G oculum. Elongetur GP; occurret eidem catheto in T. Eodem modo A venit in M, mox in S, inde ad G. Elongata GS,[7] occurret catheto eidem CA[8] in V; etiam recta magnitudo videbitur, et minor quanto oculus magis elongabitur.

In convexis specillis, oculo specillo propinquo, magnitudine prope aut[9] procul posita, semper recta videbitur. PROP. 7

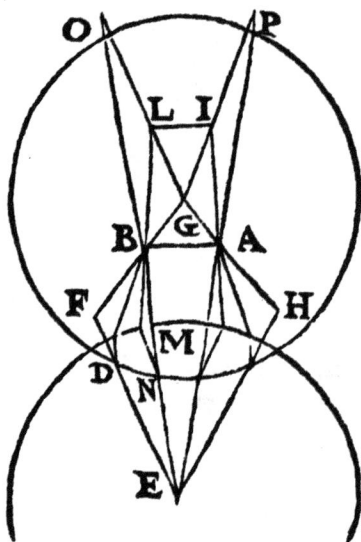

Esto magnitudo specillo propinqua AB. Veniat punctus B ad specillum C; refrangatur ad D; ex D refrangatur denuo ad E. Extenditur ED in F. Trahaturque cathetus ex G centro et B, et occurret ED in F. Eodem modo punctus A occurret catheto ex GA in H, et recta erit magnitudo. Remanente oculo in suo situ, elongetur magnitudo AB in IL; veniat punctus L ad M. Inde refrangatur ad N, et ex N ad E. Extendatur EN in O, et trahatur cathetus ex G per L. Occurret ei in O, ibi videbitur L. Eodem modo I videbitur in P. [180]

In convexis specillis, magnitudine et oculo longe positis, inversa videbitur magnitudo et propinquior. PROP. 8

Magnitudine AB longe existente et oculo E, veniat B ad C. Refrangatur ad D et ex D ad oculum E. Extendatur ED quousque coincidat cum catheto,

---

[7]"GH" in original text.
[8]"CH" in original text.
[9]"ut" in original text.

along DE. Move the eye away to G. The image of point B reaches I, then P, and then the eye at G. Extend GP, and it will intersect the same cathetus [CB] at T. By the same token, the image of point A reaches M, then S, and then G. If GS is extended, it will intersect cathetus CA at V, and so the magnitude will also appear upright, but smaller depending upon how far away the eye will lie.

PROP. 7: *In convex lenses, when the eye is near the lens and the magnitude lies near or far away, it will always appear upright.*

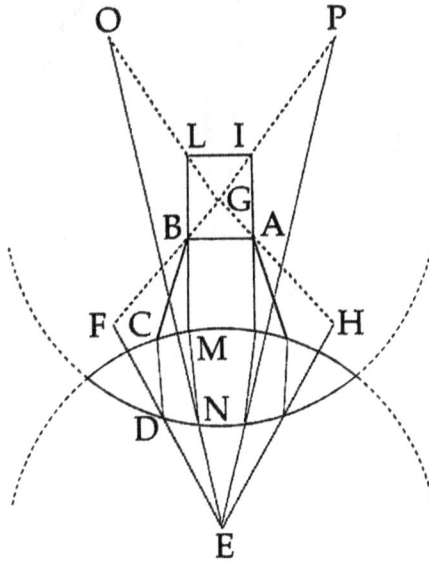

Figure 8.7

Let magnitude AB [in Figure 8.7] be near the lens. Let the image of point B reach C on the lens, let it be refracted to D; and let it be refracted once again from D to E. ED is extended to F. Draw the cathetus from center point G [of the lens's lower surface] through B, and it will intersect ED at F. Likewise, [the last refracted ray from] point A will intersect the cathetus extended along GA at H, and so the image [FH] of the magnitude will be upright. With the eye left in its original place, draw magnitude AB back to IL, and let the image of point L reach M. Let it then be refracted to N and from N to E. Extend EN to O, and drop the cathetus from G through L. It will meet EN at O, where L will be seen. By the same token, the image of I will be seen at P [and so image OP will appear upright] [180].

PROP. 8: *In convex lenses, if the magnitude and the eye lie far away, the magnitude will appear inverted and nearer than it should.*

With magnitude AB far away [in Figure 8.8] and the eye far away at E, let the image of B reach C. Let it be refracted to D and from D to the eye at E.

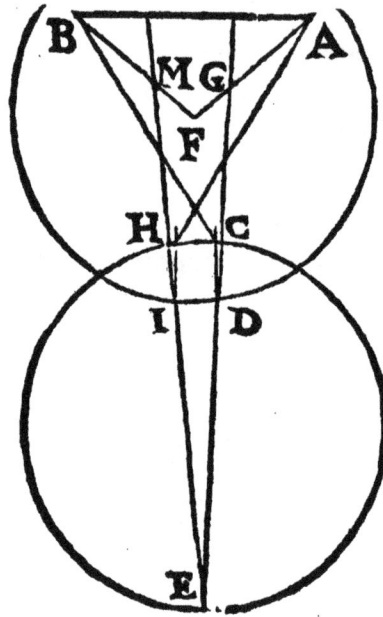

quae extendatur ex puncto B et centro circuli inferioris superficiei F, et
puncto A, et erit G. Eodem modo A punctus ad H, ad I, ad E oculum.
Extentaque EI, coincidet in M cum catheto FB; videbitur inversa, et minor,
et propinquior.

*In convexis specillis, quanto oculus et magnitudo magis elongabitur, minor
videbitur, at utroque accedente maior. PROP. 9*

Ex superiori propositione, quanto magis magnitudo AB recesserit, eo incidentiae lineae ad specillum angustabuntur, et cathetus ex centro circuli et
rei visae veniens declivius veniet, et res minorabitur. Contra vero in accessu
utriusque incidentiae lineae dilatabuntur et catheti magis rectae accedentes,
lineas formas deferentes excipient, et imagines maiores videbuntur. [181]

   Veniat prior longinquior magnitudo AB, cuius punctus B venit ad D.
Refrangitur in E, mox ad oculum F. Elongetur EF quousque coeat cum
catheto ex B et centro circuli C; erit in G. Ex alia parte, punctus A ad H,
mox ad I, mox ad F. Extenta IF coincidet cum catheto ex AC in L. Erit

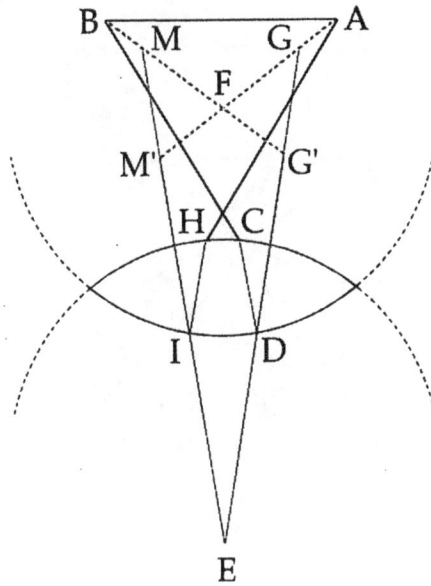

Figure 8.8

Extend ED until it meets with the cathetus extended from point B through center point F of the lower surface of the lens, as well as the cathetus from point A through F, and the intersection will be at point G. Likewise, the image of point A reaches H, then I, and then the eye at E. When it is extended, EI will intersect cathetus FB at M, and so the image of magnitude AB will appear inverted, as well as smaller and nearer than it should.[12]

PROP. 9: *In convex lenses, the more the eye and the magnitude withdraw away from the lens, the smaller the magnitude will appear to be, whereas the closer both get, the larger the magnitude will appear to be.*

According to proposition [7] above, the farther magnitude AB withdraws from the lens, the narrower the angle formed by the lines of incidence will get, the more slanted the cathetus reaching from the center of the circle to the visible object will become, and the more the image of the object will shrink.[13] On the other hand, when the two lines of incidence approach the lens, they will expand, and as the catheti become increasingly straight, the lines conveying the forms will draw apart, and the images will appear enlarged.[14] **[181]**

   Let magnitude AB [in Figure 8.9] from before [in proposition 8] be drawn farther away, and let the image of point B reach D. It is refracted to E, then to the eye at F. EF will be extended until it intersects the cathetus drawn through B and center point C of the circle; and the intersection will be at G. On the other side, the image of point A reaches H, then I, and then F. When extended, IF intersects the cathetus through AC at L. The image

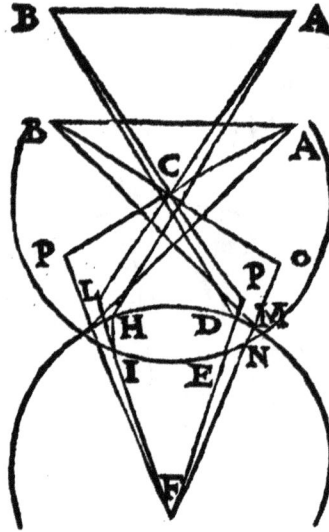

magnitudo AB, GL, ut in praecedenti. Accedat magnitudo AB propius.[10] B veniet ad M, ad N, ad F. Extenta NF, coincidet in O cum catheto ex BC. Sic A in P; maior est magnitudo OP, quam GL.

In specillis convexis, magnitudo intra centrum circulis,[11] *quanto ei propinquior, tanto maior. PROP. 10*

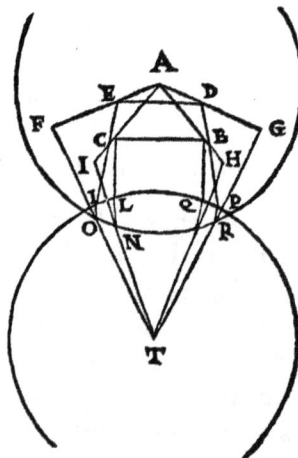

Sit magnitudo remotior a circuli centro A ipsa BC. Veniat punctus C ad L, inde ad N, mox ad oculum T. Extendatur TN usquequo coeat cum catheto

---

[10]"proprius" in original text.
[11]"oculis" in original text.

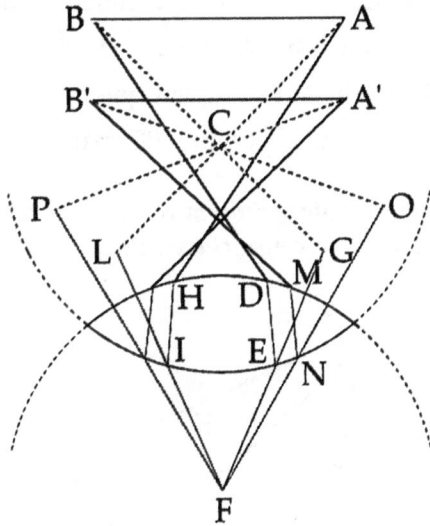

Figure 8.9

of magnitude AB will be GL, as [G'M' was] in the previous proposition. Let magnitude A'B' be drawn nearer the lens. The image of B' will reach M, then N, and then F. When extended, NF will intersect the cathetus through B'C at O. Likewise, the image of A' will be seen at P, and OP > GL.

PROP. 10: *In convex lenses, if the magnitude lies inside the center point of the circle [of the lens's lower surface], the nearer it is to that point, the larger its image will be.*

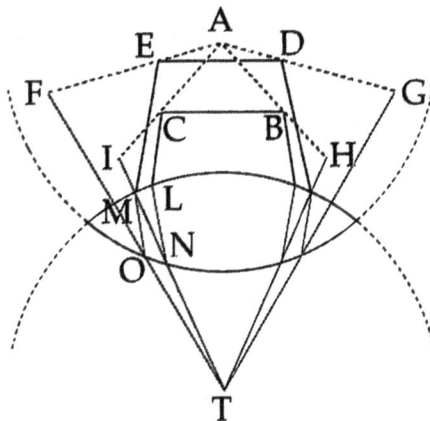

Figure 8.10

Let BC [in Figure 8.10] be a magnitude farther from center point A of the circle [forming an arc on the lower surface of the lens]. Let the image of point C reach L, then N, and then the eye at T. Extend TN until it intersects

ex A et puncto rei visae C, et coibit in I. Sic punctus alter B videbitur in H. Sit vero magnitudo propior[12] ipsi A ipsa DE. Veniat E ad M, inde ad O, tandem ad oculum T. Extendatur TO quousque cum catheto ex A et E, et erit in F; ex altera parte erit G. Erit igitur FG maior ipsa HI. [182]

Magnitudo oblique ad convexum specillum veniens, longe a suo loco videbitur, evenitque interdum ut utraque conspiciatur. PROP. 11

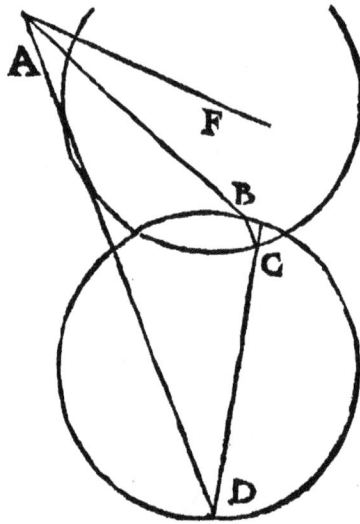

Sit magnitudo A veniat ad specilli partem B. Refrangitur ad C; venit ad oculum D.[13] Extendatur DC in E; mox ducatur cathetus per A et F centrum circuli. Occurret lineae imaginem deferenti in E. Punctus igitur A longe ab A videbitur, nempe in E. Trahatur ab oculo D linea DA. Extra specillum utramque videbis et magnitudinem A et magnitudinis imaginem E. [183]

In convexis specillis, magnitudine binis oculis opposita, duplex videbitur. PROP. 12

Magnitudo A binis oculis opposita H, E duplicata videbitur. Veniat punctus A ad C, ex C refrangatur ad D, [184] et refrangatur denuo ad oculum E dextrum. Idem punctus veniat ad specillum F, mox frangatur ad G, et refrangatur ad oculum H sinistrum. Trahatur cathetus ex A puncto rei visae et specilli centrum B; erit ABLI. Occurrit extensae ED in I et HG in L. Ab oculo

---

[12]"proprior" in original text.
[13]"B" in original text.

the cathetus drawn from A through point C of the visible object, and it will intersect at I. Accordingly, the other point B will appear at H. Now let DE be a magnitude [of the same size] nearer center point A. Let the image of E reach M, then O, and finally the eye at T. Extend TO until it intersects the cathetus drawn through A and E, and the intersection will be at F, whereas on the other side it will be at G. Accordingly, FG > HI. [182]

PROP. 11: *If the image of a magnitude reaches a convex lens from off to the side, it will appear far from where it actually is, and it sometimes happens that both it and its image are seen.*

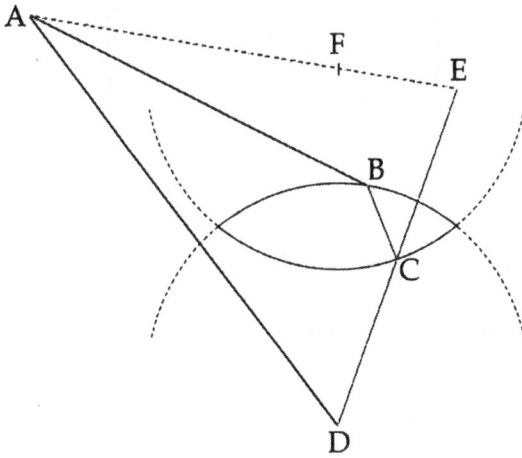

Figure 8.11

Let the image of magnitude A [in Figure 8.11] reach spot B on the lens. It is refracted to C, and it reaches the eye at D. Extend DC to E, and then draw the cathetus through A and center point F of the circle. It will intersect the line conveying the image at E. Consequently, the image of point A will appear far away from point A, namely, at E. Draw line DA from the eye at D. You will see both magnitude A outside the lens and image E of the magnitude. [183]

PROP. 12: *In convex lenses, when a magnitude faces both eyes, it will appear double.*

Magnitude A [in Figure 8.12] facing both eyes at H and E will appear double. Let the image of point A reach C, let it refract from C to D, [184] and let it refract again to the right eye at E. Let the image of the same point reach F on the lens, then refract to G, and then refract to the left eye at H. Draw the cathetus through point A on the visible object and center point B of the [lower surface of the] lens, and it will be ABLI. It intersects the extension of ED at I, and the extension of HG at L. Hence magnitude A will be seen

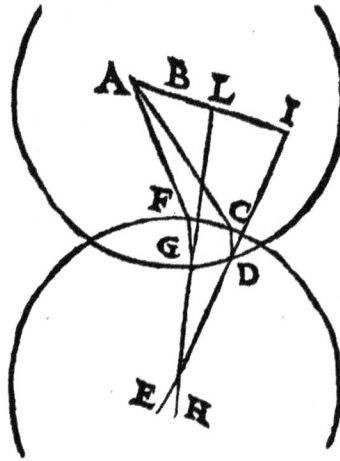

igitur dextro E videbitur magnitudo A in I; ab oculo sinistro H videbitur in L. Clauso dextro peribit imago I; clauso sinistro L.

Convexis specillis Soli oppositis ignis accenditur. PROP. 13

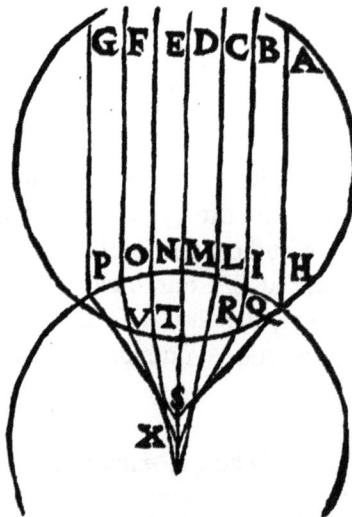

Corpus solare a quo radii profluunt aequidistantes ABCDEFG. Labuntur ex specilli superficie extima HILMNOP ad imam se refrangentes QRTV. Inde exeuntes ad perpendicularem omnes coeunt DMSX prope puncta S, X; ibi ignis accenditur, ut diximus in pila chrystrallina.

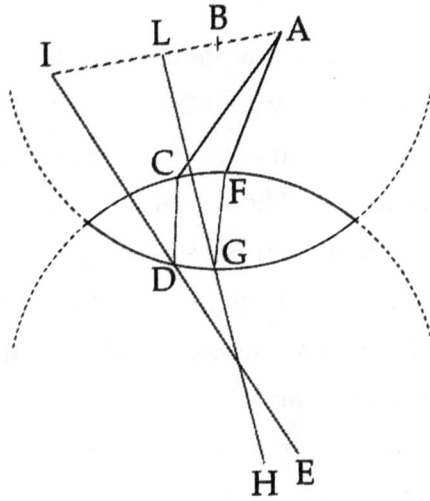

Figure 8.12

at I by the right eye at E, and it will be seen at L by the left eye at H. If the right eye is closed, image I will disappear, and if the left one is closed, image L will disappear.

PROP. 13: *Fire is ignited by convex lenses facing the Sun.*

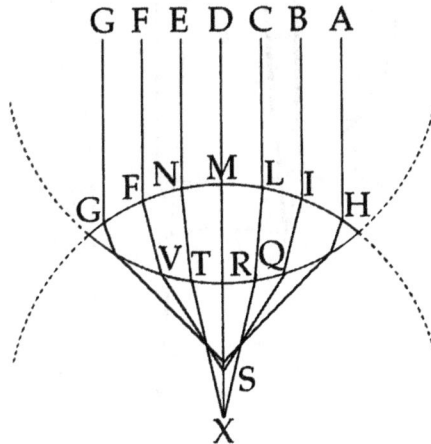

Figure 8.13

Let parallel rays at ABCDEFG [in Figure 8.13] emanate from the solar body. They are diverted by refraction from the upper surface HILMNOP to the lower surface QRTV of the lens. Passing out from there, all of them intersect axis DMSX around points S and X, where fire is lit, as we said in our discussion of the glass sphere.[15]

Senes convexis specillis clarius vident. PROP. 14

Duplex est causa cur senes convexis specillis clarius et perfectius cernant. Primo quia in senectute relaxatur pupilla—nec solum pupilla, sed omnia membra et membrorum [185] retinacula, ut laxius lotium et faeces retineant. Ex laxitate igitur pupillae radii liberius vagantur, et rem laxam, et minus certam crystallino reddunt; at convexis specillis simulachri radii coeunt et arctius pyramis colligitur, ut vidimus in praecedenti, unde naturae vitium rependunt congregando specilla convexa simulachra. Altera causa est quod senibus vitreus humor faeculentior et impurior redditur, ut in prima propositione huius libri probavimus. Intro permeans lux per crystallum clarior fulgidiorque redditur, et naturae defectus alter ex pituita resarcitur.

In concavis specillis res semper minor videbitur. PROP. 15

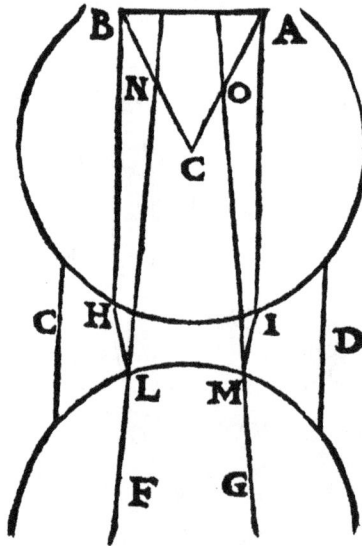

Sit magnitudo AB. Venit A punctus ad concavum specillum DE in punctum I, et ex I frangitur ad M, per tertium quinti nostri. Ex M refrangitur ad G per quartum supradicti nostri in contrariam scilicet partem. Extendatur GM. Occurrit catheto ex C, circuli extimi centro et puncto A in O. Eodem modo B ad H venit, frangitur ad L, refrangitur ad F; extenditur ad N, ubi occurrit

PROP. 14: *The elderly see more clearly through convex lenses.*

There are two reasons why the elderly see more clearly and perfectly through convex lenses. The first is that the pupil loosens in old age—and not just the pupil, but all the components and what connects the components, [185] just as older people retain urine and feces more loosely. Therefore, because of the looseness of the pupil, the incoming rays wander more freely, and they render the image of an object loose and less clear to the crystalline humor. In convex lenses, though, the rays composing the image unite and the radiant cone is bound more narrowly, as we saw in the preceding proposition, so convex lenses compensate for the shortcoming of nature by concentrating the images. Another reason is that in the elderly the vitreous humor becomes murkier and more impure, as we showed in the first proposition of this book.[16] Light spreading into crystal is rendered clearer and brighter, and for that reason the other defect of nature due to the viscosity of the humor is rectified.

PROP. 15: *An object will always appear smaller in concave lenses.*

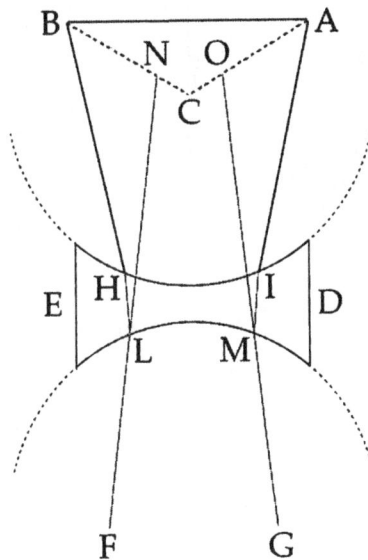

Figure 8.15

Let AB [in Figure 8.15] be a magnitude. The image of point A reaches concave lens DE at point I and is refracted to M from I, according to the third proposition of the fifth book of our treatise.[17] According to the fourth proposition of the aforementioned book, it is refracted from M to G, that is, in the opposite direction.[18] Extend GM. It intersects the cathetus [AC] drawn through the center of the upper circle and point A at O. Likewise, the image of B reaches H, is refracted to L, and is refracted to F; and [refracted ray FL] extends to N, where it intersects the cathetus drawn through CB at

catheto ex CB in N. Semper ON erit minor AB, quia franguntur intus. [186]

Longe oculo a concavo specillo se posito, magnitudo visa minor evadit, at oculo propius[14] admoto, maior priore, sed non ipsa magnitudine. PROP. 16

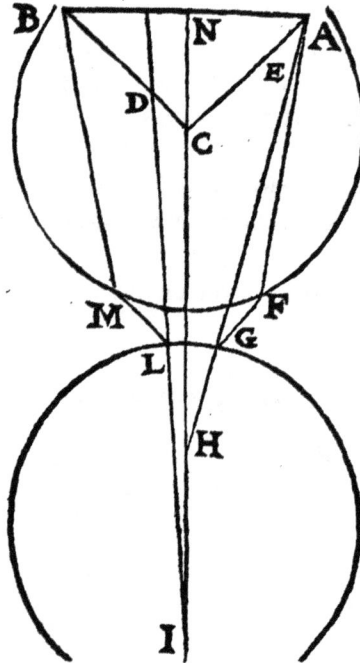

Magnitudo AB, cuius pars NB. Veniat punctus B ad M. Refrangitur ad L; labitur ad I oculum longe se positum. Elongetur IL. Venit ad D; occurrit catheto ex C centro et B in D. Igitur B in D videbitur, satis minor ipsa NB. Accedat propius[15] oculus in H. Veniat alter punctus A in F. Fugit in G; refrangitur in H. Elongetur HG in E. Occurret catheto in E; maior erit NE ipsa ON.[16] [187]

In specillis concavis magnitudo oblique veniens[17] longe a suo loco videtur, et interdum utraque conspicitur. PROP. 17

Sit magnitudo G; veniat ad specillum DF. Frangitur ad C; refrangitur ad

---

[14]"proprius" in original text.
[15]"proprius" in original text.
[16]"OB" in original text.
[17]"venient" in original text.

N. ON will invariably be smaller than AB because [the rays from AB] are refracted inward in the lens. [186]

PROP. 16: *If the eye is placed far away from a concave lens, the visible magnitude ends up looking smaller, but when the eye is moved nearer, the magnitude appears larger than before, but not the actual size of the magnitude itself.*

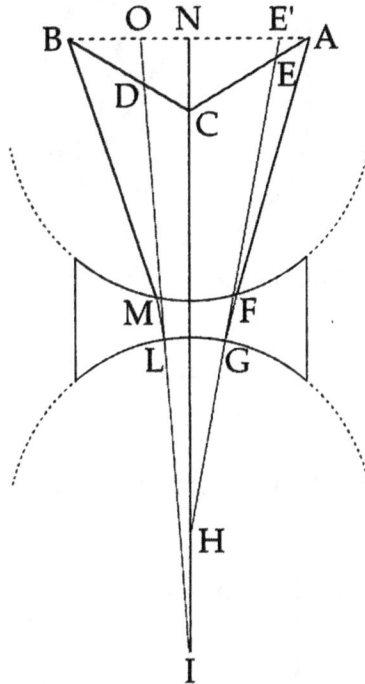

Figure 8.16

AB [in Figure 8.16] is a magnitude whose half is segment NB. Let the image of point B reach M. It is refracted to L, and it inclines to the eye placed far away at I. Extend IL. It reaches D [and O] and intersects the cathetus through center point C and B at D. B will therefore be seen at D [and its projection ON on AB is] quite a bit smaller than NB. Let the eye move nearer to the lens at H. Let the image of the other point A reach F. It diverts off to G and is refracted to H. Extend HG to E'. It will intersect the cathetus [CA] at E, and NE' > ON.[19] [187]

PROP. 17: *In concave lenses, when the image of a magnitude arrives from off to the side, it appears far away from its proper place, and sometimes it is seen double.*

Let G [in Figure 8.17] be a magnitude, and let its image reach [point B on] lens DF. It is refracted to C, and then refracted to the eye at H. Extend HC

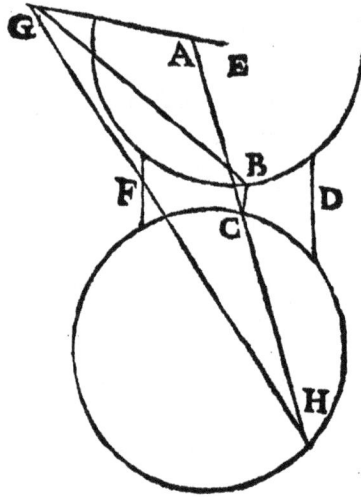

oculum H. Elongetur HC quousque occurrat catheto ex G magnitudine et E centro, et occurrit in A. Ibi igitur videbitur G. Ex oculo etiam H videbitur G magnitudo extra specillum DF[18] in G. Utramque igitur magnitudinem videt[19]: veram et verae simulachrum.

Specillo concavo Soli opposito ignis non accenditur. PROP. 18

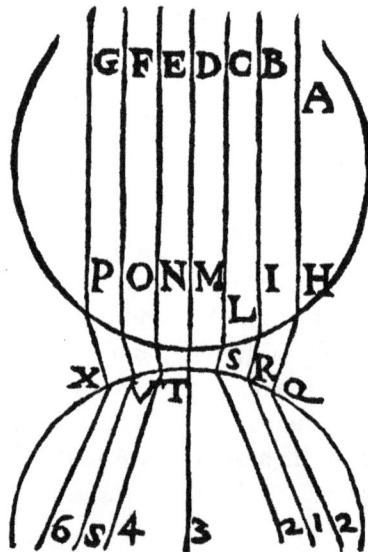

Esto Sol, cuius radii aequidistantes et perpendiculares ABCDEFG cadant recti ad concavam superficiem specilli, AH, BI, CL, DM, EN, FO, GP. Et H

---

[18] "DE" in original text.
[19] "vident" in original text.

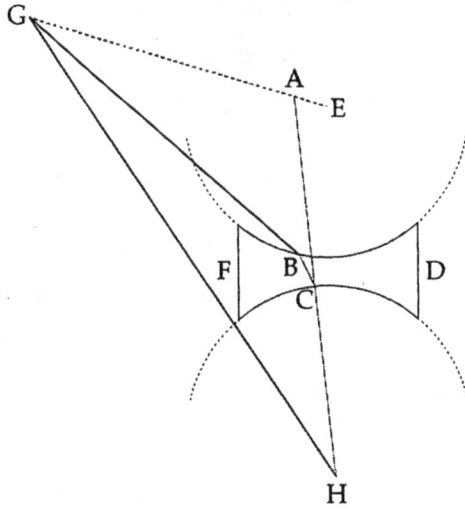

Figure 8.17

until it intersects the cathetus drawn through magnitude G and center point E [of the lens's top surface], and it intersects at A. This, then, is where G will be seen. Magnitude G will also be seen by the eye at H outside lens DF at G. Hence, the eye sees both magnitudes: the real one and the image of the real one.

PROP. 18: *Fire is not ignited by a concave lens facing the Sun.*

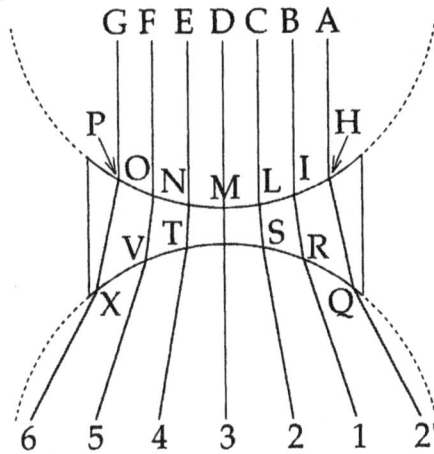

Figure 8.18

Let ABCDEFG [in Figure 8.18] be [points illuminated by] the Sun, and let parallel and perpendicular rays AH, BI, CL, DM, EN, FO, and GP fall directly

frangatur ad Q, I ad R,[20] L ad S, N ad T, O ad V, et P ad X. Et refrangatur Q ad [188] 2, R ad I, S ad 2. DM mansit directe, T ad 4, V ad 5, X ad 6. Non igitur uniuntur, sed in contrariam partem dilatantur.

Visu debiles concavis specillis acutius vident. PROP. 19

Iuvenes qui arcta sunt pupilla, si[21] vitreo humore qui in oculo continetur non claro, duo requirerent: et quae simulachra dilatarent, ut resarciretur vitium pupillae et quodammodo unirent, et quod lucem clariorem redderent. Duo haec praestat concavum specillum, nam et simulachrum quodammodo unit, ut ex refractionibus intra viri soliditatem apparet, et quodammodo aperiret, ut videmus lineis in adversam partem refugientibus, et lux pertransiens visum multiplicatur.

---

[20]"K" in original text.
[21]"se" in original text.

on the concave surface of the lens. Let the light at H be refracted to Q, that at I to R, that at L to S, that at N to T, that at O to V, and that at P to X. Then let the light at Q be refracted to [188] 2', that at R to I, and that at S to 2. DM remains straight, the light at T is refracted to 4, that at V to 5, and that at X to 6. The rays are therefore not united but spread out in the opposite direction.

PROP. 19: *Those who have weak sight see more clearly through concave lenses.*

Young people who have a narrow pupil and a cloudy vitreous humor enclosed within the eye may have needed two things: first, that the images dilate and somehow unite in order to overcome the defect of the pupil, and second that the light be rendered brighter. A concave lens accomplishes these two things, for by refraction it somehow unites the image to make it appear within the mass of the humor, and it opens it out somewhat, as we see with the lines fanning out in the opposite direction, and thus the light that passes through the eye is multiplied.[20]

NOTES

1. See *Perspectiva*, X, 48, in Witelo (1572), 443–44.

2. *Aeneid*, VIII, lines 22–25, in Virgil (1579), 162r; translation from Virgil (2005), 190–91.

3. Della Porta obviously has a glass mirror in mind, reflecting surface GF being the top of the glass plate, and reflecting surface HI being the metal backing.

4. Contrary to the claim made in the previous sentence, the light striking C does not reflect along CB to reach the eye at B. Instead, it reflects along CD and is refracted away from the normal at surface GF to reach the eye along DB.

5. In fact, according to this analysis, the eye at B would see three, not four, images of the candle flame because the image reflected from C would not reach the eye along straight line CB; instead, it would reach the eye along line DB after refraction from the glass into the air. The same holds for the image reflected from E; it, too, would reach the eye along a line refracted at surface GF of the glass. Consequently, the number of images will be determined by the number of reflections: one, and only one, from surface GF of the glass, and two more from surface HI of the metal backing to the mirror.

6. In claiming that light is brightened when it passes through extremely transparent media, such as crystal, Della Porta takes a peculiar stand against tradition (as well as his argument in book 1, proposition 20), according to which refraction weakens light to a greater or lesser extent, depending on how transparent the medium through which it passes is. Refraction was in fact attributed to this weakening so that, for instance, light passing obliquely from almost perfectly transparent aether into the atmospheric shell of air is refracted far less than light passing from air into less-transparent water. Likewise, light passing obliquely from air into glass, which is even less transparent, is refracted even more, each stage of refraction being due to the increased weakening of light; see note 21 to book 2.

7. As we saw in book 2, Della Porta brought the point of inversion into the analysis in several propositions dealing with light passing through spherical lenses; see, for example, propositions 16–20, which refer either explicitly or implicitly to the point of inversion. Because there is no actual point of intersection of rays passing through a concave lens, the point of inversion in such lenses is virtual, where the imagined continuation of the actual rays dispersing outward from the concave surface would meet on the other side.

8. Although this analysis is clearly based on the discussion of refraction through glass spheres in book 2, it does not comport with that analysis as it pertains to proposition 5 of the second book, which addresses precisely the same case as Della Porta does in this proposition. Accordingly, in Figure 2.5 from the former proposition, chord FL is the side of an inscribed hexagon, and in Figure 8.2 from this proposition, chord EC represents the same length. However, as is obvious from Figure 2.5, ray AF supposedly refracts along FR to some point on axis EL below surface FL of the glass according to the principle of reciprocity that Della Porta

applies there. In Figure 8.2, on the other hand, the equivalent ray DE refracts within the glass to point C on axis AB, which lies on, not below, surface AC of the glass. The conflict between the two accounts lies in Della Porta's treating the hemisphere as if it were a sphere, in which case the light entering the topmost surface will indeed refract to point C according to his model. In this case, he is treating the refraction of ray DE as if it were perfectly equivalent to reflection from a concave spherical mirror along EC, which in fact it is not, according to his model in book 2. Thus, Della Porta's analysis in this proposition is qualitatively but not quantitatively equivalent to the analysis that follows from his discussion in book 2. It is worth noting, as well, that Della Porta is not taking the critical angle of ~42° into account in this proposition, which means that ray DE incident at 60° will undergo total internal reflection, as will any ray between E and C incident at ~42° or more.

9. Because the positions of points A and B are improperly reversed in the diagram in the original text, I have restored them to their proper locations in order to accommodate the diagram to the demonstration.

10. A couple of points are worth mentioning here. First, it is clear that, despite his attempt to define the point of inversion for convex lenses in the previous proposition on the basis of a plano-convex lens, his analysis here is based on biconvex lenses. Second, and perhaps more interesting, Della Porta conceives of such lenses as composed of two intersecting spherical lenses of equal curvature, which allows him to treat refraction through biconvex lenses as a composite of discrete refractions through two distinct spherical lenses. In addition, although Della Porta implies that the point of inversion where the eye is placed is fixed, it in fact varies with the location of the object; the closer that object lies to the lens, the farther the point of inversion lies from the lens on the other side. It is also worth noting that from this point on Della Porta will ignore the refraction of refracted rays such as LM and OM into the eye, presumably because that refraction is too slight to make a meaningful difference where the resulting lines of sight will intersect their relevant catheti.

11. Because either Della Porta or, more likely, the printer failed to provide a diagram for this proposition, I constructed Figure 8.6 according to the dictates of the proposition and the description of lines and points in it. In the process, I was obliged to make a couple of obvious corrections in the text.

12. The conclusion of this proposition, as illustrated in the figure provided in the original text, is based on the incorrect assumption that image G of point B in Figure 8.8 lies at the intersection of the extension of refracted ray ED with the cathetus dropped from point A through center point F of the lens's lower surface. According to the same incorrect assumption, image M of point A lies where the extension of refracted ray EI intersects the cathetus dropped from B through center of curvature F. With the cathetus rule appropriately applied, however, the image of B would be at G', where refracted ray ED intersects cathetus BF, and the image of A would be at M', where refracted ray EI intersects cathetus AF. Consequently, Della Porta is correct in claiming that image will be inverted and smaller than its object, even though his "proof" of that claim is logically invalid.

13. In claiming that the cathetus becomes "more slanted" (declivius) as the object is withdrawn from the lens beyond center of curvature G in Figure 8.7, Della

Porta presumably means that the slope of the cathetus becomes increasingly steep along the vertical and therefore approaches perpendicularity.

14. Although it is unclear what Della Porta means by claiming that, as the object approaches the lens, the catheti "become increasingly straight" (magis rectae accedentes), he may be alluding to the fact that, as the object gets closer to the lens in front of center of curvature G in Figure 8.7, the catheti approach perpendicularity, just as they do when the object is withdrawn from the lens beyond G. On the other hand, according to the conditions illustrated in Figure 8.9, which accompanies the proposition that follows this paragraph, the object still lies beyond the center of curvature, so when it approaches that point, the catheti verge increasingly toward coincidence along the horizontal, thus approaching straightness.

15. See book 2, proposition 22, above.

16. A far more appropriate citation would have been to book 3, proposition 6, specifically the beginning of the fifth paragraph, where Della Porta observes that "in old people the humors thicken and become increasingly adulterated, and so it takes more light for them to see."

17. This reference is inapposite; far better would be book 1, proposition 1, where Della Porta shows empirically that a ray of light passing into a denser transparent medium is refracted toward the normal.

18. Again, this reference is inapposite; far better would be book 1, proposition 2.

19. Although the original diagram provided in the Latin text lacks a designation for point O in Figure 8.16, and although the text of the proposition itself is a bit garbled, it seems fairly clear that Della Porta meant to show that projection NO of image DC seen from far away is smaller than projection NE' of image CE seen at a closer distance, thus proving that the image of a magnitude of the same size will increase in size as the eye approaches the lens. As before, this illustration does not take into account the refraction of LI and GH into the eye.

20. This explanation of how concave lenses correct myopia, or nearsightedness, is based on a misunderstanding of how light refracts into such lenses. This misunderstanding is reflected in the diagrams that accompany propositions 15–18 in the original Latin text, where the rays entering the lens are represented as if they were refracted away from the normal rather than toward it, as they should when entering the lens. This is especially clear in the diagram accompanying proposition 18 in the Latin text, where the rays striking points P, O, N, L, I, and H all appear to converge rather than diverge, as they should according to the correct representation in Figure 8.18. As misrepresented by the diagram to prop. 18 in the original Latin text, therefore, the rays entering the lens do appear to "unite" by convergence, just as they do when entering the convex lens according to the diagram accompanying proposition 13 in the Latin text. Consequently, Della Porta's explanation of the lenticular correction of myopia assumes that the lens brightens the incoming image by bringing all of its constituent rays toward convergence, while at the same time narrowing that image for the constricted pupil. Then, when the image passes out of the lens, it is magnified by radial dispersal, the result being that the image in the eye is enlarged and brightened.

LIBER NONUS

PROOEMIUM

Loquemur hoc libro de iride et de coloribus qui ex diaphani et fulgidi mistura exoriuntur; in quorum investigatione plus quadraginta annis toto animo insudavimus, et dii faxint ut aliquid boni nacti simus. Res difficilis, admirabilis, et humanum captum excedens, ob id veteres poetae Taumantis filiam vocaverunt. Sed negotium auspicemur.

Lucem coloris expertem esse. PROP. 1

Lux sui natura splendida, tenuis, et coloris expers est. Splendida quia ea splendente cuncta nitent et videntur. Tenuis est, quia nullo temporis momento pellucidam raritatem transmittit. Nullo colore intaminata est, quod colorum omnium hipostasis futura erat. Rationes quibus nosmetipsos persuademus ex colorem esse hae sunt. Recipiens debet esse denudatum a natura recepti. Materiam primam quia quamque formam acceptura erat, ideo omni forma orbatam dicitur.[1] Chamaeleon nullo [190] colore spectatur, ut omni momento versi coloris mutationis esse possit, nam subinde colores mutat, totoque corpore reddit quoscunque attigerit. Sic et polypus.

Superiori libro de visione, diximus lucem omnem visibilis obiecti colorem indui, eundemque nostris obtutubus deffere, unde si aliquo colore conspurcaretur, quomodo sincera, non spuria, adulterataque ex corpore fluentia simulachra oculis remittere potuisset? Lux enim haudquaquam inoffensae perspicuitatis per rerum colores illabens seseque eis permiscens falsa colorum specie visum nostrum deludit. In sepum accensa, pro caeruleo viride, pro viridi flavum, pro flavo album, et omnia denique sua citrinitate defaedat. Pictores et textores qui colores lucernis seligunt falluntur, alterum pro altero substituentes. In officinis aerariis, lumen albedinem fuscam roseum squalli-

---

[1]"dicit" in original text.

BOOK NINE

INTRODUCTION

In this book we will discuss the rainbow and the colors that arise from the interaction of a light source and a transparent medium. For over forty years we have sweated over the investigation of this with our entire soul, and the gods have engineered things so that we might stumble onto something good from it. This is a difficult and wonderful subject, and one that exceeds human understanding, for which reason the ancient poets called the rainbow the daughter of Thaumas.[1] But let us broach the business at hand.

PROP. 1: *Light is colorless.*

Light is naturally bright and attenuated, and it is without color. It is bright because by its shining all things are illuminated and rendered visible. It is attenuated so that it passes through anything rare and pellucid in no time at all. It is tinged with no color because it was destined to be the substrate of all colors. Here are the reasons according to which we ourselves are convinced that it is devoid of color. Whatever receives something must be devoid of the nature of what it receives. Prime matter [is such a recipient] because it was destined to take on every form, so it is said to lack any form.[2] The chameleon [190] is judged to be of no [particular] color so that it can be altered at any moment according to a change of color, for it changes colors immediately, and its entire body takes on whatever colors it may have come in contact with. The same holds for the octopus.

Earlier, in book [four] on vision, we said that all light is imbued with the color of a visible object, and this is transmitted to our eyes, so if light were tainted with some color, how could it transmit the images emanating from a body to the eyes in pure rather than in false and adulterated form? For when it shines on the colors of things, light whose clarity is not entirely unhampered mixes with those colors and deceives our vision with a false representation of color. In the flame of tallow, blue is taken for green, golden yellow for green, and white for golden yellow, and everything is ultimately tainted by its yellowness. Painters and weavers who choose colors in candle-light make mistakes, substituting one for another. In copper smelters, the

dum, et tandem spectantium ora diro pallore effusa repraesentat. Idemque in accenso sulphure evenit.

Altera ratio. Lux est colore viduata quia tenuissima—ne dicam incorporea—nam color corporis est accidens, quomodo igitur colorata esse potest si est corporis expers, etsi colorata esset videretur? Lux per duo foramina transiens intra cubiculum, ut nil in medio attingat, non videtur; igitur sine colore. Sed ex Sole unde emanat, sorditudinis nescia per caelestes orbes pura transmittitur, at sublunari hoc mundo per elementa fluens, tingitur inficiturque, ut oculis nunquam sincera occurret nostris. Non igitur lux alba, ut Aristoteli visum est, nisi eius affectatores pro alba excolorem interpretarint. Sed debent qui docent castigate loqui, ne ipsi et legentes in praecipites salebras incidant, nullibi enim alba, pallidaque lux conspicatur nisi accenso in sulphure.

Colores fulgidos haudquaquam ex luce et opaco fieri. PROP.[2] 2

Lucem fulgorem tantum ex colorem diximus, nunc autem quomodo ex luce fulgidi colores oriantur vestigare tentandum. Sed primum maiorum placita discutiamus. Putabant veteres ex opaco et varia lucis positione colores fulgidos fieri. In exemplum adducebant columbarum cervices, quae utcunque deflectantur varias colorum species demonstrant. "Disertissime [191] Nero Caesar dicebat, 'Colla cytheriacae splendent agitata columbae,' et[3] variis quoque coloribus pavonum ceruix, quoties aliquo deflectitur nitet . . . ex varia lucis positione quam prout rectam vel obliquam receperint ita coloratur." Lucretius:

> Qualis enim caecis poterit color esse tenebris
> Lumine quin mutatur[4] in ipso, propterea quod
> Recta, aut obliqua percussus luce refulget?
> Pluma columbarum quo pacto in Sole videtur:
> Quae sita cervices circum, collumque coronat.
> Nanque alias fit uti claro sit rubra[5] pyropo:
> Interdum quoddam sensu fit, uti videatur
> Inter caeruleum virideis miscere smaragdos.
> Caudaque pavonis, larga cum luce repleta est,
> Consimili mutat ratione obversa colores:
> Qui,[6] quoniam quodam gignuntur luminis ictu,
> Scilicet id sine eo fieri non posse putandum est.

Et Cicero *Academicarum quaestionum* secundo ait, "In columba plures videri colores; non esse plus uno."[7]

---

[2]"Cap." in original text.
[3]"ex" in original text.
[4]"quin mutatur" = "qui mutantur" in original text.
[5]"rubro sit clara" in original text.
[6]"Quin" in original text.
[7]"imo" in original text.

light makes dusky white look dingy rose, and, in diffusing outward, it eventually paints the silhouettes of onlookers with a dreadful pallor. The same thing happens with burning sulfur.[3]

Here is another argument. Light is devoid of color because it is extremely attenuated—indeed, I should say incorporeal—but color is a characteristic of body, so how can light be colored if it lacks corporeality, even if it may appear to be colored? Light passing through two openings into a room and not touching anything in between is not visible; so it is colorless. Undefiled when it emanates from the Sun, however, it is transmitted in pure form through the celestial orbs, but when it flows through the elements in this sublunar world, it is tinged and tainted so that it never reaches our eyes in pure form. Light is therefore not white, as it seemed to be to Aristotle, unless his followers were to take "white" to mean uncolored. But those who propound this view should speak correctly so that they themselves and their readers do not fall headlong into ruts, for nowhere is light seen as white or pale except in burning sulfur.[4]

PROP. 2: *Iridescent colors are definitely not produced from the interaction of light and opacity.*

We have spoken only of light shining apart from color, but now we must endeavor to investigate how colors arise from brightly shining light. First, however, let us discuss the opinions of our predecessors. The ancients believed that iridescent colors are produced from a combination of opacity and shifting light. As an illustration they brought up the necks of doves, which show different kinds of colors whenever they are turned. "Nero Caesar said [191] most elegantly, 'The collar of the Cytheran's dove shines when it moves,' and the peacock's neck gleams with different colors whenever it turns; it is colored by the angle of the light according to whether it falls on them vertically or obliquely."[5] Here is what Lucretius says:

> For what colour can there be in blind darkness?
> Why, a color is changed by the light itself, according as
> The brightness responds to a direct or oblique impact of light;
> In this way the dove's plumage shows itself in the Sun,
> Lying about the nape and encircling the neck;
> For at times it is red as the blazing carbuncle,
> Again view it in a certain way, and it comes to appear
> A fusion of emerald green with blue.
> And the peacock's tail, when it is suffused with plenteous light,
> In like manner changes the colours as it turns;
> And since these colours are caused by a certain impact of light,
> Assuredly you must not think that they can be produced without it.[6]

In the second book of his *Academic Questions* Cicero says, "In [the neck of] a dove there appear to be many colors, but there is no more than one."[7]

Sed puerile profecto putandum est splendidas columbarum penintulas unius esse coloris et variis lucis ictibus illustratas differenti specie sese oculis exhiberi, nam et diversi inibi sunt colores, ut si aliquas ex collo pennulas extraxeris clare contueberis. In dextra enim parte halurgum, in sinistra viridem, infra rubidum, supra nigricantem colorem pennularum partes radiantes; etsi stet vario etiam lucis circumflexus eosdem semper conspicies. At si collum contorqueat, pristinus color, qui prius nobis erat conspicuus obliteratus delitescit; inde ad suum locum rediens, vernaculus denuo reviviscit. Si collum supra obvertit, purpurascit; si convertit, opacatur.

Non igitur ex opaco a luce colores effici poterunt. Aristoteles et Peripatetici omnes ex opaco et luce fulgidos colores gigni confitentur, sed hoc falsum est, quia opacum lucem repercutit, non recipit, nec cum eo lux poterit se commisceri. Sed error insimulandus, quum per opacum eius affectatores diaphanum intelligant. Sed cur diaphonum luci pervium, opacum impervium definiunt? [192]

Colores fulgidos ex luce et densiore aeris diaphano generari. PROP. 3

Colores alii ex mutua elementorum commixtione nascuntur, et hi dignum per se tractatu studium habent, et alibi de eis locuti sumus, nam et fulgidi non sunt. Sunt et quasi fulgidi colores alii, qui ex vaporibus nubibusque ex luce gignuntur quum densae coactaeque, et temere ex diversis partibus compositae, non pellucent, nec lucem transmittunt, sed summa cute recipiunt, ut in Solis ortu eius lucis occursu multos colores trahi, alios rubeos, alios virides, alios aureos, vel nubis partem rubei, et partem ignei coloris infici. Sed de his posterius loquemur.

Sed et alii colores ex lucis et densioris aere diaphani refractione orientes, et hi pulcherrimi sunt, fulgentes, floridi, hilaresque, et adeo nitent ut a pictoribus fingi non possint, gignuntur enim ex Sole et substantia pervia, et ars illos aemulari nequit, quia ex terrestri et obscura substantia suos parat colores. Diximus ex densioris aere diaphano nasci, non ex quovis, nam ubi Sol per tenuissimos caelestes orbes et eiusdem materiae totos illabitur, purus sincerusque resilit; ac ubi ad ima per impura elementa, per roscidas aspergines, per vitrum, et eiusmodi densioris aeris perspicui corpora, variis coloribus conspurcatur faedaturque, et ex nimia lucis et diaphani densioris mistura varia oriuntur colorum discrimina.

It is truly childish, however, to suppose that the shimmering feathers of doves are of a single color and that they show themselves in a different way to the eyes according to how they are illuminated by impinging light, for the various colors are actually in them, as you will clearly observe if you pull clumps of feathers from the neck. On the right side you will observe purple, on the left green, and red in between, all these areas shining upon the black color of the feathers, and even though the feather may stand in differently disposed light, you will still see the same colors. If the dove turns its neck, however, the original color that had first been seen by us disappears, but when the neck returns from there to its original place, the color is once again restored. If it raises its neck upward, it turns purple; but if it turns its neck back down, it turns dark.

Hence, colors cannot be produced by the interaction of opacity and light. Aristotle and all the Peripatetics avow that iridescent colors arise from the interaction of opacity and light, but this is false because opacity does not absorb but repels light, and therefore light could not mingle with it. The error to be charged, however, is that his followers take transparency for opacity.[8] But why do they define transparency as permeable to light and opacity as impermeable to it? [192]

PROP. 3: *Iridescent colors are produced from the interaction of light and a transparent medium denser than air.*

Some colors arise from the intermixing of elements, and these are worth a study in themselves, but we have spoken of them elsewhere, for they are not iridescent colors. There are others as well that are like iridescent colors, and they arise from the interaction of light and vapors or clouds when they are dense and concentrated, and being randomly gathered from different directions, they are not clear and do not transmit light but absorb it on the outer layer, just as at dawn, with the advent of the Sun's light, many colors are brought out, some red, some green, others golden, or part of a cloud is tinged red and part the color of fire. We will discuss these colors later, though.

There are other colors, however, that arise from the refraction of light in transparent bodies denser than air, and they are extraordinarily beautiful, gleaming, lurid, and delightful, and they shine so splendidly that they cannot be imitated by painters, for they arise from an interaction between the Sun and a permeable substance, and art cannot emulate them because it prepares its colors from earthly and dingy substance. We have said that colors arise from transparency of a greater density than air, but not just anywhere, for where the Sun shines through the extremely attenuated celestial orbs and everything of like substance, it comes out pure and pristine. On the other hand, where it shines down here below through impure elements, through scattered dew, through glass, and through bodies of this sort whose transparency is denser than air, it is adulterated and tainted with various colors, and distinct varieties of colors are produced from an excess of light and its mingling with a denser transparent medium.

Diximus refractione, nam quum radii ad perpendiculum feruntur, tantum est illis virium ut infracti permeent, nec faedari patiuntur, et si inficiuntur, saltim colores non nisi ex visus refractione spectari necesse est. At si per inaequale densioris[8] aere diaphanum ferantur, qui plus intimius penetrarint obscuriori colore inficiuntur; qui minus clariori. His experimentis nos rem ita nobis persuademus.

Lux per crystallinum prisma permeans minus diaphanum densius passa dilucidior nitet, plus vero meraciore colore languet. Idem in phiala aquae plena; idem et sub aquis posito speculo. Idem quoque "fistula aliquo loco rupta, aquam per tenue foramen elidi, quae sparsa contra Solem," ut media sit inter Solem et aspersam aquam, "faciem arcus repraesentat." Idem etiam accidit, ut Plutarcus admonet, si quis aquam ore conceptam [193] in cubiculo efflet contra Solem, sicut, "fullones, quum os aqua implevere et vestimenta tendiculis diducta leviter aspergunt, apparet varios edi colores in illo aere asperso, quales in arcu fulgere solent." Causam igitur esse in diaphono aere crassiore non dubitabis, nunquam enim fit arcus nisi in nubilo.

Maiorum opiniones de variis iridis coloribus. PROP. 4

Sed cur ex luce et densiore[9] aere diaphano colores tam varii et multiplices exoriantur acriter diuque priscorum philosophorum animos torsit. Anaximenes colores fieri autumavit varios

> radios Sole suos in nubem ejaculante densam, crassam, atramque, cogi enim eos super eam nubem, quam penitus dirimere non valeant. Anaxagoras Solis irradiationem densa nube repercuti . . . Metrodorus[10] quum trans nubem Sol splenduerit, nubem quidem caeruleam reddit, splendorem vero rutilat;

Plutarco referente. Senece haec:

> Quidam esse aiunt aliqua stillicidia quae Solem transmittant, quaedam magis coacta, quam ut transluceant; itaque ab illis fulgorem reddi, ab his umbram, et sic utrius intercursu effici arcum, in quo pars fulgeat quae Solem excipit, pars obscurior sit quae exclusit et ex se umbram proximis fecit.

---

[8]"densionis" in original text.
[9]"densione" in original text.
[10]"Metrodorum" in origina text.

We have said this happens by refraction because, when rays are conveyed along the perpendicular, they are so powerful that they can pass through unbent and do not suffer deformation, but if they are tinged, their colors must be observed according to the refraction of sight alone. And if they are transmitted through a transparent medium of different density than air, the rays that penetrate more deeply are tinged with a darker color, whereas those that penetrate less deeply are tinged with a brighter color. By means of the following experiences we persuade ourselves that the phenomenon is like this.

Light passing through a glass prism into a less dense transparent medium shines more brightly, but its unmixed color may be fainter. The same thing happens in a flask full of water; and so, too, with a mirror placed under water. Likewise, "when a pipe bursts at some point, we see water forced out through the tiny hole; when it is sprayed against the Sun, it displays the form of a rainbow," as long as there is an interval between the Sun and the spouting water.[9] As Plutarch reminds us, the same thing also happens if someone in a room, [193] who has a mouthful of water, blows it out against the Sun, just as "when fullers have filled their mouths with water and lightly spray the clothes that are spread out on stretching-frames, you can see various colors produced in the air that is filled with spray, like the ones that normally shine in a rainbow."[10] You will therefore have no doubt that the cause lies in a transparent medium denser than air, for the rainbow is never produced except in a cloud.

PROP. 4: *The opinions of our predecessors concerning the different colors of the rainbow.*

Long ago, the question of why so many and such varied colors are produced from the interaction of light and a transparent medium denser than air racked the brains of early natural philosophers. According to Plutarch, Anaximenes maintained that different colors are produced

when the Sun projects its rays on a thick, heavy, black cloud, for those rays concentrated on that cloud are scarcely able to penetrate. Anaxagoras claimed that the solar radiation is reflected from a dense cloud. Metrodorus argued that when the Sun shines through a cloud, it certainly renders the cloud blue, but the light is reddened.[11]

Seneca has this to say:

Some people say that there are certain raindrops that let the Sun's light through, and others that are too compacted to transmit light; and so the first sort produce brightness, the second sort shadow, and hence, by the intermingling of the two, a rainbow is generated, in which part is bright, welcoming the Sun, and part is darker, excluding it and casting a shadow over the immediate surroundings.[12]

Sed hoc falsum existimat:

> Poterat umbra et lux causa videri, si arcus duos tantum haberet colores . . . Quid ergo istic[11] duo colores faciunt, luminis atque umbrae, si innumberabilium ratio reddenda sit? . . . Sed varietas non aliam ob causam fit quam quia pars coloris a Sole est, pars a nube illa, humor autem modo caeruleas[12] lineas, modo virides, modo purpurae similes, et luteas, aut igneas ducit, duobus coloribus hanc varietatem efficientibus, remisso, et intento . . . Hinc apparet duas causas esse arcus, Solem nubemque, quia nec sereno unquam fit, nec ita nubilo ut Sol lateat. Ergo nunquam ex his est, quorum sine altero non est.

Parianus igneum colorem a Sole, caeruleum a nube, reliquos ex horum misturis esse. Nicolaus Peripateticus, "Colorum diversitatem ex diversitate partium nubis evenire," dixit, "nam ubi terrestris pars combusta cum humido misto, rubeus oritur color, ut coloratur flamma in lignis viridibus; ubi vero humidum aquosum multum, apparet viror; ubi natura tenuis et aqua, albus." [194] Albumasar ex aqua caeruleum colorem, rubeum a Sole, viridem et purpureum ex horum mistione. Sunt qui dicant lucem coloris semina secum ferre, eaque ex elementis per quae transit, ex igne ruborem, ex terra viriditatem, ex aere caeruleum, ex aqua lacteum colorem.

Albertus partim[13] ex his, partim ex se multas affert insanias. Dicit enim in iride duplicem humorem reperiri, alterum gravem, et terrestrem, faetulentum, alterum levem, fumidum, et subtilem. Hic in sublimem nubis partem scandit; ille gravis in imo nubis subsidens, ut gravium natura. Aer autem humectus ac spissus in meditullio sistitur, est et nubs prima supremaque. Si Sol his quatuor non se immiscuerit, non generabitur iris. Ex humore in quo aliquid fumidi vaporis est, fit rubeus vinosusque color, ut in parte suprema nubis. Ex humido aquoso indigesto terrestri inferior pars viret. In medio, ubi aer mistus, fit color ex rubeo et viridi mistus, scilicet croceus. Et hic color variatur, nam si nubs erit subtilis, minus habens de subtili terreo, erit color pallens, si vero de utroque plus, erit magis croceus sive puniceus. Sed praecipuum colorem caeruleum omisit, et caetera in sequentibus refellentur.

Flavum colorem ex multa luce et minima densioris·diaphani parte generari.
PROP. 5

Sed generationes colorum aggrediamur. Sed ne colorum nominum ambages legentium mentes interturbent, principo semper declarabimus colorum

---

[11]"istis" in original text.
[12]"caeruleus" in original text.
[13]"partem" in original text.

But he reckons this to be false:

> It could be thought that it was due to light and shadow, if the rainbow had just two colors, [but] then what use are two colors, light and shadow, since an explanation is required for countless colors? But the variety of color occurs for the simple reason that some of the color comes from the Sun, some from the cloud. In this liquid the Sun now produces blue lines, now green, now purplish, yellow, or fiery; and two colors, subdued and intense, generate this variety. In the present case it is clear that the rainbow has two causes, Sun and cloud, because it never occurs in a clear sky, nor in a sky so cloudy that the Sun is hidden. So it certainly derives from the things without either of which it does not exist.[13]

Artemidorus of Parium says that the fiery color is due to the Sun, the blue color to the cloud, and the rest to a mixture of these.[14] Nicholas Peripateticus maintained that "the diversity of colors arises from the diversity of the parts of the cloud, for where the earthy portion that burns is mixed with moisture, a red color is produced, as flame is colored in green wood, but where watery vapor abounds, a green color appears, and where it is attenuated and watery in nature, white appears."[15] [194] Abū Ma'shar argues that the blue color is due to water, the red color to the Sun, and green and purple to a mixture of these two.[16] There are those who might say that light carries the seeds of color with it, and those seeds come from the elements through which the light passes, red from fire, green from earth, blue from air, and milky-white from water.

Albertus Magnus asserts several senseless things on the basis of these theories as well as his own. For he says that in the rainbow two types of moisture are to be found, one heavy, earthy, and impure, the other light, vaporous, and rare. This latter moisture rises to the top of the cloud, whereas the former, heavy moisture sinks to the bottom of the cloud, as is the nature of heavy things. For its part, moist, thick air holds an intermediate place, and it constitutes the first and last limit of the cloud. If the Sun does not intermingle with these four, a rainbow will not be formed.[17] From the moisture in which there is some steamy vapor, a red, wine color is produced, as happens in the top portion of the cloud. The lower part shines green from unconsumed earthy moisture thick with water. In between, where the air is mixed, a color blended from red and green is produced: that is, saffron. And this color varies, for if the cloud is thin and contains less earthy vapor, the color will be pallid, but if it contains more of both earthy and watery vapor, the color will be more intensely saffron or reddish. But especially noteworthy is that he ignored the blue color, and the rest of his claims will be refuted in the propositions that follow.[18]

PROP. 5: *A golden yellow color is produced by the interaction of intense light and a tiny portion of a denser transparent medium.*

Let us turn to the generation of colors. In order, however, to keep the ambiguities of color terms from confusing the minds of our readers, we will always clarify the names of colors at the beginning. "The appearances of

nomina. "Multiplex colorum facies, appellationes incertae et exiguae." Flavus color dicetur. "Pleraque sunt, inquit Favorinus, in oculorum sensibus quam in verbis discriminata." Flavum colorem voco non eum quem Virgilius in frondibus olearum videri dixit, nec quem Pacuvius in aqua et puluere, inquiens, "Cedo tamen pedem lymphis flavum pulverem . . . abluam."

Sed qui in paleis nondum maturis et in croco valde aqua diluto conspicitur hicque color ex robustissima luce et ex minima diaphani aere densioris parte progigni, hisque argumentis persuademus. Sed dum per tenuem aerem fertur, flavo colore [195] induitur. Dum Borea spirat, quia purgatior, et pellucidior[14] redditur, dilutiori flavo nitet; si Austro vel nebulis addensetur, meraciori rutilat. Mane exoriens Sol per crassiores vapores transiens, adeo coloratior nitet, ut lux aurea videatur in meridie dilutior.

Robustus ignis in siccis lignis accensus flavus videtur; in medio flammae, ubi simplicissima est, flava est lux, nam ab imo conspurcatur a pabulo, in superiori parte a fumo. Carbo ubi diutissime igni excanduit, ut interior nigrities[15] absumpta sit, flavet visu forti luce vehementer inspecta; si alio oculos divertes, flava primum, mox viridia, postremo caerulea omnia conspicies, quia vehemens primo lux cum obscuriori aere collata flavet; mox languescens, viridis; tandem caerulea relaxata solvitur.

Sol per crystallini prismatis[16] aciem primam pertransiens flavo colore tingitur, quod dignosces eam partem digito velando removendoque. Nullus ergo color luci proximior quam flavus, ob id vehementissima lux Solis nullum alium sibi colorem usurpat nisi flavum vel aurem. Alexander ait, "Luteus color magis ad album quam phoeniceum accedit, quod ex hoc evenit . . . quum lumen retunditur . . . in luteum, sive pallidum transit."

Reliqui Peripatetici flavum colorem non ex vehementi luce nec[17] ex pauco densioris aere diaphano gigni, sed ex rufi et viridis coloris vicinia, quod et falsissimum est. Non enim ex duobus fuscis coloribus rubro et viridi clarus exoritur, et luculentus. Praeterea, si vitrum rubrum supra viridem collocaveris ac oculis obieceris, prospectus inde color non flavus erit, neque pictores unquam ex viridi et puniceo flavum colorem efficiunt. Et si Solem dicit per fumum visum vel per nigrum puniceum videri (quod verissimum est), non deberet prima iridis circumferentia punicea videri, quia et ibi adest lux minus passa, neque in nigro, neque per nigrum videtur, transitque. Pudet

---

[14]"pullucidior" in original text
[15]"nignities" in original text.
[16]"pusmatis" in original text.
[17]"sed" in original text.

colors are manifold, their designations uncertain and scanty. There are, says Favorinus, more colors distinguished by the sense of sight than by words."[19] Let the color golden yellow be defined. I do not call golden yellow the color that Virgil claimed he saw in olive leaves, nor what Pacuvius says in respect to water and dust, when he declaims, "Give me your foot so I may wash away the golden dust with clear [golden] water."[20]

But those who observe this color in straw that is not yet mature and in saffron that is heavily diluted in water, we will convince them by the following arguments that this color arises from the interaction of extremely intense light and the smallest part of a transparent medium denser than air. When light is transmitted through thin air, it is imbued with a golden-yellow hue. [195] When the north wind blows, the light shines with a more diluted golden-yellow hue because the air is rendered cleaner and clearer, but if the air is thickened by a south wind or by clouds, it becomes a purer red. At early dawn, when the sunlight passes through extremely thick vapors, it therefore shines with a deeper tinge, such that at midday the light appears a more diluted golden color.

A roaring fire burning in dry wood appears golden yellow, and in the middle of the flame, where it is most homogeneous, the light is golden yellow, for at the bottom it is contaminated by the fuel and at the top by smoke. When coal burns in fire for so long that its inner blackness is consumed, it is golden yellow when examined closely by the eyes in strong light, but if you turn your eyes away, you will first see everything golden yellow, then green, and finally blue because at first the strong light combined with darker air turns golden yellow; then, after weakening, it turns green, and finally, when mitigated, it devolves into blue.

Sunlight passing through the sharp edge of a glass prism is tinged golden yellow, which you will discern by covering that part of the prism with your finger and then removing it. Thus, no color is closer to light than golden yellow, which is why the brightest sunlight is likened to no other color than golden yellow or gold. Alexander of Aphrodisas says that "a golden-yellow color tends more to white than to scarlet, which happens when the light is reflected in a golden yellow object or when it passes through a pale medium."[21]

The rest of the Peripatetics claim that a golden-yellow color arises not from strong light or from a medium slightly denser than air, but from the juxtaposition of the color red and the color green, which is absolutely false. For a clear, bright color does not arise from the two dark colors red and green. Besides, if you place a red piece of glass upon a green one and place the combination in front of your eyes, the color you see through it will not be golden yellow, and painters never produce a golden-yellow color from green and scarlet. Also, if Aristotle says that the Sun seen through smoke or through blackness appears red (which is absolutely true), then the first band of the rainbow should not appear red because at this point the light is less dispersed, and it is not seen against a black background, nor does it pass through a black medium or appear through it. It is shameful of modern

reffere neutericorum de sanguine cocto et incocto similitudines, quid enim commune habet color ex elementorum mixtura cum eo qui ex luce et densioris aere diaphani generatur?

Caeruleum colorem ex languidiori luce et ex multo diaphano aere densiore nasci.
PROP. 6

Diximus proximiorem luci colorem flavum, et ex ea quae vehementissima luce rutilat, et ex minimo densiore aere [196] diaphano. Contra lux quae hebet languetque et ex multo densiore aere diaphano caeruleum colorem ementitur, est enim colorum omnium opacitati proximior, ubi enim caligat lux aut extinctioni proxima, decoloratur, fuscatur, et caerulea fit. His argumentis persuademus.

Flamma quae oleo, saepo, vel viridibus lignis proxima est caerulea est, quia ibi oriri incipit, ideo imbecilla est, et ibi etiam crassius alimentum, id est tenebra; ob id ubi crassius elementum et lumen imbecillius, ibi caeruleus color sive nigricanti proximior. Praeterea, Sol diutius intuitus, oculos claudendo, flavus, mox viridis, tandem in caeruleum abit, et solvitur lumen quia Solis fulgor multis tenebris obrutus imbecilliorque factus caeruleus apparet. Postremus etiam color qui in vitreo prismate videtur ceruleus est, quia in trianguli basi—crassiore scilicet sui parte—imbecillis radius resiliens, tenebris fit proximus, nam ater color tanquam abnegatio et privatio spectandi est.

Aer cominus observatus et pellucida corpora nullum nobis colorem exhibent; eminus tamen omnia conspecta, ubi Solis illuxit splendor, caerulea videntur.[18] Si caelum intuebimur pellucidum corpus, ubi radius lumine deficitur, tenebris fere interceptus, caeruleo color offunditur, et caeruleus color ob id a caelo dicitur. Et mare a longe intuitum etiam tale videtur. Si a montibus Sole illustratus elongamur, primum flavi, mox virides, tandem caerulei videbuntur, quum longius absunt ob nimiam luminum longinquitatem acies hebescit, et caligatis aer intercedens efacit ut caeruleus conspiciatur. Addunt alii quae per refractionem spectantur a maiori visus distantia spectantur, quum curva linea recta sit longior; quare non iniuria nigriora videri. Sed id falsum, nam linearum differentia vel minima est.

Ex colorum vicinia intimos iridis colores progigni, et itidem ex flavi et caerulei mistura. PROP. 7

Quoties igitur lucidi et densioris aere diaphani misturae variae coierint innumera fere proveniunt colorum discrimina, nam unus ex alterius vicinia

---

[18]"videtur" in original text.

thinkers to refer to the appearances of cooked and uncooked blood, for what does color yielded by a mixture of elements have in common with color that arises from the interaction of light and a transparent medium denser than air?

PROP. 6: *The color blue is produced by the interaction of weaker light and a transparent medium much denser than air.*

We have asserted that the golden-yellow color is quite close to that of light, and it arises from whatever reddens the most intense light, as well as from a transparent medium slightly denser [196] than air. On the other hand, light that is weak and faint and that interacts with a transparent medium much denser than air assumes the color blue, for it is closest of all the colors to opacity because, when light weakens or is close to being extinguished, it loses color, darkens, and becomes blue. We will convince you with the following arguments.

A flame in olive oil, tallow, or green wood is nearly blue because in this case it is beginning to emerge, so it is faint, and in this case also the fuel is quite thick, that is, in darkness. Hence, where the fuel is quite thick and the light quite faint, the color is blue or approaching blackness. Furthermore, when the Sun has been looked at for a while and the eyes are closed, the afterimage goes from golden yellow, then to green, and finally to blue, and the light dissipates because, when it is overwhelmed by profound darkness and rendered weaker, sunshine appears blue. Also, the last color seen in a glass prism is blue because the weak ray that refracts out at the base of the triangle—that is, at its thicker part—becomes nearly dark, for a coal-black color should be regarded as a negation or privation of light.

Air viewed close at hand and clear, transparent bodies reveal no color to us, yet when all of them are looked at from a distance, and when sunshine illuminates them, they appear blue. If we look closely at the transparent body of the sky, then when the radiated light fails and is taken over completely by darkness, a blue color spreads out over the sky, and for this reason the color blue is said to be from the heavens. Seen from afar, the sea also looks like that. If we are far away from mountains illuminated by the Sun, the first range will appear golden yellow, the next green, and the last blue, since the farther away they get, the more the acuteness of sight diminishes because of the excessive length of the light rays, and the intervening vapor makes the air look blue.[22] Some thinkers add that things seen by means of refraction are seen at a longer distance from the eye, since a bent line is longer than a straight one, so it is no surprise that they appear blacker. But this is false, for the difference in [the lengths of the] lines is surely minimal.

PROP. 7: *The intermediate colors of the rainbow are generated from neighboring colors in the way that golden yellow and blue are mixed.*

The distinctions that arise among colors are therefore as innumerable as the various combinations of light and a transparent medium denser than air that can be blended together, for one color combined with another neighboring

cognationem contrahens tertium efficit, et ex hoc iam orto et altero alius
subinde. Si ad lucem accesserint, perluuntur; ad opacum retuso colore [197]
colligant.[19] Et adeo alter alterum sensim subit, ut indiscriminati videantur;
ut recte a pictoribus dicatur eiusmodi colorum speciem exacte nondum
exprimi quisse, quum et a similibus auspicentur, et in diversissimos abeant.
Sed eos enumerare nostri non erit instituti, nisi extremos et intermedios
aliquos.

Medius igitur inter flavum et caeruleum colores viridis est, qui non ex
mutua lucis et densioris aere diaphani mistione nascitur, sed ex utriusque
mistura. Hi soli bini colores ex lucis mistione nascuntur, caeteri ex vicinia.
Quod ex flavo et caeruleo colore[20] fiat viridis, hoc solo argumento patebit.
Si caeruleum vitrum super flavum apposueris et oculis subieceris, viridis
color aspicietur,[21] sic et in opposito pariete; et pictores ex caeruleo et[22] flavo
viridem[23] colorem conflant. Hinc clare patet error dicentium quod ex prima
luce passa fiat puniceus color, et ex puniceo et viridi flavus; nam caeruleus
et[24] puniceus viridem vel flavum non conficiunt.

*Rufum colorem non ex lucis refractione et diaphano aere densiore gigni, sed
nigrae nubis contrapositione.* **PROP. 8**

Sed declaremus primo quid per rufum colorem intelligamus. Eius multae
sunt species. Rufus color a rubore nominatus est, sed aliter in igne, aliter
in rosa, aliter in palmae fructus, aliter in croco rutilat. Has rufi varietates
latina oratio singulis, propriisque vocabulis non demonstrat, nisi ruboris
appellatione. Graeci ἐρυθρον[25] (sanguineum), πυρρόν (flammeum), ξανθόν
(croceum), et φοινκόν (phaeniceum).

Omnes hi colores vel rufum acuunt, intenduntque, vel remittunt: nam
aliquorum mistura remittuntur. Phaeniceum dicimus, et spadis coloris species, quam in fructu palmae arboris non admodum Sole incocti visitur, et
spadica alii vocant avulsos ex palma termites cum fructu. Nos rufum sive
roseum languidum vocamus quem in extimo et supremo iridis ambitu videmus ex valida radiosa incidentia et densa nubis nigredine nascitur, non ex
mistura, ut diximus, lucidi et densioris diaphani aere, sed in sola superficie,
ubi adsit nigredo. Inquit Seneca, "Rubida fit nubes Solis incursu."

Nostram opinonem his rationibus persuademus. Si niger extinctus carbo
igni proiiciatur ut extima tantum superficies igniatur, [198] tunc carbo rufo
colore rutilabit, lux enim in summa superficie vix haerens infra nigredinem
demonstrat, adeo ex luce et nigredine nascitur. At ubi carbo diutius igne

---

[19]"coligant" in original text.
[20]"corole" in original text.
[21]"auspicabitur" in original text.
[22]"ex" in original text.
[23]"viride" in original text.
[24]"ut" in original text.
[25]"ἐριθρον" in original text.

one forms a third, related one, and from the color just produced and another one yet another one immediately forms. If they approach the light, they become washed out, and they tend toward opacity by the repeated color changes. [197] Accordingly one gradually approaches the other so as to appear indistinguishable, for which reason it is said correctly by painters that colors of this kind cannot yet be described with precision, since they may begin with similar colors yet end up as different ones. But it will be our intention to account for only a few of the extreme and intermediate colors.

Accordingly, the intermediate between the golden-yellow and blue colors is green, which is not produced from a mingling of light and a transparent medium denser than air, but from a mixture of both golden yellow and blue. These two colors alone arise from the mingling of light [and a transparent medium denser than air]; the rest are mixtures of neighboring colors. That green is formed from a golden yellow and a blue color will be evident from one single proof. If you pose blue glass upon golden-yellow glass and place it up to your eyes, a green color will be perceived, and likewise on a facing wall. Also, painters form the color green from blue and golden yellow. In this case the error of saying that a red color can be produced from the first light dispersed [through a transparent body] and that golden yellow is produced from red and green is patently obvious, for blue and red do not form green or golden yellow.

PROP. 8: *The color red is not produced from the refraction of light and a transparent medium denser than air, but by contrast against a black cloud.*

Let us first explain what we mean by the color red. It has many varieties. The color red gets its name from *rubor* ["redness"], but it reddens one way in fire, another way in a rose, another way in the fruit of the palm tree [i.e., the date], and yet another way in saffron. The Latin language does not denominate these varieties of red with single, proper terms, except by the designation *rubor*. The Greeks distinguish among ἐρυθρόν (blood red), πυρρόν (flame red), ξανθόν (saffron red), and φοινικόν (date red).[23]

These colors all make "red" more definite and broaden or loosen its use, for they are loosened by the mixture of any of them. We say that date red is a form of the color *spadix* [date brown], which is seen in the fruit of the palm tree that is not fully ripened by the Sun, but others call *spadix* the branches, along with the fruit, that are picked from the palm. We call red or dull rose what we see at the outer, topmost band of the rainbow, which is yielded by intense incident rays and the concentrated blackness of a cloud, not, as we said, from the mingling of light and a transparent medium denser than air, but only on the surface where it approaches black. Seneca says that, "a cloud turns red when sunlight strikes it."[24]

We are convinced of our opinion for the following reasons. If a piece of black charcoal that is no longer smoldering is thrust into a fire so that only its outer surface is ignited, [198] then the charcoal will turn red in color, for the light adhering to the outermost surface barely shows through the blackness, so it is produced from light and blackness. But when the

steterit, ut intus extra optime ignitus, tunc croceus evadit et ad flavum tendit. Idem in ferro et aliis metallis videre licet. Fabri ferrarii qui ferrum temperant roseum colorem observant demonstrantem ignitionem in summa tantum superficie, non intima excanduisse. Flamma accensa in re crassissima nigerrimaque nigerrimum fumum efficit; flammae apex quo eiusmodi fumo commiscetur tunc rubra videbitur. Typographi qui ex combusta pice fuliginem parant, dum pix comburitur, relucet flamma sub nigerrimo fumo sanguinei coloris.

Fuliginosae omnes flammae rubrae videntur. Puniceus color qui quasi dilutus robur est, Aristoteles ait, exoritur ubi flamma viridium lignorum fumo cohaeserit. Si candela in tenui oleo accensa est, apex puniceus videtur.[26] Splendidum, inquit, si nigro aut per nigrum videatur, puniceus apparet; per caliginem et fumum visus puniceus apparet. Ex nigri et flavi mistione puniceus efficitur, quasi croceus intensus. Ubi flavus color intenditur et fit meracior, tunc fit purpureus. Sol flavus est, ob id per fumum inspectus, puniceus visitur. Et candelae flamma flava est; per fuliginem punicea videtur. Lux conspecta per foramen nigri panni rubescit; sic etiam in cubiculo nigro.

Mane Sol exoriens, si crassas nubes et aqua praegnantes nigrescentesque pervaderit, et radii eius nubium[27] ob densitatem penetrare nequeunt, roseo colore nitentes conspicies. Ob id non semper roseus color in extima iridis superficie videtur, nec est de eius essentia. Sed tunc conspicitur quum nubes ex eius regione nigerrima adfuerit et Sol eam percutiens nec penetrare valens; tunc circulus primus rubeus spectabitur. In crystallino prismate non videtur roseus color, quia ibi opposita non adest nubes nigra, sed si nigrum pannum apposueris, puniceus[28] coloratior videbitur; unde potius ex iusta positione vel oppositione quam ex lucis et diaphani aere densioris nascitur.

Sunt quidam qui quae scribant ignorant, dicentes flavum colorem esse intensiorem[29] meracioremque, puniceum vero dilutiorem remissioremque,[30] et in flavo lucem magis tenebras passam, in puniceo minus. Bone Deus, quae nubes caliginosa eorum oculos et mentem obsidet qui discernere nequeant croceum [199] colorem qui in croco est et puniceum magis esse tenebrosum quam flavum, idest paleae colorem luculentiorem! Taedet profecto et pudet eorum rationes refellere.

Halurgum colorem non ex luce et aere densioris diaphani mistione sed ex vicinia crassi et cerulei coloris nasci. PROP. 9

Halurgus coloris intelligitur qui violaceus vulgo dicitur, quique in violis spectatur. Dicitur enim vulgo pavonaceus, quod in collo pavonum reperiatur.

---

[26]"videdur" in original text.
[27]"nubis" in original text.
[28]"puceus" in original text.
[29]"intensionem" in original text.
[30]"remissionemque" in original text.

charcoal has remained long enough in the fire to burn well beyond into its interior, it verges away from saffron red and tends toward golden yellow. You may observe the same thing in iron and other metals. Blacksmiths who temper iron maintain a rose color that reveals combustion only on the outer surface, not having ignited the interior. A flame ignited in an extremely dense and black object produces very black smoke, and the tip of the flame that is mingled with this sort of smoke will then appear red. When printers prepare soot from burning pitch, while the pitch is burning, the flame below the very black smoke glows with a blood-red color.

All sooty flames appear red. The red color that dissolves out of oak, Aristotle says, is produced when flame has been applied to the smoke of green wood. If a wick is lit in fine olive oil, its tip appears red. He says that if something bright is looked at in blackness or through blackness, it appears red; so when seen through soot or smoke it appears red.[25] A scarlet color, similar to intense saffron red, is produced from a mixture of black and golden yellow. Where a golden-yellow color becomes more concentrated and purer, it becomes purple. The Sun is golden yellow, so when it is viewed through smoke, it looks scarlet. A candle flame is also golden yellow, so it appears scarlet through soot. Light reddens when looked at through the interstices between the threads of a black cloth, and likewise in a black room.

At the crack of dawn, if sunlight pervades thick, blackened clouds full of moisture, and if the Sun's rays cannot pass through because of the clouds' thickness, you will see the clouds glow with a rosy color. For this reason, a rosy color does not always appear at the very top surface of a rainbow, nor is this color an essential part of a rainbow. It is seen, however, when the cloud directly behind it is extremely black and when the sunlight striking it is not strong enough to pass through; then the first red circular band will be seen. In a glass prism the rose color does not appear because in this case there is no black cloud behind it, but if you put a black cloth behind it, a vivid reddish color will appear. Hence, it arises from a proper disposition or facing situation rather than from the interaction of light and a transparent medium denser than air.

There are some who, not understanding what they write, claim that the golden-yellow color is more intense and purer, whereas scarlet is more diluted and faded, yet they claim that in golden yellow the light has been affected more by darkness, whereas in scarlet it has been affected less by darkness. Good God! What a sooty cloud besets the eyes and mind of those who cannot discern the saffron-red [199] color that is in saffron and cannot see that crimson is darker than golden yellow, that is, the extremely bright color of straw! It is indeed wearisome and shameful to refute their arguments.

PROP. 9: *The color purple arises not from the mingling of light and a transparent medium denser than air, but from the juxtaposition of a thick cloud and a blue color.*

Of the color purple we understand what is called "violet" in the vernacular, the color that is seen by everyone in violets. In fact, this same color, which

Resplendet etiam in amethysto purpurei species, vel iacinthi,[31] et fit quum caeruleus color diluitur et in vinum abit. Hic non ex mutua lucis et diaphani refractione nascitur, sed ex vicinia, nam rufus color et caeruleus halurgum colorem componunt. Postrema iridis circumferentia, quia caerulea est, quum ei fit obviam nubs nigerrima cum Solis colore mista colorem rufum efficit, ut diximus; ex colore rufo illo et caeruleo fit halurgus. Hic semper non videtur, quia ex iridis essentia non est sicut rufus, sed quando utrique nigerrima nubs opponitur. Hoc modo probamus. Si vitrum rufum supra caeruleum posueris et per illud inspexeris, obvium parietem halurgo colore intinctum conspicies. Pictores halurgum ex rubro et caeruleo fingunt.

Iridem non tricolorem, sed bicolorem aut multicolorem dicendam. PROP. 10

Ex variegata iridis coloria, multi sunt qui varios eos colores dixerunt. Homerus πορφυρέην ἶριν dixit [*Iliad*, 17, 547], quasi sub purpureo colore rubrum, caelureum, et halurgum conclusisset, qui adhuc a multis purpureus color dicitur. Aristoteli τρίχρος est, ex puniceo, viridi, et halurgo, quia hi praecipui colores praecipue spectantur in ea. At si simplices colores enumerare voluissent, bicolorem dicere oportuerat, ex flavo et caeruleo duo extrema, unum ex multa luce, alterum ex flavo scilicet[32] et caeruleo, nam viridis ex vicinia extremorum fieri diximus. Optime [200] igitur Parianus: igneum a Sole colorem iridem trahere, caeruleum a nube, reliquos ex horum misturis; nec nisi duos simplices colores esse.

    At si omnes colores qui fere semper in iride spectantur enumerare velis, et rufus et halurgus, pentacolor dici deberet. Si colores notandi sunt qui intus in medio ex vicinia fiunt, innumerabiles prope et infiniti sunt. Nam radius solaris sensim nubem penetrans et continuo se remittens, infinitos colores causat, tanto lascivientis naturae artificio[33] et ornatu, ut colores omnes qui maxime commendantur et in floribus visuntur praecipue suo nitore provocent. Videmus in eo aliquid flammei, aliquid lutei, aliquid caerulei, et alia in picturae modum subtilibus lineis ducta ut ad dissimiles colores sint scire non possis nisi cum primis extrema contuleris. Nam commissura decipit, usque adeo mira arte naturae conspicitur[34] quod a simillimis cepit et in dissimilia definit. Ovidius:

    Qualis ab imbre solet percussis solibus arquus
    Inficere ingenti longum curvamine caelum,
    In quo diversi niteant[35] cum mille colores:

[31]"iathini" in original text.
[32]"scillcet" in original text.
[33]"attificio" in original text.
[34]"conspiciuntur" in original text.
[35]"videant" in original text.

is encountered in the neck of a peacock, is also called "peacock" in the vernacular. A kind of purple or hyacinth color also gleams in amethyst, and it is formed when the blue color is diluted and turns into wine color. It does not arise from the interaction of light and refraction in a transparent medium, but from juxtaposition, for a red color and a blue one combine to form a violet color. Because it is blue, when the lowest band of the rainbow lies in front of a very black cloud, which forms a red color when mixed with the color of the Sun, as we said, purple is formed from that red color and the blue of the rainbow band. This color does not always appear because, like red, it is not essential to a rainbow, but when a very black cloud faces both the blue and red, it does appear. We prove it as follows. If you place red glass upon blue glass and look through it, you will see an intervening wall tinged with a purple hue. Painters form purple from red and blue.

PROP. 10: *It should be said that the rainbow is not tricolored, but bicolored or multicolored.*

There are many who have discussed the various colors among those variegated hues of the rainbow. Homer called the rainbow purple, as if he had subsumed red, blue, and purple under the heading of purple, the latter color still called purple by many. The rainbow is tricolored according to Aristotle, consisting of red, green, and purple because these particular colors are the ones that are primarily seen in it. But if they had wanted to list the simple colors, they should have said it was bicolored, formed from the two extremes of golden yellow and blue, the one arising from a great deal of light, the other from golden yellow and blue, for we have said that green is formed from the juxtaposition of the extremes. [200] Artemidorus of Parium therefore put it best: the rainbow gets its flame-red color from the Sun and its blue color from the cloud, whereas the rest come from a mixture of these, so there are only two simple colors.[26]

If, however, you wish to list all the colors that are almost always seen in a rainbow, including red and purple, it ought to be described as fire colored. If the colors that are formed inside toward the middle from juxtaposed colors are to be taken into account, they are nearly uncountable and infinite. For in gradually penetrating the cloud and reflecting back, the solar radiation creates an infinite number of colors commensurate with the artifice and supply of wanton nature, so that all the colors that are most agreeable and are seen particularly in flowers stimulate us by their luster. We see in the rainbow a bit of flame color, a bit of saffron yellow, a bit of blue, and other colors in the way thin lines are drawn in a painting so that you cannot tell they consist of different colors unless you have first looked closely at the extremes. For the conjunction is deceptive, until by the marvelous art of nature it is seen that it starts with quite similar colors and ends with dissimilar ones. As Ovid puts it:

> As when after a storm of rain the Sun's rays strike through, and a rainbow, with
> its huge curve, stains the wide sky, though a thousand different colours shine in

Transitus ipse tamen spectantia lumina fallit.
Usque adeo quod tangit idem est tamen ultima distant.

Ob id Hesiodus ex multiplicitate coloris πολυώνυμον ὕδωρ vocavit, quasi
multi nominis aquam. Virgilius eum sequutus, mille pro infinito, dixit, "Mille
trahit varios adverso Sole colores."

Contra Aristotelem, in generanda iride nullam reflexionem necessariam.
PROP. 11

Iam diximus colores quomodo generentur; nunc ad alias difficultates accedamus, et primum quomodo et Sole et nube oriantur, mox ad rotundationem
perventuri. Aristoteles satis prolixe probare nititur reflexionem fieri ex oculi
nostris ad Solem per nubem, quam opinionem Plutarchus quoque sequitur.
Seneca ex Aristotele hoc modo. Nubis iam in guttas versurae, nec dum tamen
pluit, "singula stillicidia . . . singula specula esse; a singulis ergo imaginem
reddi Solis. Deinde multas imagines, immo innumerabiles, et devexas, et in
praeceps transeuntes confundi; [201] itaque arcum multarum imaginum
Solis confusionem esse." Haec sic probat.

> Pelves . . . mille die sereno pone, et omnes habeant imaginem Solis . . . At contra
> ingens stagnum non habebit nisi unam imaginem quia omnis circumscripta laevitas
> et circundata suis finibus speculum est . . . Ergo stillicidia illa infinita . . . totidem
> specula sunt, totidemque Solis facies habent; hae[36] contra intuenti perturbatae
> apparent, nec dispiciuntur intervalla quibus singula distant, spatio prohibente
> discerni; unde pro singulis apparet una facies turbida ex omnibus . . . Ab omni
> enim lenitate acies suos radios replicat; nihil laevius aqua et aere; ergo ab aere
> spisso visus noster in nos reddit . . . Longe autem visum nostrum nobis remittit,
> qui crassior est, et pervinci non potest, sed radios nostrorum luminum moratur,[37]
> et eo, unde exierint, reflectit . . . sed quia parva sunt, Solis colorem sine figura
> exprimunt. Deinde quia in stillicidiis innumerabilibus et sine intervallo cadentibus
> reddatur idem color, incipit facies esse non multarum imaginum intermissarum
> sed unius longae atque continuae.

Sed hic multa sunt refellenda. Primo quod non ex oculis per nubem ad
Solem fit reflexio, sed contrario modo, ut probavit libro *De anima*. Sed haec
abeant, quia levia sunt. Secundo dicimus ad iridem generandam non esse
necessarium reflexionem, sed refractionem, ex guttis enim, stillicidiisque et
corporibus diaphanis non fit reflexio, sed refractio; reflexio enim a densis

---

[36]"haec" in original text.
[37]"morantur" in original text.

it, the eye cannot detect the change from each one to the next; So like appear the adjacent colors, but the extremes are plainly different.[27]

For that reason, Hesiod called water πολυώνιμον ὕδωρ, [which translates] as "multi-named water," because of its multitude of colors.[28] Following him and taking a thousand to mean infinite, Virgil said that "it elicits a thousand different colors from the facing Sun."[29]

PROP. 11: *Aristotle to the contrary, no reflection is necessary for the formation of a rainbow.*

We have already discussed how colors can be generated; let us now turn to other problems, and let us examine first how rainbows arise from both the Sun and the cloud, and then move on to the round shape of rainbows. Aristotle makes a fairly longwinded effort to show that reflection occurs from our eyes to the Sun via the cloud, an opinion that Plutarch also follows.[30] Seneca draws on Aristotle as follows. In a cloud transformed into drops, while it is still raining, "the individual drops of falling rain are individual mirrors; so they individually emit an image of the Sun. Then many, or rather innumerable, images, descending and plummeting, are merged together; [201] so a rainbow is the merging of many images of the Sun."[31] Here is how he explains this:

> On a fine day put out a thousand bowls: they will all have images of the Sun. On the other hand, a large pool will have no more than one image because every smooth surface that is confined and surrounded by a boundary is a mirror. So those infinitely many drops are so many mirrors and contain so many likenesses of the Sun. To someone facing them and looking at them, they appear confused, and the spaces between individual likenesses cannot be made out, for the distance prevents them from being distinguished. As a result, instead of individual likenesses, one confused likeness is seen emerging from all of them. From every smooth surface our sight bends back its rays. Nothing is smoother than water and air; therefore our vision returns to us from dense air also. Now water is far more effective at sending our vision back to us because it is denser and cannot be overcome; it slows down our eyes' rays and bends them back to where they have come from, but because they are small, they reproduce the color of the Sun without the shape. Then since the same color is given off in countless drops falling without a gap, there begins to be the appearance, not of many separated images, but of one long, continuous image.[32]

But many things need to be refuted here. First, we say that reflection does not take place from the eyes to the Sun via the cloud, but in the opposite direction, as Aristotle proved in the book *On the Soul*.[33] But let these ideas [about the reflection of visual radiation] be dispensed with because they are frivolous. Second, we say that refraction, not reflection, is necessary for forming a rainbow, for refraction, not reflection, occurs from drops, dripping water, and transparent bodies, whereas reflection occurs from solid, polished

politisque corporibus.[38] Quomodo igitur ibi fieri potest reflexio, si nulla
adsunt densa corpora, sed tantum diaphona? Et si per reflexionem fieret iris,
neque ad nostros oculos perveniret. [202]

Ad hoc ostendendum, sit hemispherium AEC[39]; vapores perpendiculariter ascendentes sint DFE; Sol in oriente A. Radius AF feriat nubis roscidae

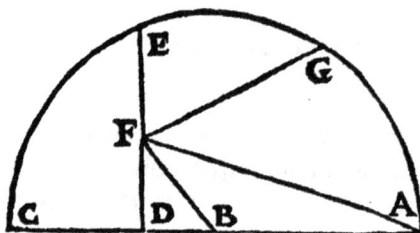

partem F; radius reflexus nunquam perveniet ad oculos B. Sed supra reflectitur, est enim lex refractionis ut ad pares angulos reflectatur; erit reflexa linea
FG, quia angulus AFD, est aequalis GFE. Idem eveniet in convexa nube. In
concava autem, si circulus nubis cavae aequidistaret lunari circulo, impossi-
bilis esset reflexio nam linea quae angulos discriminaret aequales reflexionis
et incidentiae esset linea visus, non reflexionis. Si igitur per reflexionem fieri
oporteret, necesse esset ut angulus quoque nubem ad idoneum locum locaret.
Recte igitur Philippus Menaedeus scripsit, qui primus "comperit materiam
iridis in profundo irradiari," non ex superficie reflecti. Necessaria sola igitur
refractio est. Afferemus exemplum.

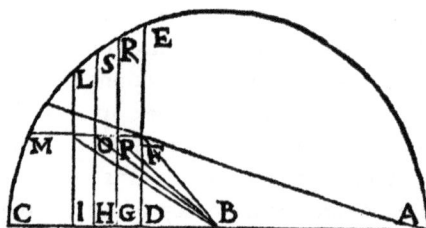

Sit hemisphaerium AEC, nubes roscida EDLI, Solis radius orientis AFT.[40]
Sed perveniens ad F superficiem nubis, quia densius diaphanum, refrangitur
FM. Dividatur nubes in tres partes secundum[41] tres colores iam dictos: FDPG

---

[38]"coporibus" in original text.
[39]"ABC" in original text.
[40]"AFL" in original text.
[41]"secunda" in original text.

bodies. In this case, then, how can reflection occur, if there are no opaque bodies but only transparent ones? Moreover, if the rainbow were formed according to reflection, its image would never reach our eyes. [202]

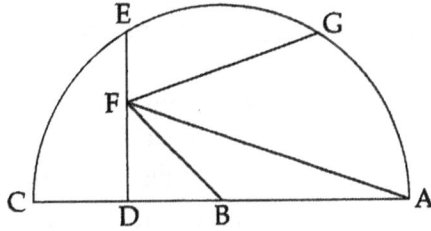

Figure 9.11a

By way of showing this, let AEC [in Figure 9.11a] be the hemisphere [of the sky], let DFE be the vapors rising straight up, and let A be the Sun in the east. Let ray AF strike spot F of the dewy cloud, and so the reflected ray will never reach the eyes at B. Instead, it is reflected upward, for the law of reflection dictates that it be reflected at equal angles, so FG will be the reflected ray because angle AFD = angle GFE. The same thing will happen in a convex cloud. In a concave cloud, however, if the circle of the concave cloud were parallel to the lunar orbit, reflection [to B] would be impossible, for the line according to which the equal angles are determined for reflection and incidence would be the line of sight, not [a separate line] of reflection.[34] If the rainbow had to be formed by reflection, then, the angle would have to be disposed in an appropriate place with respect to the cloud. Philip of Medma thus wrote correctly, being the first "to ascertain that the rainbow is based on radiation into the depth [of the cloud]," not reflection from its surface.[35] Consequently, only refraction is necessary. We will provide an example.

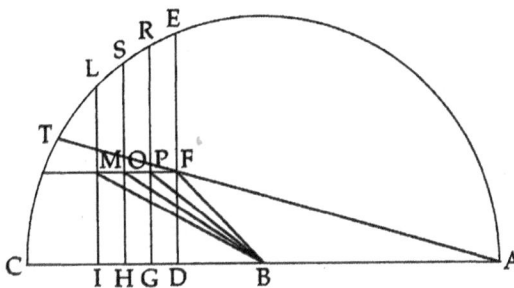

Figure 9.11b

Let AEC [in Figure 9.11b] be the hemisphere of the sky, EDLI the dewy cloud, and AFT a ray from the eastward Sun. Reaching point F on the surface of the cloud, however, it is refracted along FM because the cloud's transparency is denser [than that of the clear air]. Divide the cloud into three parts according to the three colors already discussed: that is, FDPG

flavus, PGOH viridis, OHMI caeruleus. Et ad oculum B trahantur radii a punctis divisionum FB, PB, OB, MB. In refracto igitur Solis radio videbitur pars nubis FP flava, PO viridis, OM caerulea.

Maiorum sententiae de iridis rotundationis. PROP. 12

Difficillimam arcus rotundationis aggredimur rationem, nam iridis speculationem inter naturae mira neminem praeterit. Ob id maiores nostri varia et admodum [203] diversa dixerunt, et ubi discordia ibi doctissimorum virorum caligant intellectus. Plato in *Theaeteto* irim Thaumantis, idest admirationis, filiam esse tradidit, non absurde eius originem explicans. Plutarchus ab eo, "a Thaumante genus eam ducere mortales rei miraculo adductos finxisse, adludente videlicet ad nominis originem fabula." Cicero libro *De natura deorum*: "Cur arcus species non in deorum numerum reponitur, est enim pulcher, et ob eius speciem, quia causam habet admirabilem, Thaumante dicunt esse nata." Plato in *Cratylo*: "ἶρις ab eo quod εἴρειν,⁴² idest loqui, quod nuntia est." "Ab Hesiodi interprete, irim dici quo προφορικόν λόγον, ab εἴρω, quod est dico ... videntes enim eum admirantesque orationem proferimus." Albumasar autem arcum daemonis vocat, pro daemone idem quod angelus intelligens, "idest arcus angelicae considerationis, tanquam ad eius notitiam humanus intellectus non perveniat." Plinius nubis cavitatem causam esse circulationis dixit: "Manifestum est radium Solis immissum cavae nubi, repulsa acie in Solem refringi."

> Parianus Artemidorus adjicit, quale genus nubis esse debeat quod talem Soli imaginem reddit. Si speculum, inquit, concavum feceris, quod sit sectae pilae pars, si extra medium constiteris, quicunque juxta te steterint, universi a te videbuntur propiores⁴³ tibi quam speculo. Idem, inquit, evenit, quum rotundam et cavam nubem intuentur a latere, ut Solis imago a latere ... discedat, propiorque⁴⁴ nobis sit, et in nos magis conversa.

Contra haec Seneca:

> De speculis duae opiniones sunt. Alii enim in his simulachra cerni putant, idest corporum nostrorum figuras a nostri corporibus emissas ac separatas⁴⁵; alii imagines aiunt non esse in speculo sed ipsa aspici corpora retorta oculorum acie et in se rurus deflexa. Sed nil ad rem haec pertinent quomodo res videamus, sed quomodo

---

⁴²"εἴριν" in original text.
⁴³"propriores" in original text.
⁴⁴"propriorque" in original text.
⁴⁵"temperatas" in original text.

golden yellow, PGOH green, and OHMI blue. Then extend rays FB, PB, OB, and MB to the eye at B from the points of division. Along the refracted solar ray, therefore, part FP of the cloud will appear golden yellow, part PO green, and part OM blue.[36]

PROP. 12: *Opinions of our predecessors concerning the round shape of the rainbow.*

We turn to the extremely difficult issue of the round shape of the bow, for the marvel of nature that is the rainbow escaped the contemplation of no one. Accordingly, our predecessors made various and altogether [203] different claims, and wherever there is disagreement, the intellects of learned men grope about blindly. In the *Theaetetus*, Plato says that the rainbow is the daughter of Thaumas, that is, Wonder, explaining its origin in a not irrational way.[37] Following him, Plutarch claimed that, "led by the wonder of the thing, men made up the story that it takes its descent from Thaumas by playing upon the name [i.e., "wonder"].[38] In his book *On the Nature of the Gods*, Cicero asks, "Why should not the glorious rainbow be included among the gods? It is beautiful enough, and its marvelous loveliness has given rise to the legend that Iris is the daughter of Thaumas."[39] In the *Cratylus*, Plato says that "ἶρις comes from εἴρειν, which means 'to speak,' because she is a messenger."[40] "According to the translator of Hesiod, 'iris' is called such from προφορικόν λόγον ["uttered word"], from εἴρω, which means 'I speak,' for when we see it and wonder at it, we break out in speech."[41] Abū Ma'shar calls it "demon's arc," taking "demon" to be the same as "angelic intellect": "that is, an arc of angelic contemplation, insofar as human intellect cannot attain a knowledge of it."[42] Pliny maintained that the concavity of the cloud causes the rainbow's circularity: "It is obvious that a solar ray projected to a hollow cloud is reflected back to the Sun when its leading edge is repelled."[43] [Seneca goes on to say that]

> Artemidorus of Parium even says what kind of cloud is required to produce such an image for the Sun. If you make a concave mirror, he says, that is a segment of a cutaway sphere, then, if you stand outside its center, anybody standing next to you will appear upside down and nearer to you than to the mirror. The same thing, he says, happens when we look at a round, concave cloud from the side, namely, that the Sun's image is detached from the cloud and is nearer to us and turned more toward us.[44]

Seneca argues against this as follows:

> There are two opinions about mirrors: some people think that representations are seen in them, that is, forms of our bodies that are emitted from our bodies and are distinct from them; others say that not images but the bodies themselves are seen in a mirror, as the eye-beam is twisted round and bent back upon itself. For the present it makes no difference how we see what we undoubtedly do see, but,

imago similis debet ex speculo reddi . . . Deinde si velis speculum aeri comparare, des oportet mihi eandem laevitatem corporis, eandem aequalitatem, eundem nitorem. Atqui nullae nubes [habent] similitudinem speculi. Per medias saepe transimus, nec in illis non cernimus; qui montium summa conscendunt spectant nubem nec tamen imaginem suam in illa cernunt.

Possidonius contra eos qui irim imaginem Solis dicunt, cava nube conceptam dicunt, ait:

Id si ita esset, deberet imago se ingerere nostris obtutubus ex profundo nubis per tantum interstitii quantum [204] ab occidente est ad ortum, hoc est a Sole ad nubem, quae speculum Solis dicitur. Id vero esse falsum evidens comprobat experimentum, quum saepius visatur iris longe ante nubis superficiem in qua apparet.

Praeterea, extra speculi centrum imagine apposita cavo speculo obversa videtur, ideo rotunda Solis circumferentia sursum versa nobis infra versa videri deberet. Sed hoc absurdissimum est. Contraria huic opinioni Nicolai Peripatetici est, ait enim aer, terra, et aqua accedendo, et aer recedendo semper in orbes, suas curvabunt superficies, et elatos vapores a frigore in nubem conversos, Sol feriens eius imaginem ut concavum[46] speculum arcualem ostendit. Sed hoc etiam falsissimum est, nam ascendentes vapores recta a terra ad caelum tendunt, ut ascendentes videamus, nec totum hemisferium complent, ut concavam[47] superficiem recipiant, sed partem ad nos proximam. Alia quoque accidunt absurda, quae pudet refellere.

Iridem circularem apparere. PROP. 13

Quia iris in sublimi apparet, iter nobis ad eam perscrutandam interciditur; ideo ex his quae in his inferioribus videmus ad ea quae in sublimi fiunt ascendere conabimur. Sed causas rotundationis aggrediemur. Sed ut facilius haec difficillima intelligantur, aliquas praepositiones praemittamus oportet.

    Primo quod radii a Sole in pyramidis modum iaculantur. Non illos dico qui a centro se diffunduntur, nam hi nos [205] perpendiculariter feriunt, et si vere ita non fit,[48] a mathematicis tamen ita accipitur, quum minima sit differentia. Sed qui a lateribus moventur, nam diximus libro de lumine ex omni luminosi parte lumen progredi. Exemplum: Esto Sol AB; a lateribus radii veniant AI, BH decussantes se in foramine CD. In opposito pariete rotundam efficiunt figuram[49] BCH, ADI quae sit HI.

    Secundo, pupilla nostra rotunda est, et in puncto coeunt radii crystallini.

---

[46]"convexum" in original text.
[47]"convexam" in original text.
[48]"sit" in original text.
[49]"fifiguram" in original text.

however it happens, an image that is similar must be given off from the mirror. . . .
Then, when you require air to have the properties of a mirror, you must show me
the same smooth body, the same evenness, the same brightness. Yet no clouds
bear any resemblance to a mirror: we often pass through the middle of them, but
do not see ourselves in them; those who climb mountain tops look at clouds, yet
do not see their image in them.[45]

Against those who claim that the rainbow is an image of the Sun and that
it is formed in a concave cloud, Posidonius says:

If this were true, its image ought to follow along with our sight from the depth
of the cloud according to the eastward distance [204] from the source, that is,
the distance from the Sun to the cloud, which is called the mirror of the Sun.
That this is false, however, is clearly attested to by experience, since a rainbow is
quite often seen far ahead of the surface of the cloud in which it appears.[46]

Besides, when an image is set before a concave mirror outside the center of
the mirror, it appears inverted, so the round circumference of the Sun, which
is oriented upward in relation to us, ought to appear oriented downward in
relation to us.[47] But this is totally absurd. Contrary to this opinion is that
of Nicholas Peripateticus, for he says that when air, earth, and water approach
and air draws away, always in spheres, their surfaces will curve, and when
the rising vapors are turned into a cloud by the cold, the sunlight striking
it yields an image that looks like an arc on a concave mirror.[48] This is also
completely false, however, for vapors rising from the Earth to the sky incline
straight up, as we see them ascend, and they do not fill an entire hemisphere
so as to take on a concave surface, but only the portion near us. Other
absurdities follow as well, but it would be disgraceful to refute them.

PROP. 13: *A rainbow appears circular.*

Because the rainbow appears up high, the path to a close examination of it
is cut off to us, so we will try to ascend from what we see down here to
what happens up high. Now we will proceed to what causes the rainbow's
roundness. In order to make better sense of this extremely difficult subject,
however, we need to set forth some presuppositions.

First, we say that the Sun's rays are emitted in the form of a cone. I am
not speaking of those that propagate from the center, for these strike us
[205] orthogonally, and although this is not how it actually occurs, it is
still accepted as such by mathematical opticians, since the difference is
minimal.[49] Rather, I am speaking of those that are propagated at the side,
for in the book on light we have said that light proceeds from every spot
on a luminous source.[50] As an illustration, let AB [in Figure 9.13a] be the
Sun, and let rays AI and BH on the sides proceed out and intersect at the
aperture CD. BCH and ADI form round shape HI on a facing wall.

The second presupposition is that our pupil is round, and the incident
rays converge toward a point in the crystalline lens. Since they are propagated

Et quum per rotundum foramen a puncto se diffundantur, pyramidem efficiunt habentem conum in oculo in amplo spatio basim circularem, ut iam
clarius diximus libro de visione.

Tertio, et quod valde refert, quod non sub quocunque angulo videntur
quae refracta videntur, sed stato constitutoque, nam res quae sub aquis sunt
sub angulo stato refracto videntur, ut libro diximus de refractione. Lucidioris
gratia cognitionis exemplum apponere non gravabor.

Sit nubes ex adverso ex addensato aere asperginosa iam iam in guttas
versura et in pluviam casura EDMI. Aut cava, aut convexa, aut recta nil
refert, sed sit perpendicularis. Sit fulgens astrum A ex adverso feriens nubem
in puncto E; quae si eiusdem diaphanitatis cum aere esset, transiret in C.
At quia densioris, refrangitur, et relabitur in M, et erit refractus radius EM.
Hic continet differentias colorum. Sit oculus noster respiciens in B; dividatur
in tres aequales partes EM, et tota nubs EDPG, PGOH, OHMI. Igitur radius
EP in prima nubis parte se intingens. Eum colorem indui diximus qui luci
proximior, [206] idest flavus. Ultima pars nubis OHMI, a radio debilissimo
confossa, OM in caeruleum, quasi abnegationis et extinctionis lucis se intingit.

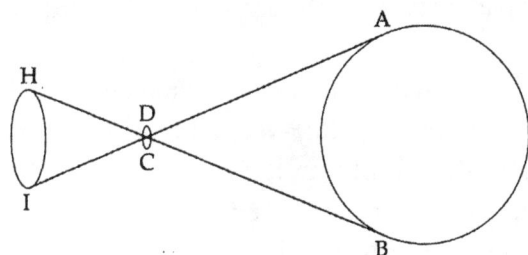

Figure 9.13a

from a point through a round opening, they form a cone that has its vertex in the eye and its circular base in the open space [outside the eye], as we discussed more clearly in the book on vision.[51]

The third presupposition, and this one has particular bearing, is that what is seen in refraction is not seen at just any angle but at a particular, given angle, for, as we said in the book on refraction, things that are under water are seen at a specific, refracted angle.[52] It is no inconvenience for me to add an illustration for the sake of clearer understanding.

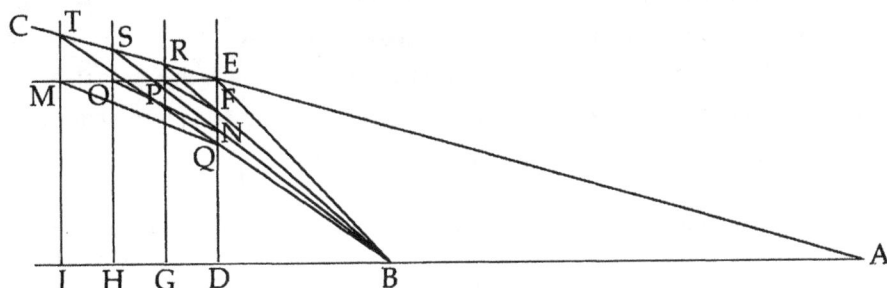

Figure 9.13b

Let EDMI [in Figure 9.13b] be a facing cloud formed from thickened, water-sprinkled air that is at this moment converted into falling raindrops. Whether it is concave, convex, or flat is irrelevant, but let it be perfectly vertical. Let A be a brightly shining heavenly body that faces the cloud, and let its light strike point E on the cloud.[53] If the cloud were of the same transparency as the air, the ray of light would pass through to C. But since it is denser, the ray is refracted and inclines toward M, and so EM will be the refracted ray.[54] This line encompasses different colors. Let our eye lie at B, let EM be divided into three equal parts [EP, PO, and OM], and let the whole cloud be divided into equal portions EDPG, PGOH, and OHMI. Accordingly, ray EP in the first portion of the cloud is tinged. We have said that the color it assumes is quite close to that of light, that is, [206] golden yellow.[55] Because the last portion OHMI of the cloud is penetrated by extremely weak radiation, portion OM is tinged blue, as a sort of negation

Medius POGH aeque valens, colorabitur eo colore qui ex utroque componitur, idest viridi.

Sed penitus[50] videamus colorationis modum et quantitates angulorum. Punctus E libere ad oculum venit. P vero ad F veniens, refrangitur, et ad B venit. Extensa igitur BF, relabitur in R. Perpendicularis igitur ex PG veniet; elongetur quousque coeat cum BF, et erit in R. Punctus igitur P videbitur in R. Sic punctus O venit in N; accedit ad[51] B. Elongetur NB; relabitur in S, ubi coit cum perpendiculari HO. Punctus igitur O videbitur in S. Sic punctus M venit per Q in B. Relabitur in T, ibi coit cum IM,[52] et punctus M erit in T.

Sed si universa nubs EDMI a Sole illustratur, cur igitur sola puncta E, P, O, M videntur? Dicimus radium nostrum BE construere angulum cum catheto ED—scilicet BED—qui fere semirectus est, vel minimum ab eo distans, nam quum linea horizontis in qua Sol A, oculus B, et centrum iridis D semper sit, erit angulus BDE rectus, et si recte metimur distantiam BD,[53] semper est semidiameter iridis, ut patet in visione iridis in caelo apparentis et in iride, quam paulo post dicemus. Est ergo BD aequalis DE; angulus igitur BED semirectus. Idem et dicendum de angulo BRG, BSH, BTI; etsi aequales non sunt, tamen parum a tali angulo recedunt.

Docuit me Plutarchus, "Si quis Soli adversus aquam ore sumat et ita insputet, ut stillicidia repercussum in Solem habeant actutum, comperit caelestis arcus imaginem factam." Et Seneca:

> Videmus[54] quum fistula aliquo loco rupta est, aquam per tenue foramen elidi, quae sparsa contra Solem oblique positum, faciem arcus repraesentat. Idem videbis accidere, si quando volueris observare fullonem, quum os aqua impleverit, et vestimenta tendiculis diducta leviter aspergit, apparet varios edi colores in illo aere asperso, quales in arcu fulgere solent.

Ego igitur cum in cubiculo ore aquam efflarem ex oppositio Solis, ex asperginoso aere illo iris videbatur circularis distans ab oculo meo per iridis semidiametrum, et si oculos in altum attollebam, attollebatur iris, et si deprimebam, deprimebatur. Sic ex lateribus, oculique motum sequebatur, et per totum illum asperginosum aerem oculos movendo, movebatur iris. Astantes qui [207] mecum erant illam non videbant, nec nisi qui post tergum erant in linea quae per Solis, oculi, et iridis centrum transiret. Intrans quoque Sol

---

[50]"penitius" in original text.
[51]"ab" in original text.
[52]"CM" in original text.
[53]"AB" in original text.
[54]"Vidi" in original text.

and extinction of the light. Halfway between the middle portion, POGH will be tinged with the color that is composed of the other two, that is, green.

But let us take a close look at the kind of coloring and the sizes of the angles. The image of point E reaches the eye uninterrupted. On the other hand, in reaching F, the image of P is refracted and then continues to B. Hence, if BF is extended, it ends up at R. The cathetus from PG will continue, so extend it until it intersects with [the extension of] BF, and it will do so at R.[56] Point P will thus be seen at R. So, too, the image of point O reaches N and then reaches B. Extend NB, and it ends up at S, where it intersects normal HO. Point O will thus be seen at S. Likewise, the image of point M reaches B through Q. [When extended, the resulting refracted ray BQ] ends up at T, where it intersects with IM, and so the image of point M will be at T.

But if the entire cloud EDMI is illuminated by the Sun, why are E, P, O, and M the only points that are seen? We respond that our ray BE forms with cathetus ED[57] an angle—that is, BED—that is precisely half a right angle, or minimally divergent from it, for since there must always be a horizon line containing the Sun at A, the eye at B, and the center of the rainbow at D, angle BDE will be right, and if we measure distance BD correctly, it invariably forms the radius of the rainbow, as is evident in how the rainbow looks in the sky and in the actual rainbow itself, which we will discuss a bit later. Accordingly, BD = DE, so angle BED is half a right angle. The same should be said as well about angles BRG, BSH, and BTI; even though they are not exactly equal [to half a right angle] they still diverge from such an angle by very little.[58]

Plutarch made me aware that "if anyone puts water in his mouth and spits it out directly away from the Sun so that the drops reflect the light immediately, he discovers that the image of a rainbow is formed."[59] Seneca, too, said that,

> when a pipe bursts at some point, we see water forced out through the tiny hole; when it is sprayed against the obliquely angled Sun, it displays the form of a rainbow. You will see the same thing happening if you ever feel like watching a fuller: when he has filled his mouth with water and lightly sprays the clothes that are spread out on stretching-frames, you can see various colors produced in the air that is filled with spray, like the ones that normally shine in a rainbow.[60]

Accordingly, when I was in a room and blew water from my mouth while facing against the Sun, a circular rainbow became apparent from that droplet-filled air at a distance of a radius of the rainbow from my eye, and if I raised my eyes upward, the rainbow was raised up, whereas if I lowered my eyes, the rainbow was lowered. The same holds for the sides; the rainbow followed the motion of my eye, and as my eyes scanned the entire body of misty air, the rainbow moved along with them. Those who were standing [207] near me did not see it, unless they were behind me on the line that passed through the Sun, my eye, and the center of the rainbow. Moreover, when the sunlight

per fenestram quando erat obliquus maior videbatur. Praeterea, iridem flor-
idissimam pluries aspexi in Tyburtino fluvio, nam ab alto cadens circumadia-
centem aerem totum irrorat, ubi accedendo recedendoque fit maior et minor
ad oculi distantiam. Ex his concludendo iridem non videri nisi sub angulo
constituto. Cuius haec diffinitio: iris est radius Solis refractus ab aere roscido,
veniens ad oculum sub angulo constituto. Hic patet dicentium error oportere
nubem esse concavam vel convexam, quum in aere in cubiculo irrorato non
convexa vel concava nubes sed recta vel irregularis est.

Iridem non nisi sub constituto angulo videri. PROP. 14

Nos ad hoc demonstrandum aliquibus mathematicis demonstrationibus
utemur. A duobus punctis in diametro circuli sumptis, quorum alterum in
extremitate, alterum in centro, plures quam duae rectae lineae non constituentur ad easdem partes ad aliud atque aliud circumferentiae punctum.

Nos sic probabimus. Esto semicirculus cuius centrum K, diametri
extremitas G. Ab his duobus punctis ad circumferentiam M duae rectae GM,

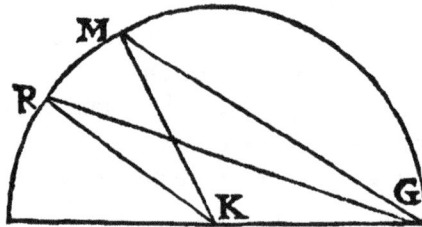

KM, dirigantur. Duae lineae dico ab eisdem signis G, K ad eandem partem
M, duae aliae recta lineae non constituentur ad aliud punctum quam M. Si
fieri poterit consituatur, et sit ad R, et sit GR, KR. Quia sicut se habet GM
ad MK sic GR, ad RK, et permutatim quoque sicut GM ad GR, sic KM ad
KR. Sunt autem KM et KR aequales (nam ambae ex centro); ergo GM ipsi
GR aequalis, quod est [208] contra 18 primi *Elementorum*, et 7 tertii. Igitur[55]
non nisi una iris videri poterit in eo aere illuminato, quia alius angulus non
aequalis erit ei.

Omnia igitur haec in unum coniungendo, si solum ab uno puncto oculi
sub stato angulo fieri posset refractio, circumvolvendo punctum illum non
nisi circulum efficiet. Rotunditas igitur iridis erit ex sectione pyramidis visualis uniformis cum sectione pyramidis radiorum Solis in superficie plana

---

[55]"Igiter" in original text.

entering the room was at a slant, the rainbow appeared larger. Several times I have also seen an extraordinarily bright rainbow in the Tiber River, for in falling from on high, it fills the surrounding air with mist, and when I approach or draw away, it becomes larger and smaller according to the distance of my eye.[61] From these observations it must be concluded that the rainbow appears only at a given angle. This is its definition: A rainbow is solar radiation refracted by misty air and reaching the eye at a particular angle. In this case the error of those who claim that the cloud must be concave or convex is clear because in the mist-filled air of the room the cloud is neither convex nor concave but straight or irregular.

PROP. 14: *A rainbow is seen only at one particular angle.*

For the purpose of proving this we will make use of some mathematical demonstrations. From two points taken on the diameter of a circle, one at the endpoint, the other at the center, more than two straight lines cannot be set up on the same sides to one point as well as to another one on the circumference.[62]

We will prove this as follows. Let there be a semicircle with center point K [in figure 9.14a] and G as the end point of the diameter. From these two

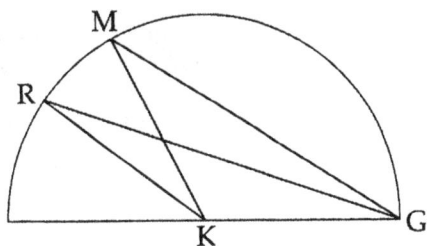

Figure 9.14a

points let the two straight lines GM and KM be drawn to M on the circumference. I say that with two lines from the same points G and K to the same side of M, two other straight lines [of equal length] cannot be set up to reach any point other than M. If such an arrangement were possible, let it be to R, and let GR and KR be the lines. Since GM:MK = GR:RK [by supposition], then *alternando* GM:GR = KM:KR. Now KM and KR are equal (for they are both drawn from the center), so GM = GR, which is [208] contrary to book 1, proposition 18, and book 3, proposition 7, of the *Elements*.[63] Consequently, only one rainbow can be seen in the illuminated air because no other angle will be equal to GKM.

To put all these points together then, if at only one point of the eye refraction can occur at a particular angle, putting that point into rotation will form nothing but a circle. Thus, the round shape of the rainbow will be due to the intersection of the uniform visual cone with the section of the cone of solar rays on the plane of the cloud's surface, since in that case the

nubis, quum tunc refractio undique aequalis fiat, existentibus Solis centro, oculi, et iridis in linea una. Id hoc modo depingemus.

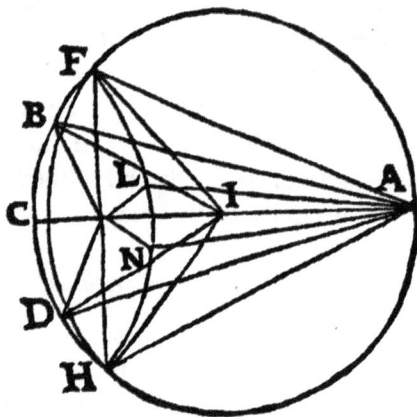

Sit meridianus AFBCDH,[56] diameter AIC, nubes opposita FMH.[57] Sol A pyramidaliter feriat eam AF, AB, AL, AD, AN, AH. Oculus in I,[58] et ex oculo pyramis radiosa concurrat cum eadem Solis nube, ut IF, IL, ID, IN, IH; quae quum undique equalis secans pyramidem Solis, erit communis sectio circulus FBDHNL[59]; anguli omnes aequales cum lineis oculi et diametro iridis IFM, IBM, IDM, ILM, INM, IHM, sic enim radii Solis. Haec demonstratio videtur Aristoteli de iridis orbitate, ut loco ubi de halone. Sed ex aliis non percipitur, sed non tam clara et facilis, et de industria occultata. [209]

Iridem apparere centro Solis, oculi, et iridis in una linea existentibus. PROP. 15

Nunquam igitur oculus iridem videre poterit nisi in una linea Solis centrum, respicientis oculus, et iridis steterint, nam si aliter starent, impossibile esset ut anguli aequales refractionis ad nostrum oculum constitui possent, quos causam dicimus iridis esse necessariam. Aristoteles pro centro oculi ponit centrum horizontis sive mundi, sed longe distat Terrae centrum ab oculo, sed oculi centrum in superficie Terrae et recte intuenti semper pars maior semicirculi iridis videbitur.

---

[56]"AFBDCH" in original text.
[57]"FEH" in original text.
[58]"L" in original text.
[59]"FBCDHNL" in original text.

refraction can occur equally on all sides when the center of the Sun, the center of the eye, and the center of the rainbow lie on one line. We will illustrate this as follows.

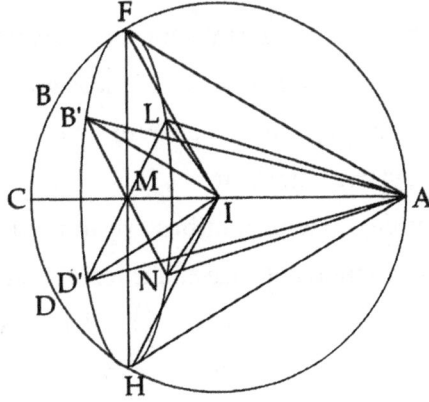

Figure 9.14b

Let AFBCDH [in Figure 9.14b] be in the plane of the meridian, let AIC be a diameter [of that circle], and let FMH be a facing cloud. Let light from the Sun at A strike it according to a cone along AF, AB', AL, AD', AN, and AH. Let the eye be at I, and let the cone of radiation from the eye intersect that same cloud illuminated by the Sun, for example, along IF, IL, ID', IN, IH, [and IB'], and since this cone cuts the cone of solar radiation uniformly throughout, circle FB'D'HNL will be the common section, and the angles IFM, IB'M, ID'M, ILM, INM, and IHM formed by the lines from the eye and the diameter of the rainbow will all be equal, and the same holds for the solar rays. This demonstration concerning the round shape of the rainbow seems equivalent to that given by Aristotle where, for instance, he deals with the halo.[64] But this demonstration [concerning the rainbow] is not understood among others; on the contrary, it is not so clear and easy, and it involves hidden diligence. [209]

PROP. 15: *A rainbow appears when the center of the Sun, the center of the eye, and the center of the rainbow lie on a single [straight] line.*

Accordingly, the eye can never see a rainbow unless the center of the Sun, the center of the viewer's eye, and the center of the rainbow lie on one line, for if they were disposed in another way, equal angles of refraction could not possibly be established to our eyes, which is why we say [that having these points on the same line] is necessary for the creation of a rainbow. Aristotle takes the center of the horizon or the Earth as the center of the eye, but the center of the Earth lies far away from the eye, and with the center of the eye on the Earth's surface and looking straight ahead, a considerable portion of the semicircle of the rainbow will always be seen.

Argumenta quae me ad id impellunt sunt experimenta visa in iride in cubiculo facta ex aqua ore insufflata et Sole: quod nunquam iris apparebat nisi Sol retro stans, oculus et iridis visae centrum in una eademque linea starent, et dum aer irroratione et guttulis madebat, elevando deprimendoque caput, elevabatur et deprimebatur iris. Et circumastantes familiares eam non videbant, nisi qui oculos prope meos haberent, et qui movebantur videbant iridem moveri, et qui starent stantem.

Idem evenit in iride, quae in fontibus apparet, qui ex alto in praeceps dalapsi, vel ex fistulis in aerem guttulas eiicientes; aerem irrorant, et ubi desit irroratus aer vel angulorum idonea positio, ibi et deest iridis circumferentia et coloris repraesentatio. Praeterea, si oculus elevabitur supra Terrae superficiem, maior pars iridis videtur, et si multum integrum circulum etiam Sole in horizonte existente. Praeterea, dum aerem irrorarem, oculum supra mensem ponebam, et non nisi circuli semissem videbam, et secundum proportionem elevantionis circuli a tabula pars iridis videbatur. Unde si quis supra montem ascenderit, clarum est quod aut totum iridis circulum, aut maiorem semisse videbit. Memini dum Hispaniis essem in ecclesia super Montem Serratum dictum, moveri infra nebulas videbam et pluere quum ego in sereno caelo supra essem, et vidi irides plusquam semicirculares, sed non perfecte memoror an perfectas viderem.

Esto Sol in horizonte, oculus supra montem; [210] esto A fulgens astrum, oculus supra montem B. Erit iridis centrum supra horizontem, et

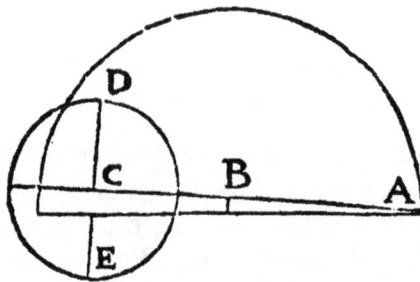

erit C; axis semidiameter DC. Et quanto erit maior super axem DC, tanto erit maior semicirculi portio. Nec putetis obsecro quod omnes oculi unam

The reasons that drive me to this conclusion are based on the following observations about the rainbow formed in my room by [the interaction between] water blown through my mouth and sunlight. The rainbow never appeared unless the Sun lay behind me, the center of the eye and of the observed rainbow had to lie on one and the same line, and as long as the air was moistened with mist and water droplets, when I raised or lowered my head, the rainbow was raised or lowered. In addition, friends who were standing around me did not see it unless they kept their eyes near mine, and those who moved saw the rainbow move, whereas those who stood still saw it stay still.

What happens in the case of the rainbow is the same as what happens in the case of fountains whose spray falls rapidly from high up, or in the case of pipes shooting droplets into the air; they fill the air with mist, and where the air lacks mist or where an appropriate arrangement of angles is lacking, the band of the rainbow fails to appear, as does the representation of color. Furthermore, if the eye is raised above the surface of the Earth, a larger portion of the rainbow appears, and yet a much more complete circle appears when the Sun is at the horizon. While I was filling the air with mist, moreover, I placed my eye level with a table, and I saw only half a circle, and the portion of the rainbow that appeared depended on the amount of the circle raised above the table. Consequently, if someone climbs a mountain, it is obvious that he will see the entire circle of the rainbow, or else more than half. I remember that when I was in Spain in a church on a mountain called Serratus, I saw clouds moving below and rain falling while I was in clear sky above, and I saw rainbows that were larger than a semicircle, but I do not remember perfectly whether I might have seen completely circular ones.[65]

Let the Sun lie at the horizon and the eye on a mountain, [210] and let A [in Figure 9.15] be the shining star and B the eye on the mountain.

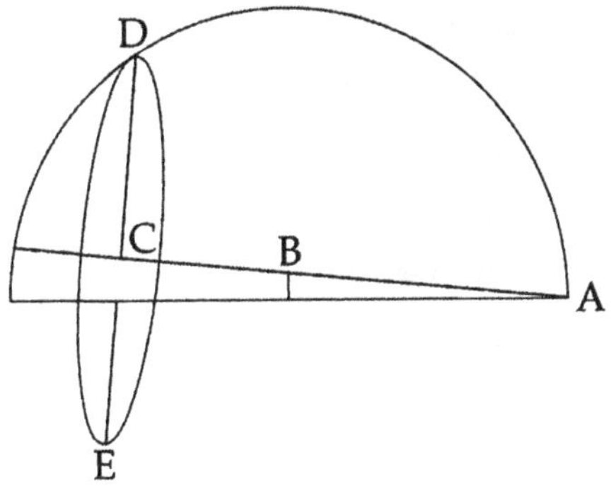

Figure 9.15

The center of the rainbow will lie above the horizon, and it will be C, and let axis CD be its radius. The higher axis CD lies, the greater the visible portion of the semicircle will be seen. And I urge you not to believe that all

iridem videant et quod omnes oculi sint unus oculus, nam quot oculi sunt, tot sunt iridis visiones, et quisque suam singularem videt, etsi putent omnes eandem videre. Et dum movetur oculus, movebitur et iris, etsi in longa distantia differentia est parva, vel minima, ut inferius dilucidius exponetur.

Querit Seneca, "Cur arcus non impleat orbem, sed dimidia eius pars videatur, quum plurimum porrigitur incurvaturque?" Dicunt aliqui: "Quum Sol sit multo altior nubibus, a superiore tantum illas percutit parte. Sequitur ut inferior pars earum non tangatur lumine; ergo quum ab una parte Solem accipiant, unam partem eius tantum imitantur, quae nunquam dimidia maior est." Sed hoc falsum, quia si Sol nubem a superiori parte sit, totam percutit nubem, et si superior sit Sol, superiori parti nubium affunditur; numquam Terra tenus descendit arcus, atqui usque ad humum demittitur. Ipsi cavam nubem faciunt et sectae[60] pilae partem, quae non potest totum orbem reddere quia ipsa est pars orbis. Sed hoc quoque falsum est quia nubem semper cavam quaerunt, quum sint aliquando planae et tumentes.

Quod moto oculo, movetur iris. PROP. 16

Hoc maxime animaduertendum quod qui ambulant et iridem aspiciunt singulis passibus singulae fiunt novae et videbuntur, unde semper secum ambulantem vident.[61]Multoties [211] expertus sum dum Italiam peragrarem et plano in agro obequitarem; ubi iridem in caelo conspicabar, additis calcaribus, citato cursu currebam, et mecum currentem iridem non nisi magna cum voluptate videbam. Amici qui mecum erant idem facientes idem experiebantur. Philippus sodalis Platonis, ut Alexander refert, ad quodcunque latus se verteret ipsa iris transferebatur, et ambulans iris mutari videbatur, ut recedentem sequeretur, et insequentem aufugeret crocodili more. Legitur apud geometras quod Philippus Menedaeus "observavit opticum illud admirabile quod insequentes fugiat[62]; fugientes insequatur." Cuius causa non alia est, quam diximus, quod in aere toto roscido, et a radiis Solis verberato, refractoque, ubi anguli constituti in oculo fiant, ibi et videbitur iris.

Ex Solis elevatione et depressione elevatur et depremitur iris. PROP. 17

De iridis elevatione et depressione Aristotelis demonstrationes omittemus, difficillime enim percipiuntur ab iis qui mathematicas ignorant, et mihi

---

[60]"sestae" in original text.
[61]"videt" in original text.
[62]"fuigiat" in original text.

eyes see a single rainbow or that all eyes constitute one eye, for there are as many views of the rainbow as there are eyes, and each one sees its unique rainbow, even though all the viewers suppose they are seeing the same one. And when they move their eyes, the rainbow will also move, even though at a great distance the difference [in separation among the individual rainbows] is slight, or minimal, as will be explained more clearly below.

Seneca asks, "Why does a rainbow not form a complete circle, but only a semicircle is seen, even when it is at its fullest extent and curvature?" Some reply that, "since the Sun is much higher than the clouds, its light strikes them only from above; consequently, their lower part is not colored by the light. So, since they receive the Sun's light at one portion, they copy only one part of it, and that is never more than half." [66] This, however, is false because if the Sun stands higher than the cloud, it strikes the entire cloud, and if the Sun is higher, the higher part of the cloud is flooded with light, and so the span of the arc never reaches the Earth and is missing all the way to the ground. These same thinkers make the cloud concave and a portion of a spherical segment that cannot yield a full orb because it is a portion of an orb. But this, too, is false because they invariably base their inquiries on a cloud that is concave, whereas clouds are sometimes flat or swollen outward.

PROP. 16: *When the eye moves, the rainbow moves.*

It is particularly noteworthy that for those who walk and look at a rainbow, with every single step new rainbows are formed and will appear, so those walkers see the rainbow continually keep pace with them. [211] I often experienced this when I traveled through Italy and rode in a flat field; there I saw a rainbow in the sky and, putting the spurs to the horse, ran at a fast clip, and it was not without considerable pleasure that I saw the rainbow running along with me. Friends who were with me had the same experience when they did the same thing. As Alexander of Aphrodisias reports, Plato's associate Philip of Medma saw the rainbow shift to whatever direction he turned, and as he walked he saw the rainbow change position such that when he gave ground to it, it followed him, whereas when he chased after it, it fled the way a crocodile does. [67] One reads in works of geometers that Philip of Medma "observed the wonderful optical effect that the rainbow flees those who chases after it and chases those who flee from it." [68] As we said, the only reason for this is that, if totally misty air is struck by the Sun's rays, which are then refracted, and if the appropriate angles are formed in the eye, a rainbow will be seen.

PROP. 17: *The rainbow is raised or lowered according the Sun's elevation [with respect to the horizon].*

We will pass over Aristotle's demonstrations concerning the raising and lowering of the rainbow, for they are grasped with great difficulty by those

videntur aliqua praesupponere quae aliter se habent. Sed his exemplis rem dilucidiorem reddemus.

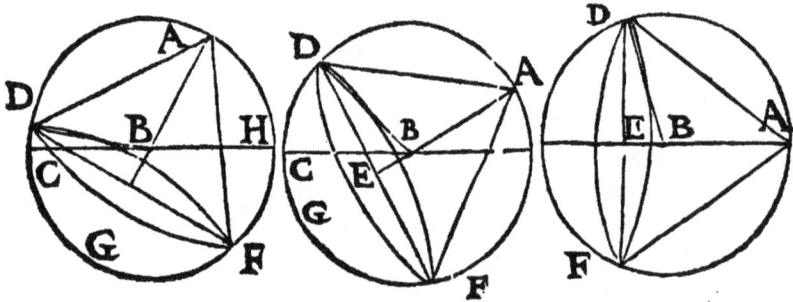

In prima figura, sit Sol oriens A, eius radius AD, AF. Erit iridis diameter DEF, oculus B. Sub angulo ADB erit iris, et observavi abstrolabio in maxima altitudine occidente Sole in fine Augusti 24 gradibus elevatum. In secunda figura, elevatus Sol in A; circulus iridis nobis apparebit minor, etsi pyramis eadem [212] ADF. Et quantum erigitur A (scilicet AH) supra horizontem, tanto iris deprimitur sub horizonte (scilicet CF[63]), et idem erit angulus ADB qui primus. In tertia, Sol maxime elevatus, ut minima videatur nisi DC, et reliqua pars sub horizonte CF,[64] angulus quoque idem erit ADB.

Non tamen hoc mirum omittam. Existente Sole ex latere et oculo in centro iridis, ita etiam ut centrum Solis, oculi, et iridis in eadem sunt linea, ad latus oculos revolvendo, tunc non videtur iris circularis, sed crus unum descendens, et colores a latere videntur ex alto descendentes, ut prima tenia flava, post viridis, et tertia caerulea videatur. Id non solum in cubiculo, ore aquam efflaudo semper video, sed fontibus, ubi stillis aerem asperginosum reddunt, ubi paulatim ad iridis centrum accedendo, oculis tamen animaduer-tendo ne emphasis evanescat, et in centro circuli constitutus ad dextra revolvendo, videtur iridis crus fere rectum prope oculos. Commorans ruri multoties accidit eiusmodi visionem conspicere in iride a nube facta, ut minus vigintiquinque passibus a me longe abesset. Multoties quoque in quadam valle nebulosa, quum ego essem in sublimi, vidi idem crus valde infra descendere. Et ego potius rationes ad sensum distorqueo, quam sensum

---

[63]"EG" in original text.
[64]"FG" in original text.

who are unfamiliar with mathematics, and they seem to me to presuppose things that are contrary to the way things actually are.[69] We will, however, clarify the matter with these illustrations.

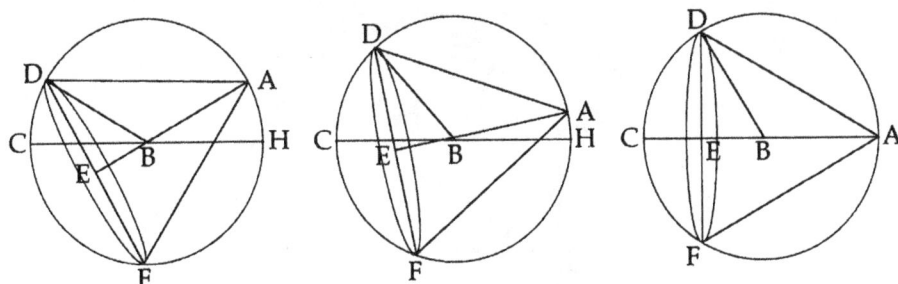

Figure 9.17a

In the first diagram [to the right of Figure 9.17a], let A be the rising Sun, and let AD and AF be rays from it. DEF will be the diameter of the complete rainbow, and B will be the eye. The rainbow will lie under angle ADB, and at the end of August, I have observed with an astrolabe that a rainbow is elevated to a maximum of 24° above the horizon at sunset.[70] In the second diagram [in the center of Figure 9.17a], the Sun has risen to A, and the circle of the rainbow will appear smaller to us, even though the cone of radiation ADF is identical. [212] The higher A gets above the horizon (i.e., along AH), the more the rainbow sinks below the horizon (i.e., along CF), but angle ADB will remain the same as it was at the beginning.[71] In the third diagram [on the left of Figure 9.17a], the Sun is at its highest so that, only the minimal amount of the rainbow may be seen along DC, whereas the remainder lies below the horizon along CF, and angle ADB will remain the same.

I should not, however, pass over this amazing thing. If the Sun lies to the side and the eye is at the center of the rainbow, so that the centers of the Sun, the eye, and the rainbow also lie on the same line, when the eyes are rotated to the side, the rainbow does not appear circular, but one leg appears to drop down, and the colors are seen at the side, descending from on high, so that the first band appears golden yellow, the following one green, and the third blue. Not only do I invariably see this in the room when I blow water from my mouth, but also in fountains, where they render the air misty with droplets and where, as I gradually approach the center of the rainbow, still turning my eyes so that the intensity does not disappear, and standing at the center of the circle while rotating my eyes to the right, the leg of the rainbow looks almost straight near the eyes. Lingering in a field, I often happened to see the same sort of image in a rainbow formed from a cloud, such that it stood less than twenty-five feet from me. In addition, several times in a certain misty valley, when I stood high above, I saw the same leg drop down far below. But rather than bend sense to reason, I prefer

ad rationem, quod video enim certum est, quod disputo incertum.

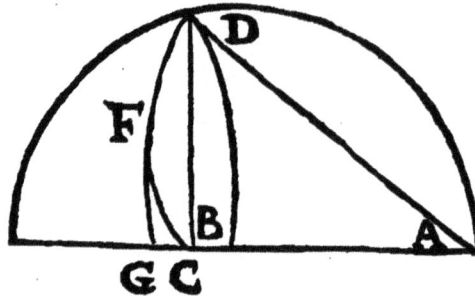

Esto astrum A, centrum iridis B; oculus vero C extra centrum iridis; obliquum crus iridis DFG, eius diameter DBE. Oculus C respiciens crus iridis FG, radii CF, CG, suis coloribus spendent. Nec audiendi sunt blacterones, dicentes non esse necessarium oculi, centrum Solis, et iridis esse in una linea, etsi pro exemplo addant iridem lucernae videri extra lineam connectentem centra. Sed exemplum est falsum, nam iris lucernae alteram causam habet, quam ipsi ignorant et iam nos supra manifestavimus. [213]

Maiorum opiniones cur iridis portiones quo minores, eo maiores videri.
PROP. 18[65]

Diximus oriente vel occidente astro, iridis semiorbem videri, eo vero ab horizonte sublato minorem videri. Sed portiones maiores semicirculi semper in meridie videri prope horizontem quam dum in sublimi apparent. Aristoteles problema solvere spopondit, set promissum non absolvit. Sui clarissimi affectatores variis rationibus id tentarunt, ut dicemus.

Olympiodorus asserit, quando iris semiorbis apparet, is in tenuissimo aere est super Terram in sublimi, at ubi prona Terram tangit, ubi ingens vaporum spiratio adest, amplioris formae videtur.

Omnia enim per aquam videntibus, longe esse maiora. Literae quamvis minutae et obscurae per vitream pilam aqua plenam maiores clarioresque cernuntur. Poma

---

[65]"17" in original text.

to bend reason to sense, for what I see is certain, whereas what I reason about is uncertain.

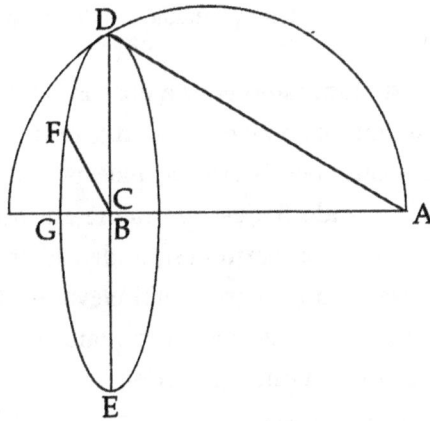

Figure 9.17b

Let A [in Figure 9.17b] be the celestial body and B the center of the rainbow, but let the eye at C lie outside the center of the rainbow. Let DFG be a leg of the rainbow at a slant, and let the rainbow's diameter be DBE. When the eye at C looks at leg FG of the rainbow, rays CF and CG shine with its colors.[72] Those bleaters who claim that the centers of the eye, the Sun, and the rainbow need not be on a single line should not be listened to, even though they may add that the rainbow around a candle is seen outside the line connecting the centers. The example is false, though, for the rainbow around a candle has another cause, which they do not know and which we have already elucidated above.[73] **[213]**

PROP. 18: *The opinions of our predecessors on why rainbows forming smaller sections than a semicircle appear enlarged.*

We have said that when the heavenly body lies to the east or west at the horizon, the rainbow appears as a semicircle, whereas when that body has risen above the horizon, the rainbow appears smaller. Yet the segments of the semicircle always appear larger at midday near the horizon than they appear when they are up high. Aristotle promised to resolve this puzzle, but he did not fulfill his promise.[74] His renowned followers tried to fulfill that promise with various arguments that we will now discuss.

Olympiodorus maintains that when the rainbow appears as a semicircle, it lies high above the Earth in extremely thin air, whereas when it turns back to touch the Earth, where a plethora of vapors are exuded, it looks larger in appearance.[75]

Everything is much larger to viewers looking through water: writing, however tiny and difficult, is seen larger and clearer through a glass sphere full of water; fruits

formosiora quam sint videntur, si innatant vitro. Sydera ampliora per nubem aspicienti cernuntur, quia acies nostra in humido labitur, nec apprendere multum fideliter potest, quod manifestum fit, si poculum adimpleverit aqua, et in id conieceris anulum . . . quicquid igitur videtur per humorem, longe amplius vero videtur.

Altera ratio: quando in sublimi est longius a nostris oculis abducta est, modo quae longius videntur, minora videntur; at iris quum horizonti proxima videtur, proximior et nostris oculis erit, et quae proximiora maiora videri. Suessanus dicit quod,[66] quando Sol est in oriente aut occidente, ob multos vapores horizontis lumen minus clarum est, unde iridem minus apparentem et visibilem facit. Contra vero in meridie ubi elevatus et a vaporibus liber, plus illustrat et facit ut res maiores videantur quam sint. Sec haec postrema ratio satis erronea et contra rationem experientiae.

Olympiodori rationes, etsi verae sint, non tamen ad rem faciunt, nam apparens arcus a summo usque ad imum semper aequalis et congruus apparet, et quae in vaporibus mersa sunt eodem modo ut summum videntur. Neque rationes attingunt scopum, nam etsi quae sub aquis cernuntur maiora conspiciantur, in hac portione in imo conspecta videtur non solum maior sed aliter disposita quam est in summo. Cardanus dicit quod oculus iudicat iridem quo humilior est [214] eo maior videtur quoniam distantiam intellegit. Quibus verbis nihil absurdius, quum sensui tribuat quod est ratiocinationis. Plinius: "Idem sublimes humili Sole, humilesque sublimi, et minores[67] occidente vel oriente, sed in latitudinem diffusi; meridie exiles, verum ambitu maiores."

Opinio propria quare iridis portiones breviores maiores videantur. PROP. 19[68]

Sed ratio vera a mathematicis petenda est, nam teniolae illae sive fasciae quae in iridis circumferentia circumductae sunt, dum in sublimi videntur, oblique videntur. Secus autem evenit quando sunt circa terram, nam recte videntur, et ubi in sublimi contractae videbantur, hic in imo explicatae. Quod exemplo clarius patebit.

Esto Sol A in hac prima figura saucians nubem roscidam ECFD supra horizontem erectam. Radius A feriens nubis partem E veniret in G, sed labens propter refractionem, in F pervenit. Oculus existens in B videt puncta extremarum teniarum; erit BE, BF. Et sub angulo EBF videbit tres iridis

---

[66]"quo" in original text.
[67]"maiores" in original text.
[68]"18" in original text.

appear more beautiful than they are if they are swimming in a glass bowl; the stars themselves seem larger when one looks at them through a cloud, because our eyesight falters in moisture and cannot reliably grasp what it wants to. This is plain if you fill a cup with water and drop a ring in it. Thus, anything seen through moisture appears much larger than in reality.[76]

Another explanation is that, when the rainbow is high, it is extended far away from our eyes, and the farther away things appear, the smaller they appear. But when the rainbow is seen near the horizon, it will be nearer our eyes, and things that are nearer appear larger. Agostino Nifo says that, when the Sun is at the eastern or western horizon, its light is less bright because of the multitude of vapors at the horizon, which therefore makes the rainbow less apparent and visible. On the other hand, when it has risen at midday and is free from vapors, it illuminates more intensely and causes things to appear larger than they are.[77] But this last argument is pretty much off track and runs counter to reason drawn from experience.

Even if they are true, Olympiodorus's arguments do not yet address the issue, for the arc that appears from top to bottom always looks uniform and consistent, and things that are immersed in vapor look the same as they do above it. Nor do his arguments achieve their goal, for although things that are seen under water appear magnified, they appear not only magnified according to how far below they lie, but they also appear disposed in a different way than is the case at the top. Girolamo Cardano says that the lower the rainbow is [214] the larger the eye judges it to be according to our apprehension of distance.[78] Nothing could be more ridiculous than what he says with these words, for he attributes to sense what pertains to reason. As Pliny says, "rainbows are high when the Sun is low, and low when the Sun is high, and they are smaller at the western and eastern horizon, but they are spread out more widely; at midday they are narrow but of a more ample circumference."[79]

PROP. 19: *The correct account for why smaller segments of the rainbow appear larger.*

The real reason is to be sought from mathematical opticians, for those narrow ribbons or bands that extend around the circumference of the rainbow are seen at a slant when they are seen up high. On the other hand, when they are near the ground, what occurs is different, for they are seen [more or less] directly, and what had appeared shrunken when seen up high appears spread out in this case, when seen lower down. This will show more clearly with an illustration.

Let light from the Sun at A in the first diagram [to the right in Figure 9.19] strike a moisture-laden cloud ECFD standing upright upon the horizon. The ray from A reaching point E on the cloud would continue to G, but being diverted by refraction, it reaches F. An eye lying at B sees points at the edges of the bands, and it will be along rays BE and BF. It will see the three colors of the rainbow along line EF under angle EBF.[80] However,

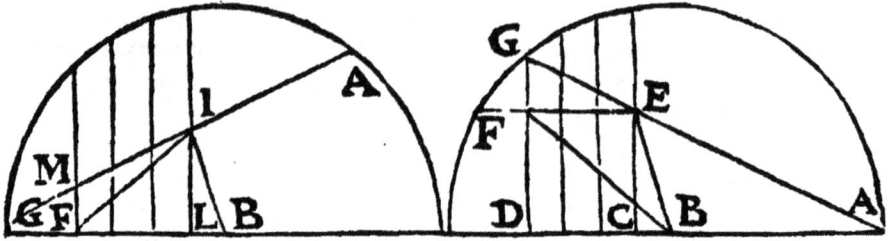

colores in linea EF. Secus autem eveniet in secunda figura, nam Sol ascendens puta ad 42 gradus in A. Et sit nubs in eadem distantia a B oculo in L, et eiusdem latitudinis LFIM. Feriat Sol nubem in I. Radius non procedit in G quoniam nubs densior; refrangitur in F. Trahatur linea a puncto B in I, et altera ad punctum F, finis refractionis. Videbitur igitur linea colorata IF sub angulo IBF, qui maior est superiori angulo EBF, ibi enim oblique, hic directe. Ut patuit in superioribus, ut quae oblique viderentur longe minora quam quae directe, si omnia proportione fiant. [215]

In iride caelesti sunt apparentes non veri colores. PROP. 20[69]

Inquit Plutarchus meteoricarum affectionum alias esse ex substantia, ut imber, grando; alias ex specie nullam habere substantiam, ut arcus. Possidonius non putat veros colores, nam "si ullus esset in arcu color, permaneret et videretur eo manifestius, quo propius.[70] Sed imago arcus ex longinquo clara est; interit quum ex vicino est ventura." Sed his non consentit Seneca, sed dicit:

> nubem colorari, et eius colorem non undique videri, nam ne ipsa quidem nubes undique apparet, nubem enim nemo qui in ipsa est videt. Nec mirum si eius color non videtur ab eo, a quo ipsa non videtur. Atqui etsi ipsa non videtur, est tamen color; itaque non est argumentum falsi coloris, etsi ipsa accedentibus apparere desinit. Idem enim evenit nubi, nec ideo falsa est . . . Praeterea, cum[71] dicitur tibi nubem Sole suffectam, non dicitur immixtum esse colorem, velut duro corpori . . . et manenti, sed fluido et vago . . . Insuper dicis omnes fulgores paulatim discuti; huius repentina facies et interitus. Respondetur proprium esse speculi quod non per partes struitur quod apparet sed statim totum fit; aeque cito omnis

---

[69]"19" in original text.
[70]"proprius" in original text.
[71]"dum" in original text.

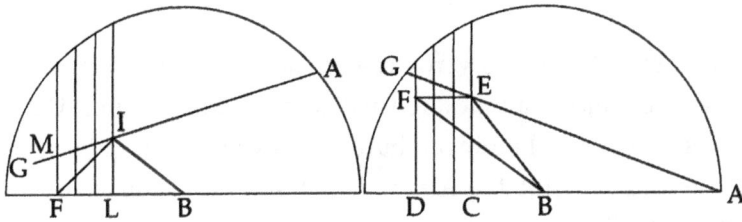

Figure 9.19

something different will occur in the second diagram [to the left in Figure 9.19], for suppose that the Sun at A rises to 42°. Let [the base of] cloud LFIM at L lie the same distance from the eye at B, and let the cloud be the same thickness. Let the light from the Sun reach the cloud at I. The ray does not continue to G because the cloud is denser [than the air outside the cloud], and so it is refracted to F. Draw a line from point B to I and another one [from B] to point F, the endpoint of refraction [along ray IF]. The colored line IF will therefore be seen under angle IBF, which is larger than the previous angle EBF, for in the previous case the line was seen at a greater slant, whereas in this case it is seen more directly. As was shown earlier, what is seen obliquely appears considerably smaller than what is see directly, if everything else is made proportionate. [215]

PROP. 20: *The colors in a heavenly rainbow are apparent, not true.*

Plutarch claims that some meteorological phenomena, such as rainstorms and hailstorms, are material in origin, whereas others are of a kind that has no material cause, for example, the rainbow.[81] Posidonius does not believe that the colors are real, for, "if there were any color in the rainbow, it would last and could be viewed more distinctly the closer one got. As it is, the image of a rainbow is clear from a distance, but it disappears when one gets up close"[82] Seneca, however, does not agree with these points, and in opposition he says that

> the cloud is colored, but its color is not visible from all sides. For in fact the cloud itself is not visible from all sides; for no one who is inside a cloud can see it. It is no surprise if its color cannot be seen by someone by whom the cloud itself cannot be viewed. Yet, although it cannot be seen, the cloud still has color. So it is not an argument for the color not being real that it stops being visible as people approach it. For the same thing happens with the cloud itself, and it is not unreal. Besides, when you are told that the cloud is dyed by the Sun, you are not being told that the color has been united to it as if on a hard, stable, enduring body, but as on a fluid, changeable one. You say, moreover, that every effulgence dissipates gradually, and you make the suddenness of [the disappearance of colors in the rainbow] a counter [to the argument that the colors are really there]. One responds that is a peculiarity of the mirror that what appears in it is not assembled bit by

imago in illo aboletur quam ponitur.

Pythagorici dicunt "non esse veros colores quod diu non permaneant . . . neque id evenire subjecti ratione quod in dilapsu sit constitutum." His respondetur non veriorem esse hominem diuturnioris vitae quam qui statim productus commoritur. Aristoteles, Alexander, et alii, non veros colores vocant sed σὺμφασης qui enim ex refractione sunt colores, mendaces colores sunt, non enim ubique nec a quocunque oculo videntur, ut diximus; sed in eo angulo oculo constituto, et si quis esset in eo loco ubi colores sunt, eos non videret. Veri colores dici debent, qui ex elementorum misturis componuntur. Sunt etiam veri colores qui ex refractione Solis in crystallino prismate videntur, qui etsi refractione lucis et diaphani aere densiores causentur, videntur in pariete, et ab omni oculo et ab omni situ conspiciuntur. [216]

De duplici iride nostrorum maiorum opiniones. PROP. 21[72]

Nec minus difficultatis habet geminatae iridis disquisitio; recensebimus maiorum opiniones, quam variae et a veritate aberrantes. Observavimus anno 1590, Sole occidente, superioris iridis elevationem 42 graduum, et inferioris 38, ut quatuor graduum esset utriusque interstitium.[73] Nicolaus Peripateticus dicebat binas irides aliquando videri, et esse a Sole et Luna genitas, ut Sol unam, alteram Luna gigneret. Sed quam futilis sit haec opinio arguit experientia quod absente Luna a nostro hemisphaerio duplex iris visa est. Alia fuit opinio opticorum temporis Alexandri, quod haec secunda iris esset reflexio primae.

Albertus exemplum affert, probatque hanc opinionem eiusmodi experimento. Fiat vitreus arcus in tres partes divisus, cuius suprema circumferentia rubea, media flava, tertia viridis; et decenter aptatus teneatur contra Solem, ut ab eo in opposito pariete iidem colores reflectantur. Imaginatur nanque colores primae iridis transfundi vel reflecti in alia nube a prima distanti, et sic duplicatam iridem videri. Sed opinio multipliciter falsa est. Primo si colores ex prima transmitterentur, colores essent eodem ordine constituti ut primi, sed quia contrario modo. Eadem fuit Cardani opinio, sed latius refellitur a Scaligero.

Dicunt alii geminatas irides ex geminata nubium oppositione, quia non plus quam duae esse possunt, altera in media aeris regione, altera in ima;

---

[72] "20" in original text.
[73] "intestitium" in original text.

bit but comes into existence instantly in its entirety. Every image in a mirror vanishes just as quickly as it is put together.[83]

The Pythagoreans claim that "they are not real colors because they do not last very long, and this failure to last happens because the color is in something that dissolves."[84] In response to them one remarks that a human who lives a long time is no more real than one who dies as soon as he is born. Aristotle, Alexander of Aphrodisias, and others call these colors not real but σὺμφασης [conj"oining together"], for colors that come from refraction are false colors because, as we have said, they are not seen everywhere or by every eye; instead they are seen by an eye at a particular angle, and if one were to be where the colors are actually manifested, he would not see them. Colors composed from the mixture of elements must be said to be real. The colors that are seen in the refraction of sunlight in a glass prism are also real, for even though they are due to the refraction of light and to transparent media denser than air, they are seen on a wall, and they are discerned by every eye from every location. [216]

PROP. 21: *The opinions of our predecessors concerning the double rainbow.*

The investigation of twin rainbows is no less fraught with difficulty, and so we will survey the opinions of our predecessors, which are various and which depart from the truth. In the year 1590, when the Sun was setting in the west, we observed that the height of an upper rainbow was 42° and that of a lower one was 38°, so that the interval between both was 4°.[85] Nicholas Peripateticus said that two rainbows are sometimes seen, and they are formed by the Sun and Moon such that the Sun would generate one and the Moon the other.[86] But experience demonstrates how fatuous this opinion is because when the Moon is missing from our hemisphere of visible sky, a double rainbow has been seen. Another opinion was that of optical thinkers at the time of Alexander of Aphrodisias, who supposed that this second rainbow was a reflection from the first.[87]

Albertus Magnus offers an illustration, and he demonstrates this opinion with an experiment of the following sort. Form a glass arc divided into three parts, its top circumferential band red, its middle one golden yellow, and its third one green, and hold it appropriately toward the Sun so that the same colors are reflected in a facing wall. Now imagine that the colors of the first rainbow are transferred or reflected to another cloud separated from the first one, and so the rainbow will appear doubled.[88] But this opinion is false on several grounds. First and foremost is that, if the colors were transmitted from the first bow, the colors in the second bow would stand in the same order as in the first, but they stand in the opposite way. Girolamo Cardano's opinion was the same, but it is refuted extensively by Julius Caesar Scaliger.[89]

Others argue that double rainbows are due to two facing clouds because there can be no more than two rainbows, one in the middle region of the

inferior habet guttas maiores, quia crassior materies, superior[74] minores, quia subtilior, ob id languidior color. Et haec est quoque Alexandri opinio, aliter quam Aristoteles verba sonant [217] interpretando. Dicit enim quod a secunda iride fit refractio a prima ad Solem, quod in textu abest, sed refractio[75] a prima. Sed si id esset, colores eodem modo ordinarentur, ut in subiecto exemplo.

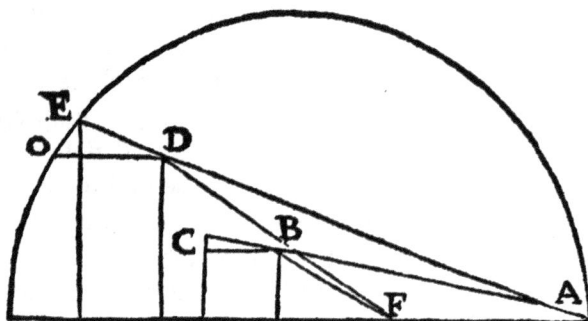

Sol A ferit primam nubem BC in B. Oculus F videt BC colores. Sic Sol idem feriens superiorem nubem DE, et oculus F videt[76] refractionis lineam DO. Sic duplex colorum ordo.

Duplicem iridem posse apparere. PROP. 22[77]

Nunc nostram afferemus opinionem, et fortasse Aristotelis germanam, quam ex obscurissimis involucris vix elicimus. A Sole enim binae apparent irides, una quidem interior, altera exterior interiorem comprehendens, sectioque maioris circuli est quae certe languidiores colores ostendit; etiam situ adversa primae. Interior iris superiorem circumscriptionem puniceam, secundam viridem, tertiam porro caeruleam habet. Sed exterior et secunda iris exteri-orem quidem circumscriptionem caeruleam, secundam viridem, intimam vero puniceam, ut ambarum iridum propinquae invicem positae sint puni-ceae. Et hanc secundam irim fieri refracto visu ex nube exteriore ad primam irim, ob id etiam languidiores esse colores ferunt, utpote a secunda facta refractione. Amplior enim a minori et interiori periferia incidit ad Solem. Propinquior enim visus existens refrangitur a periferia propinquissima primae iridi; propinquissima autem in exteriori iride, minima periferia est, quare haec habet colorem puniceum, contiguae autem et tertia secundum proportionem.

Nos exemplo rem obscuram clariorem reddemus. Dicimus ergo solarem radium vehementissimum facillime primam nubem penetrare, et si nubs tenuis, et adhuc validus erit; etiam in oppositam nubem resilire idoneam

---

[74]"inferior" in original text.
[75]"refactio" in original text.
[76]"vidit" in original text.
[77]"21" in original text.

air, the other at the bottom, and the lower region has larger drops because its matter is crasser, whereas the upper region has smaller [drops] because its matter is less crass, for which reason its color is weaker. This is also the opinion of Alexander of Aphrodisias, interpreted [217] differently from what Aristotle's words say. For he says that refraction occurs at the second rainbow from the first to the Sun, but "refraction . . . from the first" is missing in the text.[90] If this were the case, however, the colors would be arranged in the same order, as in the example below.

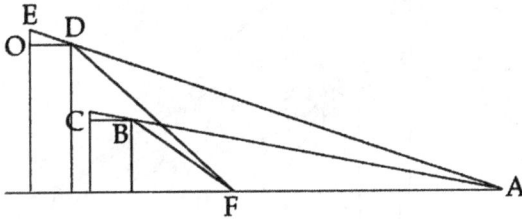

Figure 9.21

[Light from] the Sun at A [in Figure 9.21] strikes the first cloud BC at B. The eye at F sees the colors along [refracted ray] BC. Likewise, [light from] the same Sun strikes the higher cloud DE, and the eye at F sees the colors along line of refraction DO. There is thus a twofold arrangement of colors [in the same order].

PROP. 22: *A double rainbow can be seen.*

Now we will set forth our opinion, as well, perhaps, as the true opinion of Aristotle, which we draw out with difficulty from its obscure shrouding. Indeed, twin rainbows do appear from the Sun, one inside, the other outside and encompassing the interior one, and it forms the section of a larger circle that definitely shows weaker colors; also, it is disposed in an opposite way from the first one. The interior rainbow has a red outer band, a green second band, and, in turn, it has a blue third band. The exterior and second rainbow, however, has a blue outer band, a green second band, and a red lowest band, so that the bands posed nearest to each other on both rainbows are red.[91] And this second rainbow is due to sight refracted from a cloud beyond the first rainbow, which is also why the colors that the secondary rainbows display are extremely weak, insofar as they are due to a second refraction. For the larger bow reaches from the circumference of the smaller, interior bow to the Sun. Indeed, the one lying nearer the eye is refracted from the nearer band of the first rainbow, whereas the nearest band of the outer rainbow is the smaller one, since this has the red color, which is contiguous and proportionate to the third band.

We will render this obscure matter clearer with an illustration. Accordingly, we say that an extremely bright solar ray penetrates the first cloud easily, and if the cloud is thin, it will still be intense. We also say that the

generare. Hoc tamen intersit inter primam et secundam refractionem, ut colores inferioris iridis infra spectentur, superioris vero superius, et sic contrarii erunt colores. [218]

Esto Sol A; percutiat[78] nubem BCDEFGHI Solis radius AF. Erit FG prima pars flava, GH secunda viridis, HI tertia pars caerulea. Erit oculus M. Ergo FMG videt flavum, GMH viridem, HMI[79] caeruleum, et si volvatur nubs circum, circa faciet tres circulos, flavum, viridem,[80] et caeruleum. Radius igitur prior AFL refractus FI, adhuc validus et robustus, transit nubem alteram occurrentem NOPQRSTV, et percutit punctum R. Transiret in X, sed relabitur in V. Dividitur nubs in tres partes, prior RSNO flava, STOP secunda viridis, TVPQ caerulea. Et quia linea divisa, RS, ST, TV ex inferiori parte videri non potest, ut in prima; ab oculo M videbitur ex superiori. Unde prior pars et non firma RS videbitur, secunda ST, tertia TV, ut angulo superiori VMT videatur caerulea pars TVQP, angulo SMT videtur media pars STOP, et angulo SMR[81] videtur prima pars RSNO flavam. Si igitur circumducantur puncta R, S, T, V, faciet quatuor circulos tres tenias continentes, quarum maior ultima

---

[78]"percutiar" in original text.
[79]"GMI" in original text.
[80]"viviridem" in original text.
[81]"SMT" in original text.

light passes on to a properly facing cloud by refraction, and it penetrates into the cloud and produces colors in the same way [as in the primary bow]. Nevertheless, the difference between the first and second refraction is such that the colors of the lower rainbow may be observed in order from below, whereas those of the upper rainbow may be observed in order from above, and so the colors will be opposite in order. [218]

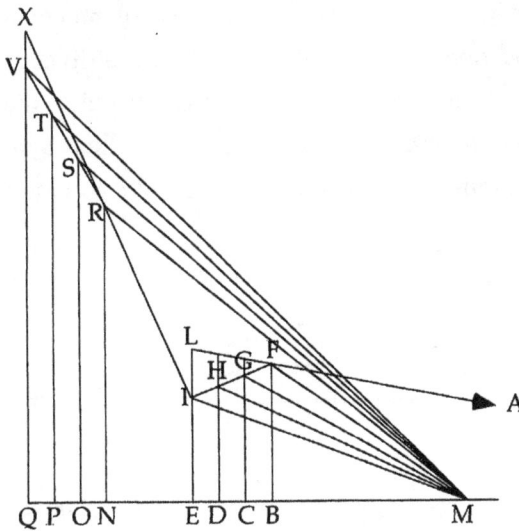

Figure 9.22

Let A [in Figure 9.22] be the Sun, and let ray AF from the Sun strike cloud BCDEFGHI. The first part FG [of ray FI refracted into the cloud] will be golden yellow, the second part GH will be green, and the third part HI will be blue. M will be the eye. Thus, it sees golden yellow [under angle] FMG, green [under angle] GMH, and blue [under angle] HMI, and if the cloud is rotated, it will form three circles in circumference, golden yellow, green, and blue.[92] Intense and strong to this point, the earlier ray AFL is refracted along FI and passes through to the second cloud NOPQRSTV, and it strikes point R. It would continue to X, but it is diverted to V. The cloud is divided into three parts, the first one, RSNO, golden yellow; the second one, STOP, green; [and the last one,] TVPQ, blue. Moreover, since the line is divided [in order from bottom to top], RS, ST, and TV cannot be seen as if from underneath by the eye at M, as in the primary bow, so it will be seen by the eye at M as if from above. Hence, RS is the first part that will appear [from the bottom up], but lacking intensity; ST is the second part, and TV the third, so under the upper angle VMT part TVQP should appear blue, under angle SMT the middle part STOP should appear [green], and under angle SMR the first part RSNO appears golden yellow. If, therefore, points R, S, T, and V are rotated, the rotation will form four circles enclosing three bands, the last and largest of which is blue, the second green, and the third

caerulea, secunda viridis, tertia flava. Et maior flava, secunda viridis, tertia caerulea, ut concava superioris et convexa inferioris flavae sint teniae. [219]

Interstitium inter duas irides esse excolorem. PROP. 23[82]

Diutorsit Alexandrum doctissimum Peripateticum difficillima quaestio, cur media nubes quae inter utramque iridem iacet excolor sit et non tota punicea. Idque evenit quod non putabat duas irides in duas diversas nubes fieri, in cuius medio aer est, et non nubes asperginosa. Evenit quoque alia ratione, quia anguli qui colores videre faciunt non sunt illius quantitatis quae ad colorum visionem requiritur, nam radii illi oculares eo modo nubem feriunt,

quo inferiorem, sed contrarii primis, et iidem sunt. Nam supremam iridem videbamus angulo VAR, infimam vero FAI; trianguli similes et proportionales, ita ut binas irides videre non refragratur primae positioni.

Iridem ex lunari radio generari posse. PROP. 24[83]

Seneca ait, "Arcum in nocte non fieri, aut admodum raro, quia Luna non habet tantum virium ut nubes transeat et illis colorem suffundat qualem accipiunt a Sole perstrictae.[84]" "Aristoteles fieri posse non infitiatur, sed Luna plena et in occasu vel ortu posita, dixitque annorom spatio quinquaginta bis apparuisse." Neuterici nulla ratione hoc dictum putant, nam in uno anno bis visam scribunt, qui Aquilone flante et Luna in meridie constituta. Plinius: "Non fieri noctu, quamvis Aristoteles prodat aliquando visum, quod tamen fatetur idem [non] nisi quartadecima Luna posse." Ego autem noctu fieri posse non infitior, nam aqua in Lunae lumine noctu ore insufflata iridem generabat, sed exilem et pene perspicuam, unde multo maius a lunaribus

---

[82]"22" in original text.
[83]"23" in original text.
[84]"Perscriptae" in original text.

golden yellow. Meantime, [in the interior bow] the largest band is golden yellow, the second band green, and the third one blue, so that the concave arc [through point R] of the upper bow and the convex arc of the lower bow [through point F] form golden-yellow bands. [219]

PROP. 23: *The space between the two rainbows is colorless.*

The extremely difficult question of why the interval in the cloud that lies between the two rainbows is without color rather than being entirely red tied the learned Peripatetic Alexander of Aphrodisias in knots.[93] This comes about because he did not believe that the two rainbows are produced in two different clouds between which there is clear air rather than moisture-laden clouds. It also comes about for another reason, because the angles that cause the colors to appear are not of the size that is required for colors to be seen in that interval, for the lines of sight from the eyes reach the cloud such that the lower rainbow is opposite the first, and they are identical. For we

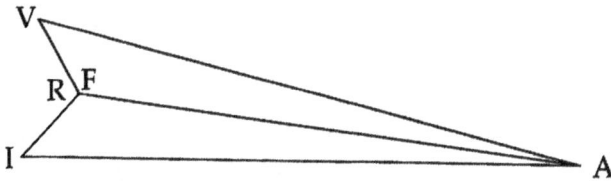

Figure 9.23

saw the upper rainbow according to angle VAR [in Figure 9.23], whereas we saw the lower rainbow according to FAI, and the triangles are similar and proportional, so that the two rainbows do not appear to depart from the first disposition.[94]

PROP. 24: *A rainbow can be generated by lunar radiation.*

Seneca says that "a rainbow does not occur at night, or only very rarely, because moonlight is not powerful enough to pass through clouds and tinge them with the kind of color they receive when struck a glancing blow by sunlight."[95] "Aristotle does not deny that a lunar rainbow can be formed, but only when the Moon is full and setting or rising, and he claimed that it appeared twice in a space of fifty years."[96] Modern thinkers suppose that this claim has no rational foundation, for they write that they saw it twice in one year when the north wind was blowing and the Moon stood in midsky.[97] Pliny says that "the rainbow is not formed at night, although Aristotle argues that it is sometimes seen then, but he nonetheless acknowledges that it can only happen on the fourteenth day of the lunar cycle."[98] I, however, do not deny that it can be formed at night, for I have produced a rainbow by blowing water from my mouth in moonlight, but the rainbow was weak and scarcely visible, so I have become all the more inclined to

radiis fieri assentirer. "Arati interpres Theon Anaximinis[85] [220] referens opinionem super affectione huiusmodi; 'et noctu,' inquit, 'fieri astruit, sed perraro, quoniam non semper plenilunium.'"

Quibus temporibus videri soleat. PROP. 25[86]

[Secundum Senecam]:

> Aristoteles ait, post autumnale aequinoctium qualilibet hora diei arcum fieri, aestate[87] non fieri, nisi incipiente aut inclinante iam die. Cuius rei causa manifesta est. Primum medio die Sol calidissimus nubes evincit, nec potest ab his imaginem suam accipere, quas scindit. At matutino tempore, et vergens ad occasum, quum minus virium[88] habeat, minus calidus est, et ideo a nubibus et substineri et repercuti potest. Deinde quum arcum facere non soleat[89] nisi adversus his in quibus facit nubibus. Quum breviores dies sunt, semper obliquus est. Itaque qualibet diei parte, etiam quum altissimus est, habet aliquas nubes, quas ex adverso ferire potest. At temporibus aestivis supra nostrum verticem fertur, itaque medio die excelsissimas terras rectiore aspicit linea, quam ut ullis[90] nubibus possit occurri, omnes enim tunc sub se habet.

Plinius:

> Fiunt autem hyeme ab aequinoctio autumnali die decrescente; quo rursus crescente ab aequinoctio verno[91] non existunt, nec circa solstitium longissimis diebus, bruma vero hoc est brevissimis diebus frequenter . . . Aestate per meridiem non cernuntur, post autumni aequinoctium quacunque hora.

Iridem serenitates et pluvias posse praesagiri. PROP. 26[92]

Strabus inter sacros interpretes non obscuros retulit in iride divinae clementiae praesagia inesse, "duos habere colores, caeruleum et rubrum, qui duo exprimunt indicia, unum aquae, quod praeteriit, alterum ignis, quod venturum creditur in fine seculi, ob id caeruleus color extrinsecus, rubeus vero sive igneus introrsum visitur." Et Deum opt. max. dixisse Moysi, "Arcum meum ponam in nubibus, et erit signum faederis inter me et Terram." Sed hoc falsum esse "scripturarum periti inquiunt, et signum illud ad [221] placitum fuisse ex divina institutione . . . ut esset iris attestatio faederis ipsius cum[93] Noe post diluvium, quo in reliquum Terra non fore peritura." Inquiunt praeterea Deum

> mundo cladem neque igni neque aqua mundo illaturam, quod inibi color rubeus visatur, qui fit ex humido inflammato, quo significatur humectam naturam non

---

[85]"Theon Anaximinis" = "Thevi et Anaximeni" in original text.
[86]"24" in original text.
[87]"astate" in original text.
[88]"vinum" in original text.
[89]"solent" in original text.
[90]"illis" in original text.
[91]"nemo" in original text.
[92]"25" in original text.
[93]"in" in original text.

agree that a rainbow is formed by lunar rays. "Theon of Alexandria, the commentator on Aratus, [220] cites the opinion of Anaximenes about a relationship of this sort, saying that 'at night conditions can be ripe for this to occur, but very rarely, because it is not always full Moon.'"[99]

PROP. 25: *At what times they are usually seen.*

[According to Seneca]

> Aristotle says that after the autumnal equinox a rainbow can form at any time of day, but in summer it can form only when the day is either beginning or drawing to a close. The explanation for this is obvious: First, because in the middle part of the day the Sun is very hot and overwhelms the clouds, and cannot receive its image back from clouds that it is breaking up; but in the early morning or when sinking toward sunset, it has less power and is less hot, and so the clouds can withstand it and send it back. Again, the Sun does not usually form a rainbow except when opposite to the clouds in which it forms it; therefore, when the days are shorter, it is always at a low angle, and so at any point in the day, even when at its highest, it finds some clouds that it can strike from the opposite side of the sky. But in summertime it travels above our heads; and so in the middle of the day it is extremely high up and looks down on Earth at too vertical an angle for an encounter with any clouds to be possible; for it then has them all beneath itself.[100]

Pliny [argues that]

> rainbows are formed in winter when the day shortens after the autumn equinox; when the day lengthens again after the vernal equinox, they do not occur, nor in the longest days near the [summer] solstice, but they occur frequently at winter solstice, during the shortest days. In summer they are not seen during midday, but after the autumn equinox they are seen at any time."[101]

PROP. 26: *A rainbow can presage clear weather or rain storms.*

Among the interpreters of scripture who are not obscure, Walafrid Strabo says that there is a portent of divine mercy in the rainbow, which "has two colors, blue and red, that represent two signs, one being of water, which is in the past, the other of fire, which is believed to be coming at the end of time, wherefore the blue color is on the outside, whereas the red or fiery color is seen on the inside."[102] And most mighty and perfect God said to Moses, "I set my bow in the clouds, and it will be a sign of the covenant between me and the Earth."[103] But this is false, "scriptural experts say, insofar as that sign [221] was for the sake of divine instruction so that the rainbow would be a testament of this very covenant with Noah after the flood, according to which the Earth will not perish for the rest of time."[104] They say, moreover, that God

> will bring about the end of the world neither by fire nor by water because, where there is a rainbow a red color is seen, which is created by kindled moisture; this

praevalere, quum ignis in arquum suum vim praeferat. Caeruleus vero color, qui et ipse inbidem[94] visitationem facit constatque ex aqueo indigesto, igneam vim haud ita praecellere videtur praemostrare. Addunt etiam irim non apparere, nisi ubi partim[95] disserenavit, partim humoris persistat densitas. Serenitatem vero ex calido provenire ac sicco, quod igneae congruit virtuti; humoris densitatem ex frigido humidoque, quae aquae proprietas est. Haec vero natura diversa quum simul fiant, victoriam in aequilibrio positam plane insinuant.

Sunt qui probent apparente iride XXXX annis non venturum iudicium, quia in eo sunt duo colores, rubeus et viridis, rubeus ex humido inflammato, et significat aqueam naturam igneae non praevalere, quum ignea in iride nonnihil sui habeat viridis color ex aqueo indigesto. Significatque igneam naturam aquae non praevalere ex simili causa, quare ea apparente, nec diluvium nec conflagrationem posse esse. Non enim diluvium nisi summa siccitas praecesserit, et ea existente non fit iris. Simile ratione, conflagratio futura non erit, nisi in signis humidis praecesserit, et si humiditas sit, non erit arcus; non deest pars ignea, scilicet rubeus color.

Sed haec evanida sunt, nam sicut diximus, rubeus color iridis et caeruleus non causatur ex humido imflammato vel indigesto, ut putabant veteres, et ignotis causis eius quoque praesagia ignorantur. "Quocirca," inquit Plutarchus, "fabulantur illum capite taurino sorbere flumina." Virgilius:

Et bibit ingens
Arquus

quum adventat imber. Non easdem ubicunque apparuerit minas affert: a meridie magnam vim aquarum trahet, vinci enim [non] poterunt vehementissimo Sole, tantum est illis virium; si circa occasum refulsit, rorabit et leviter impluet; si ab ortu circaue surrexerit, serena promittit." Plinius contra haec, arcum "ne pluvios quidem, aut serenos dies cum fide portendit." Poetae alludentes ad id quod divinae mentis praesagium iris esset, deorum nunciam finxerunt. Unde Virgilius: [222]

Irim de caelo misit Saturnia Iuno.

Et humoris ancillam dicebant quod aera per Iunonem figurabant quod aerea impressio esset, nomenque ab irim deduxere, quod est nunciare. Extat et

[94]"iridem" in original text.
[95]"parum" in original text.

means that the humid nature does not prevail, since the power of fire holds forth in its bow. On the other hand, the blue color, which shows up in the same place and is consistent with unconsumed moisture, seems to predict likewise that the fiery power in no way prevails. They add as well that the rainbow does not appear except where it has partially cleared and where a partial thickness of moisture persists. Calm weather, in fact, is due to heat and dryness because it is suited to the power of fiery matter; thickness of moisture is due to cold and wetness, which is appropriate to water. Because these different natures are produced simultaneously, however, they clearly arrive at a triumph that is poised in balance.[105]

There are those who argue that the rainbow's appearing for forty years provides no presage of the future because there are two colors in it, red and green, the red from kindled moisture, and it means that the watery nature will not prevail over the fiery nature, since what is fiery in the rainbow has some measure of a green color from unconsumed water.[106] It also means that the fiery nature will not prevail over the watery nature for the same reason, for when it appears, neither a flood nor a conflagration can occur. In fact, a flood cannot occur unless it is preceded by extreme dryness, and when that exists, a rainbow is not formed. For a similar reason, no conflagration will be in store unless it is preceded by signs of moisture, and if there is moisture, there will be no rainbow, and the fiery portion does not disappear, namely, the red color.

But these arguments have no weight because, as we said, the red and blue color of the rainbow are not due to kindled or unconsumed moisture, as the ancients thought, and to be ignorant of its causes is also to be ignorant of its portents. "For this reason," Plutarch asserts, "certain people maintain that the rainbow is like the head of a bull that swallows up rivers."[107] [As] "Virgil" [puts it]:

And the huge rainbow
Drinks[108]

when rain is approaching. But its threats vary depending on which side it appears from: if it rises in the south, it will bring a large quantity of rainwater, for the clouds could not be overwhelmed by the very strong sunlight, so powerful are they; if it shines near the west, there will be dew and light rain; if it rises due east or thereabouts, it promises clear skies."[109] Against this, Pliny argues that the rainbow "in fact forecasts with accuracy neither rainy nor calm days."[110] Alluding playfully to the notion that the rainbow was an indication of divine intelligence, the poets imagined it was a message from the gods. Accordingly, Virgil says, [222]

Saturnian Juno sent the rainbow from heaven.[111]

They used to call it the handmaiden of moisture because they fancied the rainbow was formed by Juno, and they deduced the name from "iris," which

"rusticanum praesagium . . . quod si fuerit color rubens in albido significari fertilitatem, cuius ratio est quia is color ex materiae raritate causatur ubi radius Solis fortius agat; rara autem materia facile a Sole resolvi potest, quo eveniente non insequuntur iuges pluviae, quibus impediri solet fertilitas," quod etiam falsum reiicimus ex superiori ratione. Ptolemaeus in *Quadripartito*: "Iris etiam quae quandoque videtur, quum in serenis temporibus apparuerit aerem hyemalem praesignabit; quumque in temporibus hyemalis aeris formabitur, serenum nunciabit."[96] Sed quod aqua vel serenitatem nunciet etiam imperiti sciunt qui eius generationem cogoscunt ex materia roscida. Quippe primo elevatur vapor, deinde[97] a frigore densatur in nubem, mox residentes partes convertuntur in aquam, nec diu post insequitur pluvia, eo tempore concipitur iris. Pluvia dum ad ima fertur, evanescere incipit; vertitur in pallorem, visunturque in nubibus virgae intermistae descendentes; unde quum visitur arcus, dicitur vulgo ex mari plurimum bibere. Plautus iocosissime "de anu vini vas exiccante ait, 'Ecce autem bibit arcus; pluet hercle hodie.'" Sed si ex serenitatis et pluviarum vicissitudine aliquid iudicii nancisci poterimus, post pluvias serenitatis praesagium, quia vires aquarum plenitudine exoneratae plus lucis habent ad Solem concipiendum. Vespertina sequentis diei serenitatem spondet, quia vesperi siccior est. Matutina futuram pluviam ostendit, quia vas nocturni humoris fert plenum. Sed Aratus "geminatam irim pluviarum evidens prognosticum," dicit:

Ἡ διδύμη ἔξωσε διὰ μέγαν οὐρανόν ἴρις

Confirmat et illius interpres Theonis, inquiens, quum multa fuerit nubium[98] constipatio ex priore iride altera fit circumscriptione consimili, imbrem portendit plurimum.[99]

In prismate cristallino iridis colores videri. PROP. 27[100]

Difficillimam rem aggrediemur. Virgula quidem fieri solet stricta et triangula. Nascitur crystallus exagona, [223] quae, si ex transverso Solem recipit, colorem talem qualis in arcu videri solet, reddit. Ratio est quod Sol tenuissimam aciem transiens ubi subtilior pars vitri flavo colore induitur; postremus autem basim transiens totus obscuratur, ob id caeruleo colore faedatur. Sed hoc difficile captu, quia corpus vitreum inaequaliter Soli opponitur; refrangitur et in diversam partem colores iaculatur. Exemplum.

    Sit vitreum prisma ACDEF; sit Sol G.[101] Radius GM primum et tenuem aciem pertransit et reflectitur in R. Alius radius GL percutiens obliquum

---

[96]"nunciabunt" in original text.
[97]"dein" in original text.
[98]"imbrium" in original text.
[99]"pluvium" in original text.
[100]"26" in original text.
[101]"C" in original text.

is "to announce." There is also "a rural prognostication [according to] which, if a rosy color appears in white, it means fertility, the reason being that this color arises from the subtlety of the matter where the Sun acts most forcefully. But subtle matter can easily be dissolved by sunlight, and when that happens, corresponding rainstorms do not follow, according to which fertility is usually impeded," an account we reject as false on the basis of earlier reasoning.[112] In the *Quadripartitum*, Ptolemy observes that "when it appears in calm weather, the rainbow will portend stormy air, whereas when it is formed at times of stormy air, it will indicate calm."[113] But even the unschooled know that water will actually portend calm if they understand the rainbow's generation from dewy matter. First, of course, vapor rises, then it is condensed by cold into a cloud, after which the parts that sink are converted into water, and not long after that rain follows, at which time a rainbow is formed. As the rain falls down, the rainbow begins to disappear; it becomes pale, and intermingled streaks appear to fall in the clouds. Hence, when a rainbow appears, it is often said by the common folk to drink from the sea. In an entirely joking way, Plautus says "of an old woman draining a vessel of wine dry, 'But see the rainbow drink; by Hercules, it will rain today.'"[114] But if from the alternation of calm and rainy weather we can obtain anything in the way of prediction, it is that after rain there is a harbinger of calm because on the basis of plentiful discharge, the powers of the water conduce more to the absorption of sunlight. Evening rain promises calm the following day because it is drier then than in the evening. Morning threatens rain to come because the atmospheric receptacle is full of nocturnal moisture. But Aratus says, "A double rainbow foreshadows rain:

[When] a double rainbow girds the wide sky.

His translator Theon confirms this, saying that when there is a thick gathering of clouds, another, similar bow is created from the previous rainbow, which portends several storms."[115]

PROP. 27: *The colors of a rainbow appear in a glass prism.*

We will broach an extremely difficult subject. A small, narrow, triangular rod should be formed in the usual way, and a glass hexagon is produced, [223] which renders such color as normally appears in the rainbow when it receives sunlight along its transverse side.[116] The reason is that the sunbeam that passes through the extremely narrow edge of the glass, which is the most attenuated portion, is tinged with a golden-yellow color, whereas at the other end, all the sunlight passing through the base is darkened so that it is tinged by the color blue. But this is difficult to grasp because the glass body is not evenly disposed toward the Sun, and so the sunlight is refracted and casts the colors in different directions. Here is an illustration.

Let ACD[B]EF [in Figure 9.27a] be a glass prism, and let G be the Sun. Ray GM passes through the first, narrow point and is refracted to R. Another

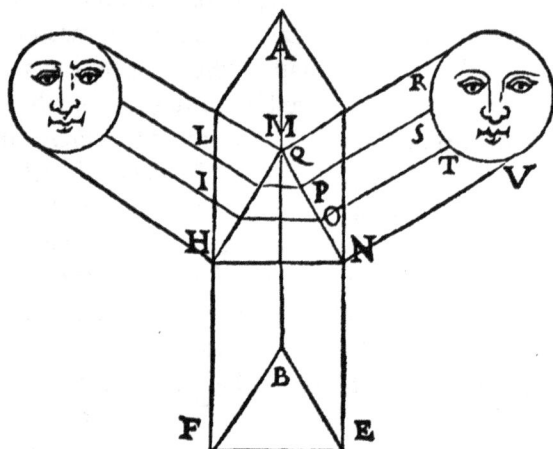

latus MH, refrangitur in LP, et ex P labitur in S ad eundem angulum. Haec tenia flava erit. Sic secundus radius GI refrangitur in O, et IO refrangitur foras OT, et erit[102] viridis tenia. Radius ultimus GH refrangitur in HN, et labitur foras NV, caerulei coloris. At si aciem Soli perfecte oppones, iaculabitur colores ex unaque parte binos. Exemplum.

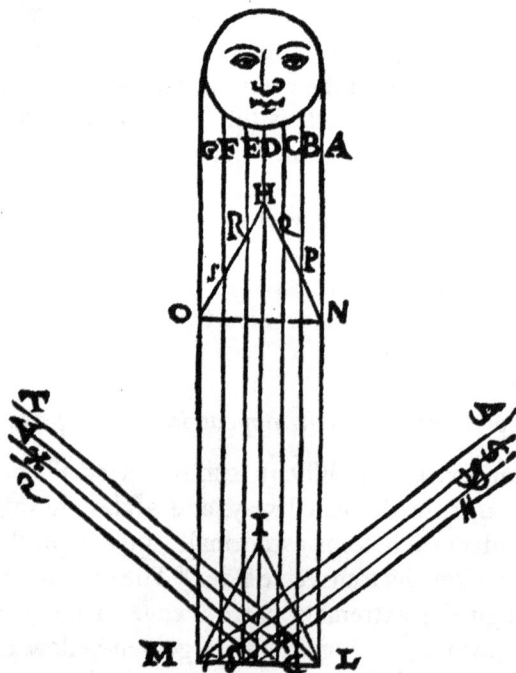

Sit crystallinum prisma directe opponens aciem Soli HNLMO; acies HI, Sol ABCDEFG. Cadat radius GM in extremam aciem. Reflectitur in oppositam

---

[102]"eris" in original text.

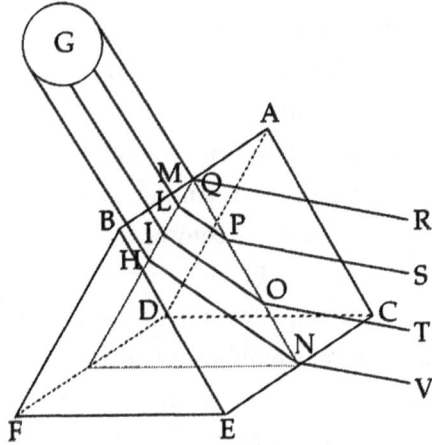

Figure 9.27a

ray GL, striking side MH at an angle, is refracted along LP, and from P it tends toward S at the same angle.[117] This band [delimited by QR and PS] will be golden yellow. Likewise, the second ray GI is refracted to O, and IO is refracted out along OT, and the band [between PS and OT] will be green. The last ray GH is refracted along HN, and it slips out along NV, which is blue in color [within the band between OT and NV]. And if you place the sharp edge of the prism directly facing the Sun, it will project the colors two by two on each side. Here is an illustration.

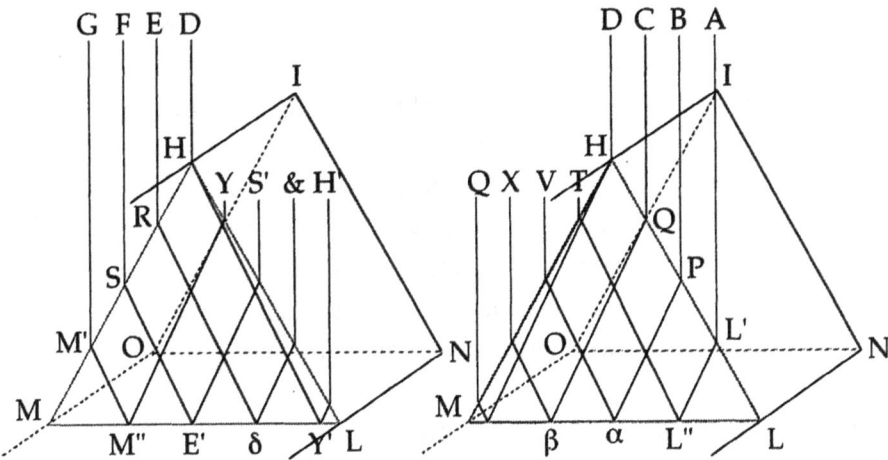

Figure 9.27b

Let HNLMO [in both diagrams of Figure 9.27b] be a glass prism directly facing the Sun, let its sharp edge be HI, and let ABCDEFG be [the endpoints of rays in] a sunbeam [represented according to a split at D within the two diagrams]. Let ray GM' [in the left-hand diagram] fall to the far [left-hand]

partem MY. Sic radius FE reflectitur in ES,[103] et tertius radius transiens circa aciem Eδ[104] reflectitur δ&, et radius DY in YH. Sic et alia parte radius [224] AL in LT, radius Bα in αV, radius Cβ[105] in βX, et ex utraque paries vel pavimentum cubiculi colorabitur.

Idem eveniet si oculis opposueris prisma, nam tres colores eosdem videbis in oppositum parietem. Sed semper extremos colores rubeum et halurgum ubi umbra aliqua, nam hi colores aliter fieri non possent; medii colores in medio plano, ubi lux. Et videntur res coloratae rotundari, nam alia ratione quam in oculo aequales anguli causantur; nam diximus non possibile iridem videri nisi in aequalibus angulis et semirectis. Idem eveniet in phiala aquae plena et in omni re in qua radius pertransiens diaphanitatem corporis inaequaliter transit, ut unus radius Solis magis altero profundetur. [225]

Virgarum colorem eisdem causis provenire. PROP. 28[106]

Aristoteles causam perscrutatus ait quum ex obliquo Solis fuerit aliqua nubes materiam habens inaequalem, ut aliqua in parte sit densa, in aliqua autem rara, et rursus [in] aliqua aquosior, in aliqua aerosior; tamen ad Solem ab eiusmodi facta refractione per pusilla quidem specula, non cernitur figura solaris, at Solis color intus apparet in parte nubis aliqua rutilans, in aliqua vero viridis, in aliqua flavus ob nubis inaequalitatem et speculorum dissimilitudinem. Vocant bacellos huiusmodi ostensiones, quod per rectas quasdam spectentur lineas. Averroes et Suessanus has virgas se nunquam vidisse fatentur. Plutarchus virgas esse dicit "affectiones quasdam quae . . . partim per substantiam, partim per speciem fiunt; quippe in nube quidem cernitur, sed colore non proprio, sed alieno, cuius speciem praeseferunt." Seneca, "virgas nil aliud esse," dicit, "nisi imperfecti arcus, colores nam facies quidem illis picta, sed nihil curvati habent; et in rectum iacent. Fiunt autem iuxta Solem fere in nube humida, etiam se spargente. Itaque idem est illis qui in arcu color; tantum figura mutatur, quia nubium quoque, in quibus extenduntur, alia est."

Mihi autem experienti duobus modis fieri contigit, et de his puto Aristotelem intellexisse. Primo, claudendo cubiculi fenestras et per foramen Sole transeunte, videntur eius radii quasi virga vel bacellus per obscuritatem

---

[103]"EZ" in original text.
[104]"ED" in original text.
[105]"CB" in original text.
[106]"28" in original text.

edge. It is refracted in the opposite direction [along] M'M"Y. So, too, ray FSE' is refracted along E'S', the third ray ERδ passing through the edge is refracted [along] δ&, and ray DHY' [is refracted] along Y'H'. Likewise, on the other side ray [224] AL' [refracts] along L"T, ray BPα [refracts along] αV, and ray CQβ [refracts] along βX, and [so] both walls or the floor of the room will be colored.[118]

The same thing will happen if you hold the prism in front of your eyes, for you will see those same three colors on the opposite wall. But the extreme colors red and violet are always where there is some shadow, for these colors cannot be formed otherwise, whereas the intermediate colors are formed in the middle plane, where there is light. Also, colored objects appear rounded because their appearances are due to a cause other than equal angles in the eye, and in fact we have said that it is impossible for a rainbow to be seen except at equal angles of 45°. The same thing will happen in a jar full of water as well as in anything else in which a ray passing through the transparency of the body penetrates nonuniformly so that one ray from the Sun will penetrate more deeply than another. [225]

PROP. 28: *The color of atmospheric streaks arises from the same causes.*

Having investigated the reason for this phenomenon, Aristotle claims that this phenomenon arises when a given cloud to the side of the Sun is composed of non-uniform substance, such that in one part it is dense, whereas it is rare in another, and on the other hand it is more moist in one part, more airy in another. Still, since reflection is made to the Sun from this sort of cloud by means of tiny mirrors, the form of the Sun is not perceived, and the color of the Sun appears inside the cloud, red in one portion, green in another portion, and golden yellow in another because of the non-uniformity of the cloud and the dissimilarity of the mirrors.[119] They call such phenomena streaks because they may be seen according to certain straight lines. Averroes and Agostino Nifo admit never having seen these streaks.[120] Plutarch says the streaks are "certain phenomena that . . . are due in part to substance and in part to mere appearance; they are certainly perceived in the cloud, but not according to its proper color; rather, they manifest a kind of extraneous color."[121] Seneca claims that

> streaks are nothing but incomplete rainbows, for their appearance is certainly colored, but not curved; they lie in a straight line. They are formed next to the Sun, generally in a cloud that is moist and already forming raindrops. So, in them there is some coloring as in a rainbow; only the shape is altered because the clouds in which they are stretched out also have a different shape.[122]

According to my experience, however, this phenomenon can occur in two ways, and I think Aristotle knew of them. First, if the windows of a room are closed and sunlight passes inside through an opening, its rays will seem to pass through the darkness like a streak or rod; and if water is blown

transire; huic aquam ore insufflando, videbitur virga, vel radius ille coloratus irinis coloribus, sed per longum eo modo quem diximus quando homo existat in centro iridis vertens se ad latus. Ore aquam proiiciendo, videtur iris dependens et irinis coloribus coloratum, teniae a summo ad imum dependentes, flava primo, viridis mox, postrema caerulea. Aristoteles ait oportet inspicientem esse a nubis latere, quod ab iride est differens, quae quaerebat Solem a tergo in eadem linea cum iridis centro. Alio quoque modo experti sumus. Bacellum ligneum accipientes (sed praestat vitreum), et aquam insufflando bullis et guttis operiemus, et Soli per longum oppositum, videbimus quasi opalum coloratum qui lapis irinis [226] coloribus decoratur, unaquaeque enim gutta dictis coloribus colorabitur. Unde si aer roscidus et radii Solis per densam nubem penetrando, eiusmodi coloribus induuntur. Ob id Aristoteles quaerebat aquosam et bullosam nubem.

Halo iisdem causis coloratus videtur. PROP. 29[107]

Ut ex Aristotele habetur, halo fit diu circa Solem, noctu circa Lunam, et ubicunque sydus est; fit ex refractione et ex nube roscida quae sit inter oculos nostros et astrum. Et videtur hic circulus circa astrum quasi astri corona. Causa rotunditatis est eadem quae in iride, nam ab astro ad nubem trahitur linea quae refracta ad nostrum oculum pervenit, nec umquam pervenit nisi anguli qui sunt incidentiae et refractionis sint aequales. Exemplum apponemus.

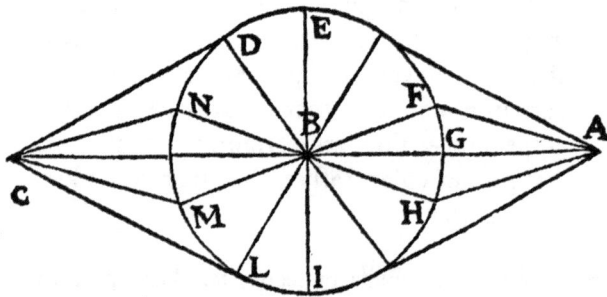

Esto Sol A, oculus noster C, nubes roscida GFEDNMLIH, radius oculum et nubem penetrans AGBC. Radius a Sole ad nubem AE, et ex E ad oculum

---

[107]"28" in original text.

through the mouth toward it, the streak will be visible and indeed the light beam within it will appear tinged with rainbow colors, but along its length in the way that we discussed when a man stands at the center of a rainbow facing to the side. As water is spewed from his mouth, a rainbow tinged with rainbow colors appears to hang down; the bands dropping from top to bottom are golden yellow first, then green, and finally blue. Aristotle claims that the viewer must be to the side of the cloud, which is different from what obtains for a rainbow, where the viewer had to be with his back to the Sun in line with the center of the rainbow. Here is another way we have experienced this phenomenon. Taking a thin wooden rod (although glass is best), we will cover it by blowing bubbles and drops of water on it, and holding it lengthwise to face the Sun, we will see it colored like opal, with the rainbow colors of the "iris" stone, [226] for each drop of water will be colored with the aforesaid colors.[123] Consequently, if the air is misty and the Sun's radiation penetrates a dense cloud, the clouds are imbued with the same sorts of colors. For that reason Aristotle looked to a moist, bubbly cloud for an explanation.

PROP. 29: *A halo appears colored for the same reasons.*

As is maintained by Aristotle, a halo is formed around the Sun during the day, around the Moon at night, and anywhere there is a star, and it is formed on the basis of refraction and a moist cloud that lies between our eyes and the heavenly body.[124] In addition, this circle around the heavenly body appears as a crown for the heavenly body. The reason for its roundness is the same as in the rainbow, for a ray extended from the heavenly body to the cloud is refracted to reach our eye, and it only arrives there when the angles of incidence and refraction are equal.[125] We will add an illustration.

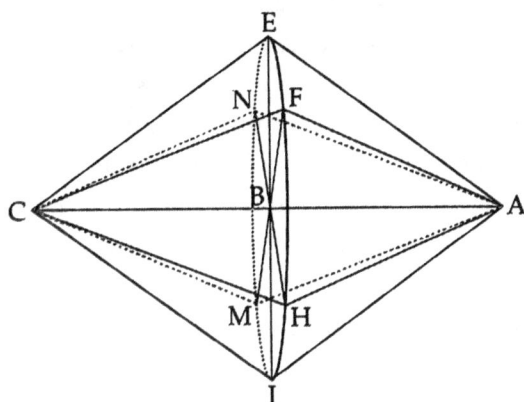

Figure 9.29

Let A [in Figure 9.29] be the Sun, C our eye, FENMIH a moist cloud, and ABC a ray through the eye and [the center of the] cloud.[126] AE is a ray from the Sun to the cloud, and it reaches from E to the eye at C. Likewise,

C. Sic AI ad oculum C. Sic AF, AM[108] ad C, et AH, AN[109] ad C. Aequales
sunt omnes lineae a Sole ad nubem, nec esse aequales possent omnes nisi
ad circulum essent. Et inde a nube ad oculum aequales, et lineae a centro
ad circumferentiam aequales EBI, FBM,[110] HBN. Necesse est igitur halonem
rotundum videri, angulus enim BIC et BEC aequales, et BNC et BMC, et
reliqui aequales; qui anguli, ut [227] diximus in iride, sunt in causa ut
colores ex refractione videantur. Aliter tota nubes colorata esset. Plutarchus:
area hunc in modum conficitur:

> Quum inter Lunam aliudve sydus visumque nostrum crassus aer nebulosusque
> intervenit, ab illo visus noster confestim reflectitur laxiusque circumfunditur, quo
> deinde in syderis orbem incurrente; duntaxat secundum circumferentiam continuo
> circulus videtur circa sydus existere. Huiusmodi autem usitatus circulus area
> propterea dicitur, quoniam illic assidue est area, esse ibi eiusmodi simulachrum
> videtur, quo nuper visus crassior inciderat.

Seneca: "Coronae ubicunque fiunt ubicunque sydus est; arcus non nisi contra
Solem . . . unde coronam, si diviseris, arcus erit." Et fiunt ab omnibus syderi-
bus. Nos halonem vidimus opposito specillo sive lenticulari portione inter
oculos nostros et candelam, vel Solem, apparent enim iridis colores circum-
circa, quia undique ibi anguli aequales. Sed color puniceus non videtur circa
flavum, quia ibi nulla adest tenebra.

Ex Solis reflexione in opaca nube etiam colores causari. PROP. 30[111]

Diximus colores qui ex refractione nascebantur; nunc autem qui ex reflexione.
In reflexione lux multiplicatur et splendescit quum densius diaphanum sive
nubes densior sit quam ut penetrari possit, quorum mistione colores nascuntur. Sed non tam pulcherrimi ut irini; sed obsoleti et obscurioris coloris.
Quum enim nubes lucem non transmittunt, extima superficies cum varia
lucis mistione colores gignit. Aureus color videtur fieri ex regesto Solis flavo
lumine in densam nubem multiplicato vel a flavae lucis continuata densitate,
multiplicatur enim lux ex reflexione, ut diximus, nam simplex incidit duplica-
tus quum resilit.

Id in stellis videre licet, nam ex ea densitate reflexo Solis lumine, color
aureus spectatur, at in nebulis quae non solidae vel caliginosae croceus color
gignitur. Color vero caeruleus videtur ubi tenuis lux in denso diaphano,

---

[108]"FM" in original text.
[109]"HN" in original text.
[110]"TBM" in original text.
[111]"29" in original text.

AI reaches the eye at C [along IC]. So, too, AF and AM [reach] C [along FC and MC], and AH and AN reach C [along HC and NC]. All of the radial lines from the Sun to the cloud are equal in length, and they could not be equal unless they were all extended to a circle. And so all the lines from the cloud to the eye are equal, and lines EBI, FBM, and HBN from the cloud's center to its circumference are equal. The halo must therefore appear round, for angles BIC and BEC, angles BNC and BMC, and all the remaining angles are equal, and, as [227] we said in the case of the rainbow, those angles are why the colors appear as they do according to refraction. Otherwise the entire cloud would be colored. Plutarch says that a halo is formed as follows:

> When thick and cloudy air intervenes between the Moon or another heavenly body and our eye, our sight is immediately reflected from it and flows wide around it, after which it reaches the globe of the heavenly body; according to the size of its circumference, a circle appears to lie continually around the heavenly body. A circle of this sort is commonly called a halo, because a halo is continuously there, and an image of that sort is seen when sight has just impinged on a denser medium.[127]

Seneca argues that "halos occur everywhere, wherever there is a heavenly body, whereas rainbows occur only facing the Sun, so if you bisect a halo, the result will be a rainbow."[128] Haloes are formed by every heavenly body. We have seen a halo with a lens posed toward the Sun, or with a lenticular section held between our eyes and a candle, for the colors of the rainbow appear circumscribed around the center because the angles to their circles are equal throughout. But the red color does not appear to surround the golden yellow one because no darkness is present at that point.

PROP. 30: *Colors also result from the reflection of sunlight in an opaque cloud.*

We have discussed the colors that arose from refraction, but now it is time to discuss those that arise from reflection. In reflection light is multiplied and shines brightly when it strikes a denser transparent medium or a cloud that is denser than it can penetrate, the resulting interaction giving rise to colors. But they are not as beautiful as in a rainbow; rather, they consist of a weak and dim color. For when the light does not penetrate the cloud, the outer surface generates colors from various interactions with the light. The golden color appears to be formed by the golden yellow radiation of the Sun reflected back and multiplied in the dense cloud or from the uniform intensity of the golden yellow light, for, as we said, light is multiplied according to reflection because when it reflects, the single light that is incident reduplicates.

This may be seen in the case of stars, for when sunlight is reflected according to their solidity, a golden color is seen, and in vapors that are not solid or dark a yellow color is generated. The color blue, on the other hand, appears where dim light shines in a dense transparent medium, and the same

idemque evenit in reflexione quod evenerat in refractione. Ideo philosophi
qui putabant iridis colores ex nubis [228] natura causati dicebant flavum
colorem in summo nubis ubi tenuis erat nubes, caeruleus in imo, ubi crassior.
Si igitur in Solis opposito nubes crassae diaphanitatis ut radium obtenebret,
caerulei coloris tincta conspicietur. Sic ex aequa utriusque potentia et lucis
et crassitudinis, viridis exoritur color, videlicet ex modice crassa nube et
modico lumine. Rubeum colorem iam diximus in iride generari ex densissimo
lumine et nigro nubis colore reflexo. Nubes caliginosa Sole suffusa alba nitet,
at in aquam conversa et iam casura, nigerrima apparet.

Aeris colores in eclipsi videri. PROP. 31[112]

Ex terra continuo vapores prodeunt, et pars ea vaporum quae tenuior levi-
orque summa petit; quae foeculentior, gravior et in imo subsidet; media in
medio residet, ut videantur quasi tres circuli terram circundantes, in supremo
tenues, in imo terrestres, in medio mediae naturae. Sed animaduertendum
non semper eiusdem rationis esse vapores, nam secundum tempora et caelestes cursus variantur. Nam aliquando tenues tenuissimi fiunt, et sic faeculentiores purgatiores. Contra autem aliquando ut qui prius terrestres
gravesque fuerit adeo gravissimi facti, ut inferioribus existentibus gravioribus,
prius levium circulus terrestres contineat; aliquando vero misti et simul
conturbati. Evenit igitur ut Sol eos feriens splendidus varios gignat colores,
ut diximus, sed nobis qui in eis versantur inconspicui. In lunari enim eclipsi,
umbram Sole in Lunae faciem immittente antequam terreni corporis umbra
ad Lunam perveniat, colores hi ex Solis lumine et vaporum causati Lunae
vultum prius feriunt, et ex eorum coloribus vaporum qualitates terram obside-
ntes cognosci possunt. Exemplum addemus. [229]

Erit Terrae globus AD, circulus primus crassos vapores continens D,
secundus medios C, supremus in quo tenuissimi B. Sit umbra vaporum
feriens Lunae faciem E, quae quia primum et supremi circuli est rubri coloris
Lune faciem foedabit. In H vero dimidium medii circuli, ut rubei et viridis
coloris Lunae facies maculetur; in F vero duorum circulorum colores subibit

---

[112]"30" in original text.

thing that happened in refraction happens in reflection. Consequently, the natural philosophers who supposed that the colors of the rainbow are due to the nature of the cloud [228] said that the golden yellow color was at the top of the cloud where the cloud was thin, whereas the blue was at the bottom, where it was thicker. If the cloud facing the Sun is of such a dense transparency as to darken the ray, then, it will appear tinged with the color blue. Likewise, if both the light and the density are of equal power, the color green is produced, namely, from a moderately dense cloud and light of moderate intensity. We have already argued that the red color is generated in the rainbow from extremely intense light and the black color reflected back from the cloud. A dark cloud suffused by sunshine glitters white, but when it is reflected in water at the verge of sunset, the water appears quite black.

PROP. 31: *The colors of the air [suffusing the entire atmospheric shell] can be seen in an eclipse.*

Vapors continually issue from the Earth, and the portion of the vapors that is thinner and lighter pushes toward the top, whereas the portion that is more adulterated and heavier sinks to the bottom, and the middle portion remains in between so that the vapors may be regarded as three circles surrounding the Earth, the thinnest vapors in the top one, the terrestrial vapors in the bottom one, and those of an intermediate nature in the middle one. But it must be noted that vapors are not always of the same nature, for they vary according to the times and the course of the heavens. Indeed, thin vapors are sometimes rendered extremely thin, and by the same token more adulterated vapors become purer. On the other hand, terrestrial vapors that were previously heavy sometimes become extremely heavy on that account, so that, according to heavier, lower vapors, the circle that previously contained lighter vapors may contain terrestrial ones; sometimes, however, the vapors are mixed and jumbled together. It therefore happens that in rendering them bright, the Sun generates various colors, as we said, but they cannot be seen by us when we are turned toward them. In a lunar eclipse, in fact, when the Sun casts its trace on the face of the Moon before the Earth's shadow reaches the moon, these colors, which are due to sunlight and vapors reach the face of the Moon earlier, and from the colors of their vapors, the qualities affecting the Earth can be known. We will add an illustration. [229]

AD [in all four diagrams of Figure 9.31] will be the globe of the Earth, the first circular band D containing the crassest vapors, the second circular band C containing the intermediate vapors, and B the topmost circular band in which the thinnest vapors exist. Let the shadow of the vapors reach the face of the Moon at E; since the first and highest circle is of a red color it will tinge the Moon red.[129] When the Moon is at H, halfway into the middle circle, the face of the moon may be stained with a red and green color, whereas at F colors of the two circles pass [fully] from red to green. At G,

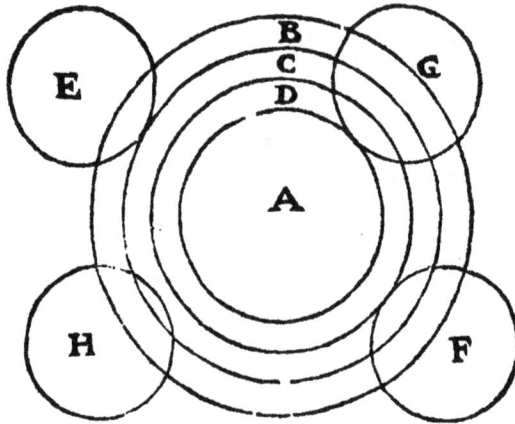

rubri et viridis. In G autem trium circulorum colores rubri, viridis, et caerulei. Ex his coloribus naturaliter futuri anni vel instantis qualitates praecognosci possunt; alibi Astrologorum opiniones refellemus.

Ex luce et densiore caeli parte lacteum circulum fieri. PROP. 32[113]

Plutarchus philosophorum opiniones referens haec habet:

> Pythagoreorum quidam stellae esse efflagrantiam a sua sede deciduae; hanc[114] autem qua pervolitarit tractum circulariter perussisse[115] quo tempore aurigante Phaethonte caelum conflagrasse dicitur. Quidam Solem iam inde ab[116] initio illac iter fecisse. Sunt qui speculare esse visum putent, Sole fulgores [230] in caelum referiente, quod in arcu caelesti et in nubibus contingit. Metrodorus Solis transmeatu fieri, esse enim hunc solarem circulum. Parmenides spissi rarique mixturam lacteum reddere colorem. Anaxagoras Terrae umbram hac caeli parte consistere; quum Sol terras subiens, caelum usquequaque illustrare non potest. Democritus stellarum collucentiam multarum, parvarum, contiguarumque inter se collustrantium propter constipationem.

Aristoteles horum multas opiniones refellit. Quod non sint stellae ibi existentes probat quod planetae omnes sub Zodiaco labuntur. Quod non sint stellae remissioris lucis quum a Sole non aspiciantur, translato Sole deberet transferri lac. Qui dicunt esse reflexionem Solis ad oculos nostros falsa dicunt, quia ibi non adest nubes, ut in iride et halone, et si esset reflexio, praesupponit

---

[113]"31" in original text.
[114]"hunc" in original text.
[115]"perusisse" in original text.
[116]"sub" in original text.

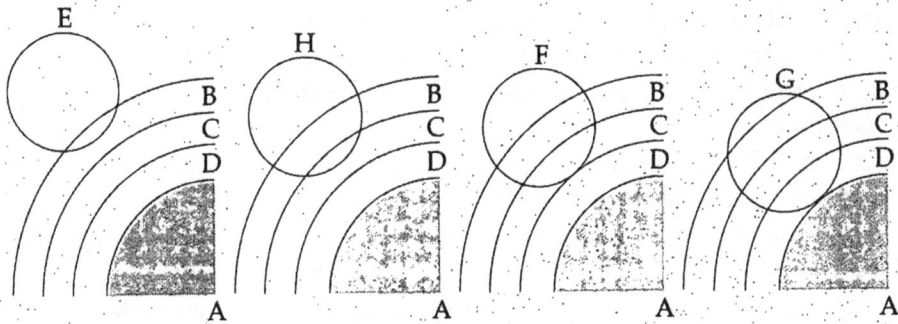

Figure 9.31

finally, the Moon will be tinged with the colors red, green, and blue of the three circles. From these colors the qualities of future years or instants can be foreknown naturally, but we will expose the opinions of astrologers elsewhere.[130]

PROP. 32: *The circle of the Milky Way is formed from light and a denser portion of the heavens.*

Citing the opinions of natural philosophers, Plutarch has this to say:

> Some Pythagoreans claim that the Milky Way is a conflagration due to a star that fell from its proper place; it is said that this star may have flown out of its circular orbit in flames at the time when the charioteer Phaeton set the heavens on fire. Certain others say that at the very beginning the Sun followed that path. There are others who may opine that it is a mirror image, the radiation from the Sun being turned back toward the heavens, which happens in the rainbow and in clouds. [230] Metrodorus argues that it arises from the transit of the Sun, for the Milky Way lies on the solar orbit. Parmenides says that the milky color comes from the mingling of thick and rare substances. Anaxagoras holds that shadow from the Earth persists in this part of the heavens because, when the Sun moves beneath terrestrial substances, it cannot illuminate the heavens at every spot. Democritus concludes that it is the combined shining from many small, continuous stars that illuminate one another according to their being tightly grouped together.[131]

Aristotle refutes many opinions of these thinkers. He demonstrates that it is not due to stars lying there because all the planets tend to lie beneath the zodiac. [He also demonstrates] that it is not due to light coming back from stars when they are not facing the Sun because [in that case], when the Sun moves, the Milky Way ought to move with it.[132] Those who claim that it consists of reflection from the Sun to our eyes speak falsely because there is no cloud at this spot, as is the case in the rainbow and the halo, and if there were reflection, it presupposes a fixed mirror, but the heavens

immobile speculum, at caelum semper movetur. "Aristoteles dicit esse aridae exhalationis flammigantiam permultae et continuae."

Sed haec opinio omnium pessima: quomodo igitur exhalationes, quae vagae, ab initio mundi semper eodem loco steterint. "Possidonius ignis coagmentationem sydere quidem rariorem, splendore vero densiorem." Cardanus dicit esse ibi "caeli densa substantia, syderum rara," contrarium dicens, nam quod lucet densius est, rarum pervium. Caelum esse rarum putet ex luminum transmissione; astra densa esse ex Lune natura, cum obiectu lumen Solis aufertur. Sed vera ratio est quod pars ea caeli densior sit aliis, ut lumen Solis recipere possit, non autem transmittere, ut Luna et caeterae stellae.

FINIS.

are always moving. "Aristotle says that it is the combustion of a copious, continuous exhalation of dry vapor."[133]

But this opinion is the worst of all; how is it, then, that vaporous exhalations, which are fluid and unstable, have remained in the same place since the beginning of the world? "Posidonius says that it is a gathering of fire, rarer than a star, to be sure, but denser than light."[134] Girolamo Cardano argues that at this location "the substance of the heavens is dense, whereas that of the stars is rare,"[135] in saying which, he lapses into contradiction, for whatever gives off light is denser and what is pervious to it is rarer. He supposes that the heavens are rare because of the transmission of light, and he supposes stars to be dense according to the nature of the Moon because when it interrupts the light of the Sun, the light disappears. But the real reason for the Milky Way is that this part of the heavens is denser than the others so that, rather than transmit the Sun's light, it can assimilate that light, as the moon and the rest of the heavenly bodies do.

THE END.

Ego Fridericus Metius legi opus Domini Ioannis Baptistae Portae Neapolitani, qui inscribitur *De Refractione Optices Parte*; et novem continet libros, quorum primus agit de refractione et eius accidentibus. 2. De pilae crystallinae refractione. 3. De oculorum partium anatome et earum muniis. 4. De visione. 5. De visionis accidentibus. 6. De cur binis oculis rem unam videamus. 7. De his quae intra oculum fiunt et foris existimantur. 8. De specillis. 9. De coloribus ex[117] refractione: scilicet de iride et lacteo circulo. In quibus nihil inveni a pietate alienum, nihil bonis moribus contrarium, sed potius auctoris doctrinam et solertiam sum admiratus, in quorum fidem propria haec scripsi manu. Romae iiii. Id. Mart. An. Domini M.D. XCIII, die scilicet festo gloriosi Gregorii Magni.

Fridericus Metius qui supra manu propria.

Imprimatur.

P. Antonius, Vicesgerens.

Imprimatur.

Fr. Bartholomaeus de Miranda. Sacri Palatii Magister.

Imprimatur.

Ardicinus Biandra, Vic. Gen. Neap.

M. Cherubinus Veronensis, August. Theol. Archiep. Neap. Vidit. Reg. fol. 3.

---

[117]"et" in original text.

I, Federico Mezio, have read the work of Master Giambattista Della Porta of Naples, which is entitled *On the Part of Optics Dealing with Refraction*; and it contains nine books, the first of which deals with refraction and its characteristics. The second [book deals] with refraction in a crystalline sphere. The third with the anatomy of the parts of the eyes and their functions. The fourth with vision. The fifth with the characteristics of vision. The sixth with why we may perceive an object as single with both eyes. The seventh with things that occur inside the eye but are judged to occur outside it. The eighth with lenses. The ninth with colors arising from refraction: that is, with the rainbow and the Milky Way. I found nothing in these books inconsistent with correct faith and nothing contrary to good morals; rather, I have been struck by the author's learning and ingenuity, trusting in which I have written this with my own hand. At Rome, 12 March, A.D. 1593, that is, on the feast day of glorious Gregory the Great.

Federico Mezio, who wrote the above by his own hand

Imprimatur.

Father Antonio [Agelli], Vicegerent

Imprimatur.

Brother Bartolomé de Miranda, Master of the Sacred Palace

Imprimatur.

Ardicino Biandra, Vicar General of Naples

Master Cherubino Veronese of the Augustinian Order, Theologian for the Archbiship of Naples, saw it. Register, folio 3

NOTES

1. See note 16 to book 1.

2. Traditionally attributed to Aristotle, the notion of prime matter applies to the completely undifferentiated stuff ultimately underlying all physical particulars. As such, it is pure potential, without qualities but capable of taking on such qualities, or "forms," as spatial boundedness, shape, size, color, texture, and so forth, which define actual objects.

3. Unlike melting copper, however, burning sulfur yields a bright-blue light.

4. Della Porta is referring here not to the brilliant blue light of the flame but, rather, the bright white core of the burning sulfur.

5. Seneca, *Natural Questions*, I, v, 6–7, in Seneca (1529), 416; translation adapted from Seneca (2010), 149. *Cytheran* refers to Aphrodite, or Venus, who was said to have been born on the island of Cythera (or Kithera), just off the southeastern tip of the Peloponnese.

6. *De rerum natura*, II, lines 798–809, in Lucretius (1570), 174–75; translation from Lucretius (1992), 159.

7. Cicero, *Academic Questions*, II, 25, in Cicero (1565), 25, line 45.

8. This notion of opacity as the limit of transparency harks back to Aristotle, *On Sense*, 3, 439a17–35.

9. Seneca, *Natural Questions*, I, v, 2, in Seneca (1529), 414; translation adapted from Seneca (2010), 144.

10. Ibid.; translation adapted from Seneca (2010), 144. The reference to Plutarch is to Pseudo-Plutarch, *On the Opinions of the Philosophers*, III, 5, in Plutarch (1530), 149v.

11. Pseudo-Plutarch, *On the Opinions of the Philosophers*, III, 5, in Plutarch (1530), 149r.

12. *Natural Questions*, I, v, 3, in Seneca (1529), 414–15; translation from Seneca (2010), 144.

13. *Natural Questions*, I, v, 4 and 12–14, in Seneca (1529), 415–16; translation adapted from Seneca (2010), 145 and 147.

14. See *Natural Questions*, I, iv, 3, in Seneca (1529), 416.

15. Albertus Magnus, *Meteororum*, III, iv, 26, in Albertus Magnus (1517a), 115rb (also [1890a], 696–97).

16. I have been unable to trace the source of this claim.

17. The four preconditions listed by Albertus are: a thick, black cloud; small drops of rain at the top of the cloud; larger, heavier drops at the bottom; and humid air; see *Meteororum*, III, iv, 14, in Albertus Magnus (1517a), 112vb–13ra [also (1890a), 682b–84a].

18. For the basis of this description of Albertus's theory, see ibid., 2–14, in Albertus Magnus (1517a), 109vb–113ra [also (1890a), 666a–84a].

19. Aulus Gellius, *Attic Nights*, II, 26, in Aulus Gellius (1560), 73.

20. Ibid., in Aulus Gellius (1560), 75; the actual quotation in the 1560 Latin edition is *Cedo tamen pedem lymphis flavis flavum ut pulverem ... abluam* ("Give me your foot so that I may wash away the golden-yellow dust with clear, golden water"). The association of *flavus* with olive fronds that Della Porta attributes to Virgil is to be found in *Aeneid*, V, line 309.

21. *Meteorologicorum*, III, in Alexander of Aphrodisias (1545), 39ra.

22. The phenomenon to which Della Porta is referring here is what Leonardo da Vinci described as aerial perspective (*prospettiva aerea*), coining the phrase in chapter 165 of his *Treatise on Painting*; see Rafaelle du Fresne, ed., *Trattato della Pittura di Lionardo da Vinci* (Paris, 1651), 45. Unlike the phrase, the concept of aerial perspective was known to artists long before Leonardo, who was the first, however, to study it closely. According to him, things appear increasingly blue as the distance increases, which is essentially what Della Porta is claiming here in arguing that their color progresses from golden yellow, which is least blue, through green, which is bluish, to true blue.

23. A likely source for this listing of Greek and Latin terms for red is Aulus Gellius, *Attic Nights*, II, 26, in Aulus Gellius (1560), 74.

24. *Natural Questions*, VII, xii, 6, in Seneca (1529), 479.

25. For Aristotelian color theory, see, *Meteorology*, III, 4, in Aristotle (1984), 601–4, and Pseudo-Aristotle, *On colors*, esp. 1–3, in Aristotle (1984), 1219–22.

26. For Artemidorus's account, see Seneca, *Natural Questions*, I, iv, 4.

27. *Metamorphoses*, VI, lines 63–66, in Ovid (1565), 122; translation from Ovid (1977), 293

28. See *Theogony*, line 785.

29. *Aeneid*, V, line 89, in Virgil (1579), 115v.

30. See Pseudo-Plutarch, *On the Doctrines of the Philosophers* III, 5, in Plutarch (1530), 148v.

31. *Natural Questions*, I, iii, 5, in Seneca (1529), 415; translation from Seneca (2010), 145.

32. *Natural Questions*, I, iii, 6–8, in Seneca (1529), 415; translation from Seneca (2010), 145-6.

33. Unlike the theory of vision implicit in Aristotle's account of the rainbow, the theory of vision he presents in *On the Soul* and *On Sense* is unequivocally intromissionist. In fact, in *On Sense*, 2, 438a26–28, he goes so far as to claim that "it is ... an irrational notion that the eye should see in virtue of something issuing from it; that the visual ray should extend itself all the way to the stars ... as some say"; translation from Aristotle (1984), 696.

34. Actually, this would be the case only if the cloud were concentric with the viewpoint at B. In specifying earlier that the cloud should be concentric with the lunar orbit, however, Della Porta located its center of curvature at the Earth's center, not at B.

35. Petrus Ramus, *Thirty-One Books of Mathematical Instruction*, I, in Ramus (1569), 20. Della Porta's misrendering of Philip's name ("Menaedeus" instead of "Medmaeus") is an apparent misrendering of Ramus's version ("Mendaeus"), which is itself a misrendering that may be traced back to Proclus (1560), 39.

36. I have followed Della Porta in representing refracted ray FM in Figure 9.11b as perpendicular to the front surface of the cloud. In reality, the ray would have to refract above that line, since it represents the normal at point F of incidence. Nonetheless, this quantitative inaccuracy does not affect the qualitative point: that is, that the three colors will be distributed as illustrated along ray FM refracted through the cloud.

37. See *Theaetetus*, 155d.

38. Pseudo-Plutarch, *On the Doctrines of the Philosophers*, III, 5, in Plutarch (1530), 148v

39. *On the Nature of the Gods*, III, 20.51, in Cicero (1565), 247; translation from Cicero (1961), 335.

40. *Cratylus*, 408b, in Plato (1551), 317.

41. Lodovico Ricchieri, *Thirty Books of Readings from the Ancients*, XXII, 32, in Ricchieri (1566), 878.

42. See Vincent of Beauvais, *Speculum naturale*, IV, 74, in Vincent of Beauvais (1591), 47vb. For the original source of the quotation, see Alfred of Sareshel (1988), 51.

43. *Natural History*, II, 60, in Pliny (1559 [as II, 59]), 26.

44. *Natural Questions*, I, iv, 3, in Seneca (1529), 416; translation from Seneca (2010), 148.

45. *Natural Questions*, I, v, 1–3, in Seneca (1529), 416; translation from Seneca (2010), 148.

46. Lodovico Ricchieri, *Thirty Books of Readings from the Ancients*, XXII, 30, in Ricchieri (1566), 876.

47. In other words, the bow should appear upside-down.

48. For this allusion to Nicolas Peripateticus's theory, see Albertus Magnus, *Meteororum*, III, iv, 26, in Albertus Magnus (1517), 115rb [also (1890a), 696].

49. I take Della Porta to mean that even the central ray does not form a true, straight line or shaft because radiation is continuous. Hence, even at its minimum, any "ray" of light forms a cone with its vertex at the point of radiation and its base on whatever it illuminates.

50. The "book on light" Della Porta mentions here presumably refers to the first book of this treatise, where he deals with the refraction of light. Nowhere in that book, however, does he state explicitly that light radiates from every point on a luminous surface, although that claim is certainly implicit in his analyses.

51. See, for example, book 4, prop. 7.

52. The "book on refraction" Della Porta mentions here is obviously book 2 of this treatise.

53. Della Porta's reference to a brightly shining heavenly body—or "star" (*aster*)—rather than to the Sun is likely due to the fact that a rainbow can be formed by moonlight, a fact he acknowledges later in proposition 25.

54. Again, I have followed Della Porta in representing refracted ray EM as perpendicular to the front surface of the cloud (see note 36 above), but of course it would be refracted along a line between EM, the normal, and EC, the continuation

of the incident ray. For the purposes of this illustration, however, all that matters is that EM lie below EC.

55. Although light, in its purest form, is perfectly colorless, according to Della Porta, it does take on a golden yellow color when it passes into the terrestrial atmosphere, which taints it somewhat. It is this slightly tainted light that passes for "white" light among optical theorists.

56. Contrary to Della Porta's claim, the cathetus in this case is not PG but, rather, the line orthogonal to ED that passes through P. The same holds *mutatis mutandis* for rays ON and MQ; their catheti will be the perpendiculars passing through points O and M, respectively, not OH and MI. Consequently, the images of those points will not be seen along line AC, although the images derived correctly will yield the same qualitative results: that is, the golden-yellow band will appear at the top of the bow, the green one in the middle, and the blue one at the bottom.

57. ED, of course, is not a cathetus; it is the refractive interface between clear air and the cloud.

58. Since angle BDE = 90°, and angle BED = 45°, angle DBE = 45°. By Della Porta's day, however, it was well established that the rainbow subtends a maximum arc of 42° (i.e., angle DBE = 42°), not 45°, as he claims; see, for example, Roger Bacon, *Opus majus*, VI, 4, in Bacon (1900), 177 (Latin), and Bacon (1928), 596 (English); see also Witelo, *Perspectiva*, X, 78, in Witelo (1572), 471; and Girolamo Cardano, *De Subtilitate*, 4, in Cardano (1551), 173. See also Carl B. Boyer, "Descartes and the Radius of the Rainbow," *Isis* 43 (1952): 95–98. Not only did Della Porta get the maximum angle wrong, but his claim that the other three angles BRG (= GBR), BSH (= HBS), and BTI (= IBT) are virtually equal to angle DBE is geometrically absurd because in that case lines of sight EB, RB, SB, and TB would all have to coincide.

59. Pseudo-Plutarch, *On the Opinions of the Philosophers*, III, 5, in Plutarch (1530), 149r.

60. *Natural Questions*, I, iii, 2, in Seneca (1529), 414; translation adapted from Seneca (2010), 144.

61. Most likely, the waterfall Della Porta has in mind here is the Grande Cascata at Tivoli. Strictly speaking, this is not on the Tiber River itself, but rather, some thirty-odd kilometers west of it on the tributary Aniene River.

62. Unstated in this enunciation is that, when added together, the two lines of any of the chosen pairs must be equal.

63. *Elements*, I, 26 is far more apposite than I, 18, as cited by Della Porta. On the other hand, *Elements*, III, 7 is perfectly apposite. See Euclid (1589), 138–39 and 334–35.

64. See *Meteorology*, III, 3, 373a5–18.

65. The mountain mentioned by Della Porta is Montserrat, and the church he alludes to is surely that of Santa Maria de Montserrat Abbey, which is poised high above the valley floor.

66. *Natural Questions*, I, viii.1, in Seneca (1529), 418; translation adapted from Seneca (2010), 153.

67. The crocodile trope to which Della Porta alludes was commonplace at the time. Likening it to glory, for instance, Juan Vives says of the crocodile that "it may flee those who pursue it but pursue those who flee it"; *Satellitium sive Symbola*, number 40, in *Ioannis Lodovici Vivis Valentini Introductio ad Sapentiam* (Paris: Simon Colinaeus, 1527), 34r.

68. See Alexander's commentary to book 3 of the *Meteorology*, in Alexander of Aphrodisias (1545), 37ra–b. For the source of the quotation, see Ramus (1569), 20.

69. For Aristotle's account, see *Meteorology*, III, 5.

70. In other words, if the Sun, the eye, and the center of the rainbow all lie on the horizon line, arc DC will be as large as possible, in which case the rainbow will encompass a full semicircle. Della Porta's figure of 24° for the size of arc DC appears to be an inversion of the size it should actually be under these conditions: that is, 42°. Note, however, that Della Porta had earlier claimed the maximum angle to be 45° (see note 58 above).

71. All the key parameters—that is, the size of the cone, the size of angle BDA, and distance BE—remain constant because B, at the center of circle ADGF, is the pivot point of rotation.

72. Suffice it to say, this geometrical analysis not only fails to demonstrate the phenomenon but also fails to illustrate it in a meaningful way.

73. See book 7, prop. 4 above, where Della Porta imputes the phenomenon to ophthalmia rather than to external circumstances.

74. See *Meteorology*, III, 2.371b28–9, where Aristotle says of the rainbow that "at sunset and sunrise the circle is smallest and the segment largest: as the Sun rises higher, the circle is larger and the segment smaller"; translation from Aristotle (1984), 599. In other words, as the Sun rises and the rainbow sinks below the horizon so that its visible portion becomes increasingly smaller than a semicircle, the circle to which its visible portion belong appears to increase in size.

75. See Olympiodorus, *Commentary on Aristotle's* Meteorology, III, 40, in Olympiodorus (1551), 71v.

76. Seneca, *Natural Questions*, I, vi, 5–6, in Seneca (1529), 418; translation adapted from Seneca (2010), 152.

77. See Nifo, *Commentaries on the Books of Aristotle's* Meteorology, III, in Nifo (1551), 100rb.

78. See *De Subtilitate*, 4, in Cardano (1551), 167.

79. *Natural History*, II, 60, in Pliny (1559 [as II, 59]), 26

80. Della Porta is speaking roughly here. In fact, as he demonstrated in proposition 13 above, the light from F will be refracted at interface EC, as will the light from every point along FE, so the angle under which FE will be seen is actually smaller than EBF.

81. See Pseudo-Plutarch, *On the Opinions of the Philosophers*, III, 5, in Plutarch (1530), 148v. The distinction lies in whether the color really inheres in a physical body, as, for example, the red in an apple—in which case it is "material" in origin—or whether the color is only apparent, as, for example, when it seems to depend on external circumstances, such as a particular juxtaposition of viewer and color, which, when changed causes the color to disappear.

82. Seneca, *Natural Questions*, I, v.10, in Seneca (1529), 417; translation adapted from Seneca (2010), 150. If the color were real, it would inhere in a drop of water but does not actually do so—in which case it is not "material" in origin.

83. Seneca, *Natural Questions*, I, v.11–12 and vi.3–4, in Seneca (1529), 417; translation adapted from Seneca (2010), 150–52.

84. Ricchieri, *Thirty Books of Readings from the Ancients*, XXII, 30, in Ricchieri (1566), 877.

85. If these angular measurements are correct, then the two rainbows Della Porta claims to have observed could not possibly have been primary and secondary because the distance between such rainbows is around 10°. What he was most likely observing was a primary bow at the maximum altitude of 42° and a supernumerary bow just below it, although a separation of 4° between the two is far too great. Note that Della Porta is using the correct figure of 42° here, not the incorrect one of 45° that he cited earlier (see note 58 above).

86. See Albertus Magnus, *Meteororum*, III, 4.26, in Albertus Magnus (1517a), 115rb [also (1890a), 697a].

87. See *Meteorologicorum*, III, in Alexander of Aphrodisias (1545), 38vb.

88. See *Meteororum*, III, 4.16, in Albertus Magnus (1517a), 113rb [also (1890a), 686a].

89. For Cardano's account, see *De Subtilitate*, 4, in Cardano (1551), 167; and for Scaliger's refutation, see *Exotericarum Exercitationum*, LXXX, 4, in Scaliger (1582), 301–4.

90. For Aristotle, see 204ra and 206vb–207ra in 1550 edition; Alexander, see 34rb–va; 38va–39rb in 1545 edition. I have been unable to trace out the textual lapse to which Della Porta refers.

91. Della Porta's designation of the color red to describe the outer and inner bands of the two rainbows seems to contradict his insistence earlier that these bands are naturally golden yellow unless the cloud behind the rainbow is extremely dark, in which case the blackness of that cloud seen through the golden yellow reddens it.

92. Here, as in proposition 19, Della Porta is speaking roughly, since the angles under which the bands would actually be seen are somewhat smaller than the two angles, FMI and VMR.

93. See *Meteorologicorum*, III, commentary to text 22, in Alexander of Aphrodisias (1545), 38v.

94. Della Porta's supposition that the two bands of the rainbows are seen under equal angles does not address the problem of why there is a colorless interval between primary and secondary bows, even according to his own analytic lights. The reason comes to ground in proposition 14, where he argues that rainbows are seen under a particular angle according to the intersection of the cone of solar radiation and the cone of visibility. It would therefore have been necessary to show that angles MFB, MGC, MHD, and MIE of the primary bow in Figure 9.22 are equal, respectively, to angles RMB, SMB, TMB, and VMB of the secondary bow, those being the key particular angles according to which the bands of color are seen. In fact, such equality is impossible according to his model, because it would require the secondary rainbow to lie directly behind the primary one.

95. *Natural Questions*, I, iii, 1, in Seneca (1529), 414; translation adapted from Seneca (2010), 144.

96. Ricchieri, *Thirty Books of Readings from the Ancients*, XXII, 27, in Ricchieri (1566), 875. For Aristotle's actual account, see *Meteorology*, III, 2, 372a21–8.

97. See Albertus Magnus, *Meteororum*, III, iv, 11, in Albertus Magnus (1517a), 112ra–b (also [1890a], 679. Cardano claims that in Germany lunar rainbows have been seen twice in a single summer; see Cardano, *De Subtilitate*, 4, in Cardano (1551), 170.

98. *Natural History*, II, lx, in Pliny (1559 [as II, 59]), 26.

99. Lodovico Ricchieri, *Thirty Books of Readings from the Ancients*, XXII, 27, in Ricchieri (1566), 875.

100. *Natural Questions*, I, viii, 6–7, in Seneca (1529), 418–19; translation adapted from Seneca (2010), 154.

101. *Natural History*, II, lx, in Pliny (1559 [as II, 59], 26.

102. Walafrid Strabo, *Glossa ordinaria*, to Genesis 9:13, in Bible (1545), 56raA.

103. Genesis 9:13, in ibid.

104. Lodovico Ricchieri, *Thirty Books of Readings from the Ancients*, XXII, 24, in Ricchieri (1566), 874

105. Ibid., 873.

106. Della Porta seems to be referring here to century I, quatrain 17, of Nostradamus's prophecies, which implies that, when rainbows appear every day for forty years, great floods will result; see *Les Vrayes Centuries et Propheties de Maistre Michel Nostradamus . . . .Reveuës et Corrigées suyvant les Premieres Editions [de]1556 [et] 1558* (Leyden, 1650), 3. The counter to this prognostication, according to Della Porta, is based on the fiery nature of the rainbow (indicated by the red band), which prevents the watery nature from prevailing. Della Porta himself dismisses this argument in the next paragraph.

107. Pseudo-Plutarch, *On the Opinions of the Philosophers*, III, 5; in Plutarch (1530), 148v.

108. Virgil, *Georgics*, I, 380–81, in Virgil (1579), 23v.

109. Seneca, *Natural Questions*, I, vi, in Seneca (1529), 417; translation from Seneca (2010), 151.

110. *Natural History*, I, 59 [60 in modern edition], in Pliny (1559), 26.

111. *Aeneid*, V, 606, in Virgil (1579), 124v.

112. Lodovico Ricchieri, *Thirty Books of Readings from the Ancients*, XXII, 29, in Ricchieri (1566), 876.

113. *Quadripartitum*, II, 13, in Ptolemy (1551), 36.

114. Lodovico Ricchieri, *Thirty Books of Readings from the Ancients*, XXII, 29, in Ricchieri (1566), 876. Quotation of Plautus from *Curculio*, act 1, scene 2.

115. Ibid., 878. Greek quotation from Aratus, *Phaenomena*, 940; translation from Aratus (1921), 455.

116. The allusion to a glass hexagon makes no obvious sense here, since Della Porta bases the subsequent analysis on a prism with a triangular rather than hexagonal cross-section. In *Perspectiva*, X, 83, however, Witelo describes the formation of a spectrum when light passes through a hexagonal crystal, and this may be the source of

Della Porta's allusion, although Bacon also mentions the prismatic effect of hexagonal crystals in *Opus majus*, part 6, ch. 2; see Witelo (1572), 473, and Bacon (1900), 173.

117. This is so because, by the principle of reciprocity, the angle of exit will be equal to the angle of entry in refraction through glass.

118. Unstated in this analysis is that the ray along DH is actually a narrow shaft of light, so when it enters the prism to the right and left of the actual apex point H, it will be refracted in contrary directions. Thus, as represented in the left-hand diagram of Figure 9.27b, it refracts along HY', reflects from the prism's bottom surface, and refracts back out to H'. In the right-hand diagram, the light refracts to a point near M, then reflects and finally emerges toward Q. Consequently, the respective bands within H'& and QX ought to be golden yellow, those within &' and XV green, and those within S'Y and VT' blue. Where the two spectrums will be projected depends, of course, on how acute or obtuse angle MHL is. One obvious problem with this analysis is that all the rays pass through a considerable portion of the glass prism via refraction and reflection, so according to Della Porta's theory, the resulting colors should all tend toward the blue end of the spectrum because of the light's weakening.

119. See *Meterology*, III, 6, 377b3–28.

120. See Averroes's commentary on *Meteorology*, III, 6 [or III, 4 in Latin version], in *Aristotle* (1550b), 210rab. See Nifo's commentary on the same text, in Nifo (1551), 120rb.

121. Pseudo-Plutarch, *On the Opinions of the Philosophers*, III, 6, in Plutarch (1530), 149r.

122. *Natural Questions*, I, ix, 1–2, in Seneca (1529), 419, translation from Seneca (2010), 154.

123. The iris stone to which Della Porta refers is presumably iris agate, which diffracts light according to the spectrum in an eye-catching way.

124. For Aristotle's account, see *Meteorology*, III, 3. Unlike Della Porta, Aristotle attributes haloes to reflection, not refraction.

125. Here Della Porta is presumably referring to the principle of reciprocity. Accordingly, a given solar, lunar, or stellar ray enters a drop in the cloud, is refracted through it, and then emerges at an angle equal to the original angle of entry into the drop.

126. For the sake of clarity and simplicity, I have ignored points G, L, and D in the original Latin text and diagram. In particular, Della Porta's designation of AGBC as a ray is confusing because ray AG in the original diagram coincides with the axial ray ABC. Moreover, even without points G, L, and D, the point of Della Porta's argument is eminently clear.

127. Pseudo-Plutarch, *On the Opinions of the Philosophers*, III, 18, in Plutarch (1530), 150r.

128. Seneca, *Natural Questions*, I, x, 1–2. in Seneca (1529), 419, translation adapted from Seneca (2010), 154.

129. According to Della Porta's theory, the color formed by light passing through the top band of vapor ought to be golden yellow because that vapor is extremely thin. On the other hand, since the eclipse is occurring against a black sky, the golden

yellow will be rendered red, just as the golden yellow of the top band of the rainbow is rendered red against a black cloud.

130. Della Porta may be referring here to his *De Aeris Transmutationibus Libri IV*, which was published in Rome seventeen years after the publication of *De refractione* but the lineaments of which Della Porta may have already had in mind by 1593.

131. Pseudo-Plutarch, *On the Opinions of the Philosophers*, III, 1, in Plutarch (1530), 148r.

132. See Aristotle, *Meteorology*, I, 8, 345a11–345b30.

133. Pseudo-Plutarch, *On the Opinions of the Philosophers*, III, 1, in Plutarch (1530), 148r. For Aristotle's own wording, see *Meteorology*, I, 8, 345b31–346b9.

134. Ibid.

135. *De Subtilitate*, 4, in Cardano (1551), 155.

BIOGRAPHICAL SKETCHES

For each of the thinkers listed here I have generally suggested a single source to consult for further details. All of these sources are printed, nearly all of them encyclopedic, and some of them quite brief. There are, however, several trustworthy online sources that can be consulted for more extensive, and sometimes more intensive, coverage. The following four, I think, stand out for quality and scope. First and foremost, is the *Stanford Encyclopedia of Philosophy* (http://plato.stanford.edu/), which is remarkable for its thorough and scholarly approach. Equally useful, though perhaps a bit less scholarly than its Stanford counterpart, is the *Internet Encyclopedia of Philosophy* (http://www.iep.utm.edu/). Somewhat broader in topical scope is *Oxford Bibliographies* (http://www.oxfordbibliographies.com/), which gives capsule accounts with some useful bibliographical references. And finally, of equal topical breadth but limited to "Italian" thinkers is the online *Enciclopedia Italiana di Scienze, Lettere ed Arti* (http://www.treccani.it/).

ALPHABETICAL LISTING

Abū Ma'shar (787?–886? CE). Known in the Latin West as Albumasar, Abū Ma'shar Ja'far ibn Muḥammad ibn 'Umar al-Balkhī was a Persian scholar, astrologer and astronomer from Balkh, in modern-day north central Afghanistan. Best known for his astrological works, he synthesized material from many different traditions, including Hellenistic Greek, Syrian, Persian, and Indian, having studied astrology ten years in Varanasi, located in the modern state of Uttar Pradesh. His introductory work, *Kitāb al-mudkhal al-kabīr* (*Introduction to Astrology*), was disseminated widely in the West after its twelfth-century translation into Latin and influenced many medieval Latin scholars, Albertus Magnus among them. For further details, see David Pingree, "Abū Ma'shar Al-Balkhī, Ja'far Ibn Muḥammad," in *The Complete Dictionary of Scientific Biography*, eds. Charles Gillispie et al. (Detroit: Scribner's, 2008), 1: 32–39.

Aelian, Claudius (c. 170–235 CE). From Praeneste in Central Italy, Aelian is known best for compiling anthologies from earlier sources. His two most famous works are *De Natura Animalium* (*On the Characteristics of Animals*) and *Varia Historia* (*Various Histories*). The former served as a key resource

for natural histories and bestiaries of the Middle Ages. The latter is a semi-ethnographic collection of anecdotes, biographies, and descriptions of far-away lands and customs, with an accent on moralizing. For further details, see Donald Zeyl, *Encyclopedia of Classical Philosophy* (New York: Routledge, 2013), 5–6.

Aëtius of Amida (fifth or sixth century CE). A Greek-speaking physician and writer of medical treatises from southeastern Anatolia (near modern-day Diarbakir, Turkey), Aëtius studied in Alexandria and incorporated a great deal of Egyptian pharmacology, as well as spells, charms, and talismans into his works. Heavily indebted to Galen and Oribasius, his magnum opus *Sixteen Books on Medicine* was one of the crucial medical sources from late antiquity. For further details, see Fridolf Kudlien, "Aëtius of Amida," in *The Complete Dictionary of Scientific Biography*, eds. Charles Gillispie et al. (Detroit: Scribner's, 2008), 1: 68–69.

Alexander of Aphrodisias (fl. c. 200 CE). From Caria in western Anatolia, Alexander was a Peripatetic scholar and renowned commentator on Aristotle. He authored several commentaries on Aristotle, as well as a number of original treatises. His *Meteorologica* is among the extant commentaries, and his *Problemata* among the original works. Both remained in circulation throughout the Renaissance. For further details, see Philip Merlan, "Alexander of Aphrodisias," in *The Complete Dictionary of Scientific Biography*, eds. Charles Gillispie et al. (Detroit: Scribner's, 2008), 1: 117–20.

Al-Farghānī (c. 800–870 CE). Known as Alfraganus in the Latin West, Abū al-'Abbās Aḥmad ibn Muḥammad ibn Kathīr al-Farghānī was a celebrated Persian astronomer and natural philosopher from Farghāna in modern-day Uzbekistan. His most recognized work is the commentary on Ptolemy's *Almagest*, known usually in English as *Elements of Astronomy on the Celestial Motions*. Upon translation into Latin in the twelfth century, it was widely circulated through the Renaissance. For further details, see Abdelhamid I. Sabra, "Al-Farghānī, Abū'l-'Abbās Aḥmad Ibn Muḥammad Ibn Kathīr," in *The Complete Dictionary of Scientific Biography*, eds. Charles Gillispie et al. (Detroit: Scribner's, 2008), 4: 541–45.

Apollonius of Rhodes (fl. third century BCE). Apollonius was an epic poet of the Homeric style and chief librarian at Alexandria. Little is known for certain about his life or his connection with the Aegean island of Rhodes. His main work, *Argonautica*, a retelling in four books of the journey of Jason and the Argonauts with contemporary literary devices, was clearly a significant influence on Virgil. In addition to his Homeric scholarship, Apollonius was known for his works on geography, incorporating descriptions of distant lands and peoples. For further details, see Doris Meyer, "Apollonius as Hellenistic Geographer," in *Brill's Companion to Apollonius*

Rhodius, eds. T. Papaghelis and A. Rengakos (2nd ed; Leiden: Brill, 2011), 267–85.

Aratus of Soli (c. 315–c. 240 BCE). From Soli, in the region of Cilicia along the Mediterranean coast of Anatolia, Aratus was a Greek didactic poet. His sole surviving major work, *Phaenomena* (*Appearances*), is a long poem concerning constellations, astronomy, and meteorology. Quite popular in Antiquity, Aratus's *Phaenomena* was the subject of several commentaries/translations in Greek and Latin. For further details, see Leonardo Tarán, "Aratus of Soli," in *The Complete Dictionary of Scientific Biography*, eds. Charles Gillispie et al. (Detroit: Scribner's, 2008), 1: 204–05.

Artemidorus of Parium (first century BCE–first century CE). Known only through his mention in Seneca's *Natural Questions*, Artemidorus, a native of Mysia on the Anatolian side of the Hellespont, is credited with an astronomical treatise on the courses of planets and stars, as well as the effects of visible light, a treatise with which Seneca strongly disagreed. For further details, see Trevor Curnow, *The Philosophers of the Ancient World: An A–Z Guide* (London: Bristol Classical Press, 2006), 50.

Arziganensis, Oliverius (fl. late fifteenth century). From Arzignano near Vicenza, Oliverius Arziganensis (or Olivero di Arzignano) is known solely through his well-received 1487 commentary on the *Memorable Deeds and Sayings* of Valerius Maximus. For further details, see Marijke Crab, "Josse Bade's *Familiaris Commentarius* on Valerius Maximus (1510): A School Commentary?" in *Transformations of the Classics via Early Modern Commentaries*, ed. Karl Enenkel (Leiden: Brill, 2013), 155.

Aulus Gellius (125/28–180 CE). Most likely from Rome, Aulus Gellius staked his claim to fame on his *Attic Nights*, a grab bag of essays about random topics that is generously laced with direct quotations from classical authors. As such, it serves as an important source for our knowledge of the intellectual context of his time. Influential in late antiquity (it was drawn on heavily by Macrobius), the *Attic Nights* became quite fashionable in the Renaissance. For further details, see Stephen Beall, "Gellius, Aulus," in *The Oxford Encyclopedia of Ancient Greece and Rome*, ed. Michael Gagarin (Oxford: Oxford University Press, 2010), 3: 278.

Campanus of Novara (c. 1220–1296). Recognized by Roger Bacon as one of the leading mathematicians of his time, Campanus, who was from the Piedmont region of northwest Italy, composed a Latin edition of Euclid's *Elements* (in fifteen books), which served as the main source of Euclidean scholarship until the sixteenth century. His *Theorica planetarum* provided an advanced description of the planetary system based on Ptolemy's *Almagest* but, because of its technical nature, was not widely read during the Middle

Ages. For further details, see G. J. Toomer, "Campanus of Novara," in *The Complete Dictionary of Scientific Biography*, eds. Charles Gillispie et al. (Detroit: Scribner's, 2008), 3: 23–29.

Cardano, Girolamo (1501–76). From Pavia, Cardano was active in many scholarly fields, most notably mathematics, in which he laid the foundations for the elaboration of algebra. He was also an educated physician and astrologer, practicing and teaching in Padua and Milan throughout his professional career. Among his more immediately popular publications was *De Subtilitate Rerum* (*Concerning the Subtlety of Things*), a compendium of scientific experiments and observations, philosophical puzzles, and collected anecdotes. For further details, see Mario Gliozzi, "Cardano, Girolamo," in *The Complete Dictionary of Scientific Biography*, eds. Charles Gillispie et al. (Detroit: Scribner's, 2008), 3: 64–67.

Dee, John (1524–1608/9). An English polymath, known primarily as a mathematician, astronomer, and occultist, Dee maintained close ties with the monarchy through his studies on navigation and cartography. Primarily because of his keen interest in the occult, he became a recognized figure in popular culture through the seventeenth century. For further details, see Nicholas Clulee, *John Dee's Natural Philosophy: Between Science and Religion* (London: Routledge, 1988).

Demosthenes Philalethes (c. first century CE). Demosthenes Philalethes was a Greek physician in Asia Minor specializing in the anatomy and pathologies of the eye. No longer extant, his *Ophthalmicus* was the single most influential text on diseases and surgery of the eye from antiquity through the high Middle Ages; it is cited in the works of Paul of Aegina, among others. For further details, see Heinrich von Staden, *Herophilus: The Art of Medicine in Early Alexandria* (Cambridge: Cambridge University Press, 1989), 69–70.

Favorinus of Arelate (first-second century CE). From Arles, Favorinus was noted as a rhetorician with a style of teaching that encouraged understanding all sides of an argument. For further details, see Ernst-Günther Schmidt, "Favorinus," in *Brill's New Pauly Encyclopedia of the Ancient World*, eds. Hubert Cancik et al. (Leiden: Brill, 2004), 5: 376.

Ficino, Marsilio (1433–99). By the early 1460s, Ficino had settled in Florence under the patronage of Cosimo de' Medici and established the so-called Platonic Academy there. Although it is questionable whether such an "academy" ever existed in the formal sense, there is no doubt that Marsilio began at that time to apply himself seriously to translating "Platonic" works from Greek into Latin. Accordingly, in 1471 he published a Latin translation of the Hermetic corpus under the title *Mercurii Trismegisti Liber De Potestate et Sapientia Dei* (*Book of Hermes Trismegistus on the Power and Knowledge of God*). Thirteen years later, in 1484 he published a Latin translation of the

complete works of Plato, with accompanying commentaries. And eight years after that, he rounded things out with a Latin translation of Plotinus's *Enneads*. With these three translations-cum-commenaries, Marsilio laid the textual ground for a resurgence of interest in Platonism/Neoplatonism during the Renaissance and early modern period. For further details, see Paul O. Kristeller, "Marsilio Ficino as a Beginning Student of Plato," *Scriptorium* 20 (1966): 41–54.

Gentile da Foligno (c. 1280–1348). Gentile had a distinguished career as a physician and professor of medicine at Siena and Perugia. Although he wrote a number of texts on original material, his crowning achievement was an extensive commentary on the *Canon* of Avicenna. A comprehensive synthesis and analysis of Greek, Arabic, and contemporary medical sources, this commentary included his own personal experiences and observations. For further details, see Joël Chandelier, "Gentile da Foligno," in *Medieval Science, Technology and Medicine: An Encyclopedia*, eds. Thomas Glick et al. (New York: Routledge, 2005), 185–86.

Heliodorus of Larissa (third century CE?). Sometimes known as Damianos, Heliodorus of Larissa, a city located on mainland Greece in Thessaly, was a Greek mathematician of whom little is known beyond the fact that he postdated Ptolemy, who died around 180 CE. His brief *Optical Hypotheses* is largely a commentary on Euclid's *Optics* with a nod to Ptolemy. For further details, see Fabio Acerbi, "Damianus of Larissa," in *The New Dictionary of Scientific Biography*, ed. Noretta Koertge (Detroit: Charles Scribners, 2007), 2: 233–34.

Hero of Alexandria (fl. c. 60 CE). Author of several works on mechanical or practical subjects, as well as on mathematics, Hero is best known for his *Pneumatics* and *Mechanics*, both of which reveal his practical, engineering bent. To some extent this same practical bent can be discerned in his *Catoptrics*, which focuses less on reflection theory than on how to arrange plane mirrors for startling visual effects. This work seems to have exerted little influence in its own day or, for that matter, after its translation into Latin in 1269 with attribution to Ptolemy. It did, however, see print in 1518. For further details, see A. G. Drachmann and Michael S. Mahoney, "Hero of Alexandria," in *The Complete Dictionary of Scientific Biography*, eds. Charles Gillispie et al. (Detroit: Scribner's, 2008), 6: 310–15.

Ḥunayn ibn 'Isḥāq (809–73 CE). A Syriac-speaking Nestorian Christian, Ḥunayn was the head of a tiny group of translators in Baghdad during the Abbasid caliphate who were instrumental in rendering certain Greek works into Arabic. Key among these works, which include many of Aristotle's, were those of Galen, well over a hundred of which Ḥunayn is credited with having translated. He was also the author of *Ten Treatises on the Eye*, which was

rendered into Latin in the late eleventh century and falsely attributed either
to Galen or to Constantine the African. As such, it became a canonical text
on ophthalmology during the Middle Ages. For further details, see G. C.
Anawati, "Ḥunayn ibn 'Isḥāq Al-'Ibādī, Abū Zayd," in *The Complete Dictionary
of Scientific Biography*, eds. Charles Gillispie et al. (Detroit: Scribner's, 2008),
15: 230–34.

Ibn Muʿādh al-Jayyānī (d. after 1079 CE). From Jaén in Spain, Ibn Muʿādh
was a *qāḍī*, or judge, by vocation and an astronomer by avocation. Among
his extant works, which include the *Tabulae Jahen*, was a clever attempt to
determine the height of the sunlight-capturing atmosphere—just under 52
miles by his calculation. This work survives in both a Hebrew and Latin
version, the latter entitled *De crepusculis et nubium ascensionibus* (*On Twilight
and the Rising of Clouds*). This brief work was commonly misattributed to
Alhacen from the later thirteenth century until quite recent times. For further
details, see A. Mark Smith, "Ibn Muʿādh al-Jayyānī," in *Encyclopedia of the
History of Science, Technology, and Medicine in Non-Western Cultures*, ed.
Helaine Selin (Dordrecht: Springer, 2016), 1: 1110–11.

Isigonus (third century BCE?). Isigonus was a Greek writer, possibly from
Nicaea, although neither that nor his place in time is certain; however, since
he is cited by Sotion of Alexandria, he most likely predates the late to mid-
second century BCE. None of his works survive, but he appears to have
been either a historian or paradoxographer. Both Pliny the Elder and Aulus
Gellius treated his work as of high repute. For further details, see Leonard
Schmitz, "Isigonus," in *A Dictionary of Greek and Roman Biography and
Mythology*, ed., William Smith (London: John Murray, 1880), 2: 630.

Johannes de Sacrobosco (fl. first half of thirteenth century). Also known
as John of Holywood, Sacrobosco was a mathematician, astronomer, and
instructor at the University of Paris, although he may have been English by
birth. His *Tractatus de sphera* (*Treatise on the [World] Sphere*), which was
composed sometime around 1230, became a canonical text for basic university instruction in astronomy for the next three centuries and more. For
further details, see Lynn Thorndyke, *The Sphere of Sacrobosco and its Commentators* (Chicago: University of Chicago Press, 1949).

Lucretius Caro, Titus (c. 95–55 BCE). Probably from Rome, Lucretius is
known through his only work, *De rerum natura* (*On the Nature of Things*),
an extended Latin poem explaining atomism and its implications for the
world structure and our place as humans within it. The primary source for
his account of atomism and its ramifications appears to be Epicurus (341–270
BCE), who drew on earlier thinkers, including Democritus, for his atomic
theory. Atomism failed to gain much of an intellectual foothold during later
antiquity and the Middle Ages, but the discovery of a manuscript of *De rerum
natura* in 1417 by Poggio Bracciolini (1380–1459) stimulated considerable

interest in atomism during the Renaissance. For further details, see David Furley, "Epicurus" and "Lucretius," in *The Complete Dictionary of Scientific Biography*, eds. Charles Gillispie et al. (Detroit: Scribner's, 2008), 4: 381–82 and 8: 536–39, respectively.

Macrobius, Ambrosius Aurelius Theodosius (fl. early fifth century C.E.). Macrobius was a Roman writer known for two influential works. The first of these, his *Commentarii in Somnium Scipionis* (*Commentaries on* [*Cicero's*] *Dream of Scipio*), was widely disseminated and served as a key source for Platonism in Medieval Europe. Second, and no less influential, was his *Saturnalia*, a collection of historical, antiquarian, and scholarly curiosities from earlier antiquity. Some of the materials within *Saturnalia*, and the form itself, are taken from Aulus Gellius's *Attic Nights*. For further details, see William H. Stahl, "Macrobius, Ambrosius Theodosius," in *The Complete Dictionary of Scientific Biography*, eds. Charles Gillispie et al. (Detroit: Scribner's, 2008), 9: 1–2.

Meletius Monachus (ninth century CE). From Tiveriopoli in Macedonia, Meletius was, as his name suggests, a monk, known for compiling medical documents. His text *De natura hominis* (*On the Nature of Man*) includes a chapter devoted to ocular anatomy, visual function, and the perception of color, based on information gleaned from various sources, including Galen. For further details, see J. Lascaratos and M. Tsirou, "Ophthalmological ideas of the Byzantine author Meletius," in *History of Ophthalmology*, eds. Harold Henkes and Claudia Zrenner (Dordrecht: Kluwer. 1990), 3: 31–35.

Metrodorus of Chios (fl. fourth century BCE). A Greek natural philosopher of the school of Democritus, Metrodorus is often cast as a skeptic, although he accepted atomism and authored a treatise on nature based on its principles. For further details, see István Bodnár, "Metrodorus of Chios," in *Brill's New Pauly Encyclopedia of the Ancient World*, eds. Hubert Cancik et al. (Leiden: Brill, 2006), 8: 835–36.

Nicholas the Peripatetic (first century BCE). Also known as Nicholas of Damascus, he was a Greek natural philosopher, historian, and prolific author, who wrote commentaries on Aristotle as well as a 144-book universal history composed relatively late in his life. He also served as tutor to the children of Marcus Antonius and Cleopatra. Apart from fragments, none of his corpus is extant today, although we know about aspects of his thought from later sources citing him, Albertus Magnus among them. For further details, see Klaus Meister, "Nicolaus of Damascus," in *Brill's New Pauly Encyclopedia of the Ancient World*, eds. Hubert Cancik et al. (Leiden: Brill, 2006), 9: 726–27.

Nifo, Agostino (c. 1470–c. 1538). A highly esteemed Italian natural philosopher, Nifo worked to reconcile his Averroist influences with Church orthodoxy. Originally from Sessa Arunca, near Naples, he authored numerous weighty commentaries on Aristotle that saw wide distribution, including his

commentary on the *Meteorology*. For further details, see Edward P. Mahoney, *Two Aristotelians of the Italian Renaissance. Nicoletto Vernia and Agostino Nifo* (Aldershot: Ashgate 2000); see also Mahoney, "Nifo, Agostino," in *The Complete Dictionary of Scientific Biography*, eds. Charles Gillispie et al. (Detroit: Scribner's, 2008), 10: 122–24.

Nymphodorus (fl. c. 300 BCE). Whether the Nymphodorus to whom Pliny refers is Nymphodorus of Syracuse, in Sicily, or Nymphodorus of Amphipolis, in Macedonia, or whether the two are one and the same, is uncertain. The Syracusan Nymphodorus is credited with a *periplus*, or description of an itinerary along the coast of Asia, whereas the Amphilopolitan Nymphodorus is credited with a work on the laws and customs of Asia. Only fragments are left of either work, so it is impossible to tell which (if either) Pliny drew on for his description of witchcraft in *Natural History*, VII, 2. For further details, see Hans Armin Gärner, "Nymphodorus," in *Brill's New Pauly Encyclopedia of the Ancient World*, eds. Hubert Cancik et al. (Leiden: Brill, 2006), 9: 928.

Olympiodorus of Alexandria (sixth century CE). Also called Olympiodorus the Younger, he was a Neoplatonist philosopher and astrologer. Besides his works on natural philosophy, he wrote two commentaries on Aristotle, one on the *Categories* and the other on the *Meteorology*. For further details, see Christina Viano, "Olympiodorus of Alexandria," in *The Complete Dictionary of Scientific Biography*, eds. Charles Gillispie et al. (Detroit: Scribner's, 2008), 23: 338–40.

Pacuvius (220 BCE–c. 120 BCE). Nephew and student of renowned poet Ennius, Pacuvius was a Roman tragic poet of great repute in his own right. He is known for introducing speculative questions and innovative language to Latin literature. For further details, see Peter Lebrecht Schmidt, "Pacuvius," in *Brill's New Pauly Encyclopedia of the Ancient World*, eds. Hubert Cancik et al. (Leiden: Brill, 2007), 10: 326.

Paul of Aegina (seventh century CE). A Greek physician and medical scholar from the island of Aegina in Greece's Saronic Gulf, Paul of Aegina is best known for his encyclopedic *De re medica libri septem* (*Seven Books on Medical Matters*, commonly known in English as the *Medical Compendium in Seven Books*). Though little is known of his life, his writings were instrumental in the field of surgery not only in the Latin West but also in the Muslim East after translation into Arabic by Ḥunayn ibn 'Isḥāq in the ninth century. For further details, see Peter E. Pormann, *The Oriental Tradition of Paul of Aegina's Pragmateia* (Leiden: Brill, 2004).

Petrus Ramus (1515–72). A French humanist from Picardy, Ramus (Pierre de la Ramée) was a logician, rhetorician, mathematician, and educator with an anti-Aristotelian perspective. His works centered on a pragmatic sensibility, and his mathematical interests were largely confined to engineering and

practical applications. For further details, see Michael S. Mahoney, "Petrus Ramus," in *The Complete Dictionary of Scientific Biography*, eds. Charles Gillispie et al. (Detroit: Scribner's, 2008), 11: 286–90.

Philip of Medma (fourth century BCE). A student of Plato's, Philip (also known as Philip of Opus, a town on the western coast of the northern arm of the Gulf of Euboea) was a Greek astronomer and mathematician who edited Plato's *Laws*, writing the dialogue *Epinomis* as an appendix to the same. Otherwise, Philip is known to us only at second hand through citations from such ancient writers as Vitruvius, Pliny the Elder, and Pseudo-Plutarch. For further details, see Trevor Curnow, *The Philosophers of the Ancient World: An A–Z Guide* (London: Bristol Classical Press, 2006), 214.

Plautus, Titus Maccius (fl. c. 254–184 BCE). A comic playwright originally from Sarsina, near Rimini northeast Italy, Plautus is credited with well over 100 plays, twenty-one of which have survived in complete or almost complete versions, but not in original form. Maintaining their popularity until the mid-first century BCE, Plautus's plays fell into relative obscurity by the first century CE and remained in obscurity throughout late antiquity and the Middle Ages. Rediscovered by Petrarch in the fourteenth century, Plautus's comedies set the standard for later Renaissance comedic theater, to which Della Porta was a significant contributor. For further details, see Niall W. Slater, "Plautus," in *The Oxford Encyclopedia of Ancient Greece and Rome*, ed. Michael Gagarin (Oxford: Oxford University Press, 2010), 5: 312–14.

Plotinus (204–270 CE). Most likely from Egypt, Plotinus studied with various philosophers in Alexandria, which was an intellectual center at the time and where he doubtless became a devoté of Plato. Moving to Rome at the age of around 40, he gained a considerable reputation as a teacher. One of his most devoted disciples, Porphyry, gathered his writings together into six books, each divided into nine parts—hence the title *Enneads* (from the Greek ἐννεάς = group of nine) for the entire collection. Within this context, Plotinus provided a sort of mystical systematization of Plato's thought that falls under the general rubric of Neoplatonism or Late Platonism. The resulting philosophical approach was enormously influential throughout late antiquity as well as the Arabic and Latin Middle Ages, and it gained a significant boost in popularity after Marsilio Ficino published a Latin translation of the *Enneads* in 1492. For further details, see A. C. Lloyd, "Plotinus," in *The Complete Dictionary of Scientific Biography*, eds. Charles Gillispie et al. (Detroit: Scribner's, 2008), 11: 41–42.

Posidonius (c. 135 BCE–c. 50 CE). Greek by parentage, Posidonius was from Apamea, which lies in the northwest corner of modern-day Syria. A philosophical thinker of Stoic bent, he wrote on a wide variety of topics from metaphysics and ethics to geodesy and meteorology. Despite his reputation for learning, none of Posidonius's works have survived, our knowledge of his

thought being restricted to citations from or mentions by later authors, such as Cicero, Pliny, Strabo, and Seneca. For further details, see E. H. Warmington, "Posidonius," in *The Complete Dictionary of Scientific Biography*, eds. Charles Gillispie et al. (Detroit: Scribner's, 2008), 11: 103-06.

Scaliger, Julius Caesar (1484–1558). A student of Lodovico Ricchieri and a native of Riva del Garda in the Trentino region of Italy, Scaliger wrote well-received texts in natural philosophy and medicine, publishing his scientific commentaries late in life. Perhaps his most influential work, the *Exotericarum exercitationum Liber XV* (*Book Fifteen of Exoteric Exercises*) approached natural philosophy with observational methods and attacked Cardano's *De subtilitate* fairly savagely. For further details, see Paul Lawrence Rose, "Scaliger (Bordonius), Julius Caesar," in *The Complete Dictionary of Scientific Biography*, eds. Charles Gillispie et al. (Detroit: Scribner's, 2008), 12: 134–36.

Solinus, Gaius Julius (third century CE?). A Roman geographer and linguist whose biographical details are unknown, Solinus is best known for *De mirabilibus mundi* (*On the Wonders of the World*). Later disseminated in a revised and expanded form as *Polyhistor*, this work consists of various accounts of foreign lands, customs, and curiosities of natural history, many of which were taken from Pliny. For further details, see Marina Tolmacheva, "Geography, Chorography," in *Medieval Science, Technology and Medicine: An Encyclopedia*, eds. Thomas Glick et al. (New York: Routledge, 2005), 186–90.

Strabo (64/3 BCE–c. 25 CE). Strabo was a Greek scholar, historian, and geographer who traveled far abroad from his home in Pontus, at the southeastern edge of the Black Sea. He is primarily known for his *Geographica*, a descriptive history of foreign lands and customs from his contemporary world. For further details, see E. H. Warmington, "Strabo," in *The Complete Dictionary of Scientific Biography*, eds. Charles Gillispie et al. (Detroit: Scribner's, 2008), 13: 83–86.

Suetonius (69–after 122 CE). Probably from Hippo Regius in Numidia, North Africa, Gaius Suetonius Tranquillus is best known for his *Lives of the Caesars*, in which he provides biographies of Julius Caesar (d. 44 BCE) and the succession of Roman emperors from Augustus (d. 14 CE) to Domitian (d. 96 CE). For further details, see Donna W. Hurley, "Suetonius," in *The Oxford Encyclopedia of Ancient Greece and Rome*, ed. Michael Gagarin (Oxford: Oxford University Press, 2010), 6: 397–99.

Theodosius (fl. c. 130 BCE?). From Bithynia in northwest Anatolia, Theodosius was recognized in his day as a mathematician and astronomer. Dealing with the geometry of spheres in three books, his *Spherics* would have been understood in its day as astronomical in purport, since spheres and their motions were central to astronomical theory at the time. For further details,

see Ivor Bulmer-Thomas, "Theodosius of Bithynia," in *The Complete Dictionary of Scientific Biography*, eds. Charles Gillispie et al. (Detroit: Scribner's, 2008), 13: 319–21.

Theon of Alexandria (fourth century CE). A Greek scholar resident in Alexandria, Theon studied mathematics and astronomy, publishing in both fields, as well as authoring commentaries on works of Ptolemy and Euclid. Next to his many commentaries, his most significant works are his edited arrangement of Euclid's *Elements*, and a "Theonine" recension of Euclid's *Optics* that may not be authentically his. For further details see G. J. Toomer, "Theon of Alexandria," in *The Complete Dictionary of Scientific Biography*, eds. Charles Gillispie et al. (Detroit: Scribner's, 2008), 13: 321–25.

Theophrastus (c. 371–c. 287 BCE). A disciple of Aristotle and successor to him as head of the Peripatetic School of Athens, Theophrastus followed Aristotle in the breadth of his interests. He is credited with over 200 works, of which only a few survive, including works devoted to botany, mineralogy, biology, physics, and sensation. Not widely cited or known during the Middle Ages, Theophrastus became an influential source during the Renaissance after recovery of his major works in the fifteenth century. For further details, see J. B. McDairmid, "Theophrastus," in *The Complete Dictionary of Scientific Biography*, eds. Charles Gillispie et al. (Detroit: Scribner's, 2008), 13.

Valerius Flaccus (first century CE). Gaius Valerius Flaccus Setinus Balbus, was a Roman poet of unknown provenance whose only surviving work is an incomplete epic poem, *Argonautica*, which is considered superior in presentation to the earlier *Argonautica* of Apollonius of Rhodes. The text, in its current form, ends so abruptly that the work is assumed to be unfinished. For further details, see Martin Dinter, "Valerius Flaccus," in *The Oxford Encyclopedia of Ancient Greece and Rome*, ed. Michael Gagarin (Oxford: Oxford University Press, 2010), 7: 146–47.

Valerius Maximus (fl. early first century CE). Hardly anything is known of Valerius Maximus other than that he was a writer, rhetorician, and collector of historical anecdotes and popular folktales, as exemplified in his moralizing *Factorum ac dictorum memorabilium libri IX* (*Nine Books of Memorable Deeds and Sayings*). He is most recognized for his use of language, which reflects the excesses and forced novelty of Silver Age Latin. For further details, see Hans-Friedrich Mueller, "Valerius Maximus," in *The Oxford Encyclopedia of Ancient Greece and Rome*, ed. Michael Gagarin (Oxford: Oxford University Press, 2010), 7: 147–48.

Varro, Marcus (116–27 BCE). An incredibly prolific writer, Varro was from Rieti, in central Italy not far from Rome. He is credited with having introduced the Greek encyclopedic tradition into the Latin world through his now-lost *Disciplinarum libri IX* (*Nine Books on the Subjects of Instruction*). A wide-

ranging discussion of the liberal arts, which included not only grammar, logic, and rhetoric (comprising the so-called trivium), as well as arithmetic, geometry, music, and astronomy (comprising the so-called quadrivium), but also medicine and architecture. Amply provided with references to (and quotations from) earlier thinkers, this work became a crucial source for knowledge about the classical intellectual tradition. For further details see Phillip Drennon Thomas, "Varro, Marcus Terentius," in *The Complete Dictionary of Scientific Biography*, eds. Charles Gillispie et al. (Detroit: Scribner's, 2008), 13: 588–89.

Vitruvius (d. c. 25 BCE). Vitruvius, about whose life virtually nothing is known, was author of the massive *De architectura libri decem* (*Ten Books on Architecture*), a wide-ranging study that deals with the design and execution of various buildings, from religious edifices to domestic housing. In it he describes the sort of training an architect should have in the arts (including optics) and provides descriptions of the various instruments needed by an architect. For further details, see Derek de Solla Price and G. J. Toomer, "Vitruvius Pollo," in *The Complete Dictionary of Scientific Biography*, eds. Charles Gillispie et al. (Detroit: Scribner's, 2008), 15: 514–21.

Walafrid Strabo (c. 808–849 CE). A monk from the monastery of Reichenau in southwestern Germany, Walafrid Strabo (= "squint-eyed") was author of several works, including the *Vita sancti Galli* (*Life of Saint Gall*) and *Libellus de exordiis et incrementis quarundam in observationibus ecclesiasticis rerum* (*A Brief Book concerning the Origins and Development of Certain Matters in the Liturgy*). From the end of the fifteenth to the first decades of the twentieth, he was assumed to be the author of the highly influential *Glossa Ordinaria* of the Bible, an assumption that has since been thoroughly discredited. For further details, see David A. Salomon, *An Introduction to the* Glossa Ordinaria *as Medieval Hypertext* (Cardiff: University of Wales Press, 2012). See also Scott Wells, "Walafrid Strabo," in *Encyclopedia of Monasticism*, eds. William H. Johnston and Christopher Kleinhenz (London: Routledge, 2015), 1385–86.

William of Moerbeke (c. 1215–1286). A Dominican most likely from Flanders, William made his way to Greece by 1260 at the latest and served as Latin bishop of Corinth from 1278 until his death. Fluent in Greek, he not only translated various works of Archimedes into Latin but also rendered most of the Aristotelian corpus from Greek into Latin. He also translated Hero of Alexandria's *Catoptrics* in 1269 but erroneously ascribed it to Ptolemy, who was assumed to be its author until modern times. To the best of our knowledge, he accomplished all this between 1260 and 1278, no mean feat considering the number and size of the works he rendered from Greek into Latin. For further details, see Lorenzo Minio-Paluello, "William of Moerbeke," in *The Complete Dictionary of Scientific Biography*, eds. Charles Gillispie et al. (Detroit: Scribner's, 2008), 9: 434–40.

PRIMARY SOURCES FOR THE TEXT

Aelian. 1548. *Aeliani De Varia Historia Libri XIIII*. Basel.

Aëtius. 1542. *Aetii Medici Graeci Contractae ex Veteribus Medicinae Tetrabiblos*. Basel.

Alberti, Leon Battista. 1540. *De Pictura Praestantissima . . . Libri Tres*. Basel.

Albertus Magnus. 1517a. *Divi Alberti Magni Naturalia ac Supernaturalia*. Venice. Available at http://bvpb.mcu.es/es/consulta/registro.cmd?id=452127

——. 1517b. *Parva Naturalia Alberti Magni*. Venice.

——. 1890a. Albert Borgnet, ed., *Beati Alberti Magni . . . Opera Omni*, vol. 4 (*Meteorology*). Paris: Vives.

——. 1890b. Albert Borgnet, ed., *Beati Alberti Magn . . . Opera Omni*, vol. 8 (*De sensu*). Paris: Vives.

Alexander of Aphrodisias. 1552. *Septimum Volumen Aristotelis Stagiritae Extra Ordinem Naturalium Varii Libri: Quibus Nonnulli Etiam Additi Sunt Aristoteli Ascripti: Alexandri Problematum Libri II*. Venice.

——. 1545. *Alexandri Aphrodisiensis . . . in Quatuor Libros Meteorologicorum Aristotelis Commentatio*. Venice.

Al-Farghānī. 1590. *Muhamedis Alfragani Arabis Chronologica et Astronomica Elementa*. Frankfurt.

Alfred of Sareshel. 1988. James K. Otte, ed. and trans., *Alfred of Sareshel's Commentary on the* Meteora *of Aristotle*. Leiden: E. J. Brill.

Alhacen. 1572. Friedrich Risner ed., *Opticae Thesaurus*. Basel.

——. 2001. A. Mark Smith, ed. and trans., *Alhacen's Theory of Visual Perception: A Critical Edition, with English Translation, Introduction, and Commentary, of the First Three Books of Alhacen's* De aspectibus. Philadelphia: American Philosophical Society.

——. 2006. A. Mark Smith, ed. and trans., *Alhacen on the Principles of Reflection: A Critical Edition, with English Translation, Introduction, and Commentary, of Books 4 and 5 of Alhacen's* De aspectibus. Philadelphia: American Philosophical Society.

——. 2008. A. Mark Smith, ed. and trans., *Alhacen on Image-Formation and Image-Distortion in Mirrors: A Critical Edition, with English Translation, Introduction, and Commentary, of Book 6 of Alhacen's* De aspectibus. Philadelphia: American Philosophical Society.

——. 2010. A. Mark Smith, ed. and trans., *Alhacen on Refraction: A Critical Edition, with English Translation, Introduction, and Commentary, of Book*

7 of *Alhacen's* De aspectibus. Philadelphia: American Philosophical Society.

Al-Kindī. 1912. Axel Björnbo and Sebastian Vogl, eds., *Alkindi, Tideus, und Pseudo Euklid: Drei optische Werke.* Lepizig: Teubner.

Aratus. 1570. *C. Iulii Hygini Augusti Liberti, Fabularum Liber.* Basel.

———. 1921. A. W. Mair and G. R. Mair, trans., *Callimachus, Lycopheron, Aratus.* London: William Heinemann.

Aristotle. 1550a. *Secundum Volumen. Aristotelis Stagiritae De Rhetorica et Poetica Libri cum Averrois Cordubensis in Eosdem Paraphrasibus.* Venice.

———. 1550b. *Quintum Volumen. Aristotelis Stagiritae De Coelo De Generatione et Corruptione, Meteorologicorum, De Plantis Libri.* Venice.

———. 1550c. *Sextum Volumen. Aristotelis Stagiritae Libri Omnes ad Animalium Cognitionem Attinentes cum Averrois Cordubensis Variis in Eosdem Commentariis.* Venice.

———. 1552. *Septimum Volumen. Aristotelis Stagiritae Extra Ordinem Naturalium Varii Libri: Quibus Nonnulli Etiam Additi Sunt Aristoteli Ascripti: Alexandri Problematum Libri II.* Venice.

———. 1984. Jonathan Barnes, ed., *The Complete Works of Aristotle.* Princeton, NJ: Princeton University Press.

Aulus Gellius. 1560. *Auli Gellii . . . Noctes Atticae.* Lyon.

Averroes. 1550. *Sextum Volumen. Aristotelis Stagiritae Libri Omnes ad Animalium Cognitionem Attinentes cum Averrois Cordubensis Variis in Eosdem Commentariis.* Venice.

———. 1552. *Decimum Volumen. Averrois Cordubensis Colliget Libri VII.* Venice.

Avicenna. 1508. *Avicenne Perhypatetici Philosophi ac Medicorum Facile Primi Opera in Lucem Redacta.* Venice.

———. 1556. *Avicennae Medicorum Arabum principis, Liber Canonis.* Basel.

Bacon, Roger. 1900. John H. Bridges, *The 'Opus Majus' of Roger Bacon,* vol. 2. London: Williams and Norgate.

———. 1928. Robert B. Burke, trans., *The Opus Majus of Roger Bacon,* vol. 2. Philadelphia: University of Pennsylvania Press.

———. 1983. David C. Lindberg, ed. and trans., *Roger Bacon's Philosophy of Nature: A Critical Edition, with English Translation, Introduction, and Notes of the* De multiplicatione specierum *and* De speculis comburentibus. Oxford: Clarendon.

———. 1996. David C. Lindberg, ed. and trans., *Roger Bacon and the Origins of Perspectiva: A Critical Edition and English Translation of Bacon's Perspectiva.* Oxford: Clarendon.

Bartholin, Caspar. 1612. *Astrologia seu de Stellarum Natura, Affectionibus, et Effectionibus.* Wittenburg.

Bible. 1545. *Biblia Sacra cum Glossis Interlineari et Ordinaria . . . Tomus Primus.* Lyon.

Cardano, Girolamo. 1551. *Hieronymi Cardani . . . De Subtilitate Libri XXI.* Lyon.

———. 2013. John M. Forrester, trans., *The* De Subtilitate *of Girolamo Cardano*. Tempe, AZ: Arizona Center for Medieval and Renaissance Studies.

Cicero. 1565. *Tomus Quartus Operum M. Tullii Ciceronis*. Paris.

———. 1961. H. Rackham, trans., *On the Nature of the Gods. Academics*. Cambridge, MA: Harvard University Press.

Della Porta, Giambattista. 1560. *Magiae Naturalis, sive de Miraculis Rerum Naturalium, Libri IIII*. Antwerp.

———. 1586. *De Humana Physiognomia Libri III*. Vico Equense.

———. 1589. *Magiae Naturalis Libri XX*. Naples.

Dürer, Albrecht. 1532. *Alberti Dureri Elementorum Geometricorum Libri IV*. Paris.

Euclid. 1537. *Euclidis Megarensis . . . Elementorum Geometricorum Libri XV . . . His adiecta sunt . . . Catoptrica et Optica*. Basel.

———. 1557a. *Euclidis Optica et Catoptrica ex Graeco Versa per Ioannem Penam*. Paris.

———. 1557b. *Euclidis Catoptrica . . . translatione per Conradum Dasympodium*. Strasbourg.

———. 1573. *La Prospettiva di Euclide . . . Insieme Con La Prospettiva Di Eliodoro Larisseo*. Florence.

———. 1589. *Euclidis Elementorum Libri XV . . . Auctore Christophoro Clavio*. Rome.

———. 1895. I. L. Heiberg and H. Menge, *Euclidis Opera Omnia*, vol. 7. Leipzig: Teubner.

Euripides. 1562. *Euripides Poeta Tragicorum Princeps, in Latinum Sermonem Conversus, Adiecto Eregione Textu Graeco*. Basel.

Galen. 1535. *Claudii Galeni Pergameni de Hippocratis et Platonis Decretis*. Basel.

———. 1541. *Galeni Prima Classis Humani Corporis Originem, Formationem, Dissectionem . . . Complectitur*. Venice.

———. 1550. *Tomus Primus Operum Galeni*. Lyon.

Gentile da Foligno. 1522. *Tertius Fan Avicennae cum Amplissima Gentilis Fulginatis Expositione*. Venice.

Grosseteste, Robert. 1912. Ludwig Baur, ed., *Die Philosophischen Werke des Robert Grosseteste, Bishofs von Lincoln*. Münster: Aschendorff.

Heliodorus of Larissa. 1573. *La Prospettiva di Euclide . . . Insieme Con La Prospettiva Di Eliodoro Larisseo*. Florence.

———. 1897. Richard Schöne, ed. and trans., *Damianos Schrift über Optik*. Berlin.

Hero of Alexandria. 1518. *Sphera cum Commentis in Hoc Volumine Contentis*. Venice.

———. 1900. Ludwig Nix and Wilhelm Schmidt, ed. and trans., *Heronis Alexandrini Opera Quae Supersunt Omnia*, vol. 2. Leipzig: Teubner.

Kepler, Johannes. 1611. *Dioptrice*. Augsburg.

Lucretius. 1570. *T. Lucretii Cari De Rerum Natura Libri VI*. Paris.

————. 1992. W. H. D. Rouse and Martin Ferguson Smith, trans., *Lucretius De rerum natura*. Cambridge, MA: Harvard University Press.

Macrobius. 1535. *Macrobii Aurelii . . . in Somnium Scipionis. Eiusdem Saturnaliorum libri VII.* Basel.

————. 2011. Howard Kaster, *Macrobius*, Saturnalia, *Books 6–7*. Cambridge, MA: Harvard University Press.

Meletius Monachus. 1552. *Meletii Philosophi de Natura Structuraque Hominis Opus.* Venice.

Nifo, Agostino. 1551. *Augustini Niphi . . . in Libris Aristotelis Meteorologicis Commentaria.* Venice.

Nostradamus. 1538. *Orus Apollo Niliacus De Hieroglyphicis Notis.* Venice.

Olympiodorus. 1551. *Olympiodori . . . in Meteora Aristotelis Commentarii.* Venice.

Ovid. 1565. *P. Ovidii Nasonis Metamorphoseen Libri XV.* Venice.

————. 1977. Paul J. Miller and G. P. Goold, trans., *Ovid*, Metamorphoses *Books 1–8.* Cambridge, MA: Harvard University Press.

Paul of Aegina. 1551. *Pauli Aeginetae Medici Opera.* Lyon.

Pecham, John. 1542. *Perspectiva communis.* Nuremberg.

————. 1970. David C. Lindberg, ed. and trans., *John Pecham and the Science of Optics*: Perspectiva communis, *edited with an Introduction, English Translation, and Critical Notes.* Madison, WI: University of Wisconsin Press.

Plato. 1551. *Omnia Divini Platonis Opera.* Basel.

Pliny. 1559. *C. Plinii Secundi Naturalis Historiae Libri Trigintaseptem.* Venice.

Plotinus. 1580. *Plotini Platonicorum facile coryphaei operum philosophicorum omnium libri LIV in sex enneades distribute.* Basel.

Plutarch. 1530. *Plutarchi Chaeronei . . . Opuscula.* Basel.

Proclus. 1560. *Procli Diadochi . . . in Primum Euclidis.* Padua.

————. 1992. Glenn R. Morrow, trans., *Proclus*: A Commentary on the First Book of Euclid's *Elements*. Princeton, NJ: Princeton University Press.

Ptolemy. 1541. *Claudii Ptolemaei . . . omnia quae extant opera.* Basel.

————. 1551. *Iulii Firmici Materni . . . His Accesserunt Claudii Ptolemaei . . . Quadripartitum . . . Libri IIII.* Basel.

————. 1605. *Claudii Ptolemaei Magna Constructionis Liber Primus . . . Io: Baptista Porta Neap. Inteprete.* Naples.

————. 1996. A. Mark Smith, trans., *Ptolemy's Theory of Visual Perception: An English Translation of the Optics with Introduction and Commentary.* Philadelphia: American Philosophical Society.

Ramus. 1569. *P. Ramis Scholarum Mathematicarum Libri Unus et Triginita.* Basel.

Ricchieri, Lodovico. 1566. *Lodovici Caelii Rhodigini Lectionum Antiquarum Libri XXX.* Basel.

Riolan the elder, Jean. 1598. *Universae Medicinae Compendia.* Paris.

Sacrobosco, Johannes de. 1578. *Sphaera Joannis de Sacro Bosco.* Lyon.

Scaliger, Julius Caesar. 1582. *Iulii Caesaris Scaligeri Exotericarum Exercitationum Liber XV De Subtilitate ad Hieronymum Cardanum.* Frankfurt.

Seneca. 1529. *L. Annei Senecae Opera*. Basel.

―――. 2010. Harry M. Hine, trans., *Seneca, Natural Questions*. Chicago: University of Chicago Press.

Solinus. 1520. *C. Iulii Solini Polyhistor seu Rerum Orbis Memorabilium Collectanea*. Cologne.

Suetonius. 1510. *Commentationes Conditae a Philippo Beroaldo in Suetonium Tranquillum*. Venice.

Theodosius. 1558. *Theodosii Tripolitae Sphaericorum, Libri Tres*. Paris.

Theophrastus. 1605: *Theophrasti Eresii . . . Pleraque Antehac Latine Nunquam, Nunc Graece et Latine Simul Edite*. Hanover.

Valerius Flaccus. 1548. *Caii Valerii Flacci Argonautica*. Lyons.

―――. 1938. J. H. Mozley, trans., *Valerius Flaccus*. Cambridge, MA: Harvard University Press.

Valerius Maximus. 1565. *Valerii Maximi Factorum et Dictorum Memorabilium Libri Novem*. Venice.

Vesalius, Andreas. 1568. *Andreae Vesalii . . . De Humani Corporis Fabrica Libri Septem*. Venice.

―――. 2009. William Frank Richardson and John Burd Carman, trans., *On the Fabric of the Human Body. Book VI: The Heart and Associated Organs. Book VII: The Brain*. Novato, CA: Norman Publishing.

Vincent of Beauvais. 1591. *Speculi Maioris Vincentii Burgundi . . . Tomi Quatuor*. Venice.

Virgil. 1579. *P. Vergilii Maronis Poemata Quae Extant Omnia*. Frankfurt.

―――. 2005. Stanley Lombardo, trans., *Virgil*, Aeneid. Indianapolis: Hackett Publishing.

Vitruvius. 1586. *M. Vitruvii Pollionis De Architectura Libri Decem*. Lyon.

―――. 1999. Ingrid D. Roland, trans., and Thomas N. Howe, comment. *Vitruvius, Ten Books on Architecture*. Cambridge, UK: Cambridge University Press.

Witelo. 1572. Friedrich Risner ed., *Opticae Thesaurus*. Basel.

―――. 1983. A. Mark Smith, ed. and trans., *Witelonis Perspectivae Liber Quintus*. Wroclaw, Poland: Ossolineum.

―――. 1991. Sabetai Unguru, ed. and trans., *Witelonis Perspectivae Liber Secundus et Liber Tertius*. Wroclaw, Poland: Ossolineum.

Xenophon. 1545. *Xenophontis Philosophi ac Historici Excellentisssimi Opera Quae Quidem Extant Omnia*. Basel.

GENERAL BIBLIOGRAPHY

Acerbi, Fabio. 2008. "Damianus of Larissa." In *New Dictionary of Scientific Biography*, vol. 2, edited by Noretta Koertge, 233–34. Detroit: Scribners.

Anawati, Georges C. 2008. "Ḥunayn ibn 'Isḥāq Al-'Ibādī, Abū Zayd." In *The Complete Dictionary of Scientific Biography*, vol. 15, edited by Charles Gillispie, Frederick Holmes, and Noretta Koertge, 230–34. Detroit: Scribner's.

Balbiani, Laura. 2001. *La Magia Naturalis di Giovan Battista Della Porta: Lingua, cultura, e scienza in Europa all'inizio dell'età moderna*. Bern: Peter Lang.

———. 1999. "La ricezione della *Magia Naturalis* di Giovan Battista Della Porta: Cultura e scienza dall'Italia all'Europa." *Bruniana & Campanelliana: Ricerche filosofiche e materiali storico-testuali* 5: 277–303.

Baldini, Ugo. 2001. "The Roman Inquisition's Condemnation of Astrology: Antecedents, Reasons and Consequences." In *Church, Censorship and Culture in Early Modern Italy*, edited and translated by Gigliola Fragnito and Adrian Belton, 79–110. Cambridge, UK: Cambridge University Press.

Banks, Martin S., et al. 2015. "Why Do Animal Eyes Have Pupils of Different Shapes?" *Science Advances* 1.

Barnes, Jonathan. See Aristotle (1984) in Primary Sources for the Text.

Baur, Ludwig. See Grosseteste (1912) in Primary Sources for the Text.

Beall, Stephen. 2010. "Gellius, Aulus." In *Oxford Encyclopedia of Ancient Greece and Rome*, vol. 3, edited by Michael Gagarin, 278. Oxford: Oxford University Press.

Bietenholz, Peter G., and Thomas B. Deutscher. 2003. *Contemporaries of Erasmus: A Biographical Register of the Renaissance and Reformation*. Toronto: University of Toronto Press.

493

Björnbo, Axel. See Al-Kindī (1912) in Primary Sources for the Text.

Black, Christopher. 2009. *The Italian Inquisition.* New Haven, CT: Yale University Press.

Blumberg, Harry, trans. 1961. *Averroes: Epitome of Parva Naturalia.* Cambridge, MA: Mediaeval Academy of America.

Bodnár, István. 2006. "Metrodorus of Chios." In *Brill's New Pauly Encyclopedia of the Ancient World*, vol. 8, edited by Hubert Cancik, Helmuth Schneider, Manfred Landfester, Christine Salazer, and Francis Gentry, 835–36. Leiden: Brill.

Borgnet, Albert. See Albertus Magnus (1890a and 1890b) in Primary Sources for the Text.

Borrelli, Arianna, Giora Hon, and Yaakov Zik, eds. 2017. *The Optics of Giambattista Della Porta (ca. 1535–1615): A Reassessment.* Cham, Switzerland: Springer.

———. 2014. "Thinking with Optical Objects: Glass Spheres, Lenses and Refraction in Giovan Battista Della Porta's Optical Writings." *Journal of Early Modern Studies* 3: 39–61.

Boyer, Carl B. 1959. *The Rainbow: From Myth to Mathematics.* New York: Thomas Yoseloff.

———. 1958. "The Tertiary Rainbow: An Historical Account." *Isis* 49: 141–54.

———. 1952. "Descartes and the Radius of the Rainbow." *Isis* 43: 95–98.

Bridges, John H. See Bacon (1900) in Primary Sources for the Text.

Bulmer-Thomas, Ivor. 2008. "Theodosius of Bithynia." In *The Complete Dictionary of Scientific Biography*, vol. 13, edited by Charles Gillispie, Frederick Holmes, and Noretta Koertge, 319–21. Detroit: Scribner's.

Burke, Robert B. See Bacon (1928) in Primary Sources for the Text.

Calabritto, Monica. 2011. "Tasso's Melancholy and its Treatment: A Patient's Uneasy Relationship with Medicine and Physicians." In *Diseases of the Imagination and Imaginary Disease in the Early Modern Period*, edited by Yasmin Haskell, 201–27. Turnhout: Brepols.

Cancik, Hubert, Helmuth Schneider, Manfred Landfester, Christine Salazer, and Francis Gentry, eds. 2006. *Brill's New Pauly Encyclopedia of the Ancient World.* 22 vols. Leiden: Brill.

Carman, John B. See Vesalius (2009) in Primary Sources for the Text.

Chandelier, Joël. 2005. "Gentile da Foligno." In *Medieval Science, Technology and Medicine: An Encyclopedia*, edited by Thomas Glick, Steven Livesey, and Faith Wallis, 185–86. New York: Routledge.

Clubb, Louise George. 1965. *Giambattista Della Porta Dramatist.* Princeton, NJ: Princeton University Press.

Clulee, Nicholas. 1988. *John Dee's Natural Philosophy: Between Science and Religion.* London: Routledge.

Crab, Marijke. 2013. "Josse Bade's *Familiaris Commentarius* on Valerius Maximus (1510): A School Commentary?" In *Transformations of the Classics via Early Modern Commentaries*, edited by Karl Enenkel, 155–66. Leiden: Brill.

Crew, Henry, trans. 1940. *The* Photismi de lumine *of Maurolycus: A Chapter in Late Medieval Optics.* New York: Macmillan.

Croll, Morris. 1921. "'Attic Prose' in the Seventeenth Century." *Studies in Philology* 18: 79–128.

Curnow, Trevor. 2006. *The Philosophers of the Ancient World: An A–Z Guide.* London: Bristol Classical Press.

Dinter, Martin. 2010. "Valerius Flaccus." In *Oxford Encyclopedia of Ancient Greece and Rome*, edited by Michael Gagarin, 146–47. Oxford: Oxford University Press.

Drachmann, Aage G., and Michael S. Mahoney. 2008. "Hero of Alexandria." In *The Complete Dictionary of Scientific Biography*, vol. 6, edited by Charles Gillispie, Frederick Holmes, and Noretta Koertge, 310–15. Detroit: Scribner's.

Eamon, William. 1994. *Science and the Secrets of Nature.* Princeton, NJ: Princeton University Press.

Eastwood, Bruce S. 1982. *The Elements of Vision: The Micro-Cosmology of Galenic Visual Theory according to Ḥunayn ibn 'Isḥāq.* Philadelphia: American Philosophical Society.

Encyclopedia Italiana di Scienze, Lettere ed Arti. Rome: Giovanni Treccani. http://www.treccani.it/

Enenkel, Karl, ed. 2013. *Transformations of the Classics via Early Modern Commentaries*. Leiden: Brill.

Favaro, Antonio. 1895–96. "Amici e corrispondenti di Galileo Galilei: Studi e ricerche (II. Ottavio Pisani)." *Atti del Reale Istituto Veneto di Scienze, Lettere ed Arti* 54: 411–40.

Forrester, John M. See Cardano (2013) in Primary Sources for the Text.

Fragnito, Gigliola, ed., and Adrian Belton, trans. 2001. *Church, Censorship and Culture in Early Modern Italy*. Cambridge, UK: Cambridge University Press.

Freedberg, David. 2002. *The Eye of the Lynx*. Chicago: University of Chicago Press.

Freeman, Kathleen. 1948. *Ancilla to the Pre-Socratic Philosophers*. Oxford: Blackwell.

du Fresne, Rafaelle, ed. 1651. *Trattato della Pittura di Lionardo da Vinci*. Paris: Giacomo Langlois.

Furley, David. 2008. "Epicurus." In *The Complete Dictionary of Scientific Biography*, vol.. 4, edited by Charles Gillispie, Frederick Holmes, and Noretta Koertge, 381–82. Detroit: Scribner's.

———. 2008. "Lucretius." In *The Complete Dictionary of Scientific Biography*, vol. 8, edited by Charles Gillispie, Frederick Holmes, and Noretta Koertge, 536–39. Detroit: Scribner's.

Gagarin, Michael, ed. 2010. *Oxford Encyclopedia of Ancient Greece and Rome*. 7 vols. Oxford: Oxford University Press.

Galluzzi, Paolo, ed. 1984. *Novità celesti e crisi del sapere*. Florence: Giunti Barbèra.

Gärtner, Hans Armin. 2006. "Nymphodorus." In *Brill's New Pauly Encyclopedia of the Ancient World*, vol. 9, edited by Hubert Cancik, Helmuth Schneider, Manfred Landfester, Christine Salazer, and Francis Gentry, 928. Leiden: Brill.

Gentilcore, David. 1998. *Healers and Healing in Early Modern Italy*. Manchester: Manchester University Press.

———. 1996. "Il Regio Protomediato nella Napoli Spagnola." *Dynamis* 16: 219–36.

Gillispie, Charles, Frederick Holmes, and Noretta Koertge, eds. 2008. *The Complete Dictionary of Scientific Biography*. 26 vols. Detroit: Scribner's.

Glick, Thomas, Steven Livesey, and Faith Wallis, eds. 2005. *Medieval Science, Technology and Medicine: An Encyclopedia*. New York: Routledge.

Gliozzi, Mario. 2008. "Cardano, Girolamo." In *The Complete Dictionary of Scientific Biography*, vol. 3, edited by Charles Gillispie, Frederick Holmes, and Noretta Koertge, 64–67. Detroit: Scribner's.

Goold, George P. See Ovid (1977) in Primary Sources for the Text.

Goulding, Robert. 2013. "Thomas Harriot's Optics, Between Experiment and Imagination: The Case of Mr. Bulkeley's Glass." *Archive for History of Exact Sciences* 68: 137–78.

Haskell, Yasmin, ed. 2011. *Diseases of the Imagination and Imaginary Disease in the Early Modern Period*. Turnhout, Belgium: Brepols.

Haskins, Charles Homer. *The Renaissance of the Twelfth Century*. Cambridge, MA: Harvard University Press, 1927.

Heiberg, I. L. See Euclid (1895) in Primary Sources for the Text.

Henkes, Harold, and Claudia Zrenner, eds. 1990. *History of Ophthalmology*. Dordrecht: Kluwer.

Henry, John. 2008. "The Fragmentation of Renaissance Occultism and the Decline of Magic." *History of Science* 46: 1–48.

Hine, Harry M. See Seneca (2010) in Primary Sources for the Text.

Hughes, David W. 1983. "On Seeing Stars (Especially Up Chimneys)." *Quarterly Journal of the Royal Astronomical Society* 24: 246–57.

Hurley, Donna W. 2010. "Suetonius." In *Oxford Encyclopedia of Ancient Greece and Rome*, vol. 6, edited by Michael Gagarin, 397–99. Oxford, UK: Oxford University Press.

Ilardi, Vincent. 2007. *Renaissance Vision from Spectacles to Telescopes*. Philadelphia: American Philosophical Society Press.

Internet Encyclopedia of Philosophy. 2018. http://www.iep.utm.edu/

Johnston, William H., and Christopher Kleinhenz, eds. 2015. *Encyclopedia of Monasticism.* 2 vols. London: Routledge.

Kaster, Howard. See Macrobius (2011) in Primary Sources for the Text.

Kodera, Sergius. 2014. "The Laboratory as Stage: Giovan Battista della Porta's Experiments." *Journal of Early Modern Studies* 3: 15–38.

———. 2013. "Giambattista Della Porta's Histrionic Science." *California Italian Studies* 3: 1–27.

Koertge, Noretta, ed. 2008. *New Dictionary of Scientific Biography.* 8 vols. Detroit: Scribners.

Kristeller, Paul O. 1966. "Marsilio Ficino as a Beginning Student of Plato." *Scriptorium* 20: 41–54.

Kudlien, Fridolf. 2008. "Aëtius of Amida." In *The Complete Dictionary of Scientific Biography*, vol. 1, edited by Charles Gillispie, Frederick Holmes, and Noretta Koertge, 68–69. Detroit: Scribner's.

Lascaratos, Ioannis, and M. Tsirou. 1990. "Ophthalmological ideas of the Byzantine author Meletius." In *History of Ophthalmology*, vol. 3, edited by Harold Henkes and Claudia Zrenner, 31–35. Dordrecht: Kluwer.

Lee, Raymond L., and Alistair B. Fraser. 2001. *The Rainbow Bridge: Rainbows in Art, Myth, and Science.* University Park, PA: Pennsylvania State University Press.

Leonardo da Vinci. See du Fresne.

Lindberg, David C. "Optics in Sixteenth-Century Italy." In *Novità celesti e crisi del sapere*, edited by Paolo Galluzzi, 131–48. Florence: Giunti Barbèra, 1984.

———. See Bacon (1983, 1996) and Pecham (1970) in Primary Sources for the Text.

Lloyd, A. C. 2008. "Plotinus." In *The Complete Dictionary of Scientific Biography*, vol. 11, edited by Charles Gillispie, Frederick Holmes, and Noretta Koertge, 41–42. Detroit: Scribner's.

Lombardo, Stanley. See Virgil (2005) in Primary Sources for the Text.

MacDonald, Katherine. 2005. "Humanistic Self-Representation in Giovan Battista Della Porta's *Della Fisonomia Dell'Uomo*: Antecedents and Innovation." *Sixteenth Century Journal* 36: 397–414.

Mahoney, Edward. 2008. "Nifo, Agostino." In *The Complete Dictionary of Scientific Biography*, vol. 10, edited by Charles Gillispie, Frederick Holmes, and Noretta Koertge, 122–24. Detroit: Scribner's.

————. 2000. *Two Aristotelians of the Italian Renaissance. Nicoletto Vernia and Agostino Nifo.* Aldershot: Ashgate.

Mahoney, Michael S. 2008. "Petrus Ramus." In *The Complete Dictionary of Scientific Biography*, vol. 11, edited by Charles Gillispie, Frederick Holmes, and Noretta Koertge, 286–90. Detroit: Scribner's.

Mair, A. W., and G. R. Mair. See Aratus (1921) in Primary Sources for the Text.

Mayhew, Robert S., trans. 2011. *Aristotle's* Problems, II. Cambridge, MA: Harvard University Press.

McCluskey, Stephen, ed. and trans. 1974. "Nicole Oresme on Light, Color, and the Rainbow: An Edition and Translation, with Introduction and Critical Notes, of Book Three of His *Questiones super quatuor libros meteororum*." PhD dissertation, University of Wisconsin.

McDairmid, John B. 2008. "Theophrastus." In *The Complete Dictionary of Scientific Biography*, vol. 13, edited by Charles Gillispie, Frederick Holmes, and Noretta Koertge, 328–34. Detroit: Scribner's.

McKirihan, Richard D. 2010. *Philosophy Before Socrates.* 2nd ed. Indianapolis, IN: Hackett.

McRae, Heather. 2018. "Pointing to Inclinations: Albertus Magnus's Physiognomy as a Scientific and Theological Nexus." PhD dissertation, University of Missouri—Columbia.

Meister, Klaus. 2006. "Nicolaus of Damascus." In *Brill's New Pauly Encyclopedia of the Ancient World*, vol. 9, edited by Hubert Cancik, Helmuth Schneider, Manfred Landfester, Christine Salazer, and Francis Gentry, 726–27. Leiden: Brill.

Menge, Heinrich. See Euclid (1895) in Primary Sources for the Text.

Merlan, Philip. 2008. "Alexander of Aphrodisias." In *The Complete Dictionary of Scientific Biography*, vol. 1, edited by Charles Gillispie, Frederick Holmes, and Noretta Koertge, 117–20. Detroit: Scribner's.

Meyer, Doris. 2011. "Apollonius as Hellenistic Geographer." In *Brill's Companion to Apollonius Rhodius*, edited by Theodore Papaghelis and Antonios Rengakos, 267–85. 2nd ed. Leiden: Brill.

Miller, Paul J. See Ovid (1977) in Primary Sources for the Text.

Minio-Paluello, Lorenzo. 2008. "William of Moerbeke," In *The Complete Dictionary of Scientific Biography*, vol. 9, edited by Charles Gillispie, Frederick Holmes, and Noretta Koertge, 434–40. Detroit: Scribner's.

Morrow, Glenn R. See Proclus (1992) in Primary Sources for the Text.

Mozley, John H. See Valerius Flaccus (1938) in Primary Sources for the Text.

Naldoni, Maria, ed. 1962. *Giovan Battista Della Porta: De telescopio*. Florence: Leo Olschki.

Nix, Ludwig. See Hero of Alexandria (1900) in Primary Sources for the Text.

Otte, James K. See Alfred of Sareshel (1988) in Primary Sources for the Text.

Oxford Bibliographies. 2018. http://www.oxfordbibliographies.com/

Papaghelis, Theodore, and Antonios Rengakos, eds. 2011. *Brill's Companion to Apollonius Rhodius*, 2nd ed. Leiden: Brill.

Piccari, Paolo. 2007. *Giovan Battista Della Porta il filosofo, il retore, lo scienzato*. Milan: FrancoAngeli.

Pingree, David. 2008. "Abū Ma'shar Al-Balkhī, Ja'far Ibn Muḥammad." In *The Complete Dictionary of Scientific Biography*, vol. 1, edited by Charles Gillispie, Frederick Holmes, and Noretta Koertge, 32–39. Detroit: Scribner's.

Pormann, Peter E. 2004. *The Oriental Tradition of Paul of Aegina's Pragmateia*. Leiden: Brill.

Price, Derek de Solla, and G. J. Toomer. 2008. "Vitruvius Pollo." In *The Complete Dictionary of Scientific Biography*, vol. 15, edited by Charles Gillispie, Frederick Holmes, and Noretta Koertge, 514–21. Detroit: Scribner's.

Quinlan-McGrath, Mary. 2012. *Influences: Art, Optics, and Astrology in the Italian Renaissance*. Chicago: University of Chicago Press.

Rackham, Harris. See Cicero (1961) in Primary Sources for the Text.

Reeves, Eileen. 2008. *Galileo's Glassworks*. Cambridge, MA: Harvard University Press.

Richardson William F. See Vesalius (2009) in Primary Sources for the Text.

Risner, Friedrich. See Alhacen (1572) and Witelo (1572) in Primary Sources for the Text.

Roland, Ingrid D. See Vitruvius (1999) in Primary Sources for the Text.

Ronchi, Vasco. 1954. "Du *De Refractione* au *De Telescopio* de G. B. Della Porta." *Revue d'histoire des sciences et de leurs applications* 7: 46–47.

Rose, Paul Lawrence. 2008. "Scaliger (Bordonius), Julius Caesar." In *The Complete Dictionary of Scientific Biography*, vol. 12, edited by Charles Gillispie, Frederick Holmes, and Noretta Koertge, 134–36. Detroit: Scribner's.

Rosen, Edward. 1947. *The Naming of the Telescope*. New York: Henry Shuman.

Rouse, William H. D. See Lucretius (1992) in Primary Sources for the Text.

Ryle, Stephen, ed. 2014. *Erasmus and the Renaissance Republic of Letters*. Turnhout: Brepols.

Sabra, Abdelhamid I. 2008. "Al-Farghānī, Abu'l-'Abbās Aḥmad Ibn Muḥammad Ibn Kathīr." In *The Complete Dictionary of Scientific Biography*, vol. 4, edited by Charles Gillispie, Frederick Holmes, and Noretta Koertge, 541–45. Detroit: Scribner's.

Salomon, David A. 2012. *An Introduction to the* Glossa Ordinaria *as Medieval Hypertext*. Cardiff: University of Wales Press.

Schmidt, Ernst-Günther. 2006. "Favorinus." In *Brill's New Pauly Encyclopedia of the Ancient World*, vol. 5, edited by Hubert Cancik, Helmuth Schneider, Manfred Landfester, Christine Salazer, and Francis Gentry, 376. Leiden: Brill.

Schmidt, Peter Lebrecht. 2006. "Pacuvius." In *Brill's New Pauly Encyclopedia of the Ancient World*, vol. 10, edited by Hubert Cancik, Helmuth Schneider, Manfred Landfester, Christine Salazer, and Francis Gentry, 326. Leiden: Brill.

Schmidt, Wilhelm. See Hero of Alexandria (1900) in Primary Sources for the Text.

Schmitz, Leonard. 1880. "Isigonus." *A Dictionary of Greek and Roman Biography and Mythology*, vol. 2, edited by William Smith, 630. London: John Murray.

Schöne, Richard. See Heliodorus of Larissa (1897) in Primary Sources for the Text.

Selin, Helaine, ed. 2016. *Encyclopedia of the History of Science, Technology, and Medicine in Non-Western Cultures*. 2 vols., 2nd ed. Dordrecht: Springer.

Shapiro, Alan E. 1975. "Archimedes's Measurement of the Sun's Apparent Diameter." *Journal for the History of Astronomy* 6: 75–83.

Shumaker, Wayne, ed. and trans. 1978. *John Dee on Astronomy*: Propae-duemata Aphoristica *(1558 and 1568), Latin and English*. Berkeley: University of California Press.

Siraisi, Nancy. 1987. *Avicenna in Renaissance Italy: The* Canon *and Medical Teaching in Italian Universities after 1500*. Princeton, NJ: Princeton University Press.

Slater, Niall W. 2010. "Plautus." In *Oxford Encyclopedia of Ancient Greece and Rome*, vol. 5, edited by Michael Gagarin, 312–14. Oxford: Oxford University Press.

Smith, A. Mark. 1992a. "The Latin Version of Ibn Muʿādh's Treatise 'On Twilight and the Rising of Clouds.'" *Arabic Sciences and Philosophy* 2: 38–88.

———. 1992b. "Picturing the Mind: The Representation of Thought in the Middle Ages and Renaissance." *Philosophical Topics* 20: 149–70.

———. 2007. "Le *De aspectibus* d'Alhacen: révolutionnaire ou réformiste?" *Revue d'histoire des sciences* 60: 65–81.

———. 2015. *From Sight to Light: The Passage from Ancient to Modern Optics*. Chicago: University of Chicago Press.

———. 2016. "Ibn Muʿādh al-Jayyānī." *Encyclopedia of the History of Science, Technology, and Medicine in Non-Western Cultures*, vol. 1, edited by Helaine Selin, 1110–11. 2nd ed. Dordrecht: Springer.

———. See Alhacen (2001, 2006, 2008, 2010), Ptolemy (1996), and Witelo (1983) in Primary Sources for the Text.

Smith, Martin F. See Lucretius (1992) in Primary Sources for the Text.

Smith, William, ed. 1880. *A Dictionary of Greek and Roman Biography and Mythology*. 3 vols. London: John Murray.

Staden, Heinrich von. 1989. *Herophilus: The Art of Medicine in Early Alexandria*. Cambridge, UK: Cambridge University Press.

Stahl, William H. 2008. "Macrobius, Ambrosius Theodosius." In *The Complete Dictionary of Scientific Biography*, vol. 9, edited by Charles Gillispie, Frederick Holmes, and Noretta Koertge, 1–2. Detroit: Scribner's.

Szulakowska, Urzula. 2000. *The Alchemy of Light: Geometry and Optics in Late Renaissance Alchemical Illustration*. Leiden: Brill.

Tarán, Leonardo. 2008. "Aratus of Soli." In *The Complete Dictionary of Scientific Biography*, vol. 1, edited by Charles Gillispie, Frederick Holmes, and Noretta Koertge, 204–05. Detroit: Scribner's.

Tarrant, Neil. 2013. "Giambattista Della Porta and the Roman Inquisition: Censorship and the Definition of Nature's Limits in Sixteenth-Century Italy." *British Journal for the History of Science* 46: 601–25.

Thomas, Phillip Drennon. 2008. "Varro, Marcus Terentius." In *The Complete Dictionary of Scientific Biography*, vol. 13, edited by Charles Gillispie, Frederick Holmes, and Noretta Koertge, 588–89. Detroit: Scribner's.

Thorndyke, Lynn. 1949. *The Sphere of Sacrobosco and its Commentators*. Chicago: University of Chicago Press.

Tolmacheva, Marina. 2005. "Geography, Chorography." In *Medieval Science, Technology and Medicine: An Encyclopedia*, edited by Thomas Glick, Steven Livesey, and Faith Wallis, 186–90. New York: Routledge.

Toomer, Gerald J. 2008. "Campanus of Novara." In *The Complete Dictionary of Scientific Biography*, vol. 3, edited by Charles Gillispie, Frederick Holmes, and Noretta Koertge, 23–29. Detroit: Scribner's.

———. 2008. "Theon of Alexandria." In *The Complete Dictionary of Scientific Biography*, vol. 13, edited by Charles Gillispie, Frederick Holmes, and Noretta Koertge, 321–25. Detroit: Scribner's.

Unguru, Sabetai. See Witelo (1991) in Primary Sources for the Text.

Valente, Michaela. 1999. "Della Porta e l'inquisizione. Nuovi documenti dell' Archivio del Sant'Uffizio." *Bruniana & Campanelliana: Ricerche filosofiche e materiali storico-testuali* 5: 415–34.

Viano, Christina. 2008. "Olympiodorus of Alexandria." In *The Complete Dictionary of Scientific Biography*, vol. 23, edited by Charles Gillispie, Frederick Holmes, and Noretta Koertge, 338–40. Detroit: Scribner's.

Vives, Juan. 1527. *Ioannis Lodovici Vivis Valentini Introductio ad Sapentiam.* Paris: Simon Colinaeus.

Vogl, Sebastian. See Al-Kindī (1912) in Primary Sources for the Text.

Wallace, William A. 1959. *The Scientific Methodology of Theodoric of Freiberg.* Fribourg: Fribourg University Press.

Warmington, Eric H. 2008. "Posidonius." In *The Complete Dictionary of Scientific Biography*, vol. 11, edited by Charles Gillispie, Frederick Holmes, and Noretta Koertge, 103–06. Detroit: Scribner's.

——. 2008. "Strabo." In *The Complete Dictionary of Scientific Biography*, vol. 13, edited by Charles Gillispie, Frederick Holmes, and Noretta Koertge, 83–86. Detroit: Scribner's.

Wells, Scott. 2015. "Walafrid Strabo." In *Encyclopedia of Monasticism*, edited by William H. Johnston and Christopher Kleinhenz, 1385–86. London: Routledge.

Wesseling, Ari. 2014. "Erasmus and Plagiarism." In *Erasmus and the Renaissance Republic of Letters*, edited by Stephen Ryle, 203–14. Turnhout: Brepols.

Zalta, Edward N., ed. 2018. *The Stanford Encyclopedia of Philosophy*. Stanford, CA: Stanford University.

Zeyl, Donald. 2013. *Encyclopedia of Classical Philosophy*. New York: Routledge.

INDEX OF NAMES

A

Abū Ma'shar/Albumasar, 396–97, 412–13, 475

Adamantius, xiii

Aelian, xxii, lxxix, 260–61, 284n29, 475–76

Aëtius of Amida, xxiv, 182n54, 278–79, 476

Agamemnon, 344–45

Agelli, Father Antonio, 464–65

Al-Farghānī (Alfraganus), 38–39, 476

Al-Kindī, Ya'qūb, xxi, lxxxv, 202–03

Albert of Saxony, lxiv, lxxxviin84

Alberti, Leon Battista, xxii, xxvi, 128n3, 226n60, 283n21

Albertus Magnus, xiii, xiv, xxii, xxv, lxxix, 290–91, 396–97, 436–37, 466nn17–18, 475, 481

Alcides, 346–47

Alexander of Aphrodisias, xxv, lxiv, 38–39, 164–65, 190–91, 288–89, 296–97, 323n20, 328–29, 332–35, 338–39, 342–43, 398–99, 426–27, 436–39, 442–43, 476

Alexander VI (Pope), 67n6

Alhacen, xxiii, xxxi, xxxix, lii, lxxiii, lxxix, lxxxvn53, lxxxvin69–70, lxxxviin81, xcn108, 40–41, 69n20, 70n22, 70nn25–26, 71n30, 72n32, 73n44–45, 74n49, 74n53, 104–05, 124–25, 128n3, 131nn21–23, 131n29, 158–59, 176–77, 181n43, 183n67, 200–03, 223n25, 224n37, 226nn58–59, 246–47, 292–93, 322n11, 322n13, 323n17, 324n27, 324nn29–30, 325n32, 480

Alighieri, Dante, x

Álvarez de Toledo y Zúñiga, Pedro, x

Anaxagoras, xxi, 394–95, 460–61

Anaximenes, xxi, 394–95, 444–45

Apollonius of Rhodes, xxvi, 330–31, 352n10, 476–77, 485

Aratus of Soli, xxvi, xxxvii, 60–61, 76n70, 444–45, 448–49, 477

Archimedes, 156–57, 181n38, 486

Aristotle/Aristotelian, xiii, xiv, xxii, xxv, lxiv, lxv, lxvii, lxviii, lxxiii, lxxix, lxxx, lxxxvin64, lxxxviiin92, 36–37, 44–45, 60–63, 67n10, 68n14, 69n21, 73n39, 77n73, 77n79, 94–95, 150–51, 160–65, 168–69, 180n31, 182n56, 186–89, 192–93, 196–97, 200–01, 222n2, 222n8, 223n30, 224n40, 280–81, 288–89, 322n8, 328–29, 334–43, 348–49, 353n31, 390–93, 398–99, 404–09, 422–23, 426–27, 430–31, 436–39, 442–45, 452–55, 460–63, 466n2, 467n25, 467n33, 470n69, 470n74, 472n96, 473n124, 474n133, 476, 479, 481, 482, 485, 486

Artemidorus of Parium, xxii, 396–97, 406–07, 412–13, 467n26, 477

Arziganensis, Oliverius, 353n32, 477

Augias/Augeas, 330–31, 352n10

Augustus, 330–31, 484

Aulus Gellius, xxii, 194–95, 467n23, 477, 480, 481

Averroes (Ibn Rushd/Averroist), xxv, lxxxvin64, 223n23, 330–31, 342–43, 353n31, 452–53, 481

INDEX OF TERMS

www.ingramcontent.com/pod-product-compliance
Lightning Source LLC
Chambersburg PA
CBHW081340190326
41458CB00018B/6061